Lecture Notes in Bioinformatics

Edited by S. Istrail, P. Pevzner, and M. Waterman

Subseries of Lecture Notes in Computer Science

Monika Heiner   Adelinde M. Uhrmacher (Eds.)

Computational Methods in Systems Biology

6th International Conference CMSB 2008
Rostock, Germany, October 12-15, 2008
Proceedings

 Springer

Series Editors

Sorin Istrail, Brown University, Providence, RI, USA
Pavel Pevzner, University of California, San Diego, CA, USA
Michael Waterman, University of Southern California, Los Angeles, CA, USA

Volume Editors

Monika Heiner
Department of Computer Science
Brandenburg University of Technology
Cottbus, Germany
E-mail: monika.heiner@informatik.tu-cottbus.de

Adelinde M. Uhrmacher
Institute of Computer Science
University of Rostock
Rostock, Germany
E-mail: adelinde.uhrmacher@uni-rostock.de

Library of Congress Control Number: Applied for

CR Subject Classification (1998): I.6, D.2.4, J.3, H.2.8, F.1.1

LNCS Sublibrary: SL 8 – Bioinformatics

ISSN   0302-9743
ISBN-10   3-540-88561-7 Springer Berlin Heidelberg New York
ISBN-13   978-3-540-88561-0 Springer Berlin Heidelberg New York

Springer is a part of Springer Science+Business Media

springer.com

© Springer-Verlag Berlin Heidelberg 2008

Typesetting: Camera-ready by author, data conversion by Scientific Publishing Services, Chennai, India
Printed on acid-free paper   SPIN: 12537532   06/3180   5 4 3 2 1 0

Preface

This volume contains the proceedings of the 6th Conference on Computational Methods in Systems Biology (CMSB) held in October 2008 in Rostock/ Warnemünde.

The CMSB conference series was established in 2003 to promote the convergence of (1) modelers, physicists, mathematicians, and theoretical computer scientists from fields such as language design, concurrency theory, software verification, and (2) molecular biologists, physicians, neuroscientists joined by their interest in a systems-level understanding of cellular physiology and pathology. Since this time, the conference has taken place annually. The conference has been held in Italy, France, and the UK, and we were glad to host CMSB in Germany for the first time.

The summaries of the invited talks by Hidde de Jong, Jane Hillston, Koichi Takahashi, Nicolas Le Novere, and Dieter Oesterhelt are included at the beginning of the proceedings. The 21 regular papers cover theoretical or applied contributions that are motivated by a biological question focusing on modeling approaches, including process algebra, simulation approaches, analysis methods, in particular model checking and flux analysis, and case studies. They were selected out of more than 60 submissions by a careful reviewing process. Each paper received at least three reviews from members of the Program Committee consisting of 27 renowned scientists from seven countries. We would like to thank all members of the Program Committee and the referees for the thorough and insightful reviews and the constructive discussions. Due to the number of high-quality submissions, the decision on which papers to accept or reject was not easy. Therefore, we integrated a rebuttal phase for the first time. The authors also contributed to the reviewing process by swift and detailed responses to the reviewers' comments. For this and their submission of interesting and cutting-edge research papers to CMSB 2008, we would like to thank the authors. Also for the first time, five tutorials representing different modeling, simulation, and analysis tools for, and approaches toward, computational biology were part of the conference attesting to the achieved maturity of research.

We used the conference management system EasyChair, which proved invaluable in handling the electronic submission of papers, the entire reviewing process, including discussions and rebuttal phase, and finally, the generation of the proceedings. CMSB 2008 received financial support from the DFG (German Research Foundation) and Microsoft Research, Cambridge. The financial support from Microsoft Research was used to waive the fee for PhD students. For their support in the local organization and administration we would like to thank our local team: Anja Hampel, Jan Himmelspach, Sigrun Hoffmann, and Matthias Jeschke.

The conference venue was the Neptun hotel, located directly at the Baltic sea. Constructed in the beginning of the 1970s and conceived as a hallmark of the GDR, it shed its history and emerged as a modern conference center after the German reunification.

We wish all readers of this volume an enjoyable journey through the challenging field of computational methods in systems biology.

August 2008   Monika Heiner
                                                     Adelinde Uhrmacher

Organization

The organizers and Co-chairs of the CMSB 2008 conference were Monika Heiner of the Brandenburg University of Technology at Cottbus and Adelinde Uhrmacher of the University of Rostock.

Steering Committee

Finn Drabløs	Norwegian University of Science and Technology, Trondheim (Norway)
Monika Heiner	TU Cottbus (Germany)
Patrick Lincoln	Stanford Research International (USA)
Satoru Miyano	University of Tokyo (Japan)
Gordon Plotkin	University of Edinburgh (UK)
Corrado Priami	The Microsoft Research – University of Trento Centre for Computational and Systems Biology (Italy)
Magali Roux-Rouquié	CNRS-UPMC (France)
Vincent Schachter	Genoscope, Evry (France)
Adelinde Uhrmacher	University of Rostock (Germany)

Program Committee

Alexander Bockmayr	Freie Universität Berlin (Germany)
Kevin Burrage	University Queensland (Australia)
Muffy Calder	University of Glasgow (UK)
Luca Cardelli	Microsoft Research Cambridge (UK)
Claudine Chaouiya	Ecole Superieure d'Ingenieurs de Luminy, Marseille (France)
Attila Csikasz-Nagy	Microsoft Resarch – University of Trento Centre for Computational and Systems Biology (Italy)
Finn Drabløs	Norwegian University of Science and Technology, Trondheim (Norway)
François Fages	INRIA, Rocquencourt (France)
Jasmin Fisher	Microsoft Research Cambridge (UK)
David Gilbert	University of Glasgow (UK)
Stephen Gilmore	University of Edinburgh (UK)
Monika Heiner	TU Cottbus (Germany)
Des Higham	University of Strathclyde (UK)
Hidde de Jong	INRIA, Rhône Alpes (France)
Walter Kolch	Beatson Institute for Cancer Research (UK)

Ursula Kummer	University of Heidelberg (Germany)
Wolfgang Marwan	Max Planck Institute Magdeburg (Germany)
Ion Moraru	University of Connecticut Health Center (USA)
Joachim Niehren	INRIA Futurs, Lille (France)
Nicolas Le Novère	European Bioinformatics Institute (UK)
Dave Parker	Oxford University (UK)
Gordon Plotkin	University of Edinburgh (UK)
Corrado Priami	Microsoft Resarch - University of Trento Centre for Computational and Systems Biology (Italy)
Koichi Takahashi	The Molecular Sciences Institute (USA)
Carolyn Talcott	Stanford Research Institute (USA)
Adelinde Uhrmacher	University of Rostock (Germany)
Olaf Wolkenhauer	University of Rostock (Germany)

External Reviewers

Paolo Ballarini	Jan Himmelspach	Davide Prandi
Grégory Batt	Matthias Jeschke	Nathan Price
Arne Bittig	Mathias John	Elisabeth Remy
Matteo Cavaliere	Sriram Krishnamachari	Ronny Richter
Federica Ciocchetta	Hillel Kugler	Aurélien Rizk
Lorenzo Demattè	Celine Kuttler	Christian Rohr
Emek Demir	Cedric Lhoussaine	Alessandro Romanel
Robin Donaldson	Hong Li	Peter Saffrey
Claudio Eccher	Jeremie Mary	Martin Schwarick
Paul Francois	Carsten Maus	Heike Siebert
Richard Fujimoto	Ivan Mura	Sylvain Soliman
Vashti Galpin	Gethin Norman	Marc Thiriet
David Gilbert	Alida Palmisano	Ashish Tiwari
Maria Luisa Guerriero	Michael Pedersen	Cristian Versari
Stefan Haar	Andrew Phillips	Andrei Zinovyev
Jane Hillston	Nir Piterman	

Table of Contents

Qualitative Modeling and Simulation of Bacterial Regulatory Networks

Hidde de Jong

INRIA Grenoble-Rhône-Alpes
655 Avenue de l'Europe, Montbonnot, 38334 Saint-Ismier Cedex, France
Hidde.de-Jong@inria.fr

The adaptation of microorganisms to their environment is controlled at the molecular level by large and complex networks of biochemical reactions involving genes, RNAs, proteins, metabolites, and small signalling molecules. In theory, it is possible to write down mathematical models of these networks, and study these by means of classical analysis and simulation tools. In practice, this is not easy to achieve though, as quantitative data on kinetic parameters are usually absent for most systems of biological interest. Moreover, the models consist of a large number of variables, are strongly nonlinear and include different time-scales, which make them difficult to handle both mathematically and computationally.

We have developed methods for the reduction and approximation of kinetic models of bacterial regulatory networks to simplified, so-called piecewise-linear differential equation models. The qualitative dynamics of the piecewise-linear models can be studied using discrete abstractions from hybrid systems theory. This enables the application of model-checking tools to the formal verification of dynamic properties of the regulatory networks. The above approach has been implemented in the publicly-available computer tool Genetic Network Analyzer (GNA) and has been used to analyze a variety of bacterial regulatory networks.

I will illustrate the application of GNA by means of the network of global transcription regulators controlling the adaptation of the bacterium *Escherichia coli* to environmental stress conditions. Even though *E. coli* is one of the best studied model organisms, it is currently little understood how a stress signal is sensed and propagated through the network of global regulators, and leads the cell to respond in an adequate way. Qualitative modeling and simulation of the network of global regulators has allowed us to identify essential features of the transition between exponential and stationary phase of the bacteria and to make new predictions on the dynamic behavior following a carbon upshift.

M. Heiner and A.M. Uhrmacher (Eds.): CMSB 2008, LNBI 5307, p. 1, 2008.

Integrated Analysis from Abstract Stochastic Process Algebra Models

Jane Hillston, Federica Ciocchetta, Adam Duguid, and Stephen Gilmore

Laboratory for Foundations of Computer Science,
The University of Edinburgh, Scotland

Extended Abstract

Bio-PEPA is a novel stochastic process algebra which has been recently developed for modelling biochemical pathways [5,6]. In Bio-PEPA a reagent-centric style of modelling is adopted, and a variety of analysis techniques can be applied to a single model expression. Such an approach facilitates easy validation of analysis results when the analyses address the same issues [3] and enhanced insight when the analyses are complementary [4]. Currently supported analysis techniques include stochastic simulation at the molecular level, ordinary differential equations, probabilistic model checking and numerical analysis of a continuous time Markov chain.

Process algebras are a well-established modelling approach for representing concurrent systems facilitating both qualitative and quantitative analysis. Within the last decade they have also been proposed as the basis for several modelling techniques applied to biological problems, particularly intracellular signalling pathways, e.g. [13,12,10,7,2,1].

A process algebra model captures the behaviour of a system as the actions and interactions between a number of entities, usually termed *processes* or *components*. In stochastic process algebras, such as PEPA [9] or the stochastic π-calculus [11], a random variable representing average duration is associated with each action. In the stochastic π-calculus, interactions are strictly binary whereas in PEPA and Bio-PEPA the more general, multiway synchronisation is supported.

The original motivation for the use of process algebras for modelling intracellular pathways was the recognition of the clear mapping that can be made between *molecules*, within a biochemical pathway, and *processes*, within concurrent systems [14]. The mapping is then elaborated with reactions between molecules represented by communication between processes, etc.

This mapping has been extremely influential with much subsequent work on process algebras for systems biology following its lead. It takes an inherently *individuals*-based view of a pathway or cell, and suffers the problem of individuals-based modelling, namely *state-space explosion*. When each individual within a system is represented explicitly and all transitions within or between individuals are captured as discrete events, the number of states becomes prohibitively high. This problem prohibits the use of techniques which rely on access to the state space in its entirety, such as model checking or numerical solution of a Markov chain. Essentially analysis is restricted to stochastic simulation where the state space is explored iteratively.

M. Heiner and A.M. Uhrmacher (Eds.): CMSB 2008, LNBI 5307, pp. 2–4, 2008.

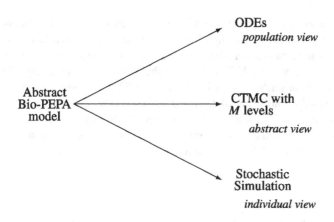

Fig. 1. Alternative modelling approaches: a single Bio-PEPA description of a system may be used to derive alternative mathematical representations offering different analysis possibilities

In contrast, biologists often take a *population*-based view of cellular systems, representing them as systems of ordinary differential equations (ODEs). These mathematical models are continuous or *fluid* approximations of the discrete state, individuals-based models of the system. In many circumstances the approximation is extremely good. In the biological context, where the exact number of molecules is often difficult to obtain but is known to be extremely large, this more abstract view is both intellectually and computationally appealing. The continuous state space models, in the form of systems of ODEs, are much more efficiently solved than their discrete state space counterparts.

In Bio-PEPA we wanted to be able to use a single model description to access both an individuals-based and a population-based view of a system. Thus we adopt an abstract style of modelling which we term, *reagent-centric*. We use the term *reagent* to mean an entity which engages in a reaction. In the basic case this will be a biochemical species such as a protein, receptor molecule or an enzyme. However it may be more abstract, capturing the behaviour of a group of entities or a whole subsidiary pathway. In this style of modelling the focus of the process algebra model is no longer the individual molecules, but rather the species or similar entities. This subtle change gives us much more flexibility in how a model may be interpreted, facilitating mappings into a number of different mathematical representations, as illustrated in Figure 1. Viewing the reagents as species, it is straightforward to use the BioPEPA description to derive the stoichiometry matrix, and the corresponding ODE model. Conversely, knowing the forms of interations which can be engaged in by the reagent, allows an individuals-based or molecular model to be derived, suitable for solution by Gillespie's stochastic simulation algorithm [8].

Moreover using the *reagents-as-processes* abstraction, together with other features of the BioPEPA language, make it straightforward to capture several characteristics of biochemical reactions which can be problematic for other process algebras. These include reactions with stoichiometry greater than one, with more than two reactants, and with general kinetic laws.

We have also been keen to investigate the extent to which "classical" process algebra analyses can be used to provide insight into system biology models. In the context of stochastic process algebras such analyses include numerical analysis of the underlying continuous time Markov chain (CTMC) and probabilistic model checking. Whilst the stochastic simulation described above is based on a CTMC, the size of the state space in most examples will prohibit any state-based analysis. Thus we have also developed a third mapping from BioPEPA models to an smaller CTMC, which we term the *CTMC with levels*. In such models the concentration of each reagent is split into discrete steps, leading to a more compact state space, more readily amenable to state-based analyses.

References

1. Bortolussi, L., Policriti, A.: Modeling Biological Systems in Stochastic Concurrent Constraint Programming. Constraints 13(1-2), 55–90 (2006)
2. Calder, M., Gilmore, S., Hillston, J.: Modelling the influence of RKIP on the ERK signalling pathway using the stochastic process algebra PEPA. In: Priami, C., Ingólfsdóttir, A., Mishra, B., Riis Nielson, H. (eds.) Transactions on Computational Systems Biology VII. LNCS (LNBI), vol. 4230, pp. 1–23. Springer, Heidelberg (2006)
3. Calder, M., Duguid, A., Gilmore, S., Hillston, J.: Stronger computational modelling of signalling pathways using both continuous and discrete-space methods. In: Priami, C. (ed.) CMSB 2006. LNCS (LNBI), vol. 4210, pp. 63–77. Springer, Heidelberg (2006)
4. Ciocchetta, F., Gilmore, S., Guerriero, M.-L., Hillston, J.: Stochastic Simulation and Probabilistic Model-Checking for the Analysis of Biochemical Systems (2008) (submitted for publication)
5. Ciocchetta, F., Hillston, J.: Bio-PEPA: an extension of the process algebra PEPA for biochemical networks. In: Proc. of FBTC 2007. Electronic Notes in Theoretical Computer Science, vol. 194(3), pp. 103–117 (2008)
6. Ciocchetta, F., Hillston, J.: Process algebras for Systems Biology. In: Bernardo, M., Degano, P., Zavattaro, G. (eds.) SFM 2008. LNCS, vol. 5016, pp. 265–312. Springer, Heidelberg (2008)
7. Danos, V., Feret, J., Fontana, W., Harmer, R., Krivine, J.: Ruled-based modelling of cellular signalling. In: Caires, L., Vasconcelos, V.T. (eds.) CONCUR 2007. LNCS, vol. 4703, pp. 17–41. Springer, Heidelberg (2007)
8. Gillespie, D.T.: Exact stochastic simulation of coupled chemical reactions. Journal of Physical Chemistry 81, 2340–2361 (1977)
9. Hillston, J.: A Compositional Approach to Performance Modelling. Cambridge University Press, Cambridge (1996)
10. Priami, C., Quaglia, P.: Beta-binders for biological interactions. In: Danos, V., Schachter, V. (eds.) CMSB 2004. LNCS (LNBI), vol. 3082, pp. 20–33. Springer, Heidelberg (2005)
11. Priami, C.: Stochastic π-calculus. The Computer Journal 38(6), 578–589 (1995)
12. Priami, C., Regev, A., Silverman, W., Shapiro, E.: Application of a stochastic name-passing calculus to representation and simulation of molecular processes. Information Processing Letters 80, 25–31 (2001)
13. Regev, A.: Representation and simulation of molecular pathways in the stochastic π-calculus. In: Proceedings of the 2nd workshop on Computation of Biochemical Pathways and Genetic Networks (2001)
14. Regev, A., Shapiro, E.: Cells as computation. Nature 419, 343 (2002)

An Exact Brownian Dynamics Method
for Cell Simulation

Koichi Takahashi

The Molecular Sciences Institute, Berkeley, USA
Computational Systems Biology Research Group, RIKEN, Yokohama, Japan
Institute for Advanced Biosciences, Keio University, Fujisawa, Japan
ktakahashi@riken.jp

1 Introduction

As we obtain better abilities to observe cellular biochemistry at the single cell / molecular levels, such as through fluorescent correlation spectroscopy and single particle tracking, evidences are accumulating that the cells may be taking advantage of intracellular spatial features to realize and optimize their functions. Computer simulation is a useful means to bridge the gap between the microscopic, physico-chemical picture of how macro-molecules diffuse and react, and the scales of time and space where biochemistry and physiology take place.

One important aspect of intracellular space is the extremely high density of macromolecules (50-400 mg/ml, compare to 1-10 mg/ml typical in vitro conditions), called intracellular macromolecular crowding[1], which results in different equilibrium points, altered reaction rates, slow and anomalous diffusion of macromolecules, and thus modified overall behaviors and dynamical characteristics of biochemical systems.

2 Computational Method

To address the pressing computational need to precisely model cellular biochemistry with microscopic details, we have been developing a high-performance Brownian Dynamics (BD) method. The new computational method I will present in this talk is called eGFRD (enhanced Greens Function Reaction Dynamics)[2], developed in collaboration with ten Wolde at AMOLF, Amsterdam and Tanase-Nicola in University of Michigan. eGFRD even further accelerates a previously proposed high-performance method called GFRD[3] by making the computation from synchronous- to asynchronous-discrete-event through introduction of the concept called first-passage processes[4]. The multi-body reaction-diffusion system that constitutes the biochemical network is decomposed into a set of one- and two-body problems, each of which are analytically solvable through Smoluchowski equation. Evolutions of the two-body problems are given by numerically evaluating the fundamental solutions to the diffusion-reaction equation called Green's functions. This new method is exact (based directly on analytical solutions of the diffusion-reaction equation), high-performance (orders of magnitude

M. Heiner and A.M. Uhrmacher (Eds.): CMSB 2008, LNBI 5307, pp. 5–6, 2008.
© Springer-Verlag Berlin Heidelberg 2008

faster than conventional BD methods), extendable (different types of diffusion and boundary conditions can be efficiently implemented) and thus opens a possibility to applications of exact particle methods to cellular-level problems which has previously been unrealistic due to high computational costs.

3 Results

I will present some of interesting results from the numerical experiments that were made possible by the advent of the very high performance particle simulation methods such as eGFRD. It will include effects of space on signaling systems that involve multisite covalent modifications such as MAPK pathways, and effects of molecular crowding on biochemical reactions in gene expression and signaling systems.

4 Other Topics

I will also briefly talk about the E-Cell Project that aims to develop cell simulation technology, some other aspects of simulation methods we are developing such as lattice- and rule-based ones, our software platform E-Cell System[5], and the critical relationship between the new simulation methods and measurement technologies such as laser spectroscopy.

Acknowledgements

The eGFRD project is being carried out in collaboration with Pieter Rein ten Wolde at AMOLF, The Netherlands, and Sorin Tanase-Nicola in University of Michigan, USA. Takahashi conducted this project as a Human Frontier Science Program Cross-Disciplinary Fellow at the Molecular Sciences Institute. The E-Cell project is partly supported by JST/CREST, Yamagata prefecture and the ministry of science (MEXT) of Japan.

References

1. Takahashi, K., Arjunan, S., Tomita, M.: Space in systems biology of signaling pathways – intracellular molecular crowding in silico. FEBS Letters 579, 8 (2005)
2. Takahashi, K., Tanase-Nicola, S., ten Wolde, P.R.: Exact Green's Function Reaction Dynamics with Analytical First-Passage Time (2008) (in preparation)
3. van Zon, J.S., ten Wolde, P.R.: Simulating biochemical networks at the particle level and in time and space: Green's function reaction dynamics. Phys. Rev. Lett. 94(12), 128103 (2005)
4. Opplestrup, T., Bulatov, V.V., Gilmer, G.H., Kalos, M.H., Sadigh, B.: First-Passage Monte Carlo Algorithm: Diffusion without All the Hops. Phys. Rev. Lett. 8, 97(23), 230602 (2006)
5. Takahashi, K., Kaizu, K., Hu, B., Tomita, M.: Multi-algorithm, multi-timescale method for cell simulation. Bioinformatics 20, 4 (2004)

Multiscale Modelling of Neuronal Signalling

Nicolas Le Novère

Computational Neurobiology, EMBL-EBI,
Wellcome Trust Genome Campus,
Hinxton, Cambridge, United-Kingdom
lenov@ebi.ac.uk

Abstract. Transduction and transmission of an input signal by a neuronal dendrite involves generation, integration and propagation of at least four kinds of information: Chemical concentration such as calcium ions, chemical modification such as phosphorylation cascades, conformational information such as allosteric modulations, and electrical signals such as membrane depolarisation. One cannot claim to understand neuronal function when focussing on a single aspect. However, developing models of the four requires using different formalisms. Furthermore running simulations implies widely different requirements in terms of compute power, storage or results and duration. I will present a few results we obtained about the synaptic function and plasticity in the striatal medium-spiny neuron, using models of signalling networks, allosteric regulations, single particle diffusion and multi-compartment electrical models. I will then discuss how we can sometimes encapsulate the results obtained at a certain level of resolution in order to increase the realism of more abstract models. I will end by outlining how one could envision to build a model striatal neuron that embodies chemical, biochemical and electrical signalling.

M. Heiner and A.M. Uhrmacher (Eds.): CMSB 2008, LNBI 5307, p. 7, 2008.
© Springer-Verlag Berlin Heidelberg 2008

Systems Biology of Halophilic Archaea

Dieter Oesterhelt

Max Planck Institute of Biochemistry,
Department of Membrane Biochemistry,
Martinsried, Germany
oesterhe@biochem.mpg.de

Abstract. Systems biology is spread over all branches of life science and attracts biologists, mathematicians, physicists, computer scientists, and engineers equally. Full of promises and visions it often signalizes that the in silico eucaryotic cell is close to realization and experimental work will be needed in the future only for confirmation. At this point science becomes fiction and destroys the great potential of interdisciplinary research aiming for added value in describing a living system or its composing modules by theoretical/simulation approaches on the basis of experimental facts. As a reliable working definition of molecular systems biology the following is useful: Modelling of cells or a modules of cells with an incomplete data set. The model (ensemble of models) must have predictive value to induce experiments which lead to falsification (verification) of subsets of models until, on the basis of available data, optimally only one model is left. The approach can be either "bottom up" or "top down". We use halophilic archaea, especially the model organism Halobacterium salinarum for systems biological experiments. These procaryotes living in concentrated brines offer biochemical features which make them very suitable for systematic analysis. A first module is signal transduction where photon absorption via two photoreceptors causes three different reactions of the target, which is the flagellar motor. The system guarantees a balanced response of the cell to light for active search of the optimal conditions for photosynthesis. Experimentally, quantitative data can be collected, which link the size of stimulus to the reaction time of the flagellar motor. Further, genome wide data on members of the network, their molecular properties and protein protein interactions were made available. Altogether a model was developed, which allows to simulate all experimental results reported so far. Bioenergetics are a second module, which is ready for modelling with a bottom up approach and the central metabolism of the cell presents an example of top down modelling with about 800 reactions in the cell. Experimental data on the course of sixteen amino acids added to the growth medium as carbon source and on the rate growth were collected and a model created which is able to quantitatively predict growth curve and carbon source usage. The lecture will give account on the details of the experimental methods used, describe the modelling approaches and summarize the results, we so far obtained.

M. Heiner and A.M. Uhrmacher (Eds.): CMSB 2008, LNBI 5307, p. 8, 2008.
© Springer-Verlag Berlin Heidelberg 2008

A Partial Granger Causality Approach to Explore Causal Networks Derived From Multi-parameter Data

Ritesh Krishna[1] and Shuixia Guo[2]

[1] Department of Computer Science, University of Warwick, Coventry, CV4 7AL, UK
[2] Department of Mathematics, Hunan Normal University, Changsha 410081,
P.R. China

Abstract. Background: Inference and understanding of gene networks from experimental data is an important but complex problem in molecular biology. Mapping of gene pathways typically involves inferences arising from various studies performed on individual pathway components. Although pathways are often conceptualized as distinct entities, it is often understood that inter-pathway cross-talk and other properties of networks reflect underlying complexities that cannot by explained by consideration of individual pathways in isolation. In order to consider interaction between individual paths, a global multivariate approach is required. In this paper, we propose an extended form of Granger causality can be used to infer interactions between sets of time series data.

Results: We successfully tested our method on several artificial datasets, each one depicting various possibilities of connections among the participating entities. We also demonstrate the ability of our method to deal with latent and exogenous variables present in the system. We then applied this method to a highly replicated gene expression microarray time series data to infer causal influences between gene expression events involved in activation of human T-cells. The application of our method to the T-cell dataset revealed a set of strong causal links between the participating genes, with many links already experimentally verified and reported in the biological literature.

Conclusions: We have proposed a novel form of Granger causality to reverse-engineer a causal network structure from a time series dataset involving multiple entities. We have extensively and successfully tested our method on synthesized as well as real time series microarray data.

1 Background

Recent advances in experimental and computational techniques have revolutionized the field of molecular biology. On the one hand experimental techniques allow us to perform experiments to produce massive amount of observation data, while on the other hand computational techniques are playing an increasing role in understanding this data and building hypothesis for understanding of the

M. Heiner and A.M. Uhrmacher (Eds.): CMSB 2008, LNBI 5307, pp. 9–27, 2008.
© Springer-Verlag Berlin Heidelberg 2008

underlying biological system. Reconstructing gene-regulatory networks is one of the key problems of functional genomics [27,17]. A gene network can be visualized as a graph in which each node represents a gene and the interactions between them is represented by the edges in the graph. The edges can represent direct or indirect interactions between the genes. Large scale monitoring of gene expression is considered to be one of the most promising techniques for reconstructing gene networks [5]. Techniques like microarrays [25] generate abundant amount of data which could be used for reconstructing gene networks. A variety of approaches have been proposed to describe gene-regulatory networks, such as boolean networks [15], difference equations [27], differential equations [31] and Bayesian networks [9,23] etc. While boolean networks, difference and differential equations are based on prior biological understanding of the molecular mechanism, Bayesian networks on the other hand have been used to infer network structures directly from the data itself. The acyclicity constraint of the Bayesian networks are addressed by Dynamic Bayesian networks [8,16] but the computational and theoretical problems arise in case of incomplete dataset which is a common problem in gene expression measurements. Relevance networks and Gaussian graphical models [28] are other commonly used methods to infer network structures from time-series data, both being simple but incapable of producing directed network structures. With each approach having its advantages and disadvantages, the field of inference of network structure from gene-expression data is still open to new techniques.

The present study is about an extension and application of a multivariate data-driven statistical technique known as Granger causality, to infer the interaction patterns among multiple time series representing a set of stochastic processes. The proposed technique relies on the statistical interdependence among multiple simultaneous time series. The interdependence could be causal in nature and therefore symmetric measures such as ordinary coherence may not be suitable for measuring it. Wiener [29] proposed a new way to measure causal influence of one time series on another by conceiving the notion that the prediction of one time series could be improved by incorporating the knowledge of the second one. Granger [12] formalized this concept in the context of the linear autoregression model of causal influences. Granger causality was extended by Geweke [11] who proposed a *measure* of interdependence between two sets of time series. We have seen a recent interest in biological community regarding application of Granger causality [19,20] for temporal Microarry data. But we realize that a straight forward application of Granger causality for biological data may not be suitable when chances of latent and exogenous variables present in the system are high. In this paper, we introduce a definition of partial Granger causality. Partial Granger causality computes the interdependence between two time series by eliminating the effect of all other variables in the system. To our knowledge, the concept of partial Granger causality is new and so is its application to a time course microarray gene-expression data. In this paper, partial Granger causality is extensively tested for various toy models representing different scenarios of interdependence between sets of time series. We then applied our approach to a highly replicated microarray time series data for T-cell activation to infer the gene network. The matlab code for implementation the partial Granger causality can be downloaded from the first author's website- http://www.dcs.warwick.ac.uk/~ritesh/pgc/index.html

2   Methods

2.1   Causal Model

First, we present Geweke's method for two univariate time series and later we introduce the concept of partial causality. Geweke's method can be explained in the following way [11]. Consider two stochastic processes X_t and Y_t. Each process admits an autoregressive representation

$$X_t = \sum_{i=1}^{\infty} a_{1i} X_{t-i} + \epsilon_{1t} \tag{1}$$

$$Y_t = \sum_{i=1}^{\infty} b_{1i} Y_{t-i} + \epsilon_{2t} \tag{2}$$

where ϵ_{1t} and ϵ_{2t} are the prediction error. According to the Granger causality theory, if the prediction of one process is improved by incorporating its own past information as well as the past information of the other process, then the second process is said to cause the first process. In other words, if the variance of prediction error for the first process is reduced by the inclusion of past measurements of the second process then a causal relation from the second process to the first process exists. A joint autoregressive representation having information of past measurements of both processes X_t and Y_t can be written as

$$X_t = \sum_{i=1}^{\infty} a_{2i} X_{t-i} + \sum_{i=1}^{\infty} c_{2i} Y_{t-i} + \epsilon_{3t} \tag{3}$$

$$Y_t = \sum_{i=1}^{\infty} b_{2i} Y_{t-i} + \sum_{i=1}^{\infty} d_{2i} X_{t-i} + \epsilon_{4t} \tag{4}$$

Equation 3 represents the prediction of the current value of X_t based on its own past value as well as the past values of Y_t. variance(ϵ_{3t}) measures the strength of prediction error. According to the definition of causality [12], if var(ϵ_{3t}) < var(ϵ_{1t}), then Y_t influences X_t.

There are three types of linear interdependence or feedback which exist for a pair of time-series. First is the causal influence form Y_t to X_t, where var(ϵ_{3t}) < var(ϵ_{1t}) and the influence can be expressed as

$$F_{Y \to X} = \ln(|\text{var}(\epsilon_{1t})|/|\text{var}(\epsilon_{3t})|) \tag{5}$$

If $F_{Y \to X} > 0$, then $Y \to X$ exists.

Second is the causal influence from X_t to Y_t defined by

$$F_{X \to Y} = \ln(|\text{var}(\epsilon_{2t})|/|\text{var}(\epsilon_{4t})|) \tag{6}$$

and the third type is instantaneous causality due to factors possibly exogenous to the (X, Y) system when $\gamma = \text{cov}(\epsilon_{3t}, \epsilon_{4t}) \neq 0$. The instantaneous causality can be expressed as

$$F_{X.Y} = \ln(|\text{var}(\epsilon_{3t})|.|\text{var}(\epsilon_{4t})|/|L|) \tag{7}$$

where

$$L = \begin{bmatrix} \text{var}(\epsilon_{3t}) & \gamma \\ \gamma & \text{var}(\epsilon_{4t}) \end{bmatrix}$$

When $\gamma = 0$, $F_{X.Y} = 0$, no instantaneous causality exists. But when $\gamma^2 > 0$, then $F_{X.Y} > 0$ and instantaneous causality exists.

The above definitions imply that the total interdependence between two time series X_t and Y_t can be defined as

$$F_{X,Y} = F_{X \to Y} + F_{Y \to X} + F_{X.Y} \tag{8}$$

2.2   Partial Causal Influence

For a network having multiple entities, various possibilities for connection among entities arise. An entity can be connected to other entities in a direct or an indirect way. This issue is of concern for network inference in order to filter out redundant channels. The benefit of multivariate model fitting is that it uses information from all the participating entities in the system, making it possible to verify whether two entities share direct causal influence while the effect of other entities are taken into account. Also, the pairwise analysis of two time series is not sufficient to reveal if the causal relationship between a pair is direct or not. The partial Granger causality between a pair can be used to analyse the strength of direct interaction between a pair of entities after eliminating the effect of other variables present in the system. By other variables, we not only mean the other observed variables in the system but also the exogenous and latent variables. Exogenous variables represent the common experimental drives present in any experimental setup, whereas, the latent variables account for the unobserved or hidden data which couldn't be captured during the experiment. The instantaneous causality proposed by Granger and Geweke represents the influence of exogenous and hidden variables on the interdependence between a pair of entities in a system. The above measurements for directed causalities depended on the effect of all the observed variables present in the system. In the proposed definition of partial causality, we compute the linear dependence between two entities by *eliminating* the effect of all other variables. As a result of this elimination, it is possible to compute the strength of direct interaction between two entities in a system. Partial Granger causality can be explained in the following way. Consider two processes X_t and Z_t. The joint autoregressive representation for X_t and Z_t can be written as

$$X_t = \sum_{i=1}^{\infty} a_{1i} X_{t-i} + \sum_{i=1}^{\infty} c_{1i} Z_{t-i} + \epsilon_{1t} \tag{9}$$

$$Z_t = \sum_{i=1}^{\infty} b_{1i} Z_{t-i} + \sum_{i=1}^{\infty} d_{1i} X_{t-i} + \epsilon_{2t} \tag{10}$$

The noise covariance matrix for the system can be represented as

$$S = \begin{bmatrix} \text{var}(\epsilon_{1t}) & \text{cov}(\epsilon_{1t}, \epsilon_{2t}) \\ \text{cov}(\epsilon_{1t}, \epsilon_{2t}) & \text{var}(\epsilon_{2t}) \end{bmatrix}$$

where var and cov represent variance and co-variance respectively. Extending this concept further, the vector autoregressive representation for a system involving three processes X_t, Y_t and Z_t can be written in the following way.

$$X_t = \sum_{i=1}^{\infty} a_{2i} X_{t-i} + \sum_{i=1}^{\infty} b_{2i} Y_{t-i} + \sum_{i=1}^{\infty} c_{2i} Z_{t-i} + \epsilon_{3t} \tag{11}$$

$$Y_t = \sum_{i=1}^{\infty} d_{2i} X_{t-i} + \sum_{i=1}^{\infty} e_{2i} Y_{t-i} + \sum_{i=1}^{\infty} f_{2i} Z_{t-i} + \epsilon_{4t} \tag{12}$$

$$Z_t = \sum_{i=1}^{\infty} g_{2i} X_{t-i} + \sum_{i=1}^{\infty} h_{2i} Y_{t-i} + \sum_{i=1}^{\infty} k_{2i} Z_{t-i} + \epsilon_{5t} \tag{13}$$

The noise covariance matrix for the above system can be represented as

$$\Sigma = \begin{bmatrix} \text{var}(\epsilon_{3t}) & \text{cov}(\epsilon_{3t}, \epsilon_{4t}) & \text{cov}(\epsilon_{3t}, \epsilon_{5t}) \\ \text{cov}(\epsilon_{3t}, \epsilon_{4t}) & \text{var}(\epsilon_{4t}) & \text{cov}(\epsilon_{4t}, \epsilon_{5t}) \\ \text{cov}(\epsilon_{3t}, \epsilon_{5t}) & \text{cov}(\epsilon_{4t}, \epsilon_{5t}) & \text{var}(\epsilon_{5t}) \end{bmatrix}$$

The partial Granger causality between X_t and Y_t by eliminating all the effect of Z_t, can be calculated by portioning the noise covariance matrices S and Σ in the following way -

$$S = \left[\begin{array}{c|c} \text{var}(\epsilon_{1t}) & \text{cov}(\epsilon_{1t}, \epsilon_{2t}) \\ \hline \text{cov}(\epsilon_{1t}, \epsilon_{2t}) & \text{var}(\epsilon_{2t}) \end{array} \right] = \left[\begin{array}{c|c} S_{11} & S_{12} \\ \hline S_{21} & S_{22} \end{array} \right]$$

$$\Sigma = \left[\begin{array}{c|c} \text{var}(\epsilon_{3t}) & \text{cov}(\epsilon_{3t}, \epsilon_{5t}) \\ \hline \text{cov}(\epsilon_{3t}, \epsilon_{5t}) & \text{var}(\epsilon_{5t}) \end{array} \right] = \left[\begin{array}{c|c} \Sigma_{XY} & \Sigma_{XYZ} \\ \hline \Sigma_{ZXY} & \Sigma_{ZZ} \end{array} \right]$$

The measure for partial causality from Y_t to X_t by eliminating the effect of Z_t can be expressed as

$$F_{Y \to X|Z} = \ln \left(\frac{S_{11} - S_{12} S_{22}^{-1} S_{21}}{\Sigma_{XY} - \Sigma_{XYZ} \Sigma_{ZZ}^{-1} \Sigma_{ZXY}} \right) \tag{14}$$

Based on the above formulation, we demonstrate in the following sections that partial causality is a good tool for inferring a network structure from a given set of time series data.

2.3   Prerequisites for Causal Models

Stationary time series: Measurement of linear dependence between multiple time series assumes the time series to be stationary. We assume our time-series t obe weakly stationary.

Linear independence among entities: Before fitting an autoregressive model on a set of processes, it is important to check that the processes are linearly independent. The check ensures that the fluctuation in the estimate of one parameter will not be compensated by the fluctuations in the estimate of other parameters. To check for linear independence among p variables, a sample variance-covariance matrix S can be calculated, which contains p variances and $\frac{1}{2}p(p-1)$ potentially different covariances. The determinant of S provides a generalized sample variance, and is equal to zero in case of linear dependence between the variables. In the case of linear dependence among variables, some of the variables should be removed from the sample. See [14] for details.

Selection of lag order: In order to find the causal relationship between variables, the equations (1-4) can be estimated using ordinary least squares which depends on a lag value, p. The model order p can be determined by minimizing the Akaike Information Criterion (AIC, [2]) defined as

$$AIC(p) = 2log(|\sigma|) + \frac{2m^2p}{n} \tag{15}$$

where σ is the estimated noise covariance, m is the dimension of the stochastic process and n is the length of data window used to estimate the model.

2.4   Bootstrap Analysis

In order to have a confidence interval for every edge present in the network, it is important to estimate the distribution of the partial causality values between different pairs of entities in a network. The confidence interval can be used as a statistical measure to separate relevant edges from the pool of all possible edges in the network. The distribution of the partial causality values in a network is determined by the bootstrap method. Consider $Y = \{E_1, E_2, \ldots, E_N\}$ to be a set of variables, where each of E_i is a time series of length l. The partial causality value between any two variables can be denoted as f_i which can be computed according to Equation 14. The set of all possible partial causality values between all possible p pairs of variables in Y can be denoted as $F = \{f_1, f_2, \ldots, f_p\}$.Following procedure can be applied to compute a bootstrap confidence interval for F using the 3σ method.

- Multiple samples of data for the system Y can be generated to create a bootstrap sample $B = \{Y_1^*, Y_2^*, \ldots, Y_L^*\}$.
- For each Y_i^* in B, compute partial causality values F_i^*. This will give a bootstrap estimates $F_1^*, F_2^*, \ldots, F_L^*$for the partial causality values obtained from the bootstrap sample B.

- A standard deviation σ_i^* for each f_i in F can be computed by the distribution of corresponding f_i^* values in $F_1^*, F_2^*, \ldots, F_L^*$.
- For 99.7% confidence level, obtain lower bound(lb) and upper bound(ub) for a f_i.

$$(lb, ub) = \{f_i - 3 \times \sigma_i^*, f_i + 3 \times \sigma_i^*\}$$

- Test the null hypothesis that the f_i values is significant by rejecting the null hypothesis if the confidence interval does not contain the value 0. So, the edges having their f_i values in F are accepted to appear in the network whose lb >0. Rest of the edges are supposed to be absent.

3 Results and Discussion

Illustrative Examples. We demonstrate the concept of partial Granger causality for network inference with the following toy models. These toy models are inspired by Baccala et al. [3].A Matlab routine was developed to compute partial causality for a given multivariate system and was tested on the following systems. The Matlab code can be found at the website mentioned in the Background section of this paper.

There are 5 entities in each of the given examples. A complete graph of 5 nodes has $10 \times 2 = 20$ possible *directed* edges (see Table 1 for edge enumeration). We computed partial causality for all the pairs (X,Y) forming an edge in the complete graph for both the directions ($X \rightarrow Y$ and $Y \rightarrow X$), eliminating the effect of all other entities in the network. The magnitude of partial Granger causality for a directed edge represents the weight associated with that edge.

Example 1. Suppose that 5 simultaneously generated time-series are represented by the equations:

$$x_1(n) = 0.95\sqrt{2}x_1(n-1) - 0.9025x_1(n-2) + w_1(n)$$
$$x_2(n) = 0.5x_1(n-2) + w_2(n)$$
$$x_3(n) = -0.4x_1(n-3) + w_3(n)$$
$$x_4(n) = -0.5x_1(n-2) + 0.25\sqrt{2}x_4(n-1) + 0.25\sqrt{2}x_5(n-1) + w_4(n)$$
$$x_5(n) = -0.25\sqrt{2}x_4(n-1) + 0.25\sqrt{2}x_5(n-1) + w_5(n)$$

where $w_i(n)$ are zero-mean uncorrelated white processes with identical variance. One can see that $x_1(n)$ is a direct source to $x_2(n), x_3(n)$, and $x_4(n)$. $x_4(n)$ and $x_5(n)$ share a feedback loop and there is no direct connection from $x_1(n)$ to $x_5(n)$. The final network structure obtained after computing the partial causality and the confidence intervals on each edge can be seen in Figure 1(a). The figure represents the equations correctly and similar network structures were obtained when tested against multiple datasets representing the above mentioned system. Figure 2(a) presents the selected edges for 20 sample datasets for the systems representing Example 1. Figure 2 also presents the results for the Examples 2 discussed below.

Example 2. The system in Example 1 is modified where $x_1(n)$ connects to $x_3(n)$ directly and also via a distinct pathway through $x_2(n)$. $x_3(n)$ directly connects to $x_4(n)$.

$$x_1(n) = 0.95\sqrt{2}x_1(n-1) - 0.9025x_1(n-2) + w_1(n)$$
$$x_2(n) = -0.5x_1(n-2) + w_2(n)$$
$$x_3(n) = 0.5x_1(n-3) - 0.4x_2(n-2) + w_3(n)$$
$$x_4(n) = -0.5x_3(n-1) + 0.25\sqrt{2}x_4(n-1) + 0.25\sqrt{2}x_5(n-1) + w_4(n)$$
$$x_5(n) = -0.25\sqrt{2}x_4(n-1) + 0.25\sqrt{2}x_5(n-1) + w_5(n)$$

The network structure found after computing partial causality for Example 2 is shown in Figure 1(b).

Example 3. We modify the system in Example 1 by adding a common exogenous input to each of the time series.

$$x_1(n) = 0.95\sqrt{2}x_1(n-1) - 0.9025x_1(n-2) + \sqrt{a_1}w_1(n) + \sqrt{1-a_1}w_6(n)$$
$$x_2(n) = 0.5x_1(n-2) + \sqrt{a_2}w_2(n) + \sqrt{1-a_2}w_6(n)$$
$$x_3(n) = -0.4x_1(n-3) + \sqrt{a_3}w_3(n) + \sqrt{1-a_3}w_6(n)$$
$$x_4(n) = -0.5x_1(n-2) + 0.25\sqrt{2}x_4(n-1) + 0.25\sqrt{2}x_5(n-1) + \sqrt{a_4}w_4(n) + \sqrt{1-a_4}w_6(n)$$
$$x_5(n) = -0.25\sqrt{2}x_4(n-1) + 0.25\sqrt{2}x_5(n-1) + \sqrt{a_5}w_5(n) + \sqrt{1-a_5}w_6(n)$$

Again $w_i(n)$, $i = 1, 2, \ldots, 6$ are zero-mean uncorrelated white processes with identical variance. a_i, $i = 1, 2, \ldots, 5$ are parameters, w_6 is the common exogenous input to the system. When $a_i = 1, i = 1, 2, \ldots, 5$, the common exogenous input is absent. The smaller the a_i, the greater is the influence of the common exogenous input on the system. In the first case, we fixed the $a_i = 0.01$ for all time series in the system and in the second case, we chose $a_i \sim U[0, 1]$ to be random variables with Uniform distribution in $[0, 1]$. In the second case, the exogenous inputs were different for different entities in the system. For both the cases, the method was able to detect the correct network structure as shown in Figure 1(a).

Example 4. Dealing with latent variables in another key issue in network inference from a given data set. To test our method in presence of latent variables in the system, we further modified the system in example 3 by adding latent variables to each of the time series.

$$x_1(n) = 0.95\sqrt{2}x_1(n-1) - 0.9025x_1(n-2) + \sqrt{a_1/2}w_1(n) + \sqrt{(1-a_1)/2}w_6(n) + 2w_7(n-1)$$
$$+ 2w_7(n-2)$$
$$x_2(n) = 0.5x_1(n-2) + \sqrt{a_2/2}w_2(n) + \sqrt{(1-a_2)/2}w_6(n) + 2w_7(n-1) + 2w_7(n-2)$$
$$x_3(n) = -0.4x_1(n-3) + \sqrt{a_3/2}w_3(n) + \sqrt{(1-a_3)/2}w_6(n) + 2w_7(n-1) + 2w_7(n-2)$$
$$x_4(n) = -0.5x_1(n-2) + 0.25\sqrt{2}x_4(n-1) + 0.25\sqrt{2}x_5(n-1) + \sqrt{a_4/2}w_4(n) + \sqrt{(1-a_4)/2}w_6(n)$$
$$+ 2w_7(n-1) + 2w_7(n-2)$$
$$x_5(n) = -0.25\sqrt{2}x_4(n-1) + 0.25\sqrt{2}x_5(n-1) + \sqrt{a_5/2}w_5(n) + \sqrt{(1-a_5)/2}w_6(n) + 2w_7(n-1)$$
$$+ 2w_7(n-2)$$

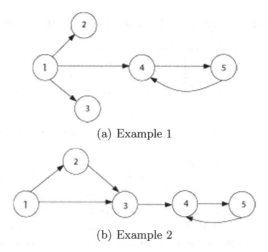

(a) Example 1

(b) Example 2

Fig. 1. The figure represents the network structures obtained for toy examples discussed in the result section

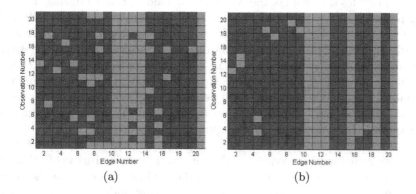

(a)   (b)

Fig. 2. The x-axis represents the edges which were expressed for the corresponding dataset on the y-axis. (a) The network in example 1 has edge number 10,11,12,13 and 20 expressed for most of the datasets. See Table 1 for relationship between edge numbers and the edges. (b) Example 2 has edges 10,11,12,15,18 and 20 expressed for most of the datasets.

$w_i(n)$, $i = 1, 2, \ldots, 7$ are zero-mean uncorrelated white processes with identical variance. a_i, $i = 1, 2, \ldots, 5$ are parameters, w_6 is the exogenous input to the system and w_7 is the latent variable. The system was tested for $a_i \sim U[0, 1]$ and the inferred network structure remained same as shown in Figure 1(a).

We used the bootstrap method to determine the significant edges forming the network. 2000 datasets were generated for each example and the final result was analysed for each dataset to check for consistency of the network structure obtained. Table 2 in the supplementary material presents the confidence interval

Table 1. Enumerating all the directed edges in the toy example

Edge Number	1	2	3	4	5	6	7	8	9	10
Edge	$(1 \leftarrow 2)$	$(1 \leftarrow 3)$	$(1 \leftarrow 4)$	$(1 \leftarrow 5)$	$(2 \leftarrow 3)$	$(2 \leftarrow 4)$	$(2 \leftarrow 5)$	$(3 \leftarrow 4)$	$(3 \leftarrow 5)$	$(4 \leftarrow 5)$
Edge Number	11	12	13	14	15	16	17	18	19	20
Edge	$(1 \rightarrow 2)$	$(1 \rightarrow 3)$	$(1 \rightarrow 4)$	$(1 \rightarrow 5)$	$(2 \rightarrow 3)$	$(2 \rightarrow 4)$	$(2 \rightarrow 5)$	$(3 \rightarrow 4)$	$(3 \rightarrow 5)$	$(4 \rightarrow 5)$

limits and the mean of causality values obtained from the bootstrap sample for toy examples 1 and 2.

Further analysis revealed that only those edges were selected by the bootstrap criteria which had significantly higher partial causality values associated with them. The edges with relatively smaller values were left out after the bootstrap confidence interval test. Figure 2 displays the selected edges for 20 sample datasets for Examples 1 and 2. As it can be seen in the figure, the majority of those datasets generated the expected network structures. The filled bright squares in the figure denote the edges which passed the bootstrap confidence interval criteria. These are also the edges having a considerably higher partial causality values than other edges in the network. Table 2 in the supplementary material provides the mean of the edges for toy models 1 and 2 during the bootstrap process. This phenomena was observed for all the toy models indicating that the edges with higher magnitude have a more significant role in detection of network structure. This is an important observation considering that bootstrap is a computationally expensive and time-consuming process. The VAR (Vector Auto Regressive) modeling of a process with q entities requires $O(q^2)$ parameters and is suitable for modeling small networks but time consuming for bigger networks. Performing bootstrapping on a bigger network using this technique will require considerable time and computational resources.

The toy models demonstrate the usefulness of partial causality for inference of network structure from synthesized datasets. This helped us verify that the network structures obtained are true to the data. Figure 2 presents a visual matrix where each row on the y-axis represents a dataset and the x-axis represents the edges which are expressed after applying our model for the corresponding dataset. We can see that similar edges are expressed for most of the datasets. There are few extra edges for some of the datasets which can be explained by the property of data, some signals in a particular dataset can be more dominant due to the introduced noise. The final verification about the selection of edges with higher causality value can be performed by looking at the Table 2. The positive lower bound for the relevant edges supports the hypothesis in the bootstrap section.

3.1 Application on T-Cell Data

The methodology was applied to a publicly available microarray gene expression data obtained from a well-established model of T-cell activation by Rangel et al.[24]. The data was collected from 2 experiments characterizing the response of a human T-cell line (Jurkat) to PMA and ionomycin treatment. The dataset comprises of recorded expression levels of 58 genes observed after 0, 2, 4, 6, 8, 18, 24, 32, 48 and 72 hours. The dataset can be downloaded from the website http://public.kgi.edu/~wild/LDS/index.htm mentioned in the publication by Rangel et al.

Fitting the VAR model on the data: The VAR model was fitted on the transformed dataset with the lag selection performed according to the AIC criteria mentioned in Equation 15. A lag value between 2 to 6 was chosen which minimized the AIC value for the system. Figure 3 represents the Q-Q plot for four genes. The plots were obtained after fitting the VAR model on the whole dataset. The linearity of the plots indicate that the actual time series values for a gene were in accordance with the predicted series. Plots in Figure 4 represent the histograms and cross-correlation measures for the standard innovations obtained for those four genes. The innovations exhibit Normal distribution. A similar pattern was observed for other genes as well after fitting the VAR model. The coefficient of determination for all 58 equations, each one representing a gene, is also presented in Figure 5. After the model fitting was done, a variance-covariance matrix for the residuals was obtained for the whole system. Partial Granger causality values were computed for each pair of genes in the dataset according to Equation 14. The distribution of calculated partial causality values can be found in the Figure 5.

Detection of the network structure: The total number of possible edges in a system of 58 entities is $\binom{58}{2} \times 2 = 3306$. Performing a bootstrap on such a big system is extremely time-consuming and computationally demanding due to the complexity of VAR models. We then relied on the observation that we made while studying the toy models, which revealed only the edges with higher partial Granger causality values compared to the rest of the edges. This was confirmed by the confidence interval tests performed on those models. Figure 2 and Table 2 support this theory for toy models. A threshold to select the most dominant edges was chosen from the tail of the empirical distribution of partial causality values for the T-cell data. Though the choice of threshold is user dependent and can vary from case to case, we use the value of 0.5743 which corresponds to the 97.5 percentile as the threshold to detect the relevant edges. A total of 83 edges were found to have partial causality value higher than the threshold. The obtained network can be seen in Figure 6.

Analysis of the inferred network structure: The threshold criteria for inference of network structure resulted in the elimination of 11 genes from the final network obtained. The elimination of nodes doesn't imply that they don't play an active role in the T-cell system, but only indicates that the interaction caused by them in our inferred causal network is weaker than the interactions caused by the entities present in the network. A complete list of the genes shown in Figure 6 along with the missing ones can be found in the Table 3 in the supplimentary material of this paper. Some key genes are listed in the caption of Figure 6.

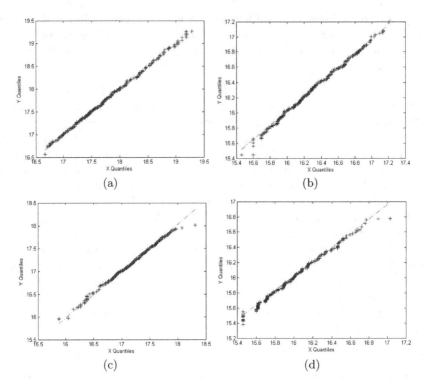

Fig. 3. Q-Q plots of actual data versus predicted data after fitting the autoregressive model

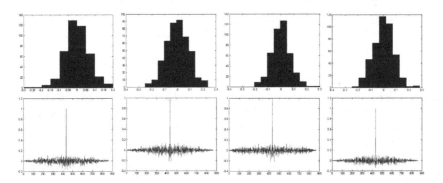

Fig. 4. Histogram and cross-correlation plot for innovations after fitting the autoregressive model

From a purely computational point of view, the network has two remarkable properties which are commonly found in most of the biological networks. The first property is the sparseness of connections in the network, and the second is the existence of hub-and-spoke structure in the network. There are several edges

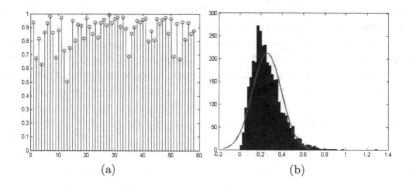

(a)   (b)

Fig. 5. (a) Plot of coefficient of determination after fitting the VAR model on tcell data.(b) Histogram plot for the partial causality between all pairs of genes in the dataset.

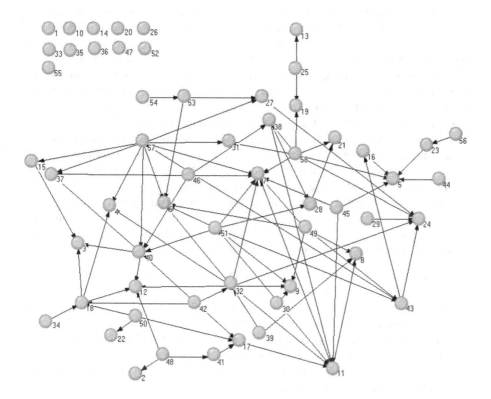

Fig. 6. The full names and complete list of all the genes is provided in the Table 3. Key genes discussed in the discussion are CD69 (gene 7),LAT (gene 57), FYB (gene 45),integrin-αM (gene 15),IL-2Rγ (gene 46),NF-κB (gene 56) and IL-16 (gene 23).

emanating from nodes 32 (superoxide dismutase 1), 57 (LAT) and 58 (v-akt murine thymoma viral oncogene homolog 1), and several edges terminating at nodes 7 (CD69), 11 (jun D proto-oncogene) and 24 (adenomatosis polyposis coli). Barabasi argues that such structures are natural for the biological systems and knocking out a hub can break down the network [4]. Among the links found in the network, we obtained a few gene-gene interactions that have been documented earlier. Zhang et al.[32] showed that LAT is required for up-regulation of CD69 in T-cells, whereas the role of IL-2Rγ for regulation of CD69 was discussed by Cheng et al.[6]. Pasquet et al. reported the activation of integrin-αM by LAT [21]. Influence of FYB on CD69 has been reported by Cambiaggi et al. [7]. A significant correlation between NF-κB activation and level of IL-16 was discovered by Takeno et al.[26] and also reported by Hidi et al. [13].

4   Conclusions

Advances in experimental techniques in molecular biology have enabled researchers to perform high-throughput experiments, enabling them to simultaneously monitor activities of numerous biological entities at different time points. Quantitative analysis of experimental data helps researchers to build hypothesis about the system and design new experiments. In this paper, we propose the use of partial Granger causality to quantitatively infer the underlying causal network structure based on microarray data. The application of this technique was first studied for various toy models and then later applied to the T-cell microarray data for deduction of causal network structure. The multivariate nature of this technique makes it useful for the systems having large number of entities engaged in cross-communication with each other. The technique is simple in nature and can be easily applied to small as well as bigger systems.

Before concluding this section, we would like to summarize the main advantages of using this technique and compare this technique with other commonly used approaches for inferring network structure from biological data. The main benefits of using our technique are the following:

- Non-availability of prior knowledge about the system is not a limitation and doesn't restrict us from studying large systems.
- Our model is inherently able to capture feedback cycles in a system which is a common feature for biological systems.
- The computation in its simplest case is very straightforward and as a result the outcome is very reliable and robust.
- In this paper we only deal with linear causality cases, but we can readily extend the concept to non-linear cases.
- When we deal with time sequence data, we are able to assess the causality relationship in the frequency domain. Such information is usually crucial when we deal with a dynamical system.

Among the commonly used methods for inference of network structure from timeseries data, relevance networks, Gaussian graphical models and Bayesian networks are the prominent and widely used ones. In the following text we present a brief overview of these methods and the problems associated with them.

Relevance networks (RN): Relevance networks are based on pairwise association score which is a correlation based method. The principle disadvantage of this method is, that inference of an interaction between two nodes is not performed in the context of the whole system. Correlation based methods are incapable of distinguishing between direct and indirect interactions and are unable to capture feedback loops.

Gaussian graphical models (GGM): GGMs - also known as undirected probabilistic graphical models are inferred by calculating partial correlation between two nodes conditional on all the other nodes in the network. Though the direct interaction between two nodes is computed in the context of the whole network, the method still suffers from some of the problems of relevance networks, namely lack of direction and feedback cycles in the inferred network.

Bayesian networks (BN): BNs are widely used directed graphical models for representing probabilistic relationship between multiple interacting entities. Although Bayesian Networks underlie a powerful technique for inferring multivariate joint probability distributions for a large amount of data, they have their own limitations. First, Bayesian networks themselves are not causal networks. Causal networks can be derived from Bayesian networks by means of interventions achieved by fixing the values of a variable to a constant value, and assuming that values of none other variables are affected by the interventions applied. Second, the technique is computationally expensive and requires a huge amount of memory and time for computation. Third, it selects from a large number of possible structures (super-exponential in number) requiring one to apply some heuristic search algorithm to decide on a single structure from the pool of structures. Heuristic algorithms suffer from their own problems such as high chances of getting struck in local minimal and time needed for computation. Heuristic algorithms also imply a bias towards putting more weight on expected structures. Another important limitation of static Bayesian Networks is their inability to capture cycles in a network. They can only capture acyclic structures. Feedback loops are a common feature of biological pathways, so static Bayesian Networks may not be a suitable tool for modeling them. The acyclicity constraints associated with the static Bayesian networks are relaxed in Dynamic Bayesian networks and the models are able to capture feedback loops in a system. In the fully observable case, learning the network structure should not be overly complicated. But computational and theoretical problems arise in case of incomplete data, which is generally a case with gene expression measurements. DBNs generally use expectation-maximization (EM) and recently more efficient structure EM (SEM) algorithm for learning network structures from a given dataset having hidden variables (unknown control/regulatory factors). However, as the EM is an optimization routine, computational costs increase and there is no guarantee to attain the global maxima. The inferred network structure is also dependent on the choice of prior and the scoring function used for evaluating networks.

We propose that our framework can be used as an alternative technique for network elucidation. It is statistically simpler and easier to implement. It also addresses the limitations associated with the above mentioned popular methods. Our approach contains a number of implicit assumptions that must be mentioned here. First, the method assumes multivariate normality and weak stationarity of the time series under consideration. This is not much of a problem because of the calibration and normalization procedures used during the preprocessing of experimental data. The second assumption is the linear relationship among the entities under consideration. Although Granger causality itself is not restricted due to the non-linear relationship between the entities, the method proposed in this paper is based on the linear regression method which can be extended to its non-linear form. Few examples of non-linear extensions of Granger causality can be found in the publications by Ancona et al. [1] and Marinazzo et al. [18]. Last, and most importantly, the proposed causality model can be most useful when experimental conditions are chosen in such a way that they activate the measured network strongly and there is minimum error in data recording.

Acknowledgements

We are thankful to Prof. JF Feng and Dr. S Kalvala for useful discussions. RK is supported by the Department of Computer Science, University of Warwick.

References

1. Ancona, N., Marinazzo, D., Stramaglia, S.: Radial basis function approach to non-linear Granger causality of time series. Physical Review E 70, 056221 (2004)
2. Akaike, H.: Fitting autoregressive models for regression. Annals of the Institute of Statistical Mathematics 21, 243–247 (1969)
3. Baccala, L., Sameshima, K.: Partial directed coherence: a new concept in neural structure determination. Biological Cybernetics 84, 463–474 (2001)
4. Barabási, A.: Linked: The New Science of Networks. Perseus Books Group, 0738206679 (2002)
5. Berkum, N.: DNA microarrays: raising the profile. Current Opinion in Biotechnology 12(1), 48–52 (2001)
6. Cheng, L., Ohlen, C., Nelson, B., Greenberg, P.: Enhanced signaling through the IL-2 receptor in CD8+ T cells regulated by antigen recognition results in preferential proliferation and expansion of responding CD8+ T cells rather than promotion of cell death. PNAS 99(5), 3001–3006 (2002)
7. Cambiaggi, C., Scupoli, M., Cestari, T., Gerosa, F., Carra, G., Tridente, G., Accolla, R.: Constitutive expression of CD69 in interspecies T-cell hybrids and locus assignment to human chromosome 12. Immunogenetics 36, 117–120 (1992)
8. Dojer, N., Gambin, A., Mizera, A., Wilczynski, B., Tiuryn, J.: Applying dynamic Bayesian networks to perturbed gene expression data. BMC Bioinformatics 7, 249 (2006)

9. Friedman, N., Linial, M., Nachman, I., Pe'er, D.: Using Bayesian Networks to Analyze Expression Data. J. Computational Biology 7, 601–620 (2000)
10. Geier, F., Timmer, J., Fleck, C.: Reconstructing gene-regulatory networks from time series knock-out data and prior knowledge. BMC Systems Biology 1, 11 (2007)
11. Geweke, J.: Measurement of Linear Dependence and Feedback Between Multiple Time Series. Journal of the American Statistical Association 77, 304–313 (1982)
12. Granger, C.: Investigating causal relations by econometric models and cross-spectral methods. Econometrica 37, 424–438 (1969)
13. Hidi, R., Riches, V., Al-Ali, M., Cruikshank, W.W., Center, D.M., Holgate, S.T., Djukanovic, R.: Role of B7-CD28/CTLA-4 costimulation and NF-kappa B in allergen-induced T cell chemotaxis by IL-16 and RANTES. J. Immunol. 164(1), 412–418 (2000)
14. Johnson, R., Wichern, D.: Applied multivariate statistical analysis. Prentice-Hall, Englewood Cliffs (1988)
15. Kauffman, S.A.: The Origins of Order. Oxford University Press, Oxford (1993)
16. Kim, S., Imoto, S., Miyano, S.: Inferring gene networks from time series microarray data using dynamic Bayesian networks. Bioinformatics 4(3), 228–235 (2003)
17. Kitano, H.: Computational System Biology. Nature 420, 206–210 (2002)
18. Marinazzo, D., Pellicoro, M., Stramaglia, S.: Nonlinear parametric model for Granger causality of time series. Physical Review E 73, 066216 (2006)
19. Mukhopadhyay, N., Chatterjee, S.: Causality and pathway search in microarray time series experiment. Bioinformatics 23, 442–449 (2007)
20. Nagarajan, R., Upreti, M.: Comment on causality and pathway search in microarray time series experiment. Bioinformatics 24(7), 1029–1032 (2008)
21. Pasque, J.M., Gross, B., Quek, L., Asazuma, N., Zhang, W., Sommers, C.L., Schweighoffer, E., Tybulewicz, V., Judd, B., Lee, J.R., Koretzky, G., Love, P.E., Samelson, L.E., Watson, S.P.: LAT is required for tyrosine phosphorylation of phospholipase cgamma2 and platelet activation by the collagen receptor GPVI. Mol. Cell Biol. 19, 8326–8334 (1999)
22. Pearl, J.: Probabilistic Reasoning in Intelligent Systems: Networks of Plausible Inference. Morgan Kaufmann, San Francisco (1998)
23. Pe'er, D., Regev, A., Elidan, E., Friedman, N.: Inferring Subnetworks from Perturbed Expression Profiles. Bioinformatics 17, S215–S224 (2001)
24. Rangel, C., Angus, J., Ghahramani, Z., Lioumi, M., Sotheran, E., Gaiba, A., Wild, D., Falciani, F.: Modeling T-cell activation using gene expression profiling and state-space models. Bioinformatics 20(9), 1361–1372 (2004)
25. Schena, M., Shalon, D., Davis, R.W., Brown, P.O.: Quantitative monitoring of gene expression patterns with a complementary DNA microarray. Science 270(5235), 467–470 (1995)
26. Takeno, S., Hirakawa, K., Ueda, T., Furukido, K., Osada, R., Yajin, K.: Nuclear factor-kappa B activation in the nasal polypepithelium: relationship to local cytokine gene expression. Laryngoscope 112(1), 53–58 (2002)
27. Van Someren, E.P., Wessels, L.F., Backer, E., Reinders, M.J.: Genetic network modeling. Pharmacogenomics 4, 507–525 (2002)
28. Werhli, A., Grzegorczyk, M., Husmeier, D.: Comparative evaluation of reverse engineering gene regulatory networks with relevance networks, graphical gaussian models and bayesian networks. Bioinformatics 22(20), 2523–2531 (2006)

29. Wiener, N.: The theory of prediction. In: Beckenbach, E.F. (ed.) Modern Mathermatics for Engineers, ch. 8. McGraw-Hill, New York (1956)
30. Yang, Y., Dudoit, S., Luu, P., Lin, D., Peng, V., Ngai, J., Speed, T.: Normalization for cDNA microarray data: a robust composite method addressing single and multiple slide systematic variation. Nucleic Acids Research 30(4), 15 (2002)
31. Yeung, M., Tegnérdagger, J., Collins, J.: Reverse engineering gene networks using singular value decomposition and robust regression. PNAS 99(9), 6163–6168 (2002)
32. Zhang, W., Irvin, B., Trible, R., Abraham, R., Samelson, L.: Functional analysis of LAT in TCR-mediated signaling pathways using a LAT-deficient Jurkat cell line. International Immunology 11(6), 943–950 (1999)

Supplementary Material

Table 2. Confidence interval bounds for Examples 1 and 2. The edge numbers correspond to the edges enumerated in Table 1.

Edge number	Example 1			Example 2		
	Mean - 3σ	Mean + 3σ	Mean	Mean - 3σ	Mean + 3σ	Mean
1	-4.03e-003	8.03e-003	2.00e-003	-4.48e-003	1.05e-002	3.02e-003
2	-3.94e-003	7.92e-003	1.99e-003	-4.48e-003	1.05e-002	3.03e-003
3	-4.16e-003	8.36e-003	2.10e-003	-4.37e-003	1.04e-002	3.03e-003
4	-3.94e-003	8.00e-003	2.03e-003	-4.44e-003	1.08e-002	3.18e-003
5	-4.25e-003	8.47e-003	2.11e-003	-4.31e-003	1.03e-002	3.04e-003
6	-4.17e-003	8.29e-003	2.05e-003	-4.64e-003	1.08e-002	3.08e-003
7	-4.00e-003	7.87e-003	1.93e-003	-4.64e-003	1.06e-002	2.97e-003
8	-6.85e-003	3.70e-002	1.51e-002	-4.39e-003	1.04e-002	3.03e-003
9	-4.84e-003	9.84e-003	2.49e-003	-4.60e-003	1.08e-002	3.12e-003
10	6.95e-002	2.04e-001	1.36e-001	6.76e-002	2.01e-001	1.34e-001
11	2.50e-001	4.50e-001	3.50e-001	4.02e-001	6.25e-001	5.13e-001
12	2.35e-002	1.24e-001	7.41e-002	1.95e-001	3.74e-001	2.84e-001
13	2.51e-001	4.50e-001	3.51e-001	-4.34e-003	1.03e-002	3.00e-003
14	-4.02e-003	8.00e-003	1.99e-003	-4.39e-003	1.05e-002	3.08e-003
15	-6.26e-003	4.19e-002	1.78e-002	9.49e-002	2.47e-001	1.71e-001
16	-3.95e-003	7.97e-003	2.00e-003	-4.53e-003	1.07e-002	3.08e-003
17	-3.96e-003	7.96e-003	1.99e-003	-4.71e-003	1.10e-002	3.15e-003
18	-4.00e-003	8.02e-003	2.00e-003	1.64e-001	3.46e-001	2.55e-001
19	-3.95e-003	8.02e-003	2.03e-003	-4.44e-003	1.05e-002	3.06e-003
20	8.72e-002	2.37e-001	1.62e-001	6.88e-002	2.01e-001	1.35e-001

Table 3. List of genes in T-cell data

Gene number	Gene name
1	retinoblastoma 1 (including osteosarcoma)
2	cyclin G1
3	TNF receptor-associated factor 5
4	clusterin (complement lysis inhibitor, SP-40,40, sulfated glycoprotein 2, apolipoprotein J)
5	mitogen-activated protein kinase 9
6	CD27-binding (Siva) protein
7	CD69 antigen (p60, early T-cell activation antigen)
8	zinc finger protein, subfamily 1A, 1 (Ikaros)
9	interleukin 4 receptor
10	mitogen-activated protein kinase kinase 4
11	jun D proto-oncogene
12	lymphocyte-specific protein tyrosine kinase
13	small inducible cytokine A2 (monocyte chemotactic protein 1, homologous to mouse Sig-je)
14	ribosomal protein S6 kinase, 70kD, polypeptide 1
15	integrin, alpha M (complement component receptor 3, alpha; also known as CD11b (p170)
16	catenin (cadherin-associated protein), beta 1 (88kD)
17	survival of motor neuron 1, telomeric
18	caspase 8, apoptosis-related cysteine protease
19	E2F transcription factor 4, p107/p130-binding
20	proliferating cell nuclear antigen
21	cyclin C
22	phosphodiesterase 4B, cAMP-specific (dunce (Drosophila)-homolog phosphodiesterase E4)
23	interleukin 16 (lymphocyte chemoattractant factor)
24	adenomatosis polyposis coli
25	inhibitor of DNA binding 3, dominant negative helix-loop-helix protein
26	Src-like-adapter
27	cyclin-dependent kinase 4
28	early growth response 1
29	transcription factor 12 (HTF4, helix-loop-helix transcription factors 4)
30	myeloid cell leukemia sequence 1 (BCL2-related)
31	cell division cycle 2, G1 to S and G2 to M
32	superoxide dismutase 1, soluble (amyotrophic lateral sclerosis 1 (adult))
33	cyclin A2
34	quinone oxidoreductase homolog
35	interleukin-1 receptor-associated kinase 1
36	SKI-INTERACTING PROTEIN
37	myeloid differentiation primary response gene (88)
38	caspase 4, apoptosis-related cysteine protease
39	transcription factor 8 (represses interleukin 2 expression)
40	apoptosis inhibitor 2
41	GATA-binding protein 3
42	retinoblastoma-like 2 (p130)
43	chemokine (C-X3-C) receptor 1
44	interferon (alpha, beta and omega) receptor 1
45	FYN-binding protein (FYB-120/130)
46	interleukin 2 receptor, gamma (severe combined immunodeficiency)
47	colony stimulating factor 2 receptor, alpha, low-affinity (granulocyte-macrophage)
48	myeloperoxidase
49	apoptosis inhibitor 1
50	cytochrome P450, subfamily XIX (aromatization of androgens)
51	CBF1 interacting corepressor
52	caspase 7, apoptosis-related cysteine protease
53	mitogen-activated protein kinase 8
54	jun B proto-oncogene
55	interleukin 3 receptor, alpha (low affinity)
56	nuclear factor of kappa light polypeptide gene enhancer in B-cells inhibitor, alpha
57	linker for activation of T cells
58	v-akt murine thymoma viral oncogene homolog 1

Functional Evolution of Ribozyme-Catalyzed Metabolisms in a Graph-Based Toy-Universe

Alexander Ullrich and Christoph Flamm

University of Vienna, Institute for Theoretical Chemistry,
Waehringerstrasse 17, A-1090 Vienna, Austria

Abstract. The origin and evolution of metabolism is an interesting field of research with many unsolved questions. Simulation approaches, even though mostly very abstract and specific, have proven to be helpful in explaining properties and behavior observed in real world metabolic reaction networks, such as the occurrence of hub-metabolites. We propose here a more complex and intuitive graph-based model combined with an artificial chemistry. Instead of differential equations, enzymes are represented as graph rewriting rules and reaction rates are derived from energy calculations of the involved metabolite graphs. The generated networks were shown to possess the typical properties and further studied using our metabolic pathway analysis tool implemented for the observation of system properties such as robustness and modularity. The analysis of our simulations also leads to hypotheses about the evolution of catalytic molecules and its effect on the emergence of the properties mentioned above.

Keywords: metabolism, evolution, simulation, enzymes, robustness.

1 Introduction

Life, in the most basic sense, constitutes of interactions between chemical compounds building complex networks which in turn can be regulated and interacted with. Living organisms adopt to the environment by means of gradual change of their internal networks and regulations. Throughout the evolutionary process, biological systems developed certain desirable properties, such as robustness and flexibility. Despite the profound knowledge of these properties and the processes within biochemical networks, the causes for the emergence of system properties are less well understood in most cases.

Metabolic networks are the best studied biochemical networks. We can reconstruct entire metabolisms because we have complete annotated genomes of model organisms at our disposal. Looking at pathways in these networks can in turn be interesting for functional genomics, e.g. gene expression data derived from DNA arrays may be better understood in terms of metabolic components, pathways or sub-networks. Insights about the metabolism of an organism can of course be useful for further biotechnological applications [1], e.g. the determination of pharmaceutical targets, metabolic engineering, changing direction and

M. Heiner and A.M. Uhrmacher (Eds.): CMSB 2008, LNBI 5307, pp. 28–43, 2008.

yield of pathways. Before we can make use of all these applications, it is essential to gain an understanding of the network properties. It is often not enough to look at textbook-like metabolisms, since they do not resemble the actual behavior of these networks. Therefore, we need means to analyze the network's topology, structure, principle, plasticity and modules. Such means exist, e.g. metabolic flux analysis [2][3], and thus we are able to make important observations about the networks and systems of interest, such as the abundance of diversity among enzymes, hub-metabolites and small world networks. But it also has to be noted that there are still some limitations to the analysis and it remains unknown how these and other possible properties emerged and further evolved. The case is especially difficult if we regard properties that cannot be sufficiently explained by looking at a static network image [4], such as robustness or evolvability.

It can be assumed and it is believed by most biologists that all organisms which we can see today are evolved from one common ancestor [5]. This ancestor would be a single cell having properties similar to those observed in cells of modern organisms. Since we do not believe this cell to have been spontaneously and magically appeared on the earth's surface, it is fair to suggest that this cell in turn gradually evolved from simpler cells. Many theories on the origin of life and scenarios for the early evolution exist, but actually we cannot say anything with certainty until the point of the common ancestor, thus all the available modern molecular techniques will fail to provide a complete account of the emergence of some of the properties we are interested in.

Models for the simulation of the emergence of network properties exist and have provided explanations for some of the properties. For example, [6] showed that gene-duplication can account for the property of a network to be scale free. So far these models of biological systems use either differential equations [7], i.e. enzymes are not modeled as actual chemical entities but only as rates, or very abstract artificial chemistries. A more complex simulation integrating more functional constraints of the metabolism should provide further insights about the metabolism itself, properties of complex networks in general and also their emergence, so that these properties may be reproducible in other applications. For instance,, artificial networks are desired to be robust and maybe even evolvable as well.

2   Model

In this section we will discuss the basic framework of the simulation -the model-, the basic ideas behind it and explain the individual components. All structures in the simulation are modeled as graphs and processes are performed through applications on the structure of graphs or the analysis of those. The choice to use graphs as the presentation for the structures in our chemical environment can be justified, firstly, by the fact that in chemistry molecules are for many years represented in graph form. Also it is the most intuitive way to regard chemical substances. Besides, networks are best understood by looking at its graph representation. And considering modern reaction classification systems[8,9,10,11], even for reactions and thus enzymes graphs can be used as appropriate models. Furthermore, it is hoped and believed that using graphs for all parts of the model results in a more realistic behavior of the entire system. Also there exist versatile applications which can be performed on graphs to analyze and transform them, such as metabolic pathway analysis or graph-rewrite systems.

The graph-based model is supported by an artificial chemistry, ToyChem[12], completing the universe in which individuals and their metabolisms can evolve. The artificial chemistry uses a graph representation of the molecule for the energy calculation. ToyChem provides the look-and-feel of a real chemistry[12] and integrates a realistically chemical behavior into our simulation, which is sufficient for our purposes and more sophisticated than has so far been at the disposal of a comparable simulation approach[13].

2.1　Genome

Every individual in the simulation population contains a genome of a fixed length and a common TATA-box sequence. Furthermore, all genes have the same length. The genome is an RNA-sequence and the single genes represent RNA-enzymes and bear the function of a particular chemical reaction from the set of reactions defined as current chemistry, as will be explained in one of the next paragraphs. In each generation, new individuals are generated from the set of optimal (with respect to metabolic yield) individuals. Those new individuals contain a copy of the parent genome to which a point mutation was applied. The mutation can occur everywhere in the genome. There can be silent mutations, i.e. the mutation takes place in a non-coding region, or neutral mutations which change a nucleotide within a gene but not the function of the corresponding RNA-enzyme, i.e. it still performs the same chemical reaction. Accordingly, there can also be missense mutations which change the structure of the RNA-enzyme in such a way that it inhabits a different function than before, and there can be mutations which either destroy a TATA-box (nonsense mutation) or build a new TATA-box and, therefore, eliminate or add a new gene to the genome, respectively.

The genome is realized by a string containing the nucleotide sequence and a list of all genes which have to be transcribed to RNA-enzymes. The sequence is treated as circular. Consequently, there are as many genes as there are TATA-boxes and some of the genes may reach over the ends or overlap. Every gene is assigned an ID. The IDs for the currently expressed genes and that of the parental genes are stored and all the genes, currently expressed or not, are listed in the genome. With this information we are able to retrace the history of every single gene and determine whether it had a single or multiple origin and if it may have disappeared for some generations before reappearing. Furthermore, the entire history of mutations is kept in the genome and we can determine the exact time of change and analyze the means of these changes. This also allows us to compare sequential and structural changes with external events, such as changes in the environment or selection for certain properties.

2.2   Metabolites

Besides the genome, individuals also include a metabolite-pool. In the first generation this pool consists only of the metabolites of the environment. The user defines the content of this environment by providing an input-file with the SMILES[14] notations of the molecules that are to be included, otherwise a predefined set of molecules is used. In each generation the newly produced metabolites are added to the metabolite-pool. To avoid redundancy, every new metabolite graph has to be checked for isomorphism against the entire pool. Since the size of the metabolite-pool can increase quickly, graph-isomorphism checking could slow down the entire simulation, therefore, we keep the graphs in a hashmap with their unique SMILES notation[15] as key, reducing the problem to a string comparison. At the end of a simulation the metabolite-pool is printed to a file, using again SMILES since it is a concise and easily interpretable way of presenting chemical molecules.

The vertices of the metabolite graph are the atoms of the respective chemical molecule and edges exist between vertices whose atoms are connected, to represent the chemical bonds of the molecule. The labels for the vertices are the atom types: hydrogen, oxygen, nitrogen or carbon (H, O, N, C). For edges the labels are the chemical bond type: single bond, double bond or triple bond $(-, =, \#)$.

2.3   Enzymes

We now turn to the most important part of the metabolism and therefore also our model. Enzymes determine the metabolic fate of a cell and almost all processes with considerable contribution to the development of a cell need enzymes. So far, simulations modeled only the reaction rates, with differential equations, but the representation of the reaction itself was rather abstract and static. We use a more flexible and realistic approach. Flexible because the set of enzymes can be adjusted by the user initially. For different purposes one might want to choose different sets of chemical reactions, e.g. sometimes only reactions working on carbon-skeletons are of interest or in another experiment only reactions involving a small number of atoms is to be observed. Realistic because many different kinds of chemical reactions are available (around 15.000 in the current simulation), but only those that are chemically valid. Due to the sheer endless number of possible chemical reactions, the set of reactions which can be chosen in the simulation is restricted here to those containing hydrogen, oxygen, carbon or nitrogen atoms. We believe that these atoms suffice to build up the most important molecules necessary for a primitive metabolism as one would expect in the early evolution of metabolism and that the simulation still resembles a realistic account for the processes at such a phase or comparable situations and does not sustain a loss of expressiveness regarding robustness and other network properties. Furthermore, we consider only the class of pericyclic reactions, which is the most important one of the three organic reaction mechanisms[16] (the other two are ioinic and radical reactions). These reactions always have a cyclic transition structure. Pericyclic reactions are used here because they are very clean, i.e. there are no unknown by-products, and they can be analyzed with frontier molecular orbital theory which is of use for the calculation of the reaction rates by the artificial chemistry. We further limit the set of reactions to those involving three to six atoms and not more than two metabolites. If we say that a certain number of atoms is involved in a reaction, then this does not mean that the metabolite which is to be worked on contains only this particular number of atoms, but rather that only the connections within a set of atoms of this particular size is changed by the reaction. Most of the already known chemical reactions lie in this range and it can be assumed that, accordingly, reactions crucial to simple metabolisms or the most basic pathways and networks underlying all metabolisms can be found there. Reactions involving more atoms or metabolites, account only for few interesting reactions and would simply add to the complexity of the computation.

The atoms and bonds of the reaction center of the chemical reaction, corresponding to the enzyme, constitute the vertices and edges of the enzyme graph. Each vertex in an enzyme graph is connected to two other vertices in such a way that the atoms build a cycle. The vertex label is equal to that of the metabolite graph, but the edge-labeling differs somewhat. In the enzyme graph, every edge has two labels for bond-types: one for the substrate molecule and the other for the product molecule. Also the bond-types for enzymes are extended by the empty symbol, indicating that two atoms are not connected. Below, the Diels-Alder reaction is shown in the GML[1][17] format which is used as the input format.

```
# ID 414141404140
rule [
  context [
    node [ id 0 label "C" ]
    node [ id 1 label "C" ]
    node [ id 2 label "C" ]
    node [ id 3 label "C" ]
    node [ id 4 label "C" ]
    node [ id 5 label "C" ]
  ]
  left [
    edge [ source 0 target 1 label "=" ]
    edge [ source 1 target 2 label "-" ]
    edge [ source 2 target 3 label "=" ]
    edge [ source 4 target 5 label "=" ]
  ]
  right [
    edge [ source 0 target 1 label "-" ]
    edge [ source 1 target 2 label "=" ]
    edge [ source 2 target 3 label "-" ]
    edge [ source 3 target 4 label "-" ]
```

---

[1] www.infosun.fim.uni-passau.de/Graphlet/GML/

```
    edge [ source 4 target 5 label "-" ]
    edge [ source 5 target 0 label "-" ]
  ]
]
```

The function of the enzyme is performed through a graph-rewriting mechanism. The graph that is transformed is that of a metabolite. The graph-rewrite rules are pairs of graphs that are gained from the enzyme graph. As explained above, an edge in the enzyme graph has two labels, one is for the substrate graph (the left side of the rule) and the other is for the product graph (the right side of the rule). First, we search for subgraphs in the metabolite that match the substrate graph and then replace it with the product graph. Note, that the atoms and the number of connections stays the same but the connections are reordered. Following, the energy of the substrate and the product are calculated as well as the reaction rate. The enzymatic reaction is only applied if the product molecule is energetically more favorable. Besides the energy calculation, metabolite graphs can be measured and assessed in terms of topological graph indices[18]. So far we use these indices (e.g. Connectivity Index[19], Platt Number[20] and Balaban Index[21]) to produce networks differing in the set of selected enzymes although starting in the same environment and with the same chemistry. We can use this to check whether the common network properties depend on lower level properties and also analyze if the selected enzymes spezialise directed or rather randomly. Furthermore, the additional use of topological indices could allow us to select for metabolites and enzymes with certain characteristics, such as very stable or very reactive metabolites and enzymes building long chains of the same molecule.

2.4 Mapping

As mentioned before, the genes encoding the enzymes are RNA-sequences and the enzymes, consequently, are modeled to act as ribozymes. To ensure a realistic behavior and evolution of our enzymes, we developed a novel genotype-phenotype mapping for the transition from the gene (genotype) to the catalytic function of the enzyme (phenotype). We use the RNA sequence-to-structure map as the basis for the mapping and consider two observations from the study of evolution and enzymes.

Firstly, it is known that neutral mutations, leading to a redundant genotype-phenotype mapping, have a considerable influence on the evolution in molecular systems. The folding of RNA-sequences to secondary structures with its many-to-one property represents such a mapping entailing considerable redundancy. Various extensive studies concerning RNA-folding in the context of neutral theory yielded to insights about properties of the structure space and the mapping itself. Thus, we will get a better understanding of some of these properties and especially of the evolution of RNA-molecules as well as their effect on the evolution of the entire molecular system.

The second observation we use is that enzymes typically have an active site where only few amino acids or bases determine its catalytic function and the remaining structure has mostly stabilization purposes. Accordingly, we extract structural and sequence information only from a restricted part of the fold. We decided to focus on the longest loop of the folded RNA since most RNA-aptamers are known to contain a loop region as their catalytic center. The idea for mapping the extracted information directly to a specific chemical reaction was inspired by the fact that many enzymes catalyze a reaction by stabilizing its transition state and the work on reaction classification systems, in particular Fujita's imaginary transition structures (ITS) approach [10,22,23].

The mapping from the structure and sequence information to the pericyclic reaction that resembles the function of the enzyme is generated as following. The length of the longest loop in the secondary structure of the RNA-enzyme determines the number of atoms that are involved in the chemical reaction to which the gene will be mapped. A statistical analysis was performed to ensure that the different reaction types occur in appropriate proportions. Further, the loop is divided in as many parts as the number of atoms involved in the reaction. The mapping to the atom types of the reaction is derived simply from the sequence information in the different parts of the loop, each corresponding to one atom. The exact mapping from sequence information to atom type here is not important since not biologically meaningful. It suffices to notice that all atom types are chosen with the same rate. The bond type of the reaction logo is derived from the structural information of the different parts of the loop, in particular, the stems contained in these parts. The number of stems in a loop region, the length of these stems, and the sequence of the first two stem pairs accounts for the decision to which bond type will be mapped. Again the exact procedure of the mapping will not be discussed because it is a rather technical detail. An example for the mapping is explained in figures 1,2 and table 1 for better understanding.

Finally, we performed several statistical tests commonly used in neutral theory. We compared it with results of approaches using cellular automatons, random boolean networks and other RNA-fold-based mappings. It exceeds all non-RNA mappings in extent and connectivity of the underlying neutral network. Further, it has a significantly higher evolvability and innovation rate than the rest. Especially interesting is the highly innovative starting phase in RNA-based mappings. This shows that the use of such a genotype-phenotype mapping contributes greatly to a more realistic modeling of evolution.

2.5   Artificial Chemistry

We will use ToyChem, an artificial chemistry with the look-and-feel of a real chemistry, for the calculation of the energy of metabolites and reaction rates. In ToyChem molecules are represented in another type of graph -the orbital graph-. From the orbital graph of a molecule all the necessary properties needed for the energy calculation can be derived. The ToyChem package also provides functionality for the computation of solvation energies [24] and reactions rates[25].

Energy Calculation. The most accurate way to calculate arbitrary properties or reaction rates for molecules is to derive a wave-function from the 3D-space

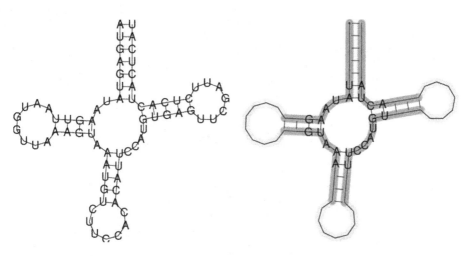

Fig. 1. Example: Reaction mapping. Left: The folded RNA. Right: Longest loop of the folded RNA and the relevant sequence and structure information (marked red).

Table 1. Example: Information derived from the longest Loop of the folded RNA-sequence. The mapped reaction contains four carbon atoms, one oxygen and one nitrogen and is unimolecular. The substrate contains one triple bond, one double bond and three single bonds, whereas the product molecule contains three double bonds and two single bonds, both are not closed.

Section	Loop	C-G pair	Neighbor > 5 bp	Bond	Valence	Seq. (loop)	Sequence
1 (red)	yes	0	yes (+1)	**1** ”–”	3	4	**4 = C**
2 (blue)	yes	1	yes (+1)	**2** ”=”	4	1	**4 = C**
3 (gray)	no	-	no	**0** ” ”	3	4	**4 = C**
4 (yellow)	yes	0	no	**0** ” ”	1	4	**4 = C**
5 (pink)	no	-	yes (+1)	**1** ”–”	2	2	**2 = O**
6 (green)	yes	1	no	**1** ”–”	3	3	**3 = N**

embedding of the molecular structures with a subsequent application of quantum mechanical methods. This approach is however rather demanding in terms of computational resources. We therefore resign to a computationally tractable approach called ToyChem [12]. This method constructs an analog of the wavefunction from the adjacency relations of a graph followed by a simplified quantum mechanical treatment called extended Hückel Theory (EHT)[26]. In the EHT the Hamilton matrix is parametrized in terms of the atomic ionization potentials and the overlap integrals between any two orbitals. The overlaps between orbitals are gained unambiguously from the molecular graph by applying the valence shell electron pair repulsion theory (VSEPR) [27]. The resulting information can conveniently be stored in the orbital graph who's vertices represent atom orbitals (labeled by atom type and hybridization state of the atom) and the edges denote overlaps of interacting orbitals. Within the ToyChem

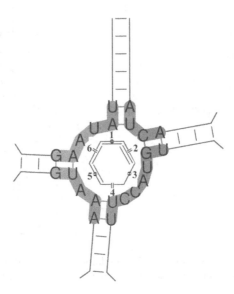

Fig. 2. Example: Extracted information maps to this particular pericyclic reaction, represented by its reaction logo. This is an example of a sigmatrophic rearrangement. The coloring indicates which information was used for the respective part of the reaction logo, the colors correspond to the labeling in the table above.

framework the atomic ionisation potentials and the overlap integrals are tabulated as functions of the atomic type and the type of the hybrid orbitals for a subset of atom which frequently occure in organic molecules. This information allows a fast construction of the Schrödinger equation from the orbital graph. Solving of the Schrödinger equation yields the eigenvectors and eigenvalues from which any physical properties of a ToyChem molecule can be calculated.

Oribtal Graph. In the orbital graph of a molecule, nodes are the atom orbitals and edges indicate overlapping orbitals. From the four atom orbitals $2p_x$, $2p_y$, $2p_z$ and $2s$, three hybrid orbitals with different geometry can be formed. The hybrid orbitals sp (linear geometry), sp^2 (trigonal geometry) and sp^3 (tetrahedral geometry) combined with the respective atom type constitute the node labels of the orbital graph. The edge labels depend on the orientation of the two interacting orbitals relative to each other. In ToyChem, three types are regarded. Therefore, there are three different edge labels, direct σ-overlap, semi-direct σ-overlap and π-overlap.

2.6   Metabolic Reaction Network

The central subject of the simulation is the metabolism, thus, we need a representation that we can easily observe and also use for analysis tools. In particular, the metabolic flux analysis but also other forms of network, graph or even grammar

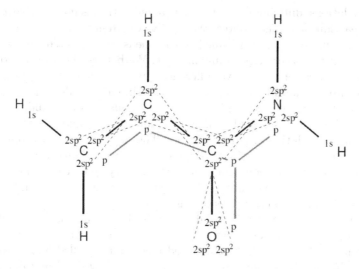

Fig. 3. Orbital graph of propenamide. Direct, semi-direct σ-overlaps, and π-overlaps are represented by solid black, dashed, and solid gray lines.

analysis as well. The most intuitive solution seems to be to use a network graph. In case of a metabolic network this could be a bidirectional and bipartite labeled graph. Bipartite because enzymes are only connected to metabolites and not to other enzymes and, vice versa, metabolites are only connected to enzymes. Bidirectional because one metabolite may at one time be the product of an enzyme and another time be the substrate of the same enzyme, therefore, the direction of a connection is important. The nodes are labeled with IDs for metabolites and enzymes. The edge labels contain information about the specific reactions in which an enzyme-metabolite pair was involved. This is necessary because we can identify the exact parts of a reaction which can be up to four metabolites and one enzyme. Further, a metabolite can be the substrate or product of an enzyme in more than one reaction. From this graph the stoichiometry matrix can easily be derived and it has the advantage that no information is lost and can be extracted in a straight forward way for almost all objectives. For example, the single reactions can be listed with substrates and products. Consequently, it is possible to analyze from which different sources a metabolite can be gained. Looking at the enzyme graph may even enable us to specify the exact regions which were joint, split or changed. The interpretability and expressiveness of this network graph, therefore, allows for a very detailed manually analysis as well as the typical computational approaches.

3 Results

We performed several simulation runs and analyzed their results to gain information about the properties of the evolved metabolic networks. Discussed will be ten simulations, differing in the topological index that is used as selection criteria for reactions and metabolites. We use five different indices and for each of them, two simulations were performed, where one is aiming to reduce the respective index and the other tries to maximize it. We believe the addition of these indices makes it easier to observe different behaviors among the networks. All simulations are initialized with a population of six individuals. The genome for the individuals is chosen randomly, but for all simulations the random number generator (RNG [28]) used for building a random genome and generate random mutations is set with the same seed number. The set of metabolites constituting the environment is the same in all simulations as well. Thus, the simulations start with equal preconditions. Furthermore, in every generation, half the population is selected and from each of the selected individuals a new individual is produced and a mutation in its genome is performed.

Metabolic networks are small world networks. Therefore, the metabolite connectivity distribution follows the power law. In other words, in a realistic metabolic network, a few highly connected metabolites, called hubs, should be observed, whereas the majority of metabolites is involved in only one or two reactions. In order to prove whether this property can be found in the networks produced by the simulation tool, the distribution of the metabolite connectivity was derived. Since the different simulations do not result in networks of equal size and it is known that in small networks, around 50 metabolites, the connectivity distribution does not follow exactly the power law and contains fewer hubs than could expected in a scale free network, we consequently group the networks into sets of networks with similar size. The values of the connectivity distribution are listed in table 2 and illustrated in figure 4.

In all networks, the majority of metabolites is involved in one or two reactions, but only larger networks ($m > 150$) contain enough highly connected metabolites to satisfy the small world property. A similar observation as in [7] can be made. In small networks, the number of hubs in the range between eight and twelve is higher than for scale-free networks, but too few hubs of higher connectivity exist. Most real world metabolic networks contain more than fifty metabolites, thus are large networks. The conclusion about the small-world property of metabolic networks, therefore, was drawn with the assumption that the network of investigation is large ($m \geq 100$). The connectivity distribution of smaller real world metabolic networks, actually, exhibit the same deviations from the power law as the networks gained from the example simulations. Accordingly, we can not state that all produced networks are scale-free, but we can assume them to resemble realistic metabolic networks.

For further analysis, one of the simulations is studied in more detail. We will discuss exemplarily the simulation that uses the minimal Balaban index [21] as selection criteria. In figure 5, 6 and 7 network graphs for generation one, two and 87 are depicted. Enzymes are drawn in light blue circles and metabolites in light gray boxes. The enzyme and metabolite indices are defined in the protocol.

From the network graph in figure 7 it can be derived, that some enzymes catalyze many reactions (e2, e12, e44, e45) and others participate in very few

Table 2. Connectivity of metabolites in networks of different sizes. Frequency in %.

Connectivity \ Metabolites	1	2	3	4	5	6	7	8 - 12	>12
avg(m)=47.5	43.91	12.61	13.03	8.61	8.82	4.83	2.31	4.41	1.47
avg(m)=104.5	50.89	17.1	6.02	5.2	5.61	5.2	3.28	5.06	1.64
avg(m)=156.5	56.21	16.36	5.85	3.56	2.92	3.29	2.1	4.02	5.67
avg(m)=249.5	58.59	12.89	8.13	6.01	2.69	2	1.37	3.67	4.64
avg(m)=570	62.8	12.1	9.12	5.49	2.89	1.2	1.46	2.34	2.63

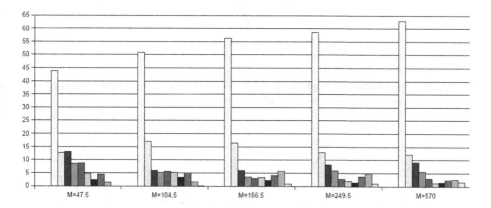

Fig. 4. Connectivity of metabolites in networks of different sizes. Frequency in %. Connectivity of 1, 2, 3, 4, 5, 6, 7, 8-12, >12, >30 for all 5 network classes.

Table 3. Specificity of enzymes in the example network

Enzyme	e37	e45	e12	e44	e18	e27	e6	e82	e4	e62	e124	e130	e2
Generation	1	1	1	2	3	3	7	10	14	42	44	51	53
Connectivity	4	5	12	9	4	2	2	2	2	2	3	1	6

reactions. In the different theories about the evolution of enzymes, it was stated that enzymes of low specificity evolve to highly specific enzymes. This is expected for the simulations as well. To study the evolution of enzymes within the simulation run, we looked at the generation in which the respective enzyme participated for the first time. This information is listed in table 3 for all enzymes in the example network in generation 87. A tendency for early enzymes to be less specific can be observed, in fact, three of the four enzymes with relatively low specificity are from generation one or two. With the exception of $e2$, all enzymes in later generations are more specific. The same observation is made for the other simulations.

The observation stated above is explained with the following scenario. In the beginning, every reaction producing valid metabolites is beneficial for the yield of the network. Since more metabolites are generated, more of the already existing enzymes find reactants. At some point, metabolites will protrude from the

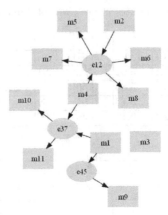

Fig. 5. Example: Network graph from simulation Balaban in generation 1. Light blue circle = enzyme, light gray box = metabolite.

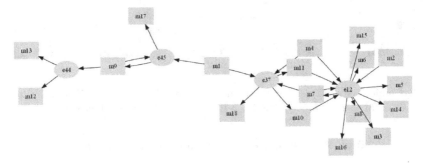

Fig. 6. Example: Network graph from simulation Balaban in generation 2. Light blue circle = enzyme, light gray box = metabolite.

metabolite pool, that is, some metabolites become more beneficial than others. This in turn means that not all of the enzymes increase the network yield. If an enzyme with low specificity overlaps in functionality with another enzyme which is specialized on the few common reactions, then the impact of the former enzyme on these reactions is very low. Since the rate of a reaction depends on how many reactions the involved enzyme performs, the remaining reactions of the lowly specific enzyme are performed on a relatively low rate. A highly specific enzyme which can catalyze the remaining non-overlapping reactions, can do so at much higher rate. Overall, it can be stated that enzymes which have a unique function have an advantage in natural selection. In later generations, most beneficial reactions are already realized by existing enzymes and only few are left. It follows that only specific enzymes can find their niche. However, sometimes lowly specific enzymes enter the network at later stages because they express a completely new functionality, e.g. different atoms, bond type or reaction type. This scenario does not comply exactly with retrograde evolution[29], since it

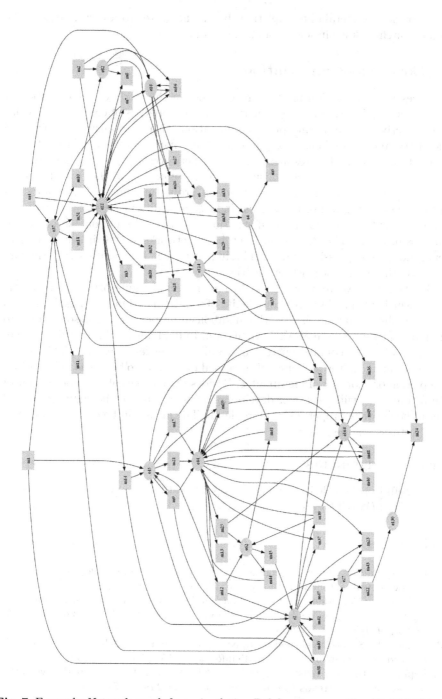

Fig. 7. Example: Network graph from simulation Balaban in generation 87. Light blue circle = enzyme, light gray box = metabolite.

does not need a metabolite depletion, but it integrates to a certain extent the idea of patchwork evolution[30] and the theory of [5].

4   Conclusions and Outlook

The presented simulation model can be a tool in the study of the evolution of metabolism and enzymes, as well as research on properties of complex networks. The underlying graph concept in combination with a sophisticated artificial chemistry and redundant genotype-phenotype map ensures a realistic behavior of the evolution. The resulting metabolic networks exhibit the characteristic properties of real world metabolisms. Various options, from the constitution of the environment and chemistry to selection properties, such as the number of descendants or the use of topological indices as additional criteria, can be adjusted. An extensive amount of information about the simulation can be gained from its protocol. The data about metabolic networks and their evolution over generations, is expressive and meaningful, so that it can be used to formulate new hypotheses or test existing theories.

For the future we are working on the integration of regulatory elements to the model which would add to the complexity of the simulated individuals, leading to a more realistic characteristic of the metabolism properties. It is also planned to interconnect the individuals of a population. Individuals would have to compete for metabolites or could cooperate and build higher-level systems. This would require a change in the modeling of the metabolite pool, so far we do not consider a depletion or shortage of metabolites. Besides the changes of the model, a lot of the future work will consist of developing ways to analyze the simulations and study the network properties in more detail. The focus will be on the emergence of robustness and flexibility.

Acknowledgements

We gratefully acknowledge financial support by the Vienna Science and Technology Fund (WWTF) project number MA05.

References

1. Liao, J.: Pathway analysis, engineering, and physiological considerations for redirecting central metabolism. Biotechnol. Bioeng. 52, 129–140 (1996)
2. Schuster, S.: Detection of elementary flux modes in biochemical networks: a promising tool for pathway analysis and metabolic engineering. Trends Biotechnol. 17, 53–60 (1999)
3. Pfeiffer, T., Sanchez-Valdenabro, I., Nuno, J., Montero, F., Schuster, S.: Metatool: for studying metabolic networks. Bioinformatics 15.3, 251–257 (1999)
4. Papin, J., Price, N., Wiback, S., Fell, D., Palsson, B.O.: Metabolic pathways in the post-genome era. Trends Biochem. Sci. 28(5), 250–258 (2003)
5. Kacser, H., Beeby, R.: Evolution of catalytic proteins. J. Mol. Evol. 20, 38–51 (1984)
6. Diaz-Mejia, J.J., Perez-Rueda, E., Segovia, L.: A network perspective on the evolution of metabolism by gene duplication. Genome. Biol. 8(2) (2007)

7. Pfeiffer, T., Soyer, O.S., Bonhoeffer, S.: The evolution of connectivity in metabolic networks. PLoS Biology 3/7, 228 (2005)
8. Arens, J.: Rec. Trav. Chim. Pays-Bas. 98, 155–161 (1979)
9. DeTar, F.: Modern approaches to chemical reaction searching. Comput. Chem. 11, 227 (1986)
10. Fujita, S.: Description of organic reactions based on imaginary transition structures. 1. introduction of new concepts. J. Chem. Inf. Comput. Sci. 26(4), 205–212 (1986)
11. Hendrickson, J.: Comprehensive system for classification and nomenclature of organic reactions. J. Chem. Inf. Comput. Sci. 37(5), 852–860 (1997)
12. Benkö, G., Flamm, C.: A graph-based toy model of chemistry. J. Chem. Inf. Comput. Sci. 43(4), 1085–1095 (2003)
13. Benkö, G.: A toy model of chemical reaction networks. Master's thesis, Universität Wien (2002)
14. Weininger, D.: Smiles, a chemical language and information system. 1. introduction to methodology and encoding rules. J. Chem. Inf. Comput. Sci. 28(1), 31–36 (1988)
15. Weininger, D.: Smiles. 2. algorithm for generation of unique smiles notation. J. Chem. Inf. Comput. Sci. 29(2), 97–101 (1989)
16. Houk, K.N., Gonzalez, J.: Pericyclic reaction transition states: Passions and punctilios, 1935-1995. Accounts of Chemical Research 28, 81–90 (1995)
17. Himsolt, M.: GML: A portable Graph File Format (Universität Passau)
18. Trinajstic, N.: Chemical Graph Theory (New Directions in Civil Engineering), 2nd edn. CRC, Boca Raton (1992)
19. Randic, M.: Characterization of molecular branching. J. Am. Chem. Soc. 97(23), 6609–6615 (1975)
20. Platt, J.: Influence of neighbor bonds on additive bond properties in paraffins. J. Chem. Phys. 15, 419–420 (1947)
21. Balaban, A.: Highly discriminating distance-based topological index. Chem. Phys. Lett. 89, 399–404 (1982)
22. Fujita, S.: Description of organic reactions based on imaginary transition structures. 2. classification of one-string reactions having an even-membered cyclic reaction graph. J. Chem. Inf. Comput. Sci. 26(4), 212–223 (1986)
23. Fujita, S.: Description of organic reactions based on imaginary transition structures. 3. classification of one-string reactns having an odd-membered cyclic reaction graph. J. Chem. Inf. Comput. Sci. 26(4), 224–230 (1986)
24. Benkö, G., Flamm, C.: Multi-phase artificial chemistry. In: The Logic of Artificial Life: Abstracting and Synthesizing the Principles of Living Systems (2004)
25. Benkö, G., Flamm, C.: Explicit collision simulation of chemical reactions in a graph based artificial chemistry. In: Capcarrère, M.S., Freitas, A.A., Bentley, P.J., Johnson, C.G., Timmis, J. (eds.) ECAL 2005. LNCS (LNAI), vol. 3630, pp. 725–733. Springer, Heidelberg (2005)
26. Hoffmann, R.: An Extended Hückel Theory. I. Hydrocarbons. J. Chem. Phys. 39(6), 1397–1412 (1963)
27. Gillespie, R.J., Nyholm, R.S.: Inorganic Stereochemistry. Quart. Rev. Chem. Soc. 11, 339–380 (1957)
28. Matsumoto, M., Nishimura, T.: Mersenne twister: a 623-dimensionally equidistributed uniform pseudo-random number generator. ACM Trans. Model Comput. Simul. 8(1), 3–30 (1998)
29. Horowitz, N.: On the evolution of biochemical syntheses. Proc. Nat. Acad. Sci. 31, 153–157 (1945)
30. Jensen, R.: Enzyme recruitment in evolution of new functions. Ann. Rev. Microbiol. 30, 409–425 (1976)

Component-Based Modelling of RNA Structure Folding

Carsten Maus

University of Rostock
Institute of Computer Science, Modelling and Simulation Group
Albert-Einstein-Str. 21, 18059 Rostock, Germany
carsten.maus@uni-rostock.de
http://www.informatik.uni-rostock.de/%7Ecm234

Abstract. RNA structure is fundamentally important for many biological processes. In the past decades, diverse structure prediction algorithms and tools were developed but due to missing descriptions in clearly defined modelling formalisms it's difficult or even impossible to integrate them into larger system models. We present an RNA secondary structure folding model described in ML-DEVS, a variant of the DEVS formalism, which enables the hierarchical combination with other model components like RNA binding proteins. An example of transcriptional attenuation will be given where model components of RNA polymerase, the folding RNA molecule, and the translating ribosome play together in a composed dynamic model.

Keywords: RNA folding, secondary structure, DEVS, model components, multi-level.

1 Introduction

Single stranded ribonucleic acids (RNA) are able to fold into complex three-dimensional structures like polypeptide chains of proteins do. The structure of RNA molecules is fundamentally important for their function, e.g. the well studied structures of tRNA and the different rRNA variants. But also other transcripts of the DNA, i.e. mostly mRNAs, perform structure formation which has been shown to be essential for many regulatory processes like transcription termination and translation initiation [1,2,3]. The shape of a folded RNA molecule can also define binding domains for proteins or small target molecules which can be found for example within riboswitches [4]. The enormous relevance for many biological key processes led to raised research efforts in identifying various RNA structures over the past decades. Unfortunately the experimental structure identification with NMR and X-ray techniques is difficult, expensive, and highly time-consuming. Therefore, many *in silico* methods for RNA structure prediction were developed which cover different requirements. Diverse comparative methods exist using alignments of similar RNA sequences to predict structures [5,6], but also many single sequence prediction algorithms work very well. Some of

M. Heiner and A.M. Uhrmacher (Eds.): CMSB 2008, LNBI 5307, pp. 44–62, 2008.

them predict the most stable RNA structure in a thermodynamical equilibrium, e.g. [7,8,9], whereas some other simulate the kinetic folding pathway over time [10,11,12,13]. The latter is also in the focus of the presented modelling approach here. Results of RNA structure predictions as well as kinetic folding simulations have reached a high level of accuracy and thus *in silico* folding became a widely used and well established technique in the RNA community. However, none of the existing tools and programs provides a flexible integration into larger system models which is also due to the fact that they are written in proprietary formalisms and do not distinguish between model description and simulation engine. To illuminate the importance of the folding processes and the possibility to integrate them into larger models, lets take a look at a concrete example of gene regulation.

2   Motivation – Modelling of Transcription Attenuation

The tryptophan (Trp) operon within bacterial genomes represents one of the best understood cases of gene regulation and has been subject to various modelling approaches [14,15]. Tryptophan is an amino acid, a building block for the production of proteins. The Trp operon includes five consecutive genes, coding for five proteins. The joint action of the proteins permits the synthesis of Trp through a cascade of enzymatic reactions. This ability is vital since the bacterium may be unable to feed on Trp from the environment. As long as Trp is obtained from the surrounding medium, its costly synthesis is impaired by a threefold control mechanism: repression of transcription initiation, transcriptional attenuation, and inactivation of the cascade of enzymatic reactions actually producing Trp. Each of these are triggered by Trp availability. Transcriptional attenuation follows if transcription starts although Trp was available. This has a small but non-negligible chance. As soon as RNA polymerase (RNAP) has transcribed the operon's leader region into an mRNA molecule, a ribosome can access this. The ribosome starts translating the mRNA content into a growing sequence of amino acids. The speed of the ribosome depends on Trp availability. The ribosome advances quickly as long as Trp is abundant, which prevents RNAP from proceeding into the operon's coding region. The attenuation is caused by the formation of a certain constellation of RNA hairpin loops in presence of a Trp molecule at a distinct segment of the mRNA molecule (figure 1). Attenuation depends on the synchronised advance of both RNAP and ribosome, and their relative positioning with respect to mRNA.

In [15] a model of the tryptophan operon was developed, in which repressors, the operon region, and the mRNA were modelled individually. However, the latter in not much detail. Only the repression of transcription initiation was included in the model. Consequently, the simulation result showed stochastic bursts in the Trp level, caused by the repressor falling off, which started the transcription. Integrating a transcriptional attenuation model would have prevented this non-realistic increase in Trp concentration, and mimicked the threefold regulation of the Trp more realistically. However, the question is how to model the

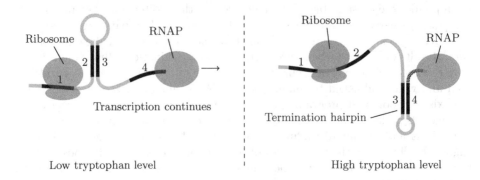

Fig. 1. Attenuation in the Trp leader sequence. With low Trp concentration the ribosome stalls at Trp codons in leader domain 1 and transcription can continue. At high Trp level, the leader domain 1 can be completely translated and thus the ribosome prevents base pairing of domain 2. Leader domains 3 and 4 can form an intrinsic transcription termination structure which causes disruption of the mRNA-RNAP complex.

attenuation. As this regulating process depends largely on structure formation, modelling of RNA folding would be a big step in the right direction for reflecting attenuation dynamics. Additionally, modelling interactions between mRNA and RNAP as well as mRNA and the ribosome are needed because both influence the kinetic folding process and RNA termination structures break up gene transcription. The focus of this paper is the RNA folding process, but at the end we will also give a detailed outlook how the composed model of tryptophan attenuation looks like and how the individual model components act together.

3   Principles of RNA Folding

3.1   Thermodynamics

The reason for RNA folding is the molecules' general tendency to reach the most favourable thermodynamical state. Complementary bases of RNA nucleotides can form base pairs by building hydrogen bonds similar to DNA double helices. Adenine (A) and Uracil (U) are complementary bases as well as Cytosine (C) and Guanine (G). In addition, the wobble base pair G-U is also frequently found in RNA structure folding. Each additional hydrogen bond of base pairs affords a small energy contribution to the overall thermodynamic stability, but there is another chemical interaction which is even more important for the RNA structure than just the number and type of base pairs. It's called base stacking and describes the interactions between the aromatic rings of adjacent bases by Van-der-Waals bonds. Base pair stacking is the cause why an uninterrupted long helix is thermodynamically more favourable than a structure of multiple single base pairs or short helices interrupted by loop regions, even if the number and type of base pairs are equal. Since the 1970s, significant progress has been done on identifying thermodynamic parameters of different base pair neighbourhoods and structural elements like hairpin loops, e.g. [16,17]. This was a precondition to develop RNA structure prediction algorithms based on energy minimisation, i.e. finding the thermodynamical most stable structure.

3.2   Primary, Secondary, and Tertiary Structure

RNA structures are hierarchically organised (see figure 2). The most simple hierarchy level is the primary structure which is nothing else than the linear sequence of nucleotides. Two nucleotides are linked over the 3' and 5' carbon atoms of their ribose sugar parts resulting in a definite strand direction. The secondary structure consists of helices formed by base pairs and intersecting loop regions. Such structural elements are formed rapidly within the first milliseconds of the folding process [18]. Interacting secondary structure elements finally build the overall three-dimensional shape of RNA molecules. Although they are formed by simple base pairs like secondary structures, helices inside loop regions are often seen as tertiary structures. Such pseudoknots and higher order tertiary interactions are, due to their complexity and analog to many other RNA structure prediction methods, not covered by our model. However, it should not retain unstated here that there are some existing tools which can predict pseudoknots quite well, e.g. [10].

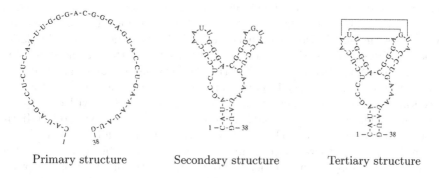

Primary structure   Secondary structure   Tertiary structure

Fig. 2. Different hierarchical levels of RNA folding

4   Modelling Formalism

As already mentioned, typically kinetic RNA folding simulations, as e.g. [10,11,12], are aimed at efficiently and accurately simulating the molecules structure formation in isolation rather than supporting a reuse of RNA folding models and a hierarchical construction of models. For approaching such model composition, we use the modelling formalism ML-DEVS [19], a variant of the DEVS formalism [20]. As DEVS does, it supports a modular-hierarchical modelling and allows to define composition hierarchies. ML-DEVS extends DEVS by supporting variable structures, dynamic ports, and multi-level modelling. The latter is based on two ideas. The first is to equip the coupled model with a state and a behaviour of its own, such that

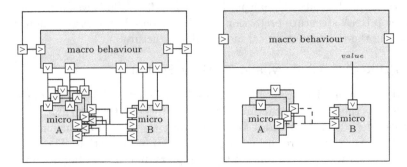

Fig. 3. Comparison of multi-level modelling with DEVS and ML-DEVS. (left) With DEVS the macro level is modelled at the same hierarchical level as the micro level models. (right) With ML-DEVS the macro dynamics are part of the coupled model. Functions for downward and upward causation reduce the number of explicit couplings needed.

the macro level does not appear as a separate unit (an executive) of the coupled model. Please recall that in traditional DEVS coupled models do not have an own state nor a behaviour. Secondly, we have to explicitly define how the macro level affects the micro level and vice versa. Both tasks are closely interrelated. We assume that models are still triggered by the flow of time and the arrival of events. Obviously, one means to propagate information from macro to micro level is to exchange events between models. However, this burdens typically modelling and simulation unnecessarily, e.g. in case the dynamics of a micro model has to take the global state into consideration. Therefore, we adopt the idea of value couplings. Information at macro level is mapped to specific port names at micro level. Each micro model may access information about macro variables by defining input ports with corresponding names. Thus, downward causation (from macro to micro) is supported. In the opposite direction, the macro level needs access to crucial information at the micro level. For this purpose, we equip micro models with the ability to change their ports and to thereby signalise crucial state changes to the outside world. Upward causation is supported, as the macro model has an overview of the number of micro models being in a particular state and to take this into account when updating the state at macro level. Therefore, a form of invariant is defined whose violation initiates a transition at macro level. In the downward direction, the macro level can directly activate its components by sending them events – thereby, it becomes possible to synchronously let several micro models interact which is of particular interest when modelling chemical reactions. These multi-level extensions facilitate modelling, figure 3 depicts the basic idea, see also [21].

5   The RNA Folding Model

The central unit in composed models using RNA structure information is an RNA folding model. Therefore, we first developed a model component which describes the folding kinetics of single stranded RNA molecules. It consists of a coupled ML-DEVS model representing the whole RNA molecule and several atomic models.

5.1 Nucleotides

Each nucleotide (nt) of the RNA strand is represented by an instance of the atomic model *Nucleotide* which is either of the type A, C, G, or U meaning its base. They are connected via ports in the same order as the input sequence (primary structure) and have knowledge about their direct neighbours. For example, the nt at sequence position 8 is connected with the nt number 7 on its 5' side and on the 3' location it is connected with the nt at position 9 (see figure 4). State variables hold rudimentary information about the neighbours, to be exact their base type and current binding partners. "Binding partner" means a secondary structure defining base pair and the term is used only in this context here and does not mean the primary backbone connections. If a partner of a nucleotide changes, an output message will be generated and the receiving (neighboured) nucleotides will update their state variables. Holding information about other atomic model states is normally not the case in DEVS models as they are typically seen as black boxes. However, here it is quite useful because of some dependencies concerning base pair stability.

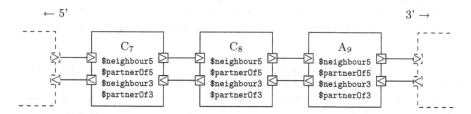

Fig. 4. Adjacent nucleotides are connected via input and output ports. A set of variables stores basic information about their neighbourhood.

Base pairs are modelled by wide range connections of nucleotides via additional interfaces. Whereas the RNA backbone bonds of adjacent nucleotides are fixed after model initialisation, the connections between base pairing nucleotides are dynamically added and removed during simulation (figure 5). Therefore, two different major states (phases) of nucleotides exist: they can be either *unpaired* or *paired*.

As already stated in section 3.1, base pair stability depends on the involved bases and their neighbourhood, especially stacking energies of adjacent base pairs provide crucial contributions for structure stabilisation. In our kinetic folding model, base pair stability is reflected by binding duration, i.e. the *time advance* function of the *paired* phase. Thus, pairing time depends on thermodynamic parameters for nucleic acids base stacking which were taken from [17] and are also be used by MFOLD version 2.3 [7]. This thermodynamic data set not only provides the free energy (Gibbs energy) for a given temperature of 37°C, but also

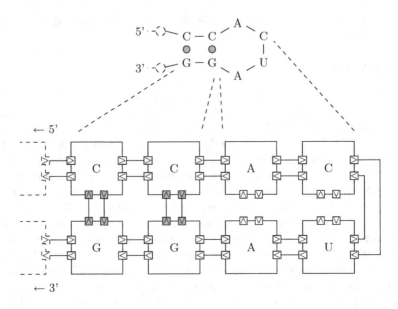

Fig. 5. Small hairpin loop structure. (top) Conventional RNA secondary structure representation. (bottom) The same structure represented by atomic DEVS models. Base pairing is modelled by dynamically coupled nucleotides (grey interfaces) with a distance of at least 5 sequence positions reflecting the minimal hairpin loop size.

the enthalpy change ΔH of various stacking situations. The enthalpy together with the free energy and the absolute temperature allows us to calculate the entropy change ΔS which allows us further to calculate the activation energy ΔE_a for base pair dissociation at any temperature T between 0 and 100°C:

$$\Delta E_a = -(\Delta H - T \Delta S) \ .$$

ΔE_a is directly used as one parameter for base pair opening time, i.e. the duration of a *paired* phase is directly dependent on the activation energy for base pair disruption. To allow RNA structures to escape from local energy minima and refold to more stable structures, the base pair dissociation time will be randomised, which leads to very short bonding times in some cases although the activation energy needed for opening the base pair is quite large.

For base pair closing an arbitrary short random time is assigned with the *unpaired* phase of nucleotides assuming that RNA base pair formation is a very fast and random process. After *unpaired* time has expired, the nucleotide model tries to build a base pair with another nucleotide randomly chosen from within a set of possible pairing partner positions. This set is determined by sterically available RNA strand regions and thus an abstraction of spatial constraints. For example, a hairpin loop smaller than 4 nucleotides is sterically impossible, but also many other nucleotide positions with larger distance can be excluded for secondary structure folding (see figure 6).

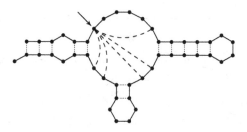

Fig. 6. Possible partners for secondary structure base pairing. The arrow-marked nucleotide is only able to pair with 5 positions within its own loop region (connected via dashed lines). All other nucleotides are excluded from base pairing with the marked position.

An *unpaired* nucleotide is not able to choose another nt for base pairing by its own. It has no global information about the RNA shape which is strongly needed here. Therefore, an implicit request by adding a new input port will be made to the coupled model that holds such macro knowledge and can therefore choose a valid position with which the requesting nt will try to pair next. For choosing this position at macro level two different model variants exist. The first and more simple one picks a position totally random from within the set of possible partners, whereas the second variant takes the entropy change into account when a pairing partner will be selected. The latter method prefers helix elongation in contrast to introducing new interior loops by single base pair formations and small loops will be more favourable than large ones [12]. Correct RNA folding with the first method is dependent on the base pair stabilities and the random folding nuclei which are the first appearing base pairs of helical regions. This last point is less important for RNA folding with model variant 2 because the chosen binding partners are more deterministic due to loop entropy consideration. A comparison of both approaches with respect to simulation results is given in section 6. Once a *nucleotide* received an input message by the macro model containing the number of a potential pairing partner, it tries to form a base pair with this nt by adding a coupling and sending a request. For a successful pairing, the partners must be of complementary base type and they must be able to pair in principle, e.g. bases can be modified so that they can not bind to others. Figure 7 illustrates the whole state flow for base pair formation and disruption of the *nucleotide* model component.

5.2   Macro Level Model

The role of the macro model and its interactions with the micro level (nucleotides) shows the schematic organisation of the whole RNA folding model in figure 8. Already mentioned in the previous section, high level information about the whole RNA molecule is needed to take sterical restrictions for base pairing into account. Therefore, the coupled model holds the overall RNA secondary structure which will be updated every time the state of a nucleotide changes

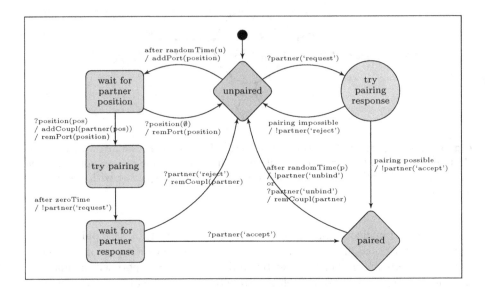

Fig. 7. *Nucleotide* state flow in a Statechart-like representation. Main states are depicted by grey diamonds. States of a nucleotide which takes up the active part of a base pair are shown as rounded rectangles and a state exclusively passed through by a reacting nucleotide is drawn as a circle. Sending and receiving events over ports have prefixed exclamation marks and question marks respectively.

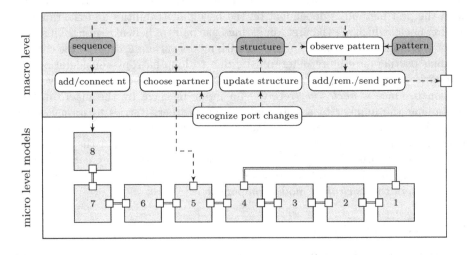

Fig. 8. Schematic overview of the whole RNA folding model. Model components are shaded in light grey and explicit couplings between them are drawn by solid double lines. Dashed lines indicate model-internal information flow and function calls.

from *unpaired* to *paired* and vice versa. This will be triggered by *nucleotide* port adding and removal recognised by the macro level. The same functionality is used to signalise the macro level the wish to try pairing with another nucleotide. The macro model detects a port adding, calculates the sterically possible partner set, chooses a position from within the set, and after all sends this position to the just now added *nucleotide* input port (figure 8, nt 5). The coupled macro model is further responsible for sequence initialisation on the micro level by adding and connecting *nucleotide* models, i.e. it generates the primary RNA structure. Another task of the macro model is to observe the current folding structure for special structural patterns. This could be for example a specific binding domain for a protein or small ligand. Also transcription termination or pausing structures can be of interest for observation. If observed structures are present during folding simulation, the macro level model can signalise this information and thus trigger dynamics to other components by adding new ports to itself (representing docking sites) or send messages over existing ports. A composed example model which uses this capability can be found in section 7.

6 Evaluation of the Folding Model

For evaluating the model's validity we simulated the folding of different RNA molecules with known structure and analysed the results. Three different types of experiments were done:

Native Structure– Correlates the majority of formed structures with the native structure after sufficient long simulation time?

Structure Distribution– Is the equilibrium ratio between minimum free energy (mfe) and suboptimal structures as expected?

Structure Refolding– Are molecules able to refold from suboptimal to more stable structural conformations?

Unfortunately, only few time-resolved folding pathways are experimentally derived and most of them treat pseudoknots [22] and higher order tertiary structure elements [22,23,24] which can not be handled by our folding model and are therefore out of the question for a simulation study. Hence, some comparisons with other *in silico* tools were also made, although we know that one has to be careful with comparing different models for validating a model as it is often unclear how valid the other models are. Because the folding model is highly stochastic, every simulation experiment was executed multiple times. Typically 100 replications were made.

6.1 Native Structure

Structural analysis of the cis-acting replication element from Hepatitis C virus revealed a stem hairpin loop conformation where the helix is interrupted by an internal or bulge loop region [25]. Figure 9 shows simulation results of its structure

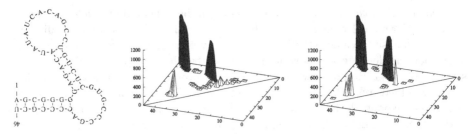

Fig. 9. Folding simulation of the Hepatitis C virus cis-acting replication element. Parameters: time 1000 ms, temperature 310.15 K, 100 replications. (left) Observed secondary structure [25]. (middle) Base pair probability matrix. Simulation with base pair formation model 1. (right) Base pair formation model 2. Peak heights indicate base pair lifetime during simulation and the native helices are shaded.

formation. The three-dimensional base pair lifetime plots indicate correct folding of both helical regions and only few misfolded base pairs. Only small differences can be seen between simulations with the two base pair formation variants described in section 5.1. Without taking entropy into account for pairing, a bit more noise of misfolded base pairs can be observed which is not surprising due to the absolutely random partner choice.

Another well known RNA structure is the cloverleaf secondary structure of tRNAs [26,27] consisting of four helical stems: the amino acid arm, D arm, anticodon arm, and the T arm. Some base modifications and unusual nucleotides exist in tRNA which stabilise its structure formation, e.g. dihydrouridine and pseudouridine. Such special conditions are not considered by our folding model as well as tertiary interactions leading to the final L-shaped form. However, folding simulations result in significant cloverleaf secondary structure formation (figure 10). Although there is much misfolded noise, the four distinct helix peaks are the most stable structural elements introduced during simulation, especially the amino acid arm. No fundamental difference can be seen between base pair formation model 1 and 2.

A third native structure validation experiment treats the Corona virus s2m motif which is a relatively long hairpin structure with some intersecting internal and bulge loops [28,29]. Simulation of the SARS virus s2m RNA folding indicates only for one of the native helix regions a conspicuous occurrence (figure 11). The other helices closer to the hairpin loop show no significant stabilisation. Competing misfolded structural elements can be observed equally frequent or even more often. Base pair formation model 2 provides a slightly better result than the first one, but it is unsatisfying too. A reason for the result can be the multiple internal and bulge loops, which destabilise the stem and thus allow locally more stable structure elements to form.

6.2   Structure Distribution

In [30] a quantitative analysis of different RNA secondary structures by comparative imino proton NMR spectroscopy is described. The results indicate that

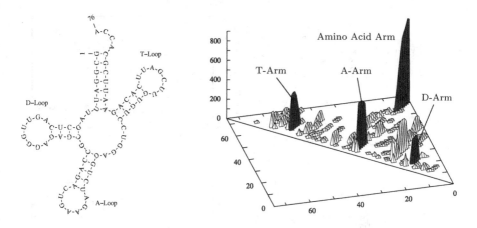

Fig. 10. Simulation of the yeast tRNA^{Phe} folding. Parameters: time 1000 ms, temperature 310.15 K, base pair formation model 2, 100 replications. (left) Native cloverleaf secondary structure. (right) Base pair probability matrix. Peak heights indicate base pair lifetime during simulation and the native helices are shaded.

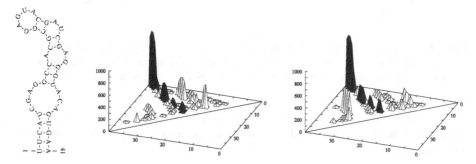

Fig. 11. Folding simulation of the SARS virus s2m motif. Parameters: time 1000 ms, temperature 310.15 K, 100 replications. (left) Known secondary structure [29]. (middle) Base pair probability matrix. Simulation with base pair formation model 1. (right) Base pair formation model 2. Peak heights indicate base pair lifetime during simulation and the native helices are shaded.

a small 34-nt RNA has two equally stable structures in thermodynamic equilibrium, one with 2 short helices and the other with a single hairpin. Folding simulations of the same RNA strand show an equal ratio of the probed structures as well (figure 12). However, both are representing just 20% of all present structures which was not detected in the NMR experiments. Many base pairs were introduced during simulation which are competing with base pairs of the two stable structures and thus reduce their appearance. This can be easily seen in the 3D matrix of figure 12 where some additional peaks show high misfolded base pair lifetimes. Simulating the RNA folding with KINFOLD [12]

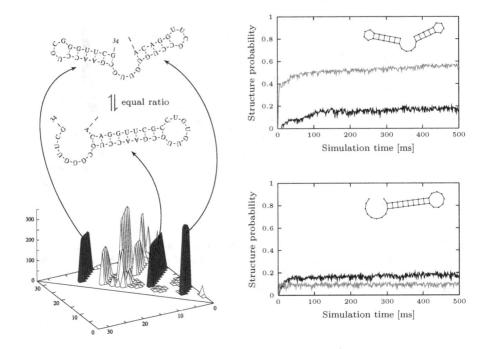

Fig. 12. Folding simulation of a designed small bistable RNA [30]. Parameters: temperature 298.15 K, base pair formation model 2, 100 replications. Grey curves describe mean structure occurrence from 100 KINFOLD simulation runs with 298.15 K.

results in a five times higher amount of the 2-helix conformation than the single hairpin, but their total sum is about 60% of all molecules and thus less misfolded structures can be observed.

6.3 Structure Refolding

Real-time NMR spectroscopy was used by Wenter et al. to determine refolding rates of a short 20-nt RNA molecule [31]. The formation of its most stable helix was temporarily inhibited by a modified guanosine at position 6. After photolytic removal of this modification a structure refolding was observed. To map such forced structure formation to relatively unstable folds at the beginning of an experiment, most RNA folding tools have the capability to initialise simulations with specified structures. We used, much closer to the original wetlab experiment, a different strategy, i.e. at first G6 was not capable to bind any other base. The time course after removing this prohibition during simulation is shown in figure 13. Wenter et al. detected structure refolding by measuring the imino proton signal intensity of U11 and U17, which show high signal intensity if they are paired with other bases. Accordingly we observed the state of both uracils over simulation time as well. After removal of G6 unpaired locking, a logarithmic decrease of structures with paired U11 and uniform increase of folds

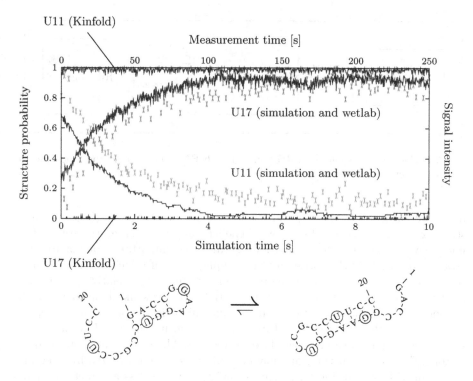

Fig. 13. Refolding of an deliberately misfolded small RNA molecule [31]. Wetlab measurements are drawn by a X. Simulation parameters: time 10 seconds, temperature 288.15 K, base pair formation model 1, 100 replications. Simulations with Kinfold were made with the same temperature and replication number but over a time period of 250 seconds.

with paired U17 can be observed reaching a complete shift of the conformational equilibrium after 4 seconds. A very similar refolding progression was experimentally measured (single spots in figure 13), but with a strong deviating time scale of factor 25. This could be a remaining model parameter inaccuracy or due to special experimental conditions which are not covered by our model, e.g. unusual salt concentrations. However, our model allows a quite realistic refolding from a suboptimal to a more stable RNA structure. Identical *in silico* experiments with KINFOLD [12] by contrast, do not show any significant refolding (figure 13, nonchanging curves). The same counts for SEQFOLD [11]. With both approaches the energy barrier seems to be too high to escape from the misfolded suboptimal structure.

6.4   Assessment of Experiments

Local optima can more easily be overcome in comparison to other traditional pure macro methods (see figure 13). We assume that even "stable" base pairs

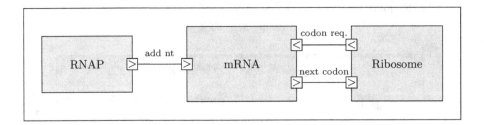

Fig. 14. Connecting a folding mRNA model with RNA polymerase and ribosome to reflect transcription attenuation

might be subject to changes, and let the nucleotides "searching" for a stable structure at micro level. This proved beneficial and emphasised the role of the micro level. However, the simulation revealed the importance of macro constraints for the folding process, and the implications of a lack of those. Macro constraints that have been considered are for example the relative positioning of the nucleotides, particularly within spatial structures like hairpin or internal loops. The interplay between macro and micro level allowed us to reproduce many of the expected structure elements, e.g. figures 9 and 10, although macro constraints have been significantly relaxed. These simplifications lead to "wrongly" formed structures and maybe could have been prevented by integrating terminal base stacking for pairing stability as well as less abstract base pair closing rules as macro constraints. A comparison of the two implemented base pair formation methods indicate only few differences. Without taking entropy into account the noise of unstable single base pairs and short helices increases, but not dramatically. The same stable structures are formed based on both rules.

7　Composed Attenuation Model

Having a working folding model we are now able to combine it with other model components that are influenced by or are influencing the RNA structure formation and come back to the motivation, the attenuation of tryptophan synthesis. At least two further models are needed to reflect transcription attenuation: the RNA polymerase and the ribosome (figure 14).

7.1　RNA Polymerase

RNA molecules are products of the transcription process which is the fundamental step in gene expression. Once the RNA polymerase enzyme complex (RNAP) has successfully bound to DNA (transcription initiation), it transcribes the template sequence into an RNA strand by sequentially adding nucleotides to the 3' end and thus elongates the molecule. To reflect this synthesising process, in the RNA model, new *nucleotide* models and their backbone connections are added dynamically during simulation. This process is triggered by the RNAP model component which interacts with RNA. This dynamic RNA elongation allows the simulation of sequential folding, where early synthesised parts of the RNA molecule can already fold whereas other parts still have to be added. Please note that this is not a unique feature of the model presented here, as kinetic folding tools typically realise sequential folding by just adding a new nt after a certain time delay. However, a component design allows to combine the RNAP with further models (e.g. the DNA template), or to model it in more detail (e.g. diverse RNAP subunits), and to exchange model components on demand. The pattern observation function of the RNA folding model, which is realised at macro level, allows us to look for an intrinsic transcription termination structure [2] during simulation. If such structure is formed, the folding model removes its elongation input port meaning the release of the RNA from the polymerase enzyme. At this time point the elongation stops, but structure folding and interactions with other components proceed.

7.2   Ribosome

The ribosome enzyme complex translates RNA sequences into protein determining amino acid sequences (peptides). Translation starts at the ribosome binding site of mRNA molecules which is reflected by a pair of input and output ports of the RNA models. The translation begins after connecting it with a ribosome model. The current ribosome position with respect to the RNA sequence is known by the RNA model. A triplet of three RNA bases (codon) encodes for one amino acid. The ribosome requests for the next codon 3' of its current RNA location when peptide chain elongation has proceeded. This is the case when the correct amino acid of the last codon entered the enzyme. The speed of the translation process depends strongly on the availability of needed amino acids. If an amino acid type is not sufficiently available, the ribosome stalls at the corresponding codon and thus pauses translation. A ribosome is quite big and thus 35-36 nucleotides are covered by its shape [32]. Therefore, a region upstream and downstream of the ribosome location is not able to form base pairs. As the RNA model knows the ribosome location, this is handled by the RNA macro level model which sends corresponding events to its nucleotide micro model components. The same counts for the helicase activity of the ribosome [32]. For sequence translation, the macro level model will disrupt a base paired structure element when it is reached by the enzyme.

Whether those additional models are realised as atomic models, or coupled models depends on the objective of the simulation study. Referring to the operon model presented in [15], the RNAP, the mRNA, and the ribosome would replace the simplistic mRNA model, to integrate the attenuation process into the model.

8   Conclusion

We presented a component-based model of RNA folding processes. Unlike traditional approaches which focus on the results of the folding process, e.g. stable structures in thermodynamical equilibrium, our objective has been different. The idea was to develop an approach that allows to integrate the folding processes into larger models and to take the dynamics into account, that has shown to be crucial in many regulation processes. Therefore, the formalism ML-DEVS was used. At macro level, certain constraints referring to space and individual locations were introduced, whereas at micro level, the nucleotides were responsible for a successful base pairing and for the stability of the structure. A model component for the nucleotides and one model component for the entire RNA molecule have been defined. The simulation results have been compared to wetlab experiments. Therefore, the model components can be parametrised for different RNA sequences (base types) as well as environmental conditions (e.g temperature). The evaluation revealed an overall acceptable performance, and in addition, insights into the role of micro level dynamics and macro level constraints. The integration of the RNA folding model into a model of transcription attenuation has been sketched. Next steps will be to realise this integration and to execute simulation experiments to analyse the impact of this more detailed regulation model on the synthesis of tryptophan.

Acknowledgments. Many thanks to Adelinde M. Uhrmacher for her helpful comments and advice on this work. I will also thank Roland Ewald and Jan Himmelspach for their instructions for using JAMES II. The research has been funded by the German Research Foundation (DFG).

References

1. Kaberdin, V.R., Blasi, U.: Translation initiation and the fate of bacterial mRNAs. FEMS Microbiol. Rev. 30(6), 967–979 (2006)
2. Gusarov, I., Nudler, E.: The Mechanism of Intrinsic Transcription Termination. Mol. Cell 3(4), 495–504 (1999)
3. Yanofsky, C.: Transcription attenuation: once viewed as a novel regulatory strategy. J. Bacteriol. 182(1), 1–8 (2000)
4. Nahvi, A., Sudarsan, N., Ebert, M.S., Zou, X., Brown, K.L., Breaker, R.R.: Genetic Control by a Metabolite Binding mRNA. Chem. Biol. 9(9), 1043–1049 (2002)
5. Torarinsson, E., Havgaard, J.H., Gorodkin, J.: Multiple structural alignment and clustering of RNA sequences. Bioinformatics 23(8), 926–932 (2007)
6. Hofacker, I.L., Fekete, M., Stadler, P.F.: Secondary Structure Prediction for Aligned RNA Sequences. J. Mol. Biol. 319(5), 1059–1066 (2002)
7. Zuker, M.: Mfold web server for nucleic acid folding and hybridization prediction. Nucleic Acids Res. 31(13), 3406–3415 (2003)
8. Hofacker, I.L., Fontana, W., Stadler, P.F., Bonhoeffer, L.S., Tacker, M., Schuster, P.: Fast folding and comparison of RNA secondary structures. Monatsh. Chem./Chemical Monthly 125(2), 167–188 (1994)
9. Rivas, E., Eddy, S.R.: A dynamic programming algorithm for RNA structure prediction including pseudoknots. J. Mol. Biol. 285(5), 2053–2068 (1999)
10. Xayaphoummine, A., Bucher, T., Thalmann, F., Isambert, H.: Prediction and Statistics of Pseudoknots in RNA Structures Using Exactly Clustered Stochastic Simulations. Proc. Natl. Acad. Sci. U.S.A. 100(26), 15310–15315 (2003)

11. Schmitz, M., Steger, G.: Description of RNA Folding by "Simulated Annealing". J. Mol. Biol. 255(1), 254–266 (1996)
12. Flamm, C., Fontana, W., Hofacker, I.L., Schuster, P.: RNA folding at elementary step resolution. RNA 6(3), 325–338 (2000)
13. Flamm, C., Hofacker, I.L.: Beyond energy minimization: approaches to the kinetic folding of RNA. Monatsh. Chem./Chemical Monthly 139(4), 447–457 (2008)
14. Santillán, M., Mackey, M.C.: Dynamic regulation of the tryptophan operon: A modeling study and comparison with experimental data. Proc. Natl. Acad. Sci. U.S.A 98(4), 1364–1369 (2001)
15. Degenring, D., Lemcke, J., Röhl, M., Uhrmacher, A.M.: A Variable Structure Model – the Tryptophan Operon. In: Proc. of the 3rd International Workshop on Computational Methods in Systems Biology, Edinburgh, Scotland, April 3-5 (2005)
16. Serra, M.J., Lyttle, M.H., Axenson, T.J., Schadt, C.A., Turner, D.H.: RNA hairpin loop stability depends on closing base pair. Nucleic Acids Res. 21(16), 3845–3849 (1993)
17. Walter, A.E., Turner, D.H., Kim, J., Lyttle, M.H., Müller, P., Mathews, D.H., Zuker, M.: Coaxial Stacking of Helixes Enhances Binding of Oligoribonucleotides and Improves Predictions of RNA Folding. Proc. Natl. Acad. Sci. U.S.A. 91(20), 9218–9222 (1994)
18. Russell, R., Millett, I.S., Tate, M.W., Kwok, L.W., Nakatani, B., Gruner, S.M., Mochrie, S.G.J., Pande, V., Doniach, S., Herschlag, D., Pollack, L.: Rapid compaction during RNA folding. Proc. Natl. Acad. Sci. U.S.A. 99(7), 4266–4271 (2002)
19. Uhrmacher, A.M., Ewald, R., John, M., Maus, C., Jeschke, M., Biermann, S.: Combining Micro and Macro-Modeling in DEVS for Computational Biology. In: WSC 2007: Proceedings of the 39th conference on Winter simulation, pp. 871–880. IEEE Press, Los Alamitos (2007)
20. Zeigler, B.P., Praehofer, H., Kim, T.G.: Theory of Modeling and Simulation. Academic Press, London (2000)
21. Uhrmacher, A.M., Himmelspach, J., Jeschke, M., John, M., Leye, S., Maus, C., Röhl, M., Ewald, R.: One Modelling Formalism & Simulator Is Not Enough! A Perspective for Computational Biology Based on James II. In: Fisher, J. (ed.) FMSB 2008. LNCS (LNBI), vol. 5054, pp. 123–138. Springer, Heidelberg (2008)
22. Bokinsky, G., Zhuang, X.: Single-molecule RNA folding. Acc. Chem. Res. 38(7), 566–573 (2005)
23. Sclavi, B., Sullivan, M., Chance, M.R., Brenowitz, M., Woodson, S.A.: RNA Folding at Millisecond Intervals by Synchrotron Hydroxyl Radical Footprinting. Science 279(5358), 1940–1943 (1998)
24. Zhuang, X., Bartley, L.E., Babcock, H.P., Russell, R., Ha, T., Herschlag, D., Chu, S.: A Single-Molecule Study of RNA Catalysis and Folding (2000)
25. You, S., Stump, D.D., Branch, A.D., Rice, C.M.: A cis-Acting Replication Element in the Sequence Encoding the NS5B RNA-Dependent RNA Polymerase Is Required for Hepatitis C Virus RNA Replication. J. Virol. 78(3), 1352–1366 (2004)
26. Rich, A., RajBhandary, U.L.: Transfer RNA: Molecular Structure, Sequence, and Properties. Annu. Rev. Biochem. 45(1), 805–860 (1976)
27. Clark, B.F.C.: The crystal structure of tRNA. J. Biosci. 31(4), 453–457 (2006)
28. Jonassen, C.M., Jonassen, T.O., Grinde, B.: A common RNA motif in the 3' end of the genomes of astroviruses, avian infectious bronchitis virus and an equine rhinovirus. J. Gen. Virol. 79(4), 715–718 (1998)

29. Robertson, M.P., Igel, H., Baertsch, R., Haussler, D., Ares, M.J., Scott, W.G.: The structure of a rigorously conserved RNA element within the SARS virus genome. PLoS Biol. 3(1), 86–94 (2005)
30. Höbartner, C., Micura, R.: Bistable Secondary Structures of Small RNAs and Their Structural Probing by Comparative Imino Proton NMR Spectroscopy. J. Mol. Biol. 325(3), 421–431 (2003)
31. Wenter, P., Fürtig, B., Hainard, A., Schwalbe, H., Pitsch, S.: Kinetics of Photoinduced RNA Refolding by Real-Time NMR Spectroscopy. Angew. Chem. Int. Ed. Engl. 44(17), 2600–2603 (2005)
32. Takyar, S., Hickerson, R.P., Noller, H.F.: mRNA Helicase Activity of the Ribosome. Cell 120(1), 49–58 (2005)

A Language for Biochemical Systems

Michael Pedersen and Gordon Plotkin

LFCS, School of Informatics, University of Edinburgh

Abstract. CBS is a *Calculus of Biochemical Systems* intended to allow the modelling of metabolic, signalling and regulatory networks in a natural and modular manner. In this paper we extend CBS with features directed towards practical, large-scale applications, thus yielding LBS: a *Language for Biochemical Systems*. The two main extensions are expressions for modifying large complexes in a step-wise manner and parameterised modules with a notion of subtyping; LBS also has nested declarations of species and compartments. The extensions are demonstrated with examples from the yeast pheromone pathway. A formal specification of LBS is then given through an abstract syntax, static semantics and a translation to a variant of coloured Petri nets. Translation to other formalisms such as ordinary differential equations and continuous time Markov chains is also possible.

Keywords: Large-scale, parametrised modules, subtyping, coloured Petri nets.

1 Introduction

Recent years have seen a multitude of formal languages and systems applied to biology, thus gaining insight into the biological systems under study through analysis and simulation. Some of these languages have a history of applications in computer science and engineering, e.g. the pi calculus [1], PEPA [2], Petri nets [3] and P-systems [4], and some are designed from scratch, e.g. Kappa [5], BioNetGen [6], BIOCHAM [7], Bioambients [8], Beta binders [9,10], Dynamical Grammars [11] and the Continuous Pi Calculus [12].

The *Calculus of Biochemical Systems* (CBS) [13] is a new addition to the latter category which allows metabolic, signalling and regulatory networks to be modelled in a natural and modular manner. In essence, CBS describes reactions between modified complexes, occurring concurrently inside a hierarchy of compartments but allowing cross-compartment interactions and transport. It has a compositional semantics in terms of Petri nets, ordinary differential equations (ODEs) and continuous time Markov chains (CTMCs). Petri nets allow a range of established analysis techniques to be used in the biological setting [14], and ODEs and CTMCs enable deterministic and stochastic simulations.

This paper proposes extensions of CBS in support of practical, large-scale applications, resulting in LBS: a *Language for Biochemical Systems*. The two main extensions are *pattern expressions* and *parameterised modules*. Patterns represent complexes and pattern expressions provide a concise way of making small

M. Heiner and A.M. Uhrmacher (Eds.): CMSB 2008, LNBI 5307, pp. 63–82, 2008.

changes to large complexes incrementally, a common scenario in signal transduction pathways. Parameterised modules allow general biological "gadgets", such as a MAPK cascade, to be modelled once and then reused in different contexts. Modules may be parameterised on compartments, rates, patterns and species types, the latter resulting in a notion of parametric type. A notion of subtyping of species types and patterns is also employed, allowing a module to specify general reaction schemes and have a concrete context provided at the time of module invocation. Modules may furthermore return pattern expressions, thus providing a natural mechanism for connecting related modules.

Other improvements over CBS include species and compartment declarations which involve scope and new name generation. While species modification sites in CBS always have boolean type, LBS does not place any limitations of modification site types, allowing, e.g., location (real number pairs) or DNA sequences (strings) to be represented. LBS also allows general rate expressions to be associated with expressions, although mass-action kinetics may be assumed as in CBS.

The syntax and semantics of CBS are outlined informally in Section 2 through some basic examples, including a MAPK cascade module drawn from the yeast pheromone pathway. In Section 3 we introduce LBS by further examples from the yeast pheromone pathway and demonstrate how the features of LBS can be used to overcome specific limitations of CBS. We then turn to a formal presentation of the language. An abstract syntax of LBS is given in Section 4. An overview of the semantics of LBS, including a static semantics, the general approach to translation, and a specific translation to Petri nets, is given in Section 5. Section 6 discusses related work and future directions. Due to space constraints, only selected parts of the semantics are given in this paper. A full presentation, together with the full LBS model of the yeast pheromone pathway, can be found in [15].

A compiler from LBS to the *Systems Biology Markup Language* (SBML) [16] has been implemented and supports the main features of the language presented formally in this paper. We have validated the compiler on the yeast pheromone pathway model by deriving ODEs from the target SBML using the Copasi tool [17]. By manual inspection, the derived ODEs coincide, up to renaming of species, with the ODEs published in [18], although there is some discrepancy between the simulation results.

2   The Calculus of Biochemical Systems

Basic examples of CBS modules with no modifications or complexes are shown in Program 1, with the last line informally indicating how the modules may be used in some arbitrary context. Module M1 consists of two chemical reactions taking place in parallel as indicated by the bar, |. The first is a condensation reaction, and the second is a methane burning reaction. In M2 the second reaction takes place inside a compartment c, and in M3 the species O2 is transported out of compartment c to whatever compartment the module is later instantiated inside. In contrast, a language such as BIOCHAM would represent the same model by

Program 1. Example of three modules in CBS

```
module M1 { 2 H2 + O2 -> 2 H2O | CH4 + 2 O2 -> CO2 + 2 H2O };
module M2 { 2 H2 + O2 -> 2 H2O | c[CH4 + 2 O2 -> CO2 + 2 H2O] };
module M3 { c[O2] -> O2 };
... | M1 | ... | M2 | ... | M3 | ...
```

an unstructured list of five reactions and explicitly specifying the compartments of each species.

Graphical representations of the two individual reactions in module M1 are shown in Figure 1a; the reader familiar with Petri nets (see [19] for an overview) can think of the pictures as such. When considering two reactions together, *in parallel*, the standard chemical interpretation is that the reactions share and compete for species which have syntactically identical names in the two reactions, in this case O2 and H2O. A graphical Petri net representation of M1 based on this interpretation is shown in Figure 1b. In the case of module M2, the species O2 and H2O are not considered identical between the two reactions because they are located in different compartments. Consequently, none of the species are merged in the parallel composition that constitutes M2, see Figures 1c and 1d.

This example illustrates how reactions, or more generally, modules, are composed in CBS, and hints at how a *compositional* semantics in terms of Petri nets can be defined. Similar ideas can be used to define compositional semantics in terms of ODEs and CTMCs, assuming that reactions are equipped with rates; see [13] for the full details.

Let us turn to a more realistic example featuring modifications and complexes, drawn from a model of the yeast pheromone pathway that will serve as a case study throughout the paper. Figure 2, adapted from [18], shows a graphical

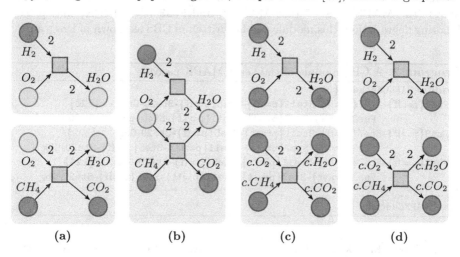

                    (a)   (b)   (c)   (d)

Fig. 1. Petri net representations of reactions and their parallel composition. Places (circles) represent species, transitions (squares) represent reactions, and arc weights represent stoichiometry.

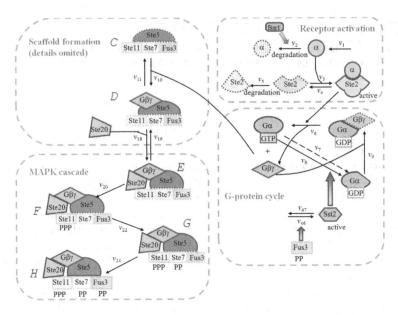

Fig. 2. Selected and reorganised parts of the informal model from [18] of the yeast pheromone pathway. Copyright © 2008 John Wiley & Sons Limited. Reproduced with permission.

representation of the model divided into several interacting modules. We do not discuss the biological details of the model but rather consider the general structure and how this can be represented formally, and we start by focusing on the MAPKCascade module. The cascade relies on a scaffolding complex holding Ste11 (the MAPK3), Ste7 (the MAPK2) and Fus3 (the MAPK) into place. Ignoring degradation, this module can be written in CBS as shown in Program 2.

Program 2. A CBS module of the yeast MAPK cascade

```
module MAPKCascade {
  Fus3{p=ff}-Ste7{p=ff}-Ste11{p=ff}-Ste5{p=ff}-Ste20-Gbg ->{k20}
          Fus3{p=ff}-Ste7{p=ff}-Ste11{p=tt}-Ste5{p=ff}-Ste20-Gbg |
  Fus3{p=ff}-Ste7{p=ff}-Ste11{p=tt}-Ste5{p=ff}-Ste20-Gbg ->{k22}
          Fus3{p=ff}-Ste7{p=tt}-Ste11{p=tt}-Ste5{p=ff}-Ste20-Gbg |
  Fus3{p=ff}-Ste7{p=tt}-Ste11{p=tt}-Ste5{p=ff}-Ste20-Gbg ->{k24}
          Fus3{p=tt}-Ste7{p=tt}-Ste11{p=tt}-Ste5{p=ff}-Ste20-Gbg
};
... | MAPKCascade | ...
```

The labels k20, k22 and k24 associated with reaction arrows represent the mass-action rates given in [18]. Fus3, Ste7, Ste11, Ste5, Ste20 and Gbg are the names of *primitive species*, i.e. non-complex species, and can exist in various states of modification. In this case all primitive species except Ste20 and

Gbg have a single modification site, p, which can be either phosphorylated or unphosphorylated, indicated by assigning boolean values (**tt**/**ff**) to the sites. For example, Fus3{p=ff} represents Fus3 in its unphosphorylated state. Complexes are formed by composing modified primitive species using a hyphen, -.

Complexes such as the above will generally be referred to as *patterns*. As in BIOCHAM, the term "pattern" reflects that modification sites in reactants can be assigned *variables* rather than actual boolean values, hence "matching" multiple physical complexes and thereby ameliorating the combinatorial explosion problem on the level of species modifications.

Two limitations of the CBS representation emerge from this example:

1. **Redundancy.** Many signalling pathways involve making small changes to large complexes. Therefore, patterns are often identical except for small changes in modification, but in CBS we are forced to write all patterns out in full.
2. **Reuse.** The MAPK cascade is a typical example of a "biological gadget" which is utilised in many different contexts but with different participating species [20]. The CBS MAPK module in our example "hard codes" the species and rates involved and hence cannot be used in another context.

The next section shows how LBS offers solutions to these limitations.

3   The Language for Biochemical Systems

We give three examples of LBS programs and informally explain their syntax. The first shows how species declarations and pattern expressions can be used to improve the yeast MAPK cascade module. The second shows how a general, reusable MAPK cascade module can be written by taking advantage of parameters and subtyping. The third shows how modules can communicate by linking an output pattern of a receptor activation module to an input pattern of a G-protein cycle module.

3.1   The Yeast MAPKCascade

Program 3 shows how the model in Program 2 can be re-written in LBS. The first difference is that all primitive species featuring in a program must be declared by specifying their modification site names and types, if any. For example, Fus3 has a single modification site named p of type **bool**, and Ste20 has no modification sites. In general, arbitrarily many modification sites may be declared.

The second difference is that we assign the first pattern to a pattern identifier called e. This identifier, and the ones that follow, correspond directly to the names given to complexes in Figure 2. We can then simply refer to e in the first reaction instead of writing out the full pattern. The product of the first reaction uses the pattern expression e<Ste11{p=tt}> to represent "everything in e, except that site p in Ste11 is phosphorylated," and subsequently assigns the resulting pattern to a new identifier f which, in turn, is then used as a

Program 3. Species declarations and pattern expressions in the yeast MAP-KCascade module

```
module YeastMAPKCascade {
  spec Fus3{p:bool}, Ste7{p:bool}, Ste11{p:bool}, Ste5{p:bool}, Ste20, Gbg;
  pat e = Fus3{p=ff}-Ste7{p=ff}-Ste11{p=ff}-Ste5{p=ff}-Ste20-Gbg;

  e ->{k20} e<Ste11{p=tt}> as f;
  f ->{k22} f<Ste7{p=tt}> as g;
  g ->{k24} g<Fus3{p=tt}> as h
};
... | YeastMAPKCascade; ...
```

reactant of the second reaction, and so on. When using such *in-line* pattern declarations, reactions are separated by semi-colons (;) rather than the parallel composition (|), indicating that the order in which reactions are written matters.

The module can be invoked in some parallel context as indicated informally in the last line. Since, e.g., Fus3 is declared locally, inside the module, multiple instances of the module would give rise to multiple, distinct instances of this species. If we prefer species to be shared between multiple instances of a module, they should either be declared globally or passed as parameters, as we see in the next subsection.

3.2 A General MAPK Cascade Module

The model in the previous subsection still suffers from a lack of reusability. From a more general perspective, a (scaffolded) MAPK cascade is a series of reactions operating on some potentially very big complex but which contains specific species serving the K3, K2 and K1 functions of the cascade. Each of these species must have at least one phosphorylation site (i.e. of boolean type) which in the general case could be be called ps. With this in mind, the K3, K2 and K1 become *species parameters* of the MAPK module. The scaffold complex containing these species becomes a pattern parameter which we will call mk4, indicating its role as an upstream initiator of the cascade. Reaction rates become rate parameters. Program 4 shows how the resulting module can be written in LBS. The species parameters follow the structure of species declarations. But the pattern parameter is different: it provides a pattern identifier together with the *type* of the pattern. A pattern type simply represents the names of primitive species in the pattern: no more is needed to determine how the pattern may be used, because the types of the primitive species are defined separately.

In the context of the yeast model, we can simply use the MAPK cascade module by declaring the specific species of interest, defining the scaffold pattern, and passing these together with the rates as arguments to the module. This is shown in Program 5 which is semantically equivalent to Program 3, i.e. the two translate to the same Petri net. A closer investigation of this program tells us that the pattern e, which is passed as an *actual* parameter, has the type Fus3-Ste7-Ste11-Ste5-Ste20-Gbg, namely the species contained in the pattern.

Program 4. Defining a general, scaffolded MAPKCascade module

```
module MAPKCascade( spec K1{ps:bool}, K2{ps:bool}, K3{ps:bool};
                    pat mk4 : K1-K2-K3; rate r1, r2, r3 ){
  mk4 ->{r1} mk4<K3{ps=tt}> as mk3;
  mk3 ->{r2} mk3<K2{ps=tt}> as mk2;
  mk2 ->{r3} mk2<K1{ps=tt}> as mk1
};
```

Program 5. Using the general MAPKCascade module

```
spec Fus3{p:bool}, Ste7{p:bool}, Ste11{p:bool}, Ste5{p:bool}, Ste20, Gbg;
pat e = Fus3{p=ff}-Ste7{p=ff}-Ste11{p=ff}-Ste5{p=ff}-Ste20-Gbg;

MAPKCascade(Fus3{p:bool}, Ste7{p:bool}, Ste11{p:bool}, e, k20, k22, k24);
```

The corresponding *formal* parameter has type K1-K2-K3, which is instantiated to Fus3-Ste7-Ste11 through the species parameters. This works because the actual parameter type contains *at least* the species required by the formal parameter type. This is all the module needs to know, since these are the only species it is going to manipulate. We say that the type of the actual parameter is a *subtype* of the type of the formal parameter, and hence the module invocation is legal. A similar idea applies at the level of species parameters, although in this example the corresponding formal and actual species parameters have the same number of modification sites. Note that the names of corresponding modification sites need not be the same for actual and formal parameters. In this example, the formal parameters use ps while the actual parameters use p.

3.3 Receptor Activation and G-Protein Cycle Modules

Our last example illustrates how modules can be linked together. If we look at the general structure of the yeast pheromone picture in Figure 2, we notice that many of the modules produce outputs which are passed on to subsequent modules: the scaffold formation module produces a scaffold which is passed on to the MAPK cascade module, and the receptor activation module produces a receptor-ligand complex (consisting of Alpha and Ste2) which is used to activate the G-protein cycle. The G-protein cycle in turn passes on a beta-gamma subunit.

In order to naturally represent these interconnections, LBS provides a mechanism for modules to return patterns. Program 6 shows this mechanism involving the receptor activation and G-protein cycle modules. The modules are named as in Figure 2. They are commented and should be self-explanatory, except perhaps for three points. Firstly, enzymatic reactions are represented using the tilde operator ($\sim$) with an enzyme (a pattern) on the left and a reaction on the right. Secondly, reversible reactions are represented using a double-arrow (<->) followed by rates for the forward and backward directions. Thirdly, the rate v46 in the G protein cycle module is defined explicitly because it does not follow

---

Program 6. Receptor activation and G-protein cycle modules

```
spec Fus3{p:bool};
module ReceptorAct(comp cyto, pat degrador, patout rl) {
  spec Alpha, Ste2{p: bool};
  (* pheromone and receptor degradation: *)
  degrador ~ Alpha ->{k1} | cyto[Ste2{p=ff}] ->{k5} |
  (* Receptor-ligand binding and degradation: *)
  Alpha + cyto[Ste2{p=ff}] <->{k2,k3} cyto[Alpha-Ste2{p=tt}] as rl;
  cyto[rl] ->{k4}
};
module GProtCycle(pat act, patout gbg) {
  spec Ga, Gbg, Sst2{p:bool};
  pat Gbga = Gbg-Ga-GDP;
  (* disassociation of G-protein complex: *)
  act ~ Gbga ->{k6} Gbg + Ga-GTP |
  (* ... and the G-protein cycle: *)
  Ga-GTP ->{k7} Ga-GDP |
  Sst2{p=tt} ~ Ga-GTP ->{k8} Ga-GDP |
  rate v46 = k46 * (Fus3{p=tt}^2 / (4^2 + Fus3{p=tt}^2));
  Fus3{p=tt} ~ Sst2{p=ff} <->[v46]{k47} Sst2{p=tt} |
  Ga-GDP + Gbg ->{k9} Gbga |
  pat gbg = Gbg; Nil
};
spec Bar1;
comp cytosol inside T vol 1.0;
ReceptorAct(cytosol, Bar1, pat link);
GProtCycle(link, pat link2);
(* rest of model ... *)
```

---

mass-action kinetics, and this is indicated by the use of square brackets around the forward rate of the reaction.

Let us consider how the two modules interface to each other. The receptor activation module takes parameters for the cytosol compartment and a pattern which degrades the pheromone. The latter has empty type, indicating that the module does not care about the contents of this pattern. The new feature is the last parameter, rl (short for *receptor-ligand* complex). This is an output pattern: it is defined in the body of the module and is made available when the module is invoked. This happens towards the end of Program 6 by first declaring the relevant species and the compartment cytosol with volume 1 inside the distinguished top level compartment T, which are then passed as actual parameters to the module. Note that a species is a special, non-modified and non-complex case of a pattern, and hence can be used as a pattern parameter. The last actual parameter is the output parameter identifier, here called link. This pattern identifier will be assigned the return pattern of the module (namely the receptor-ligand complex) and is then passed as the "activating" parameter for the G-protein cycle module. It follows that the ordering of module invocation matters, which as above is indicated by a semicolon rather than the parallel composition. In this particular example, output patterns have the empty type, but they could have arbitrary types and are subject to a subtyping mechanism similar to that of input patterns.

4   The Abstract Syntax of LBS

Having given an informal introduction to the main features of LBS and its *concrete* syntax, we now present its *abstract* syntax. This is shown in Table 1; each of the three main syntactic categories is explained further in the following subsections.

Table 1. The core abstract syntax of LBS: Pattern expressions and their types (top), programs (middle) and declarations (bottom)

$$PE ::= \beta \mid p \mid PE - PE' \mid PE\langle PE\rangle \mid PE.s \mid PE\backslash s$$

$$\beta ::= \{s_i \mapsto \alpha_i\} \quad \alpha ::= \{l_i \mapsto E_i\} \quad \tau ::= \{s_i \mapsto n_i\} \quad \sigma ::= \{l_i \mapsto \rho_i\}$$

$$P ::= \{LPE_i \mapsto n_i\} \xrightarrow{RE} \{LPE'_j \mapsto n'_j\} \text{ if } E_{\text{bool}}$$

$$\mid \mathbf{0} \mid P_1 \mid P_2 \mid c[P] \mid Decl; P \mid m(APars; \mathbf{pat}\ p); P$$

$$LPE ::= PE \mid c[LPE] \qquad APars ::= \underline{s : \sigma}; \underline{c}; \underline{PE}; \underline{RE}$$

$$RE ::= k \mid LPE \mid c \mid r(APars) \mid \log(RE) \mid RE\ \mathbf{aop}\ RE$$

$$Decl ::= \mathbf{comp}\ c : c', v \mid \mathbf{spec}\ s : \sigma \mid \mathbf{pat}\ p = PE \mid \mathbf{rate}\ r(FPars) = RE$$

$$\mid \mathbf{module}\ m(FPars; \mathbf{patout}\ \underline{p : \tau})\{P\}$$

$$FPars ::= \mathbf{spec}\ s : l : \rho; \mathbf{comp}\ \underline{c : c'}; \mathbf{pat}\ \underline{p : \tau}; \mathbf{rate}\ \underline{r}$$

4.1   Notation

Tuples $(x_1, \ldots, x_k)$ are written $\underline{x}$ when the specific elements are unimportant. The set of finite *multisets* over a set S is denoted by FMS(S), the total functions from S to the natural numbers which take value 0 for all but finite many elements of S. The *power set* of a set S is written 2^S. Partial finite *functions* f are denoted by finite indexed sets of pairs $\{x_i \mapsto y_i\}_{i \in I}$ where $f(x_i) = y_i$, and I is omitted if it is understood from the context. When the ith element of a list or an indexed set is referred to without explicit quantification in a premise or condition of a rule, the index is understood to be universally quantified over the index set I. The *domain of definition* of a function f is denoted by dom(f); the empty partial function is denoted by $\emptyset$. We write $f[g]$ for the update of f by a partial finite function g; for the sake of readability, we often abbreviate e.g. $f[\{x_i \mapsto y_i\}]$ by $f[x_i \mapsto y_i]$.

4.2   Pattern Expressions

In the abstract syntax for pattern expressions, s ranges over a given set NAMES$_s$ of *species names*, l ranges over a given set NAMES$_{\text{mo}}$ of *modification site names*, and p ranges over a given set IDENT_p of *pattern identifiers*. The simplest possible pattern expression, a *pattern* β, maps species names to lists of modifications, thus allowing homomers be be represented.

Modifications in turn map modification site names to expressions E, which range over the set $\text{EXP} \triangleq \bigcup_{\rho \in \text{TYPES}_{\text{mo}}} \text{EXP}_\rho$; here, TYPES_{mo} is a given set of modification site types, ranged over by ρ, and EXP_ρ is a given set of expressions of type ρ. We assume that TYPES_{mo} contains the boolean type **bool** with values $\{\mathbf{tt}, \mathbf{ff}\}$. Expressions may contain *match variables* from a given set X and we assume a function $\text{FV} \triangleq \text{EXP} \to 2^X$ giving the variables of an expression.

Pattern composition, $PE - PE'$, intuitively results in a pattern where the lists of modifications from the first pattern have been appended to the corresponding lists of modifications from the second pattern. This operation is therefore *not* generally commutative, e.g. $\mathbf{s\{1=tt\}}$-$\mathbf{s\{1=ff\}}$ is not the same as $\mathbf{s\{1=ff\}}$-$\mathbf{s\{1=tt\}}$. This is not entirely satisfying and will be a topic of future work. We have already encountered the pattern update expression $PE\langle PE \rangle$ in the MAPK cascade module in Program 3. The expression $PE.s$ restricts the pattern to the species s and throws everything else away, while the expression $PE \backslash s$ keeps everything except for s. So a dissociation reaction of s from PE can be written (in the concrete syntax) as `PE ->{r} PE\s + PE.s`. If there are multiple instances of `s` in `PE`, only the first in the list of modifications will be affected.

Finally, α and β represent the types of species and patterns, respectively; in the grammar, n ranges over $\mathbb{N}_{>0}$. Henceforth we let TYPES_p and TYPES_s be the sets generated by these productions.

4.3   Programs

In the abstract syntax for programs, c ranges over a given set NAMES_c of *compartment names*, m ranges over a given set IDENT_m of *module identifiers*, r ranges over a given set IDENT_r of *rate identifiers*, $n \in \mathbb{N}_{>0}$ and $k \in \mathbb{R}$. The first line in the grammar for programs is a reaction. Products and reactants are represented as functions mapping located patterns to stoichiometry. A reaction may furthermore be conditioned on a boolean expression. We encountered located patterns in Programs 1 and 6: they are just patterns inside a hierarchy of compartments. Rate expressions associated with reactions can employ the usual arithmetic expressions composed from operators $\mathbf{aop} \in \{+, -, *, /, \}$. But they can also include located patterns (which refer to either a population or a concentration, depending on the semantics), compartments (which refer to a volume) and rate function invocations.

In the second line, $\mathbf{0}$ represents the null process, and we have encountered parallel composition and compartmentalised programs in the examples. Declarations are defined separately in the next subsection. Module invocations are followed sequentially by a program since a new scope will be created if patterns are returned from modules as in Program 6.

Enzymatic reactions, reversible reactions, mass-action kinetics and inline pattern declarations (using the **as** keyword) which we encountered in the examples are not represented in the abstract syntax. They can all be defined, in a straightforward manner, from existing constructs. Take for example the following definitions of reactions from Programs 3 and 6:

- e ->{k10} e<Ste11{p=tt}> as f $\triangleq$
  pat f = e<Ste11{p=tt}>; e ->[k10 * e] f.
- Fus3{act=tt} ~ Sst2{act=ff} <->[v46]{k47} Sst2{act=tt} $\triangleq$
  Fus3{act=tt} + Sst2{act=ff} ->[v46] Fus3{act=tt} + Sst2{act=tt}
  | Sst2{act=tt} ->[k47 * Sst2{act=tt}] Sst2{act=ff}

4.4 Declarations

Compartment declarations specify the volume $v \in \mathbb{R}_{>0}$ and parent of the declared compartment. Compartments which conceptually have no parent compartment should declare the "top level" compartment, $\top$, as their parent, so we assume that $\top \in \text{NAMES}_c$. Modules and rate functions may be parameterised on species with their associated type, compartments with their declared parents, patterns with their associated type, and finally on rate expressions. Henceforth we let FORMALPARS be the set of formal parameter expressions, as generated by the $FPars$ production.

5 The Semantics of LBS

Having introduced the syntactic structure of LBS programs, we now turn to their *meaning*.

5.1 Static Semantics

The static semantics tells us which of the LBS programs that are well-formed according to the abstract syntax are also semantically meaningful. This is specified formally by a type system of which the central parts are given in Appendix A, and full details can be found in [15]. In this subsection we informally discuss the main conditions for LBS programs to be well-typed.

The type system for pattern expressions checks that pattern identifiers and species are declared and used according to their declared type. For pattern updates, selections and removals, only species which are present in the target pattern expression are allowed. If a pattern expression is well-typed, its type τ is deduced in a compositional manner. Suptyping for patterns is given by multiset inclusion, and subtyping for species is given by record subtyping as in standard programming languages [21].

For reactions, the type system requires that any match variables which occur in the products also occur in the reactants. This is because variables in the product patterns must be instantiated based on matches in the reactant patterns during execution of the resulting model. It also requires that the reactant and product located patterns agree on parent compartments; for example, the reaction c1[s] -> c2[s] is not well-typed if the compartments c1 and c2 are declared with different parent compartments. A similar consideration applies for parallel composition. For a compartment program c[P] we require that all top-level compartments occurring in P are declared with parent c. In order for these conditions on compartments to be checked statically, i.e. at time of module declaration rather than invocation, the type system must associate parent compartments with programs in a bottom-up manner as detailed in Appendix A.

For module invocations, the type system first of all ensures that the number of formal and actual parameters match. It also ensures that the type of actual species and pattern parameters are subtypes of the corresponding formal parameter types, both for standard "input" parameters but also for output pattern parameters. Finally, all formal output patterns identifiers must be defined in the body of the module.

5.2   The General Translation Framework

A key advantage of formal modelling languages for biology is that they facilitate different kinds of analysis on the same model. This is also the case for CBS which is endowed with compositional semantics in terms of Petri nets, ODEs and CTMCs. The semantics of LBS is complicated by the addition of high-level constructs such as pattern expressions, modules and declarations, but this is ameliorated by the fact that the definition of its semantics is to some extent independent of the specific choice of target semantical objects. The *general translation framework* defines these independent parts of the translation, and concrete translations then tie into this framework by defining the semantic objects associated with the following:

1. *Normal form* reactions, where pattern expressions have been evaluated to patterns, together with the types of species featuring in the reaction.
2. The **0** program.
3. The parallel composition of semantic objects.
4. Semantic objects inside compartments.

Here we only outline the central mechanisms involved; a complete account can be found in [15].

The framework evaluates pattern expressions PE to patterns β according to the intuitions set forth in Section 4.2 and by replacing pattern identifiers with their defined patterns. Located pattern expressions are evaluated to pairs of (immediate) parent compartments and the resulting patterns. Reactions are then evaluated to *normal form* reactions of the form:

$$\{(c_i^{\text{in}}, \beta_i^{\text{in}}) \mapsto n_i^{\text{in}}\} \xrightarrow{RE} \{(c_j^{\text{out}}, \beta_j^{\text{out}}) \mapsto n_j^{\text{out}}\} \text{ if } E_{\textbf{bool}}$$

where pattern expressions and rate function invocations in RE have been evaluated, and compartment identifiers have been replaced by their declared volumes. A *species type environment* $\Gamma_s :$ NAMES$_s \hookrightarrow_{\text{fin}}$ TYPES$_s$ recording the type of species in reactions is also maintained by the framework.

Modules are evaluated to semantic functions which, given the relevant actual parameters at time of invocation, return the semantic objects of the module body together with the output patterns. The definition of these functions is complicated by the need to handle species parameters and pattern subtyping: the former requires renaming to be carried out, and the latter entails handling type environments for species that are not necessarily within the scope of their declaration.

As an example, using the general translation framework on the module in Program 4 results in a function which, when invoked with the parameters given in Program 5 , computes the (almost) normal form reactions in Program 2, uses the first concrete semantics function to obtain concrete semantic objects for each reaction, and finally applies the third concrete semantic function to obtain the final result of the parallel compositions.

5.3   Translating LBS to Petri Nets

This subsection demonstrates the translation framework by giving the definitions needed for a concrete translation to Petri nets. Translations to ODEs or CTMCs are also possible, see [13] or [14] for the general approach.

Petri Nets. When modelling biological systems with Petri nets, the standard approach is to represent species by places, reactions by transitions and stoichiometry by arc multiplicities as in Figure 1. Complex species with modification sites can be represented compactly using a variant of *coloured Petri nets* where places are assigned *colour types* [22]. In our case, the colour types of places are given by pairs (c, τ) of compartments and pattern types together with a global primitive species type environment Γ_s. Then a pair (c, τ) uniquely identifies a located species, so there is no need to distinguish places and colour types. Arcs are equipped with multisets of *patterns* rather than plain stoichiometry, allowing a transition to restrict the colour of tokens (e.g. modification of a species) that it accepts or produces. Transitions are strings over the binary alphabet. This enables us to ensure that the transitions from two parallel nets are disjoint simply by prefixing 0 and 1 to all transitions in the respective nets.

In the following formal definition of *bio-Petri nets*, we shall need the sets of patterns conforming to specific types (here a type system $E : \rho$ on expressions is assumed):

$$\text{PATTERNS}_{\tau, \Gamma_s} \triangleq \{\beta \in \text{PATTERNS} \mid$$
$$\text{type}(\beta) = \tau \wedge \forall s \in \text{dom}(\beta).\text{type}(\beta(s)) = \Gamma_s(s)\}$$

$$\text{type}(\{s_i \mapsto \underline{\alpha_i}\}) \triangleq \{s_i \mapsto |\underline{\alpha_i}|\}$$

$$\text{type}(\{l_i \mapsto E_i\}) \triangleq \{l_i \mapsto \rho_i\} \text{ where } E_i : \rho_i$$

We use a standard notation $\prod_{i \in I} X_i$ for dependent sets.

Definition 1. *A bio-Petri net $\mathcal{P}$ is a tuple $(S, T, F_{\text{in}}, F_{\text{out}}, B, \Gamma_s)$ where*

- $S \subset \text{NAMES}_c \times \text{TYPES}_p$ *is a finite set of places (located pattern types).*
- $T \subset \{0, 1\}^*$ *is a finite set of transitions (reactions).*

- $F_{\text{in}} : \prod_{t,(c,\tau)\in T\times S} \text{FMS}(\text{PATTERNS}_{\tau,\Gamma_s})$ *is the* flow-in *function (reactants).*
- $F_{\text{out}} : \prod_{t,(c,\tau)\in T\times S} \text{FMS}(\text{PATTERNS}_{\tau,\Gamma_s})$ *is the* flow-out *function (products).*
- $B : T \to \text{EXP}_{\mathbf{bool}}$ *is the* transition guard *function.*
- $\Gamma_s : \bigcup\{\text{dom}(\tau) \mid (c,\tau) \in S\} \to \text{TYPES}_s$ *is the* species type *function.*

We use the superscript notation $S^{\mathcal{P}}$ to refer to the places S of Petri net $\mathcal{P}$, and similarly for the other Petri net elements. For a formal definition of behaviour (qualitative semantics) of bio-Petri nets, please refer to [15].

The Concrete Translation of LBS to Bio-Petri Nets. Following [13], the concrete translation to bio-Petri nets is given by the following four definitions.

1. Let $P = \{(c_i^{\text{in}}, \beta_i^{\text{in}}) \mapsto n_i^{\text{in}}\} \xrightarrow{RE} \{(c_j^{\text{out}}, \beta_j^{\text{out}}) \mapsto n_j^{\text{out}}\}$ **if** $E_{\mathbf{bool}}$ be an LBS reaction in normal form and let Γ_s be a species type environment. Then define $\mathcal{P}(P, \Gamma_s)$ as follows, where ϵ denotes the empty string:

   - $S^{\mathcal{P}} \triangleq \{(c_i^{\text{in}}, \text{type}(\beta_i^{\text{in}}))\} \cup \{(c_j^{\text{out}}, \text{type}(\beta_j^{\text{out}}))\}$
   - $T^{\mathcal{P}} \triangleq \{\epsilon\}$
   - $F_{\text{io}}^{\mathcal{P}}(\epsilon, (c,\tau)) \triangleq \{(\beta_h^{\text{io}} \mapsto n_h^{\text{io}}) \mid c_h^{\text{io}} = c \wedge \text{type}(\beta_h^{\text{io}}) = \tau\}$ for io $\in \{\text{in}, \text{out}\}$
   - $B^{\mathcal{P}}(\epsilon) \triangleq E_{\mathbf{bool}}$
   - $\Gamma_s^{\mathcal{P}} \triangleq \Gamma_s$

2. Define $\mathcal{P}(0) \triangleq (\emptyset, \emptyset, \emptyset, \emptyset, \emptyset, \emptyset)$
3. Let $\mathcal{P}_1$ and $\mathcal{P}_2$ be Petri nets with $\Gamma_s^{\mathcal{P}_1}(s) = \Gamma_s^{\mathcal{P}_2}(s)$ for all $s \in \text{dom}(\Gamma_s^{\mathcal{P}_1}) \cap \text{dom}(\Gamma_s^{\mathcal{P}_2})$. Define parallel composition $\mathcal{P} = \mathcal{P}_1 | \mathcal{P}_2$ as follows, where $b \in \{0,1\}$:

   - $S^{\mathcal{P}} \triangleq S^{\mathcal{P}_1} \cup S^{\mathcal{P}_2}$
   - $T^{\mathcal{P}} \triangleq \{0t \mid t \in T^{\mathcal{P}_1}\} \cup \{1t \mid t \in T^{\mathcal{P}_2}\}$
   - $F_{\text{io}}^{\mathcal{P}}(bt, p) \triangleq \begin{cases} F_{\text{io}}^{\mathcal{P}_1}(t,p) & \text{if } t \in T^{\mathcal{P}_1} \wedge p \in S^{\mathcal{P}_1} \\ F_{\text{io}}^{\mathcal{P}_2}(t,p) & \text{if } t \in T^{\mathcal{P}_2} \wedge p \in S^{\mathcal{P}_2} \\ \emptyset & \text{otherwise} \end{cases}$ for io $\in \{\text{in}, \text{out}\}$
   - $B(bt) \triangleq \begin{cases} B^{\mathcal{P}_1}(t) & \text{if } t \in T^{\mathcal{P}_1} \\ B^{\mathcal{P}_2}(t) & \text{if } t \in T^{\mathcal{P}_2} \end{cases}$
   - $\Gamma_s^{\mathcal{P}} \triangleq \Gamma_s^{\mathcal{P}_1} \cup \Gamma_s^{\mathcal{P}_2}$

4. First define $c[(c',\tau)] \triangleq \begin{cases} (c,\tau) & \text{if } c' = \top \\ (c',\tau) & \text{otherwise} \end{cases}$

   Let $\mathcal{P}'$ be a Petri net. Then define the compartmentalisation $\mathcal{P} = c[\mathcal{P}']$ as follows:

   - $S^{\mathcal{P}} \triangleq \{c[p] \mid p \in S^{\mathcal{P}'}\}$
   - $T^{\mathcal{P}} \triangleq T^{\mathcal{P}'}$
   - $F_{\text{io}}^{\mathcal{P}}(t, c[p]) \triangleq F_{\text{io}}^{\mathcal{P}'}(t, p)$ for io $\in \{\text{in}, \text{out}\}$
   - $B^{\mathcal{P}}(t) \triangleq B^{\mathcal{P}'}(t)$
   - $\Gamma_s^{\mathcal{P}} \triangleq \Gamma_s^{\mathcal{P}'}$

Observe that for programs where species have no modification sites and do not form complexes, the above definitions collapse to the simple cases of composition illustrated in Program 1 and Figure 1 for standard Petri nets.

6   Related Work and Future Directions

Compared to the other languages for biochemical modelling mentioned in the introduction, CBS is unique in its combination of two features: it explicitly models *reactions* rather than individual agents as in process calculi, and it does so in a *compositional* manner. BIOCHAM, for example, also models reactions explicitly using a very similar syntax to that of CBS, but it does not have a modular structure.

Whether the explicit modelling of reactions is desirable or not depends on the particular application. Systems which are characterised by high combinatorial complexity arising from complex formations are difficult or even impossible to model in CBS and LBS; an example is a model of scaffold formation which considers all possible orders of subunit assembly, where one may prefer to use Kappa or BioNetGen. But for systems in which this is not an issue, or where the combinatorial complexity arises from modifications of simple species (which CBS and LBS deal with in terms of match variables), the simplicity of a reaction-based approach is attractive. It also corresponds well to graphical representations of biological systems, as we have seen with the yeast pheromone pathway example.

To our knowledge, no other languages have abstractions corresponding to the pattern expressions and nested declarations of species and compartments of LBS. The notion of parameterised modules is however featured in the *Human-Readable Model Definition Language* [23], a draft textual language intended as a front-end to the *Systems Biology Markup Language* (SBML) [16]. The modules in this language follow an object-oriented approach rather than our functional approach, but there is no notion of subtyping or formal semantics.

Tools for visualising LBS programs and, conversely, for generating LBS programs from visual diagrams, are planned and will follow the notation of [24,25] or the emerging *Systems Biology Graphical Notation* (SBGN). We also plan to use LBS for modelling large scale systems such as the EGFR signalling pathway [26], although problems with interpretation of the graphical diagrams are anticipated. While some parts of the EGFR map are well characterised by modules, others appear rather monolithic and this is likely to be a general problem with modular approaches to modelling in systems biology. In the setting of *synthetic* biology, however, systems are *programmed* rather than *modelled*, so it should be possible to fully exploit modularity there.

With respect to language development, it is important to achieve a better understanding of homomers, enabling a commutative pattern composition operation. We also anticipate the addition of descriptive features for annotating species with e.g. Gene Ontology (GO) or Enzyme Commission (EC) numbers. One may also consider whether model analyses can exploit modularity, e.g. for Petri net invariants. Results for a subset of CBS without complexes and modifications, corresponding to plain place/transition nets, can be found in [27], and extensions to LBS and coloured Petri nets would be of interest.

Acknowledgements

The authors would like to thank Vincent Danos and Stuart Moodie for useful discussions, and Edda Klipp for supplying the diagram in Figure 2. This work was supported by Microsoft Research through its European PhD Scholarship Programme and by a Royal Society-Wolfson Award. The second author is grateful for the support from The Centre for Systems Biology at Edinburgh, a Centre for Integrative Systems Biology funded by BBSRC and EPSRC, reference BB/D019621/1. Part of the research was carried out at the Microsoft Silicon Valley Research Center.

References

1. Regev, A., et al.: Representation and simulation of biochemical processes using the pi-calculus process algebra. In: Pacific Symposium on Biocomputing, pp. 459–470 (2001)
2. Calder, M., et al.: Modelling the influence of RKIP on the ERK signalling pathway using the stochastic process algebra PEPA. In: Priami, C., Ingólfsdóttir, A., Mishra, B., Riis Nielson, H. (eds.) Transactions on Computational Systems Biology VII. LNCS (LNBI), vol. 4230, pp. 1–23. Springer, Heidelberg (2006)
3. Reddy, V.N., et al.: Petri net representation in metabolic pathways. In: Proc. Int. Conf. Intell. Syst. Mol. Biol., pp. 328–336 (1993)
4. Paun, G., Rozenberg, G.: A guide to membrane computing. Theor. Comput. Sci. 287(1), 73–100 (2002)
5. Danos, V., et al.: Rule-based modelling of cellular signalling. In: Caires, L., Vasconcelos, V.T. (eds.) CONCUR 2007. LNCS, vol. 4703, pp. 17–41. Springer, Heidelberg (2007)
6. Faeder, J.R., et al.: Graphical rule-based representation of signal-transduction networks. In: Liebrock, L.M. (ed.) Proc. 2005 ACM Symp. Appl. Computing, pp. 133–140. ACM Press, New York (2005)
7. Chabrier-Rivier, N., et al.: The biochemical abstract machine BIOCHAM. In: Danos, V., Schachter, V. (eds.) CMSB 2004. LNCS (LNBI), vol. 3082, pp. 172–191. Springer, Heidelberg (2005)
8. Regev, A., et al.: BioAmbients: an abstraction for biological compartments. Theor. Comput. Sci. 325(1), 141–167 (2004)
9. Priami, C., Quaglia, P.: Beta binders for biological interactions. In: Danos, V., Schachter, V. (eds.) CMSB 2004. LNCS (LNBI), vol. 3082, pp. 20–33. Springer, Heidelberg (2005)
10. Guerriero, M.L., et al.: An automated translation from a narrative language for biological modelling into process algebra. In: Calder, M., Gilmore, S. (eds.) CMSB 2007. LNCS (LNBI), vol. 4695, pp. 136–151. Springer, Heidelberg (2007)
11. Mjolsness, E., Yosiphon, G.: Stochastic process semantics for dynamical grammars. Ann. Math. Artif. Intell. 47(3-4), 329–395 (2006)
12. Kwiatkowski, M., Stark, I.: The continuous pi-calculus: a process algebra for biochemical modelling. In: Heiner, M., Uhrmacher, A.M. (eds.) Proc. CMSB. LNCS. Springer, Heidelberg (2008)

13. Plotkin, G.: A calculus of biochemical systems (in preparation)
14. Heiner, M., et al.: Petri nets for systems and synthetic biology. In: Bernardo, M., Degano, P., Zavattaro, G. (eds.) SFM 2008. LNCS, vol. 5016, pp. 215–264. Springer, Heidelberg (2008)
15. Pedersen, M., Plotkin, G.: A language for biochemical systems. Technical report, University of Edinburgh (2008), http://www.inf.ed.ac.uk/publications/report/1270.html
16. Hucka, M., et al.: The systems biology markup language (SBML): a medium for representation and exchange of biochemical network models. Bioinformatics 19(4), 524–531 (2003)
17. Hoops, S., et al.: COPASI – a COmplex PAthway SImulator. Bioinformatics 22(24), 3067–3074 (2006)
18. Kofahl, B., Klipp, E.: Modelling the dynamics of the yeast pheromone pathway. Yeast 21(10), 831–850 (2004)
19. Murata, T.: Petri nets: properties, analysis and applications. Proc. IEEE 77(4), 541–580 (1989)
20. Garrington, T.P., Johnson, G.L.: Organization and regulation of mitogen-activated protein kinase signaling pathways. Current Opinion in Cell Biology 11, 211–218 (1999)
21. Pierce, B.C.: Types and Programming Languages. MIT Press, Cambridge (2002)
22. Jensen, K.: Coloured Petri Nets: Basic Concepts, Analysis Methods and Practical Use, vol. 1. Springer, Heidelberg (1992)
23. Bergmann, F., Sauro, H.: Human-readable model definition language (first draft, revision 2). Technical report, Keck Graduate Institute (2006)
24. Kitano, H., et al.: Using process diagrams for the graphical representation of biological networks. Nat. Biotechnol. 23(8), 961–966 (2005)
25. Moodie, S.L., et al.: A graphical notation to describe the logical interactions of biological pathways. Journal of Integrative Bioinformatics 3(2) (2006)
26. Oda, K., et al.: A comprehensive pathway map of epidermal growth factor receptor signaling. Molecular Systems Biology (2005)
27. Pedersen, M.: Compositional definitions of minimal flows in Petri nets. In: Heiner, M., Uhrmacher, A.M. (eds.) CMSB 2008. LNCS, vol. 5307, pp. 288–307. Springer, Heidelberg (2008)

A   The Type System for LBS

This section presents the central parts of the type system for LBS.

A.1   Type Environments

The type system relies on the following type environments corresponding to each of the possible declarations in the abstract syntax.

- Γ_s : $\text{NAMES}_s \hookrightarrow_{fin} \text{TYPES}_s$ for species declarations.
- Γ_c : $\text{NAMES}_c \hookrightarrow_{fin} \text{NAMES}_c$ for compartment declarations.
- Γ_p : $\text{IDENT}_p \hookrightarrow_{fin} \text{TYPES}_p$ for pattern declarations.
- Γ_r : $\text{IDENT}_r \hookrightarrow_{fin} \text{FORMALPARS}$ for rate function declarations.
- Γ_m : $\text{IDENT}_m \hookrightarrow_{fin} \text{FORMALPARS} \times \text{TYPES}_p^* \times 2^{\text{NAMES}_s}$ for module declarations.

The Γ_m environment stores the type of defined modules which includes a set of compartments in which the module may be legally instantiated according to the compartment hierarchy specified through compartment declarations. For a well-typed module this set will either be a singleton compartment, indicating that the module can only be instantiated in a particular compartment, or the entire set of compartment names indicating that the module can be instantiated inside any compartment. The type system also needs to check that output patterns are used appropriately. For this purpose, two additional environments are required.

- Γ_p^{out} : $\text{IDENT}_p \hookrightarrow_{fin} \text{TYPES}_p$. This stores the declared types of module output patterns.
- $R_p \in 2^{\text{IDENT}_p}$. This records pattern identifiers which have an entry in the Γ_p^{out} environment *and* are defined inside a module.

The R_p environment is necessary to ensure that a module does in fact define the pattern identifiers which it has declared as outputs.

The type systems to be given in the following use and modify the above type environments. In general, type judgements for a given type system may have the form $\Gamma_s, \Gamma_c, \Gamma_p, \Gamma_r, \Gamma_m \vdash \ldots$ but for the sake of readability we may rather write, e.g., $\Gamma, \Gamma_p \vdash \ldots$ where Γ represents all environments to the left of the turnstile except Γ_p.

A.2   Pattern Expressions

Subtyping is formalised by the first two rules in Table 2. The remaining rules use the environments for primitive species and patterns, so judgements are of the form $\Gamma_s, \Gamma_p \vdash PE : \tau$. The first rule furthermore relies on a type system (not given here) for basic expressions with judgements of the form $E : \rho$. We

Table 2. The type system for patterns

STSPEC
$$\frac{\text{dom}(\sigma') \subseteq \text{dom}(\sigma) \wedge \forall l \in \text{dom}(\sigma').\sigma(l) = \sigma'(l)}{\sigma <: \sigma'}$$

STPAT
$$\frac{\forall s \in \text{NAMES}_s.\tau(s) \geq \tau'(s)}{\tau <: \tau'}$$

TPAT
$$\frac{E_{i_j} : \rho_{i_j} \text{ and } \rho_{i_j} = \Gamma_s(s)(l_{i_j})}{\Gamma, \Gamma_s \vdash \{s_i \mapsto \{l_i \mapsto E_i\}\} : \{s \mapsto |\{l_i \mapsto E_i\}|\}}$$

TPATCOMP
$$\frac{\Gamma \vdash PE : \tau \text{ and } \Gamma \vdash PE' : \tau'}{\Gamma \vdash PE - PE' : \tau + \tau'}$$

TPATUPD
$$\frac{\Gamma \vdash PE : \tau \text{ and } \Gamma \vdash PE' : \tau'}{\Gamma \vdash PE\langle PE'\rangle : \tau} \quad \tau <: \tau'$$

TPATSEL
$$\frac{\Gamma \vdash PE : \tau}{\Gamma \vdash PE.s : \{s \mapsto 1\}} \quad s \in \text{dom}(\tau)$$

TPATREM
$$\frac{\Gamma \vdash PE : \tau + \{s \mapsto 1\}}{\Gamma \vdash PE\backslash s : \tau}$$

TPATIDEN
$$\frac{}{\Gamma, \Gamma_p \vdash p : \Gamma_p(p)}$$

Table 3. The type system for programs

$\mathrm{T}_{\mathrm{REAC}}$
$$\frac{\Gamma_s, \Gamma_c, \Gamma_p \vdash LPE_i : \tau_i, \gamma_i \text{ and } \Gamma_s, \Gamma_c, \Gamma_p, \Gamma_r \vdash RE \text{ and } \Gamma_s, \Gamma_c, \Gamma_p \vdash LPE'_j : \tau'_j, \gamma'_j}{\Gamma, \Gamma_s, \Gamma_c, \Gamma_p, \Gamma_r \vdash \{LPE_i \mapsto n_i\} \xrightarrow{RE} \{LPE'_j \mapsto n'_j\} \text{ if } E_{\mathbf{bool}} : \gamma'' \dashv \emptyset}$$

if 1) $\gamma'' \neq \emptyset$ and 2) $\bigcup \mathrm{FV}(LPE'_j) \subseteq \bigcup \mathrm{FV}(LPE_i)$ and

3) $\mathrm{FV}(E_{\mathbf{bool}}) \subseteq \bigcup \mathrm{FV}(LPE'_j) \cup \bigcup \mathrm{FV}(LPE_i)$ where $\gamma'' = \bigcap\{\gamma_i\} \cap \bigcap\{\gamma'_j\}$

$\mathrm{T}_{\mathrm{PAR}}$
$$\frac{\Gamma \vdash P_1 : \gamma \dashv R_p \text{ and } \Gamma \vdash P_2 : \gamma' \dashv R'_p}{\Gamma \vdash P_1 \mid P_2 : \gamma \cap \gamma' \dashv R_p \cup R'_p} \quad \gamma \cap \gamma' \neq \emptyset \text{ and } R_p \cap R'_p = \emptyset$$

$\mathrm{T}_{\mathrm{NIL}}$
$$\overline{\Gamma \vdash \mathbf{0} : \mathrm{NAMES_c} \dashv \emptyset}$$

$\mathrm{T}_{\mathrm{COMP}}$
$$\frac{\Gamma, \Gamma_c \vdash P : \gamma' \dashv R_p}{\Gamma, \Gamma_c \vdash c[P] : \{\Gamma_c(c)\} \dashv R_p} \quad c \in \gamma'$$

$\mathrm{T}_{\mathrm{DEC}}$
$$\frac{\Gamma \vdash Decl \dashv \Gamma', R_p \text{ and } \Gamma' \vdash P : \gamma \dashv R'_p}{\Gamma \vdash Decl; P : \gamma \dashv R_p \cup R'_p} \quad R_p \cap R'_p = \emptyset$$

$\mathrm{T}_{\mathrm{MODINV}}$
$$\frac{\Gamma, \Gamma_c, \Gamma_p, \Gamma_m, \Gamma_p^{\mathrm{out}} \vdash APars <: FPars \text{ and } \Gamma, \Gamma_c, \Gamma'_p, \Gamma_m, \Gamma_p^{\mathrm{out}} \vdash P : \gamma'' \dashv R''_p}{\Gamma, \Gamma_c, \Gamma_p, \Gamma_m, \Gamma_p^{\mathrm{out}} \vdash m(APars; \mathbf{pat} \; \underline{p^{o'}}); P : \gamma' \cap \gamma'' \dashv R'_p \cup R''_p}$$

if 1) $\gamma' \cap \gamma'' \neq \emptyset$ and 2) $R'_p \cap R''_p = \emptyset$ and 3) $|\underline{p^o}| = |\underline{p^{o'}}|$ and

4) if $p_i^{o'} \in \mathrm{dom}(\Gamma_p^{\mathrm{out}})$ then $\tau_i^{o'} <: \Gamma_p^{\mathrm{out}}(p_i^{o'})$ where

$\Gamma'_p \triangleq \Gamma_p[p_i^{o'} \mapsto \tau_i^{o'}]$

$\tau_i^{o'} \triangleq \tau_i^o \Theta$ where $\Theta \triangleq \{s_i \mapsto s'_i\}$

$\gamma' \triangleq \gamma \Theta$ where $\Theta \triangleq \{c_i \mapsto c'_i, c''_i \mapsto \Gamma_c(c_i)\}$

$R'_p = \{p_i^{o'} \mid p_i^{o'} \in \mathrm{dom}(\Gamma_p^{\mathrm{out}})\}$

$(FPars, \underline{\tau^o}, \gamma) \triangleq \Gamma_m(m)$

spec $s : \underline{l} : \rho;$ **comp** $\underline{c : c''};$ **pat** $\underline{p : \tau};$ **rate** $\underline{r} \triangleq FPars$

$\underline{s' : l' : \rho'; \underline{c'}; \underline{PE}; RE} \triangleq APars$

$\mathrm{T}_{\mathrm{AFPAR}}$
$$\frac{\Gamma_s, \Gamma_p \vdash PE_l : \tau'_l \text{ and } \Gamma_s, \Gamma_c, \Gamma_p, \Gamma_r \vdash RE_m}{\Gamma_s, \Gamma_c, \Gamma_p, \Gamma_r \vdash APars <: FPars}$$

if 1) $|\underline{x}| = |\underline{x'}|$ for $x \in \{s, \underline{l_i}, c, p, r\}$ and 2) $s'_i \in \mathrm{dom}(\Gamma_s), c'_k \in \mathrm{dom}(\Gamma_c)$ and

3) $\rho_{ij} = \rho'_{ij}, \Gamma_s(s'_i) <: \{l'_{ij} \mapsto \rho'_{ij}\}$ and 4) $\tau'_i <: \tau''_i$ where

$\tau''_i \triangleq \tau_i \Theta$ where $\Theta \triangleq \{s_i \mapsto s'_i\}$

spec $s : \underline{l} : \rho;$ **comp** $\underline{c : c''};$ **pat** $\underline{p : \tau};$ **rate** $\underline{r} \triangleq FPars$

$\underline{s' : l' : \rho'; \underline{c'}; \underline{PE}; RE} \triangleq APars$

use pattern types as multisets with the obvious extension to $\mathbb{N}$ and the standard multiset operation $+$.

A.3   Programs

The type system for programs in Table 3 tells us if a program is well-typed and, if so, gives the compartments where a program can legally reside according to the compartment hierarchy specified in compartment declarations. Judgements are of the form $\Gamma_s, \Gamma_c, \Gamma_p, \Gamma_r, \Gamma_m, \Gamma_p^{out} \vdash P : \gamma \dashv R_p$ where $\gamma \in 2^{\text{NAMES}_c}$, and thus rely on all available environments. Rules for rate expressions, located patterns and declarations are omitted.

The conditions on compartment hierarchies are enforced in the TREAC rule by using standard set theoretic notation, and the condition on free match variables uses the obvious extension of FV to located pattern expressions. The TPAR rule imposes similar conditions on compartments, and also ensures that parallel compartments do not define the same return pattern identifiers. The declaration rule, TDEC, relies on the declaration type system which is omitted. If the declaration is a pattern flagged for return, the pattern identifier will be in R_p.

Module invocation is checked by the final rule, TMODINV which relies on a separate rule for checking that actual and formal parameters match. The first two conditions are similar to the TPAR rule, and the third checks that actual and formal output parameters have matching length. The fourth condition requires that any of the actual output patterns which are also declared as outputs at a higher level are subtypes of the declared outputs at that higher level. The program following module invocation is evaluated in a pattern environment updated with entries for the actual output patterns. There is however one catch: pattern types may contain the names of formal species parameters, and these must be substituted for the corresponding actuals using a substitution Θ. Similar ideas apply to the set of legal parent compartments.

The Attributed Pi Calculus

Mathias John[1], Cédric Lhoussaine[2,4],
Joachim Niehren[3,4], and Adelinde M. Uhrmacher[1]

[1] University of Rostock
Institute of Computer Science, Modeling and Simulation Group
[2] University of Lille 1, LIFL, CNRS UMR8022
[3] INRIA, Lille, Mostrare project
[4] BioComputing project, LIFL, Lille

Abstract. The attributed pi calculus ($\pi(\mathcal{L})$) forms an extension of the
pi calculus with attributed processes and attribute dependent synchronization. To ensure flexibility, the calculus is parametrized with the language $\mathcal{L}$ which defines possible values of attributes. $\pi(\mathcal{L})$ can express
polyadic synchronization as in pi@ and thus diverse compartment organizations. A non-deterministic and a stochastic semantics, where rates
may depend on attribute values, is introduced. The stochastic semantics
is based on continuous time Markov chains. A simulation algorithm is
developed which is firmly rooted in this stochastic semantics. Two examples underline the applicability of $\pi(\mathcal{L})$ to systems biology: Euglena's
movement in phototaxis, and cooperative protein binding in gene regulation of bacteriophage lambda.

1 Introduction

A plethora of formal concurrent modeling languages has been proposed for
systems biology since the seminal work of Regev and Shapiro [2,1] on the
π-calculus. These languages subscribe to two main paradigms: In the object-
centered paradigm, interaction capacities are attached to concurrent actors, as
for instance in the π-calculus [7,6,5,4,3]. Rule-based languages focus on chemical
reactions and pathways being composed thereof [8,10,9]. Deterministic models
in terms of differential equations can be obtained by averaging over possible
behaviors of non-deterministic models in concurrent languages [11,12,8].

In the following, we introduce the attributed π-calculus, an extension of the
π-calculus by attributed processes and attribute dependent synchronization. Attribute values are useful in order to define diverse properties of biological processes. Two examples on different levels of abstraction illustrate the basic idea:
First, a cell `Cell(coord,vol)` is attributed by coordinates $\text{coord} \in \mathbb{R}^3$ and a
volume $\text{vol} \in \mathbb{R}^+$. Second, protein `Prot(comp)` is attributed by the cellular
compartment $\text{comp} \in \{\,'\texttt{nucleus}',\dots\}$ in which it is located.

Whether two processes of the attributed π-calculus are allowed to interact
depends on constraints on their attribute values. Consider e.g. the binding action
of a protein `Prot(x)` of type $x \in \{\,'\texttt{A}',\,'\texttt{B}'\,\}$ to an operator `Op(y)` able to bind

M. Heiner and A.M. Uhrmacher (Eds.): CMSB 2008, LNBI 5307, pp. 83–102, 2008.

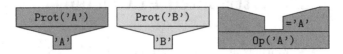

Fig. 1. Only the protein of type 'A' is permitted to bind to the operator of type 'A', but not the protein of type 'B'. Type equality is tested by the operator, once it sees the type of the protein, by applying the test function $\lambda x.x=$'A'.

proteins of type y $\in$ {'A','B'}, as for instance in the system in Fig. 1. For the binding of a protein of type x to an operator of type y, equality of their types x=y is required. In the attributed π-calculus this can be expressed by the following definitions, which are valid for all possible values of the attributes x and y:

Prot(x) $\triangleq$ bind[x]!().0
Op(y) $\triangleq$ bind[λx.x=y]?().OpBound(y)

This says, that before enabling a binding action, Prot(x) needs to provide its type, which is the value of attribute x, as specified by bind[x]. The operator receives the value of x and tests it for equality with its own type, the value of attribute y, as specified by bind[λx.x=y]; the equality constraint is expressed by the Boolean valued function λx.x=y. Consider for instance a system with two actors Prot('A') and Op('A'):

Prot('A') | Op('A') $\rightarrow$ OpBound('A')

The interaction constraint for these two actors is composed by function application (λx.x='A')'A'. It evaluates to the truth value of 'A'='A' which is true. This enables the above binding action, by which the operator turns into its bound state. The slightly more complex system Prot('A') | Prot('B') | Op('A') where only the first protein is permitted to bind to the operator is illustrated in Fig. 1. Assuming constraints that evaluate to positive real numbers, e.g. stochastic rates not only Booleans, also specific affinities of interaction partners can be expressed.

Attribute values are defined in a language $\mathcal{L}$. They subsume possible reaction rates and constraints as in higher order logic. The language $\mathcal{L}$ forms a parameter of the attributed π-calculus, which we call therefore $\pi(\mathcal{L})$, similarly to constraint logic programming CLP($\mathcal{L}$) [13] or concurrent constraint programming cc($\mathcal{L}$) [14]. This avoids inventing completely independent calculi for the many reasonable choices of attribute values and constraints. E.g. encoding the π-calculus with polyadic synchronization [15] requires a constraint language $\mathcal{L}_{\wedge,=}$ with equality tests for channel names and conjunctions thereof.

The paper is structured as follows. We start with the syntax of $\pi(\mathcal{L})$. Thereafter, we present a non-deterministic and a stochastic semantics for $\pi(\mathcal{L})$, independently a concrete choice of attribute language. Two examples provide deeper insight into the modeling with $\pi(\mathcal{L})$. The first one describes the light dependent movement of Euglena, a single celled organism living in inland water. The second example regards cooperative binding, a phenomenon, which is often observed in gene regulatory systems [17,16]. Compared to previous variants of the

π-calculus, the ability to make reaction rates dependent on the attribute values of interaction partners facilitates modeling for both examples.

Afterward, we present a stochastic simulator for $\pi(\mathcal{L})$ independently of the attribute language. In contrast to previous simulators for the stochastic π-calculus [5,4,3], we show how to infer the simulator directly from the definition of the stochastic semantics in terms of continuous time Markov chains.

We complete the paper with encoding $\pi@$ [15] in $\pi(\mathcal{L}_{\wedge,=})$ with respect to the non-deterministic semantics[1]. The language $\pi@$ was proposed as a unifying extension of the π-calculus, in which to express different compartment organizations as in BioAmbients [20] or Brane Calculi [21]. By encoding $\pi@$ in $\pi(\mathcal{L}_{\wedge,=})$, this feature can be inherited, while supporting richer classes of constraints and values.

2 Extending the π-Calculus

We present the attributed π-calculus $\pi(\mathcal{L})$ which extends the π-calculus with respect to some attribute language $\mathcal{L}$. This sublanguage provides expressions for describing values of process attributes. Values subsume constraints, for specifying communication abilities, and stochastic rates.

Keeping the attribute language $\mathcal{L}$ parametric avoids us reinventing independent calculi for the many useful choices in practice. We present two semantics for $\pi(\mathcal{L})$, a non-deterministic semantics and a stochastic semantics, such that the latter refines the former (for error-free programs). Both semantics are defined independently of the concrete choice of $\mathcal{L}$. The stochastic semantics leads to a stochastic simulator which also is independent of the concrete choice of $\mathcal{L}$, see Section 4. Fixing the attribute language is an orthogonal issue, which is easy in practice, for instance by using the implementation language of the simulator.

2.1 Languages of Attribute Values

Let $\mathbb{B} = \{0, 1\}$ be the set of Booleans, $\mathbb{N}$ the set of natural numbers starting from 1, and $\mathbb{R}^+$ the set of non-negative real numbers.

An attribute language is a functional programing language that provides expressions by which to compute values. Expressions are built from constants for numbers, functions, relations, such as 0, $+$, and $\geq$, and channel names, such as x and y. Whenever there exists any ambiguity in concrete syntax, we write constants with quotes such as 'val' in order to distinguish them from channel names without quotes such as val.

Besides defining communication abilities, channel names behave like variables, whose values are defined by the environment. The value of a channel name x can be accessed by applying the function constant 'val' to x. Constraints can be understood as expressions of type Boolean, as for instance $('val' \, x) + ('val' \, y) \geq 0$.

---

[1] It remains unclear how to add reaction rates to $\pi@$ in a general manner as needed for defining stochastic semantics. This has only be done for particular adaptations of $\pi@$ [18,19].

A constraint is successful in environments in which it evaluates to the successful Boolean value 1. In the stochastic setting, successful values are relaxed to stochastic rates in $\mathbb{R}^+ \cup \{\infty\}$.

More formally, we start from an infinite set *Chans*, whose elements are called channel names and ranged over by x, y. An attribute language over *Chans* is a triple $\mathcal{L} = (\textit{Consts}, \textit{Succ}, \Downarrow)$. The first component is a finite set *Consts*, whose elements are called constants and ranged over by c. The set *Exprs* of $\mathcal{L}$ is defined as the set of all λ expressions with constants in *Consts* and variables in *Chans*. The set *Vals* of $\mathcal{L}$ is defined as the set of all values, i.e. channel names, constants, and λ abstractions.

$$e \in \textit{Exprs} ::= x \mid c \mid \lambda x.e \mid e_1 e_2$$
$$v \in \textit{Vals} ::= x \mid c \mid \lambda x.e$$

The second component of $\mathcal{L}$ is a finite set $\textit{Succ} \subseteq \textit{Vals}$, whose elements are called successful values and ranged over by r. In examples, *Succ* will cover the Boolean 1 only or all stochastic rates $r \in \mathbb{R}^+ \cup \{\infty\}$.

The third component of $\mathcal{L}$ is a big-step evaluator $\Downarrow$, a black box algorithm, which evaluates expressions to values. Its behavior depends on environments $\rho : \textit{Chans} \to \textit{Vals}$ mapping channel names to values (including stochastic rates). Let *Env* be the set of all such environments. The big-step evaluator is a partial function of type: $\Downarrow : \textit{Env} \times \textit{Exprs} \rightharpoonup \textit{Vals}$. Instead of $\Downarrow(\rho, e) = v$ we write $\rho \vdash e \Downarrow v$ and call v the value of e wrt. ρ. If the special constant `'val'` belongs to *Consts* then we assume that it satisfies for all channel names $x \in \textit{Chans}$:

$$\rho \vdash \text{'val'}\; x \Downarrow \rho(x)$$

When considering stochastic semantics, we have to choose an attribute language in which *Succ* is a subset of stochastic rates, i.e., $\textit{Succ} \subseteq \mathbb{R}^+ \cup \{\infty\}$.

Values of $\Downarrow(\rho, e)$ may be undefined for two possible reasons: program errors or non-termination. Typical examples for program errors are type errors, e.g. applying constants to real numbers instead of functions. Non-termination is notorious, when the big-step semantics is defined by a small-step evaluator, as usual. We treat non-termination of expressions in processes as program errors. In this case execution blocks, since it is running into an unproductive infinite loop. The assumption of termination is essential for stochastic simulation, where we have to evaluate all expressions in redexes (i.e. pairs of sender and receiver on the same channel) before performing any communication step. One could try to exclude non-termination statically by imposing a simple type system on the level of processes. Instead, we prefer to treat attempts to access undefined values $\Downarrow (\rho, e)$ as program errors dynamically.

An example for an attribute language is $\mathcal{L}_{\wedge,=}$, which is used in Section 5 to encode the π-calculus with polyadic synchronization and a non-deterministic semantics. It provides constraints by which to test a conjunction of equalities between channel names $\wedge_{i=1}^n x_i = y_i$. We use the following set of constants: $\textit{Consts} = \mathbb{B} \cup \{\text{'and'}, \text{'equal'}, 0, 1\}$ of which there is a single successful value in $\textit{Succ} = \{1\}$. The types of the function constants are $\text{'and'} : \mathbb{B} \to \mathbb{B} \to \mathbb{B}$

Processes	$P ::= A(\tilde{e})$	defined process
	$\mid P_1 \mid P_2$	parallel composition
	$\mid (\nu x{:}v)\, P$	channel creation
	$\mid \pi_1.P_1 + \ldots + \pi_n.P_n$	summation of alternative choices
	$\mid \mathbf{0}$	empty solution
Prefixes	$\pi ::= v[e]?\tilde{y}$	receiver
	$\mid v[e]!\tilde{v}$	sender
Definitions	$D ::= A(\tilde{x}) \triangleq P$	parametric process definition

Fig. 2. Syntax of $\pi(\mathcal{L})$ where $v, \tilde{v} \in Vals$, $x, \tilde{y} \in Chans$, and $e \in Exprs$

$$fn(\pi_1.P_1 + \ldots + \pi_n.P_n) = \bigcup_{i \in \{1,\ldots,n\}} fn(\pi_i.P_i) \quad fn(\mathbf{0}) = \emptyset$$
$$fn(v[e]?\tilde{y}.P) = fn(v) \cup fn(e) \cup (fn(P) \setminus \{\tilde{y}\}) \quad fn(P_1 \mid P_2) = fn(P_1) \cup fn(P_2)$$
$$fn(v[e]!\tilde{v}.P) = fn(v) \cup fn(\tilde{v}) \cup fn(e) \cup fn(P) \quad fn(A(\tilde{e})) = fn(e)$$
$$fn((\nu x{:}v)\, P) = (fn(P) \cup fn(v)) \setminus \{x\} \quad fn(A(\tilde{x}) \triangleq P) = fn(P) \setminus \{x\}$$

Fig. 3. Free channel names

and `'equal'` : $Chans \to Chans \to \mathbb{B}$. Therefore, the evaluator is defined for all $x, y \in Chans$ and $c_1, c_2 \in \mathbb{B}$ and undefined for all other values:

$$\rho \vdash \text{'equal'}\; x\; y \Downarrow \begin{cases} 1 \text{ if } x = y \\ 0 \text{ otherwise} \end{cases} \qquad \rho \vdash \text{'and'}\; c_1\; c_2 \Downarrow \begin{cases} 1 \text{ if } c_1 = c_2 = 1 \\ 0 \text{ otherwise} \end{cases}$$

2.2 Syntax of $\pi(\mathcal{L})$

Let $\mathcal{L}$ be an attribute language over some infinite set of channel names $x \in Chans$. The vocabulary for building attributed processes in $\pi(\mathcal{L})$ consists of an infinite set of process names $A \in Proc$, each of which comes with a fixed arity in $\mathbb{N} \cup \{0\}$, and the set of expressions $e \in Exprs$ and values $v \in Vals$ of $\mathcal{L}$.

We use tuple notation in many places. We write $\tilde{e}$ for tuples of expressions, $\tilde{v}$ for tuples of values, and $\tilde{x}$ for tuples of channel names. Substitutions $[\tilde{v}/\tilde{x}]$ apply to tuples of the same length. The same holds for sequences of expressions and values in tuple evaluation $\rho \vdash \tilde{e} \Downarrow \tilde{v}$.

The syntax of the attributed π-calculus $\pi(\mathcal{L})$ is defined in Fig. 2. There are three syntactic categories, (attributed) processes P, prefixes π, and definitions D. The expressions of these categories are as usual, except that we permit λ expressions e in places, where previously, only channel names x or rates r could be found.

A defined process is a term $A(\tilde{e})$ where the arity of tuple $\tilde{e}$ is equal to the arity of A.

Channel creation $(\nu x{:}v)\, P$ creates a new channel name x and assigns a value v to it. The scope of x ranges over v and P. The assignment of v to x is stored in the current environment ρ, such that it can be accessed later on by using the function constant 'val'. This mechanism is useful in order to assign stochastic rates to channels, as in the usual stochastic π-calculus. More generally, it makes channel names behave like variables or memory blocks which refer to values.

Prefixes $v[e_1]!\tilde{v}$ for receiving inputs and $v[e_1]?\tilde{x}$ for sending outputs over channels are generalized in two aspects. First of all, we add expressions $[e_1]$ and $[e_2]$ to senders resp. receivers, which impose constraints on attribute values to be satisfied before communication. Second, we permit values in non-binding positions, where usually only channel names can be used. This way, arbitrary values can be send and received over channels. Prefixes in which v is not a channel name are considered to be erroneous. Such errors are difficult to exclude statically since channel names can be substituted by values dynamically. Even more permissive would be to allow expressions $\tilde{e}$ instead of values $\tilde{v}$. This does not raise any problems as long as they are evaluated before communication. For sake of simplicity, we stick to the slightly more restrictive setting.

The free channel names $fn(P)$ are defined as usual in Fig. 3. They account for free channel names in lambda expressions e denoted by $fn(e)$, i.e., those occurring out of the scope of all λ binders in e. Bound channel names $bn(P)$ are defined as before, except that λ-binders in expressions $e \in \textit{Exprs}$ are included too. The structural congruence on processes $\equiv$ remains the least congruence satisfying α conversion, the axioms of commutative monoids w.r.t. $(|, \mathbf{0})$, associativity and commutativity of summation $+$, and the usual scoping rules of ν-binders:

$$(\nu x{:}v)\ (P_1 \mid P_2) \equiv (\nu x{:}v)\ P_1 \mid P_2 \quad \text{if } x \notin fn(P_2)$$
$$(\nu x{:}v)\ (\nu y{:}u)\ P \equiv (\nu y{:}u)\ (\nu x{:}v)\ P \quad \text{if } x \notin fn(u) \text{ and } y \notin fn(v)$$

2.3   Non-deterministic Semantics

The non-deterministic operational semantics of the attributed π-calculus is given in Fig. 4. It is defined in terms of a small step reduction relation between processes w.r.t. to an environment $\rho : \textit{Chans} \to \textit{Vals}$:

$$\rho \vdash P \to P'$$

Rule (DEF_{nd}) applies the definition of a process in a call-by-value manner once all parameters have been evaluated. This will always succeed for error free programs, but may run into an infinite loop or raise an exception otherwise. The communication rule (COM_{nd}) applies to two parallel sums, which contain matching communication parts, i.e. a receiver $x[e_1]?\tilde{y}.P_1$ and a sender on the same channel $x[e_2]!\tilde{v}.P_2$ such that the constraint $e_1 e_2$ evaluates to some successful value. Reduction cancels all other alternatives of the communicating sums. Furthermore, it substitutes $\tilde{y}$ by $\tilde{v}$ in the continuation P_1 of the receiver, and keeps the continuation P_2 of the sender. The structural rule (NEW_{nd}) eliminates a binder $(\nu x{:}v)\ P$ and updates the value of x in the environment to v.

The usual π-calculus as used by BioSpi or Spim [4,3] (without stochastics) can be encoded in $\pi(\mathcal{L}_1)$ with the successful Boolean value as unique constant, i.e., $\textit{Consts} = \textit{Succ} = \{1\}$. The translation introduces dummy constraints that are

Application of definitions

$$(\text{DEF}_{nd}) \ \frac{\rho \vdash \tilde{e} \Downarrow \tilde{v} \qquad A(\tilde{x}) \triangleq P}{\rho \vdash A(\tilde{e}) \rightarrow P[\tilde{v}/\tilde{x}]}$$

Communication

$$(\text{COM}_{nd}) \ \frac{\rho \vdash e_1 e_2 \Downarrow r \in Succ}{\rho \vdash x[e_1]?\tilde{y}.P_1 + \ldots \mid x[e_2]!\tilde{v}.P_2 + \ldots \rightarrow P_1[\tilde{v}/\tilde{y}] \mid P_2}$$

Structural rules

$$(\text{PAR}_{nd}) \ \frac{\rho \vdash P_1 \rightarrow P_1'}{\rho \vdash P_1 \mid P_2 \rightarrow P_1' \mid P_2} \qquad (\text{NEW}_{nd}) \ \frac{\rho \cup \{x : v\} \vdash P \rightarrow P'}{\rho \vdash (\nu x{:}v) \, P \rightarrow (\nu x{:}v) \, P'}$$

$$(\text{STRUC}_{nd}) \ \frac{P \equiv P_1 \qquad \rho \vdash P_1 \rightarrow P_2 \qquad P_2 \equiv P'}{\rho \vdash P \rightarrow P'}$$

Fig. 4. Non-deterministic semantics of $\pi(\mathcal{L})$

Solutions	$S ::= A(\tilde{e})$	defined molecule
	$\mid S_1 \mid S_2$	parallel composition
	$\mid (\nu x{:}v) \, S$	channel creation
	$\mid \mathbf{0}$	empty solution
Molecules	$M ::= \pi_1.S_1 + \ldots + \pi_n.S_n$	sum of alternative choices
	$\mid (\nu x{:}v) \, M$	channel creation
Prefixes	$\pi ::= v[e]?\tilde{y}$	receiver
	$\mid v[e]!\tilde{v}$	sender
Definitions	$D ::= A(\tilde{x}) \triangleq M$	molecule definition

Fig. 5. Biochemical forms where $v, \tilde{v} \in Vals$, $x, \tilde{y} \in Chans$, and $e \in Exprs$

always true. In order to do so, let _ be an arbitrary channel. We rewrite all receivers $x!\tilde{y}.P$ to $x[\_]!\tilde{y}.P$ and all senders $x?\tilde{y}.P$ to $x[\lambda\_.1]?\tilde{y}.P$. Channel declarations $(\nu x)P$ are translated to $(\nu x{:}\_) \, P$, i.e.; they introduce dummy values too.

2.4 Biochemical Forms

Every process of $\pi(\mathcal{L})$ can be normalized into biochemical forms [4]. These forms are well-suited for defining the stochastic semantics and implementing simulators.

The idea is to introduce parametric process definitions for all sums in processes. What remains is a parallel composition of defined processes possibly with some local channel names. A systematic transformation replaces all nested sums by newly defined processes. The parameters of the new process definitions are the free names of the nested sums, while the sums itself are moved into the new process definitions.

Biochemical forms as defined in Fig. 5 have four categories of terms. The previous category of processes is split into two levels, *solutions* ranged over by S and *molecules* ranged over by M. Solutions can be identified with multisets of defined molecules $A(\tilde{e})$ up to some local channels. Molecule definitions are sums possibly in the scope of ν-binders. The alternatives of the sums are prefixed solutions, so that no nested sums can be produced by communication.

2.5   Stochastic Semantics

The stochastic semantics of $\pi(\mathcal{L})$ is given in Fig. 6. It defines Markov chains for solutions of $\pi(\mathcal{L})$. The states of such Markov chains are the congruence classes of solutions with respect to structural congruence $\equiv$. The stochastic rates, associated to state transitions, are obtained by evaluating constraints to successful values. Rather than saying "that" a constraint is successful, this expresses "how" successful it is. We use product notation for parallel compositions, by writing $\prod_{i=1}^{n} S_i$ instead of $S_1 \mid \ldots \mid S_n$. We use meta variables N to range over quantifier prefixes $(\nu \widetilde{y{:}v})$, write $N(S)$ instead of $(\nu \widetilde{y{:}v})\,(S)$, and $N(M)$ instead of $(\nu \widetilde{y{:}v})\,(M)$. We are looking for a stochastic semantics expressing the *Chemical Law of Mass Action* according to which the speed of a chemical reaction in a solution is proportional to the number of possible interactions of its reactants in the solution. This can be translated in $\pi(\mathcal{L})$ as follows: given a source state S, the semantics defines how many distinct interactions allow a transition to a common (i.e. structurally equivalent) target state S'. For instance, suppose that $S = A \mid A \mid B$, with definitions associated to A and B such that $A \mid B \to C$ with rate r, then one expects $S \to A \mid C$ with rate $2r$. Indeed, two distinct interactions lead to $A \mid C$ involving either the first or the second occurrence of A. To this end, we use an intermediate reduction relation $\xrightarrow[\ell]{r}$ using labels $\ell \in \mathbb{N}^4$ and rates $r \in \mathbb{R}^+ \cup \{\infty\}$, from which we build a Markov chain $\xrightarrow{r}$ by summation over labels.

A label locates a potential redex in solutions. Pairs $(i_1, j_1) \in \mathbb{N}^2$ mark the position of the j_1'th alternative in the i_1'th molecule of a solution. Labels $\ell = (i_1, j_1, i_2, j_2) \in \mathbb{N}^4$ fix a pair of alternatives in a solution S uniquely up to α-renaming. Such a pair is a redex, if it is a receiver-sender-pair on the same channel from different defined molecules. The set $\text{redex}_\ell^\rho(S)$ defined by rule (REDEX) contains all the α-variants of the redex at ℓ, if there is any. The environment ρ matters here, since all arguments of the defined molecules with numbers i_1 and i_2 need to be evaluated. The redex contains the j_1'th alternative of molecule i_1 and the j_2'th alternative of molecule i_2 as selected by rule (CHOOSE). Only alpha congruence $\equiv_\alpha$ is permitted here. Using full structural congruence instead would spoil the meaning of labels, so that redexes could not be uniquely identified by them. This would falsify summation in rule (SUM) and counting in rule (COUNT).

Going back to our previous example, and assuming that A and B are defined with a single alternative consisting of complementary prefixes, we have $A \mid A \mid B \xrightarrow[(1,1,3,1)]{r} A \mid B$ and $A \mid A \mid B \xrightarrow[(2,1,3,1)]{r} A \mid B$.

Redexes $(1 \leq j \leq m,\ i_1, j_1, i_2, j_2 \in \mathbb{N})$

$$(\text{CHOOSE}) \quad \frac{\rho \vdash \tilde{e} \Downarrow \tilde{v} \qquad A(\tilde{x}) \triangleq N(\pi_1.S_1 + \ldots + \pi_m.S_m)}{\text{choose}_j^\rho(A(\tilde{e})) = (N(\pi_j.S_j))[\tilde{v}/\tilde{x}]}$$

$$(\text{REDEX}) \quad \frac{\text{choose}_{j_1}^\rho(A_{i_1}(\tilde{e}_{i_1})) \equiv_\alpha (\nu \widetilde{y_1{:}v_1})\ (x[e_1']?\tilde{y}.S_1) = S_1' \qquad \text{choose}_{j_2}^\rho(A_{i_2}(\tilde{e}_{i_2})) \equiv_\alpha (\nu \widetilde{y_2{:}v_2})\ (x[e_2']!\tilde{v}.S_2) = S_2' \qquad i_1 \neq i_2}{(S_1', S_2') \in \text{redex}_{(i_1,j_1,i_2,j_2)}^\rho(\prod_{i=1}^n A_i(\tilde{e}_i))}$$

where $x \in Chans$, $x \notin \{\tilde{y_1}\} \cup \{\tilde{y_2}\}$ and $\{\tilde{y_1}\} \cap \{\tilde{y_2}\} = \emptyset$.

Labeled reduction $(r \in \mathbb{R}^\infty$ and $\ell = (i_1, j_1, i_2, j_2) \in \mathbb{N}^4)$

$$(\text{COM}) \quad \frac{(N_1(x[e_1']?\tilde{y}.S_1), N_2(x[e_2']!\tilde{v}.S_2)) \in \text{redex}_\ell^\rho(\prod_{i=1}^n A_i(\tilde{e}_i)) \qquad \rho \vdash e_1'e_2' \Downarrow r \in Succ}{\rho \vdash \prod_{i=1}^n A_i(\tilde{e}_i) \xrightarrow[\ell]{r} \prod_{i=1, i \neq i_1, i_2}^n A_i(\tilde{e}_i) \mid N_1 N_2(S_1[\tilde{v}/\tilde{y}] \mid S_2)}$$

$$(\text{NEW}) \quad \frac{\rho \circ [v/x] \vdash S \xrightarrow[\ell]{r} S'}{\rho \vdash (\nu x{:}v)\ S \xrightarrow[\ell]{r} (\nu x{:}v)\ S'} \quad \text{where } x \notin dom(\rho).$$

Markov chain $(r, r' \in \mathbb{R}^+)$

$$(\text{CONV}) \quad \frac{\forall \ell \in \mathbb{N}^4 \forall (N_1(x[e_1']?\tilde{y}.S_1), N_2(x[e_2']!\tilde{v}.S_2)) \in \text{redex}_\ell^\rho(S) \exists v \in Vals : \rho \vdash e_1'e_2' \Downarrow v}{\rho \vdash S \Downarrow}$$

$$(\text{SUM}) \quad \frac{\rho \vdash S \Downarrow \qquad S \equiv S_1 \qquad \sum_{\{\ell \mid \rho \vdash S_1 \xrightarrow[\ell]{r'} S_2 \equiv S'\}} r' = r \neq 0 \qquad \nexists \ell.\rho \vdash S_1 \xrightarrow[\ell]{\infty}}{\rho \vdash S \xrightarrow{r} S'}$$

$$(\text{COUNT}) \quad \frac{\rho \vdash S \Downarrow \qquad S \equiv S_1 \qquad \begin{array}{c} n = \sharp\{\ell \mid \rho \vdash S_1 \xrightarrow[\ell]{\infty} S_2 \equiv S'\} \neq 0 \\ m = \sharp\{\ell \mid \rho \vdash S_1 \xrightarrow[\ell]{\infty} S_2\} \end{array}}{\rho \vdash S \xrightarrow{\infty(n/m)} S'}$$

Fig. 6. Stochastic operational semantics of $\pi(\mathcal{L})$

Rule (COM) performs the interactions of the redex with label $\ell = (i_1, j_1, i_2, j_2)$, under the condition that the constraint of the redex is successful. Rule (NEW) states that reduction may occur under channel restriction with corresponding environment updating. We assume that restricted channels do not occur in the environment.

Markov chains are derived by summing up the rates of all labeled reductions leading from S to S' modulo structural congruence. Summation in rule (SUM) requires that all constraints in all redexes of S converge to some value, as defined by rule (CONV). It equally presupposes that no immediate reaction with rate ∞

is enabled. Rule (COUNT) does the same except that it assigns to each transition a probability. This probability represents the number of immediate located transitions leading to a common state over the total number of outgoing immediate transitions. Thanks to those probabilities, immediate transitions can be eliminated in order to obtain a true Continuous Time Markov chain preserving *probability transitions* and *sojourn times* (see [5] for details).

The usual stochastic π-calculus can be encoded in $\pi(\mathcal{L}_{\mathbb{R}^+ \cup \{\infty\}})$ with $Consts = Succ = \mathbb{R}^+ \cup \{\infty\}$. The translation only needs to access the rate of the channel in the environment. One way to do so is to rewrite sender $x!\tilde{y}.P$ to $x['val'x]!\tilde{y}.P$ and receiver $x?\tilde{y}.P$ to $x[\lambda r.r]?\tilde{y}.P$. Channel declarations with stochastic rates $(\nu x{:}r)$ P remain unchanged. Another encoding would be to rewrite output sender $x!\tilde{y}.P$ to $x[\_]!\tilde{y}.P$ and input receiver $x?\tilde{y}.P$ to $x[\lambda\_.'val'\ x]?\tilde{y}.P$. The only difference is whether sender or receiver accesses the rate of the communicating channel (which both of them know).

3   Modeling Techniques and Biological Examples

We present two examples for modeling biological systems with attribute dependent rates, in order to illustrate the advantages of $\pi(\mathcal{L})$ as a modeling language. The first example introduces a simplistic, discrete spatial model of Euglena's phototaxis [22], while the second more complex example shows how to deal with cooperative enhancement in gene regulation at the lambda switch [17].

3.1   Space

Spatial aspects of molecular systems gain increasing interest in systems biology [23]. We illustrate the use of attribute dependent rates for spatial modeling with a simple example, which is the modeling of Euglena's light dependent motion (phototaxis). More complex compartment structures can be modeled in the attributed π-calculus as well, as we will show by encoding $\pi@$ in Section 5.

Euglena is a single cell organism that lives in inland water and performs photosynthesis. Depending on the brightness, it swims up and down in order to reach a zone with just the right amount of light, [24]: if the amount of light decreases it moves towards a lighter zone, and vice verca. This behavior is specified in the attributed π-calculus model in Fig. 7. The model assumes a light source positioned over water, of which it distinguishes five discrete zones of water, of depths $\mathtt{d} \in \{0, 1, 2, 3, 4\}$.

Euglenas at depth $\mathtt{d}$ are defined by processes $\mathtt{Euglena(d)}$. There are two channels $\mathtt{up}$ and $\mathtt{down}$ for upward respectively downward motions. The speed of upward motions is $\mathtt{d}*\mathtt{(1-i)}$ where $\mathtt{i} \in [0,1]$ is the current light intensity, and the speed of the downwards movement is $\mathtt{(4-d)}*\mathtt{i}$. For $\mathtt{i} = 0.5$, Euglenas are expected to concentrate at depth level $\mathtt{d} = 2$. Note that $\mathtt{Euglena(0)}$ cannot climb further, since the value of $0*(1-i)$ is 0 and thus not successful for all $\mathtt{i}$. For the same reason, $\mathtt{Euglena(4)}$ cannot descend. The interaction partners are processes $\mathtt{Light(i)}$ modeling light sources of intensity $\mathtt{i} \in [0,1]$. In our example, we use two light sources, with different intensities $\mathtt{Light(0.5)}$ and $\mathtt{Light(0.6)}$. $\mathtt{Euglena(d)}$

Process definitions

$$Euglena(d) \triangleq up[\lambda i.d*(1-i)]?().Euglena(d-1)$$
$$+ down[\lambda i.(4-d)*i]?().Euglena(d+1)$$
$$Light(i) \triangleq up[i]!().Light(i)$$
$$+ down[i]!().Light(i)$$

Example solution

$$Euglena(2) \mid ... \mid Euglena(2) \mid Light(0.5) \mid Light(0.6)$$

Fig. 7. $\pi(\mathcal{L})$ model of Euglena's light-dependent motion: the rates for climbing and falling depend on the organism's current depth level and the light intensity

can adapt to different intensities of light. For this we use λi abstractions in its definition.

The same system can be modeled in the stochastic π-calculus, since all parameters are finitely valued: $d \in \{0,\ldots,4\}$ and $i \in \{0.5, 0.6\}$. The idea is to duplicate channels for all possible reaction rates. In the above example, we need channels $up_{d,i}$ with rates $d*(i-1)$, and channels $down_{d,i}$ with rates $(4-d)*i$, for all possible values of i and d that yield nonzero rates. This leads to independent definitions of Euglenas for all possible depths, see Fig. 8.

$$Euglena_0() \triangleq down_{0,0.5}?().Euglena_1() + down_{0,0.6}?().Euglena_1()$$
$$Euglena_1() \triangleq up_{1,0.5}?().Euglena_0() + up_{1,0.6}?().Euglena_0()$$
$$+ down_{1,0.5}?().Euglena_2() + down_{1,0.6}?().Euglena_2()$$
$$...$$
$$Euglena_4() \triangleq up_{4,0.5}?().Euglena_3() + up_{4,0.6}?().Euglena_3()$$

Fig. 8. A model of Euglena's in the stochastic π-calculus. Distinct definitions for $Euglena_d$ are used for all depth levels $d \in \{0,4\}$.

Whether arbitrary processes of $\pi(\mathcal{L})$ are expressible in the stochastic π-calculus is open, particularly for infinitely valued parameters. Even if it is possible, the size of the the stochastic π-calculus definitions may often become prohibitively larger.

3.2 Cooperative Enhancement

Cooperative binding is a frequent and often decisive aspect in gene regulatory networks, where proteins stabilize each other's binding to neighboring DNA sites by adhesive contacts. In quantitative terms, the decay rate of one DNA-protein complex decreases by the existence of another. This is an instance of cooperative enhancement of reaction rates by third partners. As shown in [16,17], cooperative enhancement can be modeled in the stochastic π-calculus. It however requires nontrivial encodings, that can be alleviated within $\pi(\mathcal{L})$.

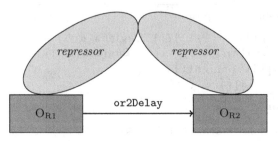

Fig. 9. The decay of the *repressor*-O_{R2} complex: in order to make the decay rate of the *repressor*-O_{R2} complex dependent on O_{R1}'s state, the two sites communicate on or2Delay before O_{R2} unbinds

A well understood instance of cooperative binding occurs during transcription initiation control at the λ switch. The λ switch is a segment of the DNA of bacteriophage λ. It contains two binding sites O_{R1} and O_{R2}, where *repressor* and *cro* proteins can bind. An unstable binding of a *repressor* molecule to O_{R2} is stabilized by the simultaneous presence of another *repressor* at the neighboring site O_{R1}. As illustrated in Fig. 9, the two proteins actually touch each other.

A $\pi(\mathcal{L})$ model of cooperative binding at O_{R2} is presented in Fig. 10. It contains the parametric definition Prot(type), which emulates the behavior of the proteins. The parameter type can be instantiated by either 'rep' or 'cro', for modeling *repressor* or *cro* proteins respectively. Proteins can bind to both sites O_{R1} and O_{R2}. Free sites are defined by processes $O_{R1}()$ and $O_{R2}()$, where proteins can attach via channel bind. As this occurs the channel release is created, and henceforth connects the protein to the site (complexation). Later communication on release breaks the complex. The reaction rate of complexation is fixed to 0.098. For decomplexation the rate is determined by the sender, i.e. the binding site, the receiving protein accepts it by applying the identity function $\lambda r.r$.

Now consider the models for the protein bound DNA sites. O_{Ri}Bound(type,release) describes the unbinding from the occupied site O_{Ri}, where type indicates the type of the bound protein. For $i = 1$ the rate of the unbinding reaction merely depends on the protein type. Using the global channel or1Delay it is specified as ['val' or1Delay type]. Recall that 'val' simply access the channel's value from the environment.

For the second site ($i = 2$) decomplexation is influenced by cooperative binding. To model this, O_{R1} and O_{R2} are linked via the channel or2Delay, illustrated in Fig. 9. Additionally, the release operation is decomposed into an interaction on channel or2Delay, with a reaction rate defining the actual unbinding delay, and an immediate communication on release. As stated in the definition of the global channel or2Delay the unbinding delay depends not only on the type of the bound protein, but also on the state of O_{R1}, which can be either 'free', bound to 'rep' or bound to 'cro'. The reaction rate is computed by applying the function λ state1.'val' or2Delay state1 type to the state of O_{R1}.

Values of global channel names

```
bind: 0.098
or1Delay: λ type.
   if type='rep' then 0.155 else
   if type='cro' then 2.45
or2Delay: λ state1. λ type2.
   if state1='rep' then
      if type2='rep' then 0.155 else % big delay (cooperative)
      if type2='cro' then 3.99        % small delay
   else 2.45 % 'cro' or 'free'
```

Process definitions

```
Prot(type) ≜ (ν release:_)bind[_]!(type,release).
                 release[λr. r]?().Prot(type)
O_R1() ≜
   bind[λ_. 'val' bind]?(type,release).O_R1Bound(type,release)
 + or2Delay['free']!().O_R1()
O_R1Bound(type,release) ≜
   release['val' or1Delay type]!().O_R1()
 + or2Delay[type]!().O_R1Bound(type,release)
O_R2() ≜
   bind[λ_. 'val' bind]?(type,release).O_R2Bound(type,release)
O_R2Bound(type,release) ≜
   or2delay[λ state1. 'val' or2Delay state1 type]?().
      release[∞]!().O_R2()
```

Example Solution

$$O_{R1}() \mid O_{R2}() \mid \prod_{i=1}^{28} Prot('rep') \mid \prod_{i=1}^{67} Prot('cro')$$

Fig. 10. $\pi(\mathcal{L})$ model of cooperative binding between O_{R1} and O_{R2} at the λ switch

A previous model [17] in the the stochastic π-calculus calculus [2] requires to keep O_{R2} constantly informed about state changes of O_{R1}, which is implemented by immediate communication steps. Keeping state information consistent in this manner is error-prone, it may easily lead to deadlocks. A subsequent model [16] in SPiCO [5], the the stochastic π-calculus calculus with concurrent objects, requires significantly fewer updates. In $\pi(\mathcal{L})$, reaction rates directly depend on the attribute values of the interaction partners. State changes are propagated without additional communication steps, preventing deadlocks.

Comments on the syntax. Biochemical forms are sometimes cumbersome for modeling, such that we prefer using the full language, as introduced in Section 2.2. In the definition of Prot(type) and O_{R2}Bound(type,release), for instance, we use sequences of communication prefixes like $\pi_1.\pi_2.P$, which are a special case of nested sums and thus excluded by biochemical forms. The usual unnesting algorithm, which replaces nested sums by newly defined processes, provides the required transformation.

We freely use `if-then-else` statements in λ-terms of the attribute language, for instance in the values of channel names `or1Delay` and `or2Delay`. This is syntactic sugar increasing the readability of the model, which can be directly replaced by λ calculus expressions[2]. In practice, however, further syntactic sugar and extensions of the lambda calculus might be useful, in particular the addition of pattern matching `case` statements. These are advantageous for typeful programming, in contrast to pure `if-then-else` expressions.

4   Stochastic Simulator

The development of the simulator, as presented in this section, closely follows the introduced stochastic semantics in terms of continuous time Markov chains (Section 2.5). This shows that a simulator for $\pi(()\ \mathcal{L})$ can be obtained independently of the choice of $\mathcal{L}$, by extending of previous simulators for the stochastic π-calculus or SPiCO [4,5,3]. Our presentation differs previous ones though. The main advantage is to firmly root the simulator in the operational semantics. Algorithmic aspects remain mostly unchanged, but become applicable in greater generality.

The stochastic semantics for $\pi(\mathcal{L})$ induces the naive stochastic simulator given in Fig. 11. A simulator's input comprises a solution S, an environment ρ, and a time point $t \in \mathbb{R}$. The next reduction step for S is chosen in a memory less stochastic manner. The sojourn time $\Delta \geq 0$ in S is inferred, and the simulator proceeds with the resulting solution at time point $t + \Delta$.

Central to this simulator is to determine the set of labeled reactions in S with respect to the current environment ρ:

$$\texttt{Reacts} = \{(\ell, r) \mid \rho \vdash S \xrightarrow[\ell]{r} S'\}$$

Labeled reactions with rate $r = \infty$ are executed with priority and without time consumption. If no reaction with rate $r = \infty$ exists, we apply Gillespie's algorithm [25] to select a reaction $(\ell, r) \in \texttt{Reacts}(S)$ with probability r/s where $s = \sum_{(\ell, r') \in \texttt{Reacts}(S)} r'$. The sojourn time in S is $\Delta = -ln(1/U)/s$ for some uniformly distributed random number $0 < U \leq 1$.

The set `Reacts` can be computed as follows. First, all expressions in defined molecules $A(\tilde{e})$ of S are evaluated, and then replaced by their definitions. Next, all pairs of alternatives are enumerated and filtered for those that use the same channel and satisfy the communication constraint. The constraint is tested by applying the evaluation algorithm for the attribute language $\mathcal{L}$ in environment ρ. It terminates by assumption for error-free processes, so that the computation of `Reacts` terminates in that case too.

Most fortunately, the Markov chain itself does not need to be computed by the simulation algorithm. This would be largely unfeasible, since the number of possible outcomes of non-deterministic interactions may grow exponentially.

---

[2] When assuming $\rho \vdash \text{'true'}\ v_1\ v_2 \Downarrow v_1$ and $\rho \vdash \text{'false'}\ v_1\ v_2 \Downarrow v_2$ we can replace conditionals `if` e `then` e_1 `else` e_2 by applications $e\ e_1\ e_2$.

Simulate−naive (S, ρ, t)
　　// *solution S, environment* $\rho : fn(S) \rightarrow Vals$, *time point* $t \in \mathbb{R}$
　　case S
　　of 0 **then skip** // *termination*
　　of $\prod_{i=1}^{n} A_i(\tilde{e}_i)$ **then** // *no* ν *binders in* S
　　　　let Reacts $= \{(\ell, r) \mid \exists S'.\ \rho \vdash S \xrightarrow[\ell]{r} S', r \in Succ, \ell \in \mathbb{N}^4\}$

　　　　if $\{(\ell, r) \in \text{Reacts} \mid r = \infty\} = \emptyset$
　　　　then
　　　　　　let $((\ell, r), \Delta) =$ Gillespie (Reacts) // (SUM)
　　　　　　let S' such that $S \xrightarrow[\ell]{r} S'$ // (COM)
　　　　　　Simulate−naive $(S', \rho, t + \Delta)$
　　　　else
　　　　　　select (ℓ, ∞) **in** Reacts **with equal probability** //(COUNT)
　　　　　　let S' such that $\rho \vdash S \xrightarrow[\ell]{\infty} S'$ // (COM)
　　　　　　Simulate−naive (S', ρ, t)
　　else // *some* ν *binder in* S
　　　　let S', x, v such that $S \equiv (\nu x{:}v)\, S'$ with $x \notin dom(\rho)$
　　　　let $\rho' = \rho \circ [v/x]$
　　　　Simulate−naive (S', ρ', t) // (NEW)

Fig. 11. Naive simulator interpreting the stochastic semantics

Furthermore, it would require to decide whether two solutions are structurally congruent (rules SUM and COUNT), which is a graph isomorphism complete problem [26].

In order to increase efficiency of the naive simulation algorithm, we apply an idea hidden already in the implementations BioSpi. The objective is to avoid the enumeration of all pairs of alternatives (and thus redexes), since there may be quadratically many in the size of S. The strategy is to *group* all reactions, using the same channel x and the same constraints $e_2 e_1$, modulo evaluation of e_1 and e_2, i.e. all reactions with $x[v_1]!\ldots$ and $x[v_2]?\ldots$ for some v_1 and v_2. We then apply the Gillespie algorithm to such *grouped* reactions.

A *label* for a grouped reaction in a solution S is a triple $L \in Chans(S) \times Vals^2(S)$. It represents the following set of reactions with respect to a solution S of the form $S = \prod_{i=1}^{n} A_i(\tilde{e}_i)$ and an environment ρ:

Reacts$(L) = \{(\ell, r) \in \text{Reacts} \mid L = (x, v_1, v_2),\ \ell = (i_1, j_1, i_2, j_2),$
$$\text{choose}_{j_1}^{\rho}(A_{i_1}(\tilde{e}_{i_1})) = x[v_1]!\ldots, \text{choose}_{j_2}^{\rho}(A_{i_2}(\tilde{e}_{i_2})) = x[v_2]?\ldots\}$$

The stochastic rate for a grouping label L is usually called propensity $\text{prop}(L) \in \mathbb{R}^+ \uplus \{\infty(n) \mid n \in \mathbb{N}\}$. It sums up all rates of the labeled reactions that it groups together, or counts the number of labels of infinite rate reactions if there are any:

$$\text{prop}(L) = \begin{cases} \infty(n) & \text{if } n = \#\{\ell \mid (\ell, \infty) \in \text{Reacts}(L)\} \geq 1 \\ \sum_{(\ell, r) \in \text{Reacts}(L)} r & \text{otherwise} \end{cases}$$

With this we can define the set of grouped reactions in S with respect to ρ. They will be used as input to Gillespie's algorithm:

$$\texttt{GReacts} = \{(L, \texttt{prop}(L)) \mid L \in Chans(S) \times Vals(S)^2\}$$

In practice, the cardinality of $\texttt{Reacts}(L)$ is often linear in the size of S. Fig. 12 gives a simulation algorithm based on grouped reactions with propensities. In contrast to the naive simulator, it first selects a label of a grouped reaction by Gillespie's algorithm, and then a label of a reaction in the group with equal distribution.

It remains to compute the propensities of all labels of grouped reactions in a solution S. These can be derived from the values below where $S = \prod_{i=1}^{n} A_i(\tilde{e}_i)$:

$$out(x, v_1) = \#\{(i,j) \mid \text{choose}_j^\rho(A_i(\tilde{e}_i)) = x[v_1]! \ldots\}$$
$$in(x, v_2) = \#\{(i,j) \mid \text{choose}_j^\rho(A_i(\tilde{e}_i)) = x[v_2]? \ldots\}$$
$$mixin(x, v_1, v_2) = \#\{(i, j_1, j_2) \mid \text{choose}_{j_1}^\rho(A_i(\tilde{e}_i)) = x[v_1]! \ldots,$$
$$\text{choose}_{j_2}^\rho(A_i(\tilde{e}_i)) = x[v_2]? \ldots\}$$

Lemma 1. $prop(x, v_1, v_2) = (out(x, v_1) * in(x, v_2) - mixin(x, v_1, v_2)) * r$ *if the solution does not contain infinite rates and* $\rho \vdash v_1 v_2 \Downarrow r$.

The computation of mixins can still produce an output of quadratic size and thus need quadratic time. The square factor, however, is in the maximal number of alternatives in sums defining molecules of S, which will be small in practice. All other needed values can be computed in linear time in the size of S, when ignoring the time for evaluating expressions, which is justified in many practical cases.

The final step toward an efficient simulator consists in computing the propensities $prop(x, v_1, v_2)$ incrementally, so that they don't have to be recomputed from scratch in every reduction step. This can be based on Lemma 1, since the values of $out(x, v_1)$, $in(x, v_2)$, $mixin(x, v_1, v_2)$ can be updated incrementally, when adding new solutions or canceling alternative choices by communication.

5   Polyadic Synchronization and Compartments

In this section, we encode a non-deterministic variant of $\pi@$ [15] in $\pi(\mathcal{L}_{\wedge,=})$, where $\mathcal{L}_{\wedge,=}$ is the attribute language providing conjunctions of equalities between channel names, as defined in Section 2.1.

The calculus $\pi@$ is an extension of the π-calculus, able to express various spatial aspects of compartment organization. There are two main ingredients in $\pi@$, polyadic synchronization and different levels of priorities for different types of reduction. Adding priorities to the $\pi(\mathcal{L})$ is straightforward. They can be defined by a total order on the successful values $Succ$ of the attribute language. The operational semantics has to be adapted such that reduction steps with greater rates are executed first. This can be done by labeling possible reduction steps with the rate values of their constraints. Indeed, the stochastic semantics of

Simulate (S, ρ, t) //solution S, environment $\rho : fn(S) \to Vals$, time point $t \in \mathbb{R}$
  case S
  of 0 then skip // termination
  of $\prod_{i=1}^{n} A_i(\tilde{e}_i)$ **then** // no ν binders in S
    let GReacts $= \{(L, \text{prop}(L)) \mid L \in Chans(S) \times Vals(S)^2\}$
    if $\{(L, r) \in \text{GReacts} \mid r = \infty(n)\} = \emptyset$
    then
       let $((L, r), \Delta) = $ Gillespie (GReacts)
       select (ℓ, r) **in** GReacts(L) equally distributed
       let S' such that $S \xrightarrow[\ell]{r} S'$
       Simulate $(S', \rho, t + \Delta)$
    else
       select $(L, \infty(n))$ **in** GReacts
         with probability n/m where $m = \sum_{(L', \infty(n')) \in \text{GReacts}} n'$
       select (ℓ, ∞) **in** Reacts(L) with equal probability
       let S' such that $\rho \vdash S \xrightarrow[\ell]{\infty} S'$
       Simulate (S', ρ, t)
  else // some ν binder in S
    let S', x, v such that $S \equiv (\nu x{:}v)\, S'$ with $x \notin dom(\rho)$
    let $\rho' = \rho \circ [v/x]$
    Simulate (S', ρ', t) // (NEW)

Fig. 12. Stochastic simulator for $\pi(\mathcal{L})$ (to be implemented incrementally)

$\pi(\mathcal{L})$ is already defined such that it gives highest priority to immediate reactions (with rate ∞). For sake of simplicity, we ignore priorities and focus on how to express polyadic synchronization in the following.

The syntax of $\pi@$ is given in Figure 13. It is similar to the syntax of the π-calculus with two exceptions. Inputs have the form $\tilde{x}?\tilde{z}.P$ and outputs the form $\tilde{x}!\tilde{y}.P$. A communication is possible for inputs and outputs that use the same sequence of channels rather than a single channel only:

$$\ldots + \tilde{x}?\tilde{z}.P \mid \tilde{x}!\tilde{y}.Q + \ldots \;\to\; P[\tilde{y}/\tilde{z}] \mid Q$$

The only values of $\pi@$ are channels. Local binders $(\nu x)Q$ do not assign any values to channels. Therefore, the semantics of $\pi@$ does not consider environments.

The encoding $[\![\_]\!] : \pi@ \to \pi(\mathcal{L}_{\wedge,=})$ is compositional; only, the encoding of communication prefixes deserves special attention:

$$[\![(x_1, \ldots, x_n)!\tilde{z}.P]\!] = x_1[\lambda y_2 \ldots \lambda y_n.$$
$$\text{'and'} \ldots \text{'and'}(\text{'equal'}\,x_2 y_2) \ldots (\text{'equal'}\,x_n y_n)]\tilde{z}.[\![P]\!]$$
$$[\![(y_1, \ldots, y_n)?\tilde{z}.P]\!] = y_1[\lambda e.e y_2 \ldots y_n]?\tilde{z}.[\![P]\!]$$

Tuple equality in $\pi@$ is translated as a conjunction of name equalities in the attribute language: an output with subject $(x_1, \ldots, x_n)$ is translated as an output with subject x_1 with a constraint that checks whether its first argument (bound to y_2) is equal to x_2 and its second argument (bound to y_3) is equal to

Processes	$P ::= A(\tilde{x})$	defined process
	$\mid P_1 \mid P_2$	parallel composition
	$\mid (\nu x)P$	channel creation
	$\mid \pi_1.P_1 + \ldots + \pi_n.P_n$	summation of alternative choices
	$\mid 0$	empty solution
Prefixes	$\pi ::= \tilde{x}?\tilde{y}$	receiver
	$\mid \tilde{x}!\tilde{y}$	sender
Definitions	$D ::= A(\tilde{x}) \triangleq P$	parametric process definition

Fig. 13. Syntax of $\pi@$ [15] where $x, \tilde{x}, \tilde{y} \in Chans$

x_3, etc. Dually, an input with subject $(y_1, \ldots, y_n)$ is translated as an input with subject y_1 with a constraint applying its argument to $y_2, \ldots, y_n$. We also have to translate the definitions of molecules in use

$$[\![A(\tilde{x}) \triangleq P]\!] = A(\tilde{x}) \triangleq [\![P]\!]$$

and consider reduction with respect to a set of such encoded definitions.

Theorem 1 (Operational correspondence)

(a) if $S_1 \to S_2$ w.r.t. $\mathcal{D}$, then $[\![S_1]\!] \to [\![S_2]\!]$ w.r.t. $[\![\mathcal{D}]\!]$
(b) if $[\![S]\!] \to S_1$ w.r.t $[\![\mathcal{D}]\!]$, then $\exists S_2$ s.t. $S_1 \equiv [\![S_2]\!]$, and $S \to S_2$ w.r.t $\mathcal{D}$.

The proof is mostly straightforward. It might be worth noticing, however, that all λ expressions terminate in linear time. This can be seen by inspecting the occurring terms.

6 Conclusion and Outlook

We presented the attributed π-calculus $\pi(\mathcal{L})$, which incorporates an attribute language $\mathcal{L}$ fixing the values and constraints. Since $\mathcal{L}$ is introduced as a parameter, a proliferation of domain specific calculi can be avoided. The central idea of our approach is to control communication by constraints on attribute values of processes. $\pi(\mathcal{L})$ can encode $\pi@$ and thus, also BioAmbients, and BraneCalculi. Furthermore, an encoding of a version of Beta-binders into $\pi@$ was proposed recently [27]. Beyond discrete spatial structures, e.g. a cell's compartments, spatial attributes like volume and position can influence reaction rates. Thus, $\pi(\mathcal{L})$ appears particularly promising for modeling spatial processes. We have illustrated the usefulness of $\pi(\mathcal{L})$ as a modeling language for systems biology by the examples of cooperative protein-DNA binding in gene regulation of bacteriophage λ, and the phototaxis of the single celled organism Euglena. Currently, a model of the Wnt-pathway with focus on membrane interactions and receptors is being developed. If constitutes a further test case for the modeling formalism.

We defined a non-deterministic and a stochastic semantics for $\pi(\mathcal{L})$. The stochastic semantics has been specified in terms of a continuous time Markov chain (CTMC) and then transferred to a simulator, independently of the choice of $\mathcal{L}$. This simulator permits to reuse all algorithmic tricks of state-of-the-art simulators for the stochastic π-calculus. An implementation of the simulator is currently under way. Identifying parameter values for the Wnt-pathway model and, in addition, performance studies along the lines of [28] will, both, challenge and help assessing the simulator's practical efficiency.

In the context of modeling, one shall explore the implication of $\pi(\mathcal{L})$'s central idea, i.e. to map attribute values to interaction affinity, on discrete modeling languages, e.g. rule-based ones. Another issue to be investigated is its potential for large scale modeling. Therefore, object-oriented language concepts such as inheritance, shall be lifted from SPiCO [5] to $\pi(\mathcal{L})$. With respect to language theory, another yet open question is whether one can encode $\pi(\mathcal{L})$ into the π-calculus.

Acknowledgements

We thank Céline Kuttler for suggesting our collaboration, and valuable comments on this paper. Part of the research was financed by the DFG in the context of the Research Training School "dIEM oSiRiS". The BioComputing activity of the LIFL in Lille is funded by the ANR Jeunes Chercheurs BioSpace.

References

1. Regev, A., Shapiro, E.: Cells as Computation. Nature 419, 343 (2002)
2. Regev, A.: Computational Systems Biology: A Calculus for Biomolecular Knowledge. Tel Aviv University, PhD thesis (2003)
3. Priami, C., Regev, A., Shapiro, E., Silverman, W.: Application of a Stochastic Name-Passing Calculus to Representation and Simulation of Molecular Processes. Information Processing Letters 80, 25–31 (2001)
4. Phillips, A., Cardelli, L.: Efficient, Correct Simulation of Biological Processes in the Stochastic Pi Calculus. In: Calder, M., Gilmore, S. (eds.) CMSB 2007. LNCS (LNBI), vol. 4695, pp. 184–199. Springer, Heidelberg (2007)
5. Kuttler, C., Lhoussaine, C., Niehren, J.: A Stochastic Pi Calculus for Concurrent Objects. In: Anai, H., Horimoto, K., Kutsia, T. (eds.) Ab 2007. LNCS, vol. 4545, pp. 232–246. Springer, Heidelberg (2007)
6. Dematté, L., Priami, C., Romanel, A.: Modelling and Simulation of Biological Processes in BlenX. SIGMETRICS Performance Evaluation Review 35(4), 32–39 (2008)
7. Ciocchetta, F., Hillston, J.: Bio-PEPA: An Extension of the Process Algebra PEPA for Biochemical Networks. ENTCS 194(3), 103–117 (2008)
8. Chabrier-Rivier, N., Fages, F., Soliman, S.: The Biochemical Abstract Machine BIOCHAM. In: Danos, V., Schachter, V. (eds.) CMSB 2004. LNCS (LNBI), vol. 3082, pp. 172–191. Springer, Heidelberg (2005)
9. Faeder, J.R., Blinov, M.L., Goldstein, B., Hlavacek, W.S.: Rule-Based Modeling of Biochemical Networks. Complexity 10(4), 22–41 (2005)

10. Danos, V., Feret, J., Fontana, W., Harmer, R., Krivine, J.: Rule-based modelling of cellular signalling. In: Caires, L., Vasconcelos, V.T. (eds.) CONCUR 2007. LNCS, vol. 4703, pp. 17–41. Springer, Heidelberg (2007)
11. Hillston, J.: Process Algebras for Quantitative Analysis. In: LICS 2005: Proceedings of the 20th Annual IEEE Symposium on Logic in Computer Science, pp. 239–248. IEEE Computer Society, Los Alamitos (2005)
12. Cardelli, L.: On Process Rate Semantics. TCS 391(3), 190–215 (2008)
13. Jaffar, J., Lassez, J.L.: Constraint Logic Programming. In: 14th ACM SIGACT-SIGPLAN Symposium on Principles of Programming Languages, pp. 111–119. ACM Press, New York (1987)
14. Saraswat, V.A., Rinard, M., Panangaden, P.: The Semantic Foundations of Concurrent Constraint Programming. In: 18th ACM SIGPLAN-SIGACT Symposium on Principles of Programming Languages, pp. 333–352. ACM Press, New York (1991)
15. Versari, C.: A Core Calculus for a Comparative Analysis of Bio-inspired Calculi. Programming Languages and Systems, pp. 411–425 (2007)
16. Kuttler, C.: Modeling Bacterial Gene Expression in a Stochastic Pi Calculus with Concurrent Objects. PhD thesis, Université des Sciences et Technologies de Lille - Lille 1 (2007)
17. Kuttler, C., Niehren, J.: Gene regulation in the pi calculus: Simulating cooperativity at the lambda switch. Transactions on Computational Systems Biology VII, 24–55 (2006)
18. Versari, C., Busi, N.: Stochastic Simulation of Biological Systems with Dynamical Compartment Structure. In: Calder, M., Gilmore, S. (eds.) CMSB 2007. LNCS (LNBI), vol. 4695, pp. 80–95. Springer, Heidelberg (2007)
19. Versari, C., Busi, N.: Efficient Stochastic Simulation of Biological Systems with Multiple Variable Volumes. ENTCS 194(3), 165–180 (2008)
20. Regev, A., Panina, E.M., Silverman, W., Cardelli, L., Shapiro, E.: BioAmbients: An Abstraction for Biological Compartments. TCS 325(1), 141–167 (2004)
21. Cardelli, L.: Brane calculi. In: Danos, V., Schachter, V. (eds.) CMSB 2004. LNCS (LNBI), vol. 3082, pp. 257–278. Springer, Heidelberg (2005)
22. John, M., Ewald, R., Uhrmacher, A.M.: A Spatial Extension to the Pi Calculus. ENTCS 194(3), 133–148 (2008)
23. Kholodenko, B.N.: Cell-Signalling Dynamics in Time and Space. Nature Reviews Molecular Cell Biology 7(3), 165–176 (2006)
24. Grell, K.G.: Protozoologie. Springer, Heidelberg (1968)
25. Gillespie, D.T.: A General Method for Numerically Simulating the Stochastic Time Evolution of Coupled Chemical Reactions. Journal of Computational Physics 22(4), 403–434 (1976)
26. Khomenko, V., Meyer, R.: Checking Pi Calculus Structural Congruence is Graph Isomorphism Complete. Technical Report CS-TR: 1100, School of Computing Science, Newcastle University, 20 pages (2008)
27. Cappello, I., Quaglia, P.: A translation of beta-binders in a prioritized pi-calculus. In: From Biology to Concurrency and back, Workshop FBTC (2008)
28. Ewald, R., Jeschke, M.: Large-Scale Design Space Exploration of SSA. In: Computational Methods in Systems Biology, International Conference CMSB 2008. LNCS. Springer, Heidelberg (2008)

The Continuous π-Calculus: A Process Algebra for Biochemical Modelling

Marek Kwiatkowski and Ian Stark

Laboratory for Foundations of Computer Science
School of Informatics, The University of Edinburgh, Scotland
{M.Kwiatkowski,Ian.Stark}@ed.ac.uk

Abstract. We introduce the *continuous π-calculus*, a process algebra for modelling behaviour and variation in molecular systems. Key features of the language are: its expressive succinctness; support for diverse interaction between agents via a flexible network of molecular affinities; and operational semantics for a continuous space of processes. This compositional semantics also gives a modular way to generate conventional differential equations for system behaviour over time. We illustrate these features with a model of an existing biological system, a simple oscillatory pathway in cyanobacteria. We then discuss future research directions, in particular routes to applying the calculus in the study of evolutionary properties of biochemical pathways.

1 Introduction

This research aims to develop computational methods for studying the Darwinian evolution of biochemical pathways. We work in the framework introduced by Regev et al. [1,2,3], who identified the *π-calculus* process algebra as a promising formalism for biological modelling. We modify it in a way that allows us to mention quantitative parameters explicitly, and makes the interaction network of the agents more flexible (see §1.1 below). To take advantage of this quantitative information, we develop a novel operational semantics through a compositional description of continuous system behaviour in terms of real vector spaces (§2.2). We illustrate these concepts with an example of a concrete biological system, a simple oscillatory pathway in cyanobacteria (§7). Finally, we discuss the possibilities of answering questions related to the evolution of pathways using process-algebraic techniques such as model checking and behavioural equivalences (§4).

Reliable models and simulations of evolution on the molecular level would have wide applications, from pure evolutionary theory to drug design. Our particular interest is in the ubiquitous phenomenon of mutational robustness, which has recently received attention as a cross-level organisational principle of biological systems [4,5]. Its understanding, especially on the level of gene regulation and cellular signalling, is an important challenge; and, moreover, one where a computational approach may give essential assistance in tackling the complexity of the systems involved.

M. Heiner and A.M. Uhrmacher (Eds.): CMSB 2008, LNBI 5307, pp. 103–122, 2008.

We motivate the use of process algebras in this context as follows: firstly, they have already been successfully used to model biochemical networks (see §1.2 below). Secondly, much of the genetic variation results in qualitative or quantitative changes in the interaction network; process-algebraic descriptions of networks enable us to express this variability easily, in syntactic terms. Finally, to model evolution we need means to express fitness and related concepts such as neutrality [6,5]; to this end we plan to use well-developed process-algebraic techniques like model checking and behavioural equivalences (§4).

The main contribution of this paper is the *continuous π-calculus* (*cπ*), a process algebra designed specifically to model biomolecular systems in an evolutionary context, with an original semantics in terms of real vector spaces. It also offers a fully modular and compositional method of generating a set of ordinary differential equations (ODEs) governing a given system. Moreover, we give a process-algebraic model of a biomolecular system of considerable interest — a primitive bacterial circadian clock recently recreated *in vitro* [7].

In the remainder of this section we introduce *cπ* by means of small examples (§1.1) and then very briefly recall related work in the fields of computational and theoretical biology (§1.2). Section 2 is a formal presentation of the calculus, while §3 contains the *cπ* model of the KaiC circadian clock [8], with graphs showing its oscillatory behaviour. In the final section we present and discuss the future directions of our research.

1.1 Key Features

The classic π-calculus is well-described in existing texts [9,10,11]. Here we focus on the key distinguishing features of our continuous π, and follow with two small examples to illustrate them.

1. Every *cπ* process is a parallel (∥) combination of *species*. Species are very similar to classic π-processes. Every species in a given process is equipped with a real number, to be thought of as the concentration of the substance described by the species. Thus, every *cπ* process represents a complete molecular system at a certain point in time.

2. As usual in the π-calculus, communication is through named channels. However, in contrast to most π derivatives, there are no co-names. Instead, any name can in principle communicate with any other — potentially more than one — and for any two names it is specified whether they can communicate and at what rate. Biologically, names are intended to model distinctive reaction sites, with the communication rate between two names corresponding to the rate constant of the biochemical reaction between sites. The relevant technical device to manage this information is an *affinity network*. One consequence of this approach is that every new name must come with the information about its communication potential. Also, whenever we consider a particular process P, we assume some given affinity network on the free names of P.

   The reason for this approach is two-fold. First, it makes sense in an evolutionary context to abandon the strict correspondence of sites and co-sites, and hence the symmetry of names and co-names. Second, the affinity networks give a convenient collection of parameters for varying the model: in particular, those important for questions of evolvability and robustness.

3. All $c\pi$ reactions have at most two substrates and are governed by the Law of Mass Action (following, for example, the observations of [12, p.298]). It is worth stressing, however, that more complex kinetic behaviour can emerge as we build up larger processes from smaller ones. Nonetheless, the system dynamics remains purely deterministic: for every process P we derive a term $\frac{dP}{dt}$, denoting the speed and direction of the temporal evolution of P.

4. To model spontaneous monomolecular reactions, such as degradation, or conformation changes, we use silent actions labelled with real numbers denoting reaction rate constants.

5. Molecular complexes are represented by parallel components within the shared scope of one or more private names, following Regev [2]. As usual for the π-calculus, communication between shared private names within the complex gives rise to silent actions; and these in turn model spontaneous actions like complex dissociation.

As a first simple example, consider two molecules, A and B, that can bind to each other (at rate k_1) and subsequently unbind (at rate k_2). As noted above, we model complexation as scope extrusion and decomplexation as interaction on a shared private channel. This gives rise to the following $c\pi$ definition of species A and B:

$$A \overset{\mathrm{df}}{=} (\boldsymbol{\nu}\, u \overset{k_2}{-\!\!\!-} v)(a\langle u\rangle.v.A) \tag{1}$$

$$B \overset{\mathrm{df}}{=} b(x).x.B \tag{2}$$

with the global affinity network of Fig. 1.

Here the public names a and b represent protein interaction sites, and a communication event between these two names models binding of these sites. The prefix $(\boldsymbol{\nu}\, u \overset{k_2}{-\!\!\!-} v)$ declares a new affinity network consisting of two private names, u and v, that can communicate with each other at the rate k_2. Species A and B can react by a communication event on the public $a \overset{k_1}{-\!\!\!-} b$ channel, with the private name u sent via a, received on b, and then substituted for x in $x.B$. This extends the scope of the network to encompass the remaining parts of A and B and so form complex C:

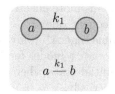

Fig. 1. A very simple affinity network and its textual rendering

$$A \mid B \overset{\tau\langle a,b\rangle}{\longrightarrow} C \overset{\mathrm{df}}{=} (\boldsymbol{\nu}\, u \overset{k_2}{-\!\!\!-} v)(v.A \mid u.B) \tag{3}$$

If we mix species A and B together in concentrations c_1 and c_2, then we obtain a $c\pi$ process $(c_1 \cdot A \parallel c_2 \cdot B)$. The formal semantics of this process reflect mass-action dynamics: the complex C is produced in proportion to the product of substrate

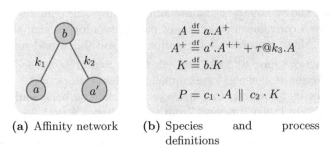

$$A \overset{\mathrm{df}}{=} a.A^+$$
$$A^+ \overset{\mathrm{df}}{=} a'.A^{++} + \tau@k_3.A$$
$$K \overset{\mathrm{df}}{=} b.K$$

$$P = c_1 \cdot A \parallel c_2 \cdot K$$

(a) Affinity network   **(b)** Species   and   process definitions

Fig. 2. A simple $c\pi$ system with a non-trivial affinity network modelling discriminative binding of the kinase K (via site b) to the molecules A and A^+ (at sites a and a', respectively). The definition of the inactive A^{++} species is omitted.

concentrations, while the substrates themselves (A and B) deplete similarly. If we compute the semantics (an appropriate $\frac{d\cdot}{dt}$ term) as described in §2.2, we see that this is indeed the case:

$$\frac{d(c_1 \cdot A \parallel c_2 \cdot B)}{dt} = k_1 c_1 c_2 \cdot C - k_1 c_1 c_2 \cdot A - k_1 c_1 c_2 \cdot B . \tag{4}$$

Similarly, when we consider a solution of complexes in concentration c_3 and derive the semantics of the process ($c_3 \cdot C$), we observe that the complex dissolves to give back substrates A and B at the expected rate:

$$\frac{d(c_3 \cdot C)}{dt} = -k_2 c_3 \cdot C + k_2 c_3 \cdot A + k_2 c_3 \cdot B . \tag{5}$$

As another example, consider a molecule A that can exists in three states: unphosphorylated, phosphorylated and doubly phosphorylated; denoted A, A^+ and A^{++} respectively. Furthermore, suppose that kinase K (the phosphorylating agent) is more effective at the initial phosphorylation step $A \to A^+$ than at the subsequent one $A^+ \to A^{++}$, having reaction rate constants $k_1 > k_2$. Finally, assume that A^+ (only) can spontaneously relax back to A at a rate k_3. Figure 2 shows a $c\pi$ model for this system, and in particular a process P representing an initial state where only A and K are present, at concentrations c_1 and c_2 respectively.

The affinity network in Fig. 2(a) indicates that site b can react with either site a or site a': this will capture the double action of the kinase. Figure 2(b) gives the definitions of the species involved in the system. The first equation states that species A can be transformed into A^+ on interaction at the site a. The second states that A^+ can either interact on a' and be transformed into A^{++} or convert back to A at rate k_3 without any external agent — here "+" models the mutually exclusive choice of alternatives. In the third equation, kinase K can interact at site b and then return to its initial state; recall that according to the affinity network, this interaction might be with site a (on A) or a' (on A^+). The final line defines the initial state of the system, with A and K present at the specified concentrations.

This is a dynamic model, with P evolving in a continuous fashion. At any time instant we can, using the methods of §2.2, formally derive the vector $\frac{dP}{dt}$ specifying the gradient of this temporal evolution. In particular, in the initial state we have:

$$\frac{dP}{dt} = k_1 c_1 c_2 \cdot A^+ - k_1 c_1 c_2 \cdot A . \tag{6}$$

1.2 Related Work

The application of the π-calculus to biology is due to Regev and Shapiro [3], who identified the fundamental correspondence between cellular processes and concurrent computations [13]. They proposed modelling molecules as π-processes, and the use of parallel composition to express the fact that such molecules act independently. In this framework, names denote molecular interaction sites and communication models interaction. A further important abstraction was the use of private names to model molecular complexes and compartments.

Further refinements of this framework addressed the introduction of quantitative information into the model and on modelling compartments more directly. This led, for example, to the development of the biochemical stochastic π-calculus [1] and BioAmbients [14]. Other process algebras, such as PEPA [15] have also been applied to model biological systems [16]. Although seen mainly as simulation engines, these formalisms have also been used to perform static analysis of the model [17].

Using process algebras as a high-level descriptive language, Calder et al. [18] have shown how PEPA models can generate both discrete (stochastic) and continuous (ODE) behavioural specifications for the same system. Unlike raw ODEs then, a process algebra model is not itself the behaviour, but can be used to generate it. We believe that this abstraction step is important in modelling variation, to identify how emergent behaviour depends on changes in a process or its parameters.

Meanwhile, the interest in mutational robustness has been growing amongst biologists for the past 15 years. A recent monograph on the subject [5] identifies the explosion of high-throughput techniques as an important factor for this interest; the other is the importance of this concept in the context of systems biology [19]. The methods applied to study this phenomenon range from pure mathematics [20] to exhaustive computations [21].

2 The Continuous π-Calculus

In this section we set out a formal syntax and mathematical semantics for the continuous π-calculus. Both syntax and semantics have two "layers": of *species* corresponding to individual molecules, and *processes* to populations of these. It is important to keep in mind, however, that none of these terms should invoke associations with their meaning in the context of evolutionary theory.

2.1   Syntax

Definition 1. Take $\mathcal{N}$ a fixed, countably infinite set of *names*, denoted by lower-case letters a, b, x, y, ... Vectors of names are denoted by $\vec{a}$, $\vec{x}$ etc.; these may be of zero length.

Definition 2. A *prefix* is a syntactic expression of the form $a(\vec{b}; \vec{y})$ (a *communication prefix*) or $\tau@k$ (*spontaneous* or *silent prefix*), where all elements of $\vec{y}$ are distinct, $\tau \notin \mathcal{N}$ is a fixed symbol and $k \in \mathbb{R}_{\geq 0}$. We use symbols like π, π', etc. to denote prefixes.

A communication prefix models the ability of one molecule to engage in an interaction with another, at site a. Details of the interaction are modelled as name-passing, where $\vec{b}$ is the vector of names to be sent and $\vec{y}$ is a vector of placeholders for names to be received (and so binds subsequent occurrences of the $\vec{y}$). This symmetry and synchrony of communication is mildly novel, and introduced to reflect the fact that molecular interaction is a synchronous and (usually) symmetric event. For readability we abbreviate when possible: $a(\vec{y})$ for $a(\ ; \vec{y})$, $a\langle \vec{b} \rangle$ for $a(\vec{b};)$, and a for $a(\ ;)$.

The silent prefix $\tau@k$ models the case where a molecule may undergo spontaneous change, without any interaction with the external environment; or, at least, no observed interaction at the level of abstraction being modelled (cf. the dephosphorylation of species A^+ in Fig. 2(b)). The rate of this transformation is recorded directly in the prefix as k.

Definition 3. An *affinity network* is a pair $\langle M, f \rangle$, where $M \subseteq_{\text{fin}} \mathcal{N}$ is a carrier set and f is a symmetric partial function $f \colon M \times M \rightharpoonup \mathbb{R}_{\geq 0}$. We often blur the distinction between a network and its carrier: for a network M we write $x \in M$ to indicate that x lies in the carrier of M, and similarly for other set-theoretic predicates. The expression $M(a, b)$ denotes the value the network assigns to a pair of names (a, b) if this is defined; we write $M(a, b)\!\downarrow$ when this is the case.

From here on we assume a distinguished *global affinity network Aff*, which must be a total relation on its carrier. We shall also use the notation $X \mathbin{\#} Y$, read X *fresh for* Y, to state that $X \cap Y = \varnothing$ for name sets X and Y; and then overload this when X or Y are terms to refer to their free name sets.

Definition 4. The set of *species* is generated by the following grammar:

$$A, B ::= \mathbf{0} \mid D(\vec{a}) \mid \Sigma_{i=0}^{n} \pi_i.A_i \mid A \mid B \mid (\boldsymbol{\nu} M)A \tag{7}$$

where $M \mathbin{\#} Aff$. These are in turn the inactive species $\mathbf{0}$, invocation of a species definition, guarded choice, parallel composition, and local name declaration. This last is also *restriction*: names declared in the affinity network M are available in A, but not elsewhere until explicitly passed out in a communication from A. Small instances of guarded choice Σ are usually written with $+$; as in Fig. 2(b). For every invocation $D(\vec{a})$ we assume a corresponding definition $D(\vec{y}) \stackrel{\text{df}}{=} A$, such that every free name of A appears in either $\vec{y}$ or Aff. Furthermore, these definitions must be *productive* in that any recursive cycle includes a prefix guard. We identify α-equivalent species: the binding operations are prefix π and restriction (νM). Finally, there is a *structural congruence* $\equiv$ on species, generated by the axioms in Fig. 3. We use S to denote the set of species modulo $\equiv$, and write $[A]$ for the equivalence class of species A up to structural congruence.

Definition 5. The set of *processes* is generated by the following grammar:

$$P, Q ::= c \cdot A \mid P \| Q \tag{8}$$

where A is a species, $c \in \mathbb{R}_{\geq 0}$ and all free names appear in Aff. Figure 3 gives a structural congruence $\equiv$ on processes.

In what follows we maintain a careful distinction between a species and its $\equiv$-equivalence class, and the same for processes, in order to precisely state the correctness results for $c\pi$ semantics.

$$
\begin{aligned}
A \mid \mathbf{0} &\equiv A \\
A \mid B &\equiv B \mid A \\
(A \mid B) \mid C &\equiv A \mid (B \mid C) \\
\Sigma_{i=0}^{n} \pi_i.A_i &\equiv \Sigma_{i=0}^{n} \pi_{\sigma_i}.A_{\sigma_i} \quad \text{perm. } \sigma \\
(\nu M)(A \mid B) &\equiv A \mid (\nu M)B \quad M \# A \\
(\nu M)A &\equiv A \quad M \# A \\
(\nu M)(\nu N)A &\equiv (\nu N)(\nu M)A \quad M \# N
\end{aligned}
\qquad
\begin{aligned}
P \| (c \cdot \mathbf{0}) &\equiv P \\
P \| Q &\equiv Q \| P \\
P \| (Q \| R) &\equiv (P \| Q) \| R \\
(c \cdot A) \| (d \cdot A) &\equiv (c + d) \cdot A \\
c \cdot A \equiv c \cdot B \qquad & A \equiv B
\end{aligned}
$$

Fig. 3. Axioms generating structural congruence on species (l) and processes (r)

2.2   Semantics

The two-layered nature of syntax (a "layer" of species, then another of processes) is reflected in the semantics: to obtain the semantics of a process, we first examine the species involved. Formally, we do this by defining a multi-transition system on species and then using this information to build the continuous semantics for processes. Both levels maintain compositionality.

The transition system for species. A *multi-transition system* is a labelled transition system that allows multiple transitions with the same source, target and label. This extension is necessary to keep track of quantitative aspects of behaviour in $c\pi$, as done for example in PEPA [15].

We present the multi-transition system for species in an *abstraction-concretion* style, following Milner [9] (or see [11] for a shorter explanation). In continuous π the use of symmetric prefixes eliminates the distinction between abstractions and concretions, and hence we use "concretion" to refer to both constructions.

Definition 6. A *concretion* is a term generated by the following grammar:

$$F, G ::= (\vec{b}; \vec{y})A \mid F \mid A \mid A \mid F \mid (\nu M)F \qquad (9)$$

where all elements of $\vec{y}$ are distinct and binding. As with species, we identify α-equivalent concretions. We build a structural congruence $\equiv$ from the axioms in Fig. 4, writing $[F]$ for the $\equiv$-equivalence class of F, and take $\mathcal{C}$ as the set of concretions modulo structural congruence.

$$
\begin{array}{llll}
F \mid 0 \equiv F & & & \\
F \mid A \equiv A \mid F & (\nu M)(A \mid F) \equiv A \mid (\nu M)F & M \,\#\, A \\
(F \mid A) \mid B \equiv F \mid (A \mid B) & (\nu M)(F \mid A) \equiv F \mid (\nu M)A & M \,\#\, F \\
(A \mid F) \mid B \equiv A \mid (F \mid B) & (\nu M)F \equiv F & M \,\#\, F \\
F \mid A \equiv F \mid B \qquad A \equiv B & (\nu M)(\nu N)F \equiv (\nu N)(\nu M)F & M \,\#\, N \\
(\vec{b}; \vec{y})A \equiv (\vec{b}; \vec{y})B \qquad A \equiv B & (\vec{b}; \vec{y})(A \mid B) \equiv A \mid (\vec{b}; \vec{y})B & \vec{y} \,\#\, A \\
\end{array}
$$

Fig. 4. Axioms generating structural congruence for concretions

A concretion can be seen as a species that has committed to take part in a specific binary interaction. When it encounters another compatible concretion that interaction may take place. We formalize this with a notion of "pseudo-application".

Definition 7. The operation of *pseudo-application* is a binary partial function $\circ$ on concretions, defined by structural induction over its arguments. For the base case, $(\vec{a}; \vec{x})A \circ (\vec{b}; \vec{y})B$ is defined if and only if $|\vec{a}| = |\vec{y}|$ and $|\vec{b}| = |\vec{x}|$, in which case the result is $A\{\vec{b}/\vec{x}\} \mid B\{\vec{a}/\vec{y}\}$. The inductive clauses are as follows:

$$
\begin{array}{ll}
(\vec{a}; \vec{x})A \circ (F \mid B) \stackrel{\mathrm{df}}{=} ((\vec{a}; \vec{x})A \circ F) \mid B & (A \mid F) \circ F' \stackrel{\mathrm{df}}{=} A \mid (F \circ F') \\
(\vec{a}; \vec{x})A \circ (B \mid F) \stackrel{\mathrm{df}}{=} B \mid ((\vec{a}; \vec{x})A \circ F) & (F \mid A) \circ F' \stackrel{\mathrm{df}}{=} (F \circ F') \mid A \\
(\vec{a}; \vec{x})A \circ (\nu M)F \stackrel{\mathrm{df}}{=} (\nu M)((\vec{a}; \vec{x})A \circ F) & (\nu M)(F) \circ F' \stackrel{\mathrm{df}}{=} (\nu M)(F \circ F') .
\end{array}
$$

For the two clauses in the bottom line we assume that M is fresh for the other concretion involved. Observe that in the presence of α-equivalence this condition can always be met, and hence the only reason for a pseudo-application to be undefined is the arity mismatch of the concretions in the base case.

Where pseudo-application is defined we write $F \circ G {\downarrow}$ and say that F and G are *compatible*. It is possible to define a type system for sites to ensure compatibility, but this complicates the calculus and we shall not do so here.

Proposition 8. *The following hold for any compatible concretions F and G.*

   (i) *Application $F \circ G$ is a species.*
  (ii) *Application $G \circ F$ is defined and $G \circ F \equiv F \circ G$.*
 (iii) *If $F' \equiv F$ and $G' \equiv G$ then $F' \circ G'$ exists and is congruent to $F \circ G$.*

Proof. Induction over the derivation of $F \circ G$ shows (i) and (ii), while (iii) uses induction over the derivations of $F' \equiv F$ and $G' \equiv G$. $\square$

Finally, Fig. 5 sets out the rules for generating the multi-transition system on species, as a structural operational semantics [22]. For a species A, we write *Trans*(A) for the associated transition multiset. These transitions are of three kinds:

1. From a species to a concretion, labelled by a name. This represents a potential for interaction; more precisely, a transition $A \xrightarrow{a} (\vec{b}; \vec{y})B$ means that the species A can interact with another by sending $\vec{b}$ on the channel a and then evolve to B, with $\vec{y}$ replaced by data received.

2. From a species to another species, labelled by $\tau@k$ where k is a real number, such as $A \xrightarrow{\tau@1.5} B$. This denotes the ability of species A to evolve into B without other interaction, at a basal rate of k (here 1.5). Examples include degradation, where B is $\mathbf{0}$, or complex dissociation, with B of the form $B' \mid B''$.

3. From one species to another, labelled by a term $\tau\langle a, b \rangle$ with names a and b, for example $A \xrightarrow{\tau\langle a,b \rangle} B$. This also denotes a potential for evolution of A into B, but now the basal rate of evolution is the affinity between a and b. This affinity will be determined by either the global affinity network *Aff* or some local network M to be introduced by restriction (νM). Thus local interaction at private names becomes visible externally as a spontaneous action, maintaining the compositionality of the semantics.

The following result states that the structural congruence of species is indeed a behavioural equivalence.

Theorem 9. *Let $A \equiv B$. There exists a bijection ϕ: Trans(A) $\rightarrow$ Trans(B) such that if $\phi(A \xrightarrow{\alpha} E) = B \xrightarrow{\alpha'} E'$ then $\alpha = \alpha'$ and $E \equiv E'$.*

Proof (sketch). We proceed by induction on the derivation of $A \equiv B$: for every transition in *Trans*(A) we exhibit a corresponding one in *Trans*(B) via a case analysis of the transition derivation tree, and then show that this association is a bijection. $\square$

The behaviour of processes. We give a compositional semantics to $c\pi$ processes in terms of real vector spaces $\mathbb{P}$ and $\mathbb{D}$, capturing respectively the immediate actions $\frac{dP}{dt}$ and the potential interactions ∂P of a process P. First, however, we need several preliminary definitions. Recall that $\mathcal{S}$ and $\mathcal{C}$ are the sets of species and concretions, respectively, modulo structural congruence.

Definition 10. A non-zero $[A] \in \mathcal{S}$ is *prime* if it is not a parallel composition of non-trivial species, i.e. if $A \equiv (B \mid C)$ implies either $B \equiv \mathbf{0}$ or $C \equiv \mathbf{0}$. We write $\mathcal{S}^{\#}$ for the set of prime species, with $\mathcal{S}^{\#} \subsetneq \mathcal{S}$.

$$\frac{0 \le j \le n \quad \pi_j = a_j(\vec{b}_j; \vec{y}_j)}{\Sigma_{i=0}^n \pi_i.A_i \xrightarrow{a_j} (\vec{b}_j; \vec{y}_j)A_j} \quad \text{CHOICE-1}$$

$$\frac{A \xrightarrow{\alpha} E}{A \mid B \xrightarrow{\alpha} E \mid B} \quad \text{PAR-LEFT}$$

$$\frac{0 \le j \le n \quad \pi_j = \tau@k}{\Sigma_{i=0}^n \pi_i.A_i \xrightarrow{\tau@k} A_j} \quad \text{CHOICE-2}$$

$$\frac{B \xrightarrow{\alpha} E}{A \mid B \xrightarrow{\alpha} A \mid E} \quad \text{PAR-RIGHT}$$

$$\frac{A \xrightarrow{a} F \quad B \xrightarrow{b} G \quad F \circ G\downarrow}{A \mid B \xrightarrow{\tau\langle a,b\rangle} F \circ G} \quad \text{COM-1}$$

$$\frac{A \xrightarrow{\alpha} E \quad \alpha \notin M}{(\nu M)A \xrightarrow{\alpha} (\nu M)E} \quad \text{RES-1}$$

$$\frac{A \xrightarrow{\tau\langle a,b\rangle} B \quad a,b \in M \quad M(a,b)\downarrow}{(\nu M)A \xrightarrow{\tau@M(a,b)} (\nu M)B} \quad \text{COM-2}$$

$$\frac{A \xrightarrow{\tau\langle a,b\rangle} E \quad a,b \notin M}{(\nu M)A \xrightarrow{\tau\langle a,b\rangle} (\nu M)E} \quad \text{RES-2}$$

$$\frac{B \xrightarrow{\alpha} E \quad D(\vec{y}) \overset{\text{df}}{=} B}{D(\vec{b}) \xrightarrow{\alpha} E\{\vec{b}/\vec{y}\}} \quad \text{DEFN}$$

Fig. 5. Transition rules for species. Here α ranges over all kinds of transition labels and E ranges over both species and concretions.

Theorem 11. *For any nonzero species A there exists a unique finite multiset $\{\![A_1], \ldots, [A_n]\!\} \subset \mathcal{S}^\#$ such that $A \equiv A_1 \mid \cdots \mid A_n$. Call this prime decomposition of A.*

Proof (sketch). We assign normal forms to species using a normalising and confluent term rewriting system respecting $\equiv$, and take the prime decomposition as the multiset of $\mid$-components of the normal form.   □

The decomposition theorem allows us to represent any $c\pi$ process as a collection of prime species weighted by real numbers.

Definition 12. The *process space* $\mathbb{P}$ is the (infinite dimensional) vector space $\mathbb{R}^{(\mathcal{S}^\#)}$.

There is a natural mapping from species into process space, with the concentration of participating prime species matching their multiplicity in the prime decomposition.

Definition 13. Define $\langle \cdot \rangle \colon \mathcal{S} \to \mathbb{P}$ inductively over the structure of its argument:

$$\langle A \rangle \overset{\text{df}}{=} \begin{cases} 0 \text{ at every position} & \text{if } [A] = [\mathbf{0}] \\ 1 \text{ at } [A], 0 \text{ elsewhere} & \text{if } [A] \text{ prime} \\ \langle B \rangle + \langle C \rangle & \text{if } [A] = [B \mid C] \text{ for } B, C \not\equiv \mathbf{0}. \end{cases} \tag{10}$$

It follows from Thm. 11 that $\langle \cdot \rangle$ is well-defined and constant inside every equivalence class of species.

In due course we shall use space $\mathbb{P}$ to capture immediate process behaviour. Although this behaviour is what we are most interested in, it is impossible to define it compositionally without further information on the possible ways a process may interact with others. We therefore define a space of interaction potentials $\mathbb{D}$ and an interaction tensor $\odot \colon \mathbb{D} \times \mathbb{D} \to \mathbb{P}$ that combines two compatible potentials into an immediate behaviour.

Definition 14. The *interaction space* $\mathbb{D}$ is the (infinite dimensional) vector space $\mathbb{R}^{S \times C \times \mathcal{N}}$ of ternary real functions with pointwise addition and scalar (real) multiplication. Note that it has a basis consisting of functions of the form $1_{[A],[F],a}$ which take the value 1 for the indicated arguments and 0 for any other set of inputs.

Definition 15. The *interaction tensor* $\odot \colon \mathbb{D} \times \mathbb{D} \to \mathbb{P}$ is the bilinear function defined by the following action on basis values:

$$1_{[A],[F],a} \odot 1_{[B],[G],b} \stackrel{\mathrm{df}}{=} \begin{cases} \mathit{Aff}(a,b)(\langle F \circ G \rangle - \langle A \rangle - \langle B \rangle) & \text{if } F \circ G \downarrow \\ 0 & \text{otherwise} . \end{cases} \tag{11}$$

Each element $\xi \in \mathbb{D}$ associates with every triple $([A], [F], a)$ a real number. When ξ describes the interaction capabilities of a process, this real number denotes the sum of concentrations of all species present in the process which can make the transition $A \xrightarrow{a} F$, or structural equivalent. The interaction tensor $\odot$ combines two such interaction potentials into an actual process behaviour. Combination of basis elements, which can be seen as the two transitions $A \xrightarrow{a} F$ and $B \xrightarrow{b} G$, gives a new species resulting from the interaction $F \circ G$, the substrate species A and B are lost, and the whole expression is weighted by the interaction rate of the $a - b$ channel in Aff. Combination of more complex $\xi, \xi' \in \mathbb{D}$ is computed by bilinear extension of this, which ensures that all combinations of interaction potentials in ξ and ξ' are considered. Moreover, bilinearity means that the result is scaled in proportion to the "amounts" of interaction potential, and so reflects the Law of Mass Action.

With the above definitions at hand, we are now in a position to define the formal semantics of $c\pi$.

Definition 16. The *complete behaviour* of a process P is a pair $(\frac{dP}{dt}, \partial P) \in \mathbb{P} \times \mathbb{D}$ of its *immediate actions* and *potential interactions* defined inductively on the structure of P as follows:

$$\partial(c \cdot A)([B], [F], a) \stackrel{\mathrm{df}}{=} c \cdot \mathrm{card}\{\!\{ C \xrightarrow{a} G \in \mathit{Trans}(A) \mid C \in [B] \wedge G \in [F] \}\!\} \tag{12}$$

$$\begin{aligned} \frac{d(c \cdot A)}{dt} \stackrel{\mathrm{df}}{=} \ & c \cdot \Sigma_{B \xrightarrow{\tau@k} C \in \mathit{Trans}(A)} k \cdot (\langle C \rangle - \langle B \rangle) \\ & + c \cdot \Sigma_{B \xrightarrow{\tau\langle a,b \rangle} C \in \mathit{Trans}(A)} \mathit{Aff}(a,b) \cdot (\langle C \rangle - \langle B \rangle) \\ & + \frac{1}{2}(\partial(c \cdot A) \odot \partial(c \cdot A)) \end{aligned} \tag{13}$$

$$\partial(P \parallel Q) \stackrel{\mathrm{df}}{=} \partial P + \partial Q \tag{14}$$

$$\frac{d(P \parallel Q)}{dt} \stackrel{\text{df}}{=} \frac{dP}{dt} + \frac{dQ}{dt} + \partial P \textcircled{0} \partial Q . \tag{15}$$

We explain briefly the intuitions behind these definitions. Because every process can be identified with a point in $\mathbb{P}$, we can view the immediate behaviour $\frac{dP}{dt}$ as a vector field over $\mathbb{P}$, associating with each process the gradient of its temporal evolution. The equations (13) and (15) reflect this interpretation. Thus in (13) we take into account the effect of all τ-transitions of species A, weighted with the interaction rates (given by the transition labels or the global affinity network) and the initial concentration c, and then add the behaviour arising from interactions between pairs of A molecules. In (15), the immediate behaviour of a composition of two processes is the sum of immediate behaviours of the components plus the behaviour that emerges from their interaction.

Computing the interaction potential ∂P for a process is more straightforward: equation (12) lifts all the appropriate transitions from the multi-transition system and multiplies them by the concentration c; while (14) reflects the fact that the interaction potential of a composition of processes is simply the sum of the interaction potentials of the components, with no cancellation or further emergent interaction.

The following theorems demonstrate that structural congruence of processes is a behavioural equivalence, and that further identification of $|$ and $\parallel$ only slightly weakens this.

Theorem 17. *Let $P \equiv Q$. Then $\partial P = \partial Q$ in $\mathbb{D}$ and $\frac{dP}{dt} = \frac{dQ}{dt}$ in $\mathbb{P}$.*

Proof. Straightforward induction on derivation of $P \equiv Q$. $\qquad\square$

Theorem 18. *Let $\equiv^+$ be the smallest congruence on processes containing $\equiv$ and satisfying the additional rule*

$$c \cdot (A \mid B) \; \equiv^+ \; (c \cdot A) \parallel (c \cdot B) \tag{16}$$

and let $P \equiv^+ Q$. Then $\frac{dP}{dt} = \frac{dQ}{dt}$ and for any $\xi \in \mathbb{D}$, $\partial P \textcircled{0} \xi = \partial Q \textcircled{0} \xi$.

Proof. By induction on derivation of $P \equiv^+ Q$. $\qquad\square$

In general we may have $P \equiv^+ Q$ but $\partial P \neq \partial Q$, because the transitions $A \xrightarrow{x} F$ and $A \mid B \xrightarrow{x} F \mid B$ give rise to different points of $\mathbb{D}$ via (12), despite being essentially equivalent as interaction potentials. We discuss this further in §4.4, with a possible remedy. Notice, though, that the property we do have of equality under $- \textcircled{0} \xi$ for all ξ is a form of *observational equivalence*: there is no way from within process space to observe any difference between ∂P and ∂Q.

3   Example

In this section we give a $c\pi$ model for a simple biomolecular system, the KaiC circadian clock. Our reference for this system is the work of van Zon et al. [8].

3.1   The System

Introduction. Circadian clocks are molecular systems that exhibit oscillatory behaviour synchronized with the 24-hour day cycle. They play an important role in many organisms by helping to regulate their cellular behaviour according to the circadian rhythm.

The system we model is a primitive circadian clock found in the cyanobacterium *Synechococcus elongatus* [23,24]. It consists of three kinds of protein: KaiA, KaiB and KaiC. In particular, KaiC forms hexamers with 6 phosphorylation sites which are phosphorylated and dephosphorylated in a cyclic manner, thus dictating the circadian rhythm.

The KaiC circadian clock has two features that make it of particular interest to the biological community. The first is its simplicity — it requires only 3 kinds of molecules to function. The other is that it does not rely on either intracellular compartments or gene regulation, which sets it apart from other circadian clocks and (remarkably) makes it possible to reproduce its self-sustaining cycle relatively easily *in vitro* [7].

The allosteric model. Although there are extensive experimental results on the activity of the various components of the KaiC system, its precise mechanism is not yet understood. In order to explain the observed behaviour, the authors of [8] propose an elegant model based on two assumptions. The first assumption is that every KaiC protein is *allosteric*, i.e. it can adopt two distinct 3D shapes (conformations), denoted active and inactive. This gives every KaiC hexamer a propensity to spontaneously undergo a phosphorylation-dephosphorylation cycle, as shown in Fig. 6. The cycles of individual KaiC are then synchronized with each other thanks to the other assumption: that the phosphorylating agent KaiA binds more strongly to weakly phosphorylated KaiC molecules. This mechanism is called *differential affinity*. The role of KaiB in this model is to stabilise the inactive form of KaiC and to increase the competition for free KaiA molecules between the differently phosphorylated active forms.

3.2   The $c\pi$ Model

Figure 7 displays our $c\pi$ model of the KaiC system proposed in [8]. It combines species declarations, a global affinity network *Aff* and local affinity networks M_i.

Species. In the first few lines (18)–(23) we define the 14 distinct species of KaiC: active and inactive forms, each in one of 7 phosphorylation states. Lines (24)–(29) then define the complexes formed by inactive KaiC with KaiB and KaiA, while (30) defines species KaiA and KaiB themselves. Finally, process (31) describes an initial state of the model. Note that several intermediate species, not explicitly defined here, naturally emerge from subsequent interactions. For example, the A-C_0 complex

$$(\nu M_0)((u_0.C_0 + r_0.C_1) \mid act_0.A) \tag{17}$$

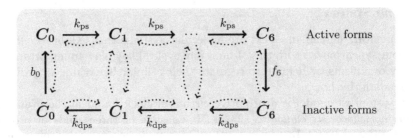

Fig. 6. The phosphorylation cycle of a single KaiC molecule. The two allosteric forms have opposite (de-)phosphorylation tendencies: phosphorylation proceeds from left to right, dephosphorylation from right to left. The potential to flip between the conformations closes the cycle. Non-dominant reactions are indicated with dotted arrows. Adapted from [8, Fig. 1(B)].

arising from an interaction between A and C_0 on the $a \overset{k_0^{\mathrm{Af}}}{\longrightarrow} a_0$ channel, communicating local names u_0 and r_0. This complex can then dissociate, triggered by the interaction between local site act_0 and either u_0 or r_0, corresponding respectively to simple unbinding of C_0 or catalysed phosphorylation to C_1.

The definitions of the species are based entirely on their interaction capabilities as postulated in [8]. For example, the active and unphosphorylated KaiC molecule (species C_0 in (18)) can either flip to the inactive state at rate f_0 (the $\tau@f_0.\tilde{C}_0$ component), spontaneously phosphorylate at rate k_{ps} (the $\tau@k_{\mathrm{ps}}.C_1$ component), or bind a KaiA molecule to form the complex of (17) above (the $a_0\langle act_0\rangle.(u.C_0 + r.C_1)$ component).

We have one minor deviation from [8], concerning the binding of KaiB and KaiA to inactivated KaiC in lines (24)–(29). Both of these bind in multiples to KaiC: based on size measurements, van Zon et al. assume that each KaiC binds two KaiB and then two KaiA. They model these with a 3-substrate reaction, which cannot be expressed directly in $c\pi$. Instead, we model the binding by two consecutive binary interactions, where the rate of the second (k_{vf}) is much greater than that of the first (k_i^{Bf} or k_i^{Af}).

This occurs, for example, in the KaiB-KaiC complex $B\tilde{C}_0$ of (24). This can either spontaneously dissociate, at rate k_0^{Bb}, into an inactivated KaiC and two KaiB, phosphorylate at rate $\tilde{k}_{\mathrm{ps}}$, or bind successively to the $\tilde{a}$ site on two KaiA molecules to form a KaiA-KaiB-KaiC complex.

The model also shows other kinetics in action, for example in the binding of KaiA to active KaiC modelled by scope extrusion shown above (17). When the extrusion is reversible, as it is here, the $c\pi$ semantics of the combined reaction generates a Michaelis-Menten kinetics. Other binding events are modelled as simple communication, which gives rise to Mass Action kinetics.

Affinity networks. We model the differential affinity mechanism with fan-like affinity networks, where a single site can interact with several others at different rates (Fig. 7(c)). For the sake of symmetry and for the ease of potential perturbation analysis of the model, we retain the fan shape even where

$$C_0 \stackrel{\mathrm{df}}{=} (\nu M_0)(\tau@f_0.\tilde{C}_0 + \tau@k_{\mathrm{ps}}.C_1 + a_0\langle act_0\rangle.(u_0.C_0 + r_0.C_1)) \tag{18}$$

$$C_i \stackrel{\mathrm{df}}{=} (\nu M_i)(\tau@f_i.\tilde{C}_i + \tau@k_{\mathrm{ps}}.C_{i+1} \tag{19}$$
$$+ \tau@k_{\mathrm{dps}}.C_{i-1} + a_i\langle act_i\rangle.(u_i.C_i + r_i.C_{i+1}))$$

$$C_6 \stackrel{\mathrm{df}}{=} \tau@f_6.\tilde{C}_6 + \tau@k_{\mathrm{dps}}.C_5 \tag{20}$$

$$\tilde{C}_0 \stackrel{\mathrm{df}}{=} \tau@b_0.C_0 + \tau@\tilde{k}_{\mathrm{ps}}.\tilde{C}_1 + b_0.b'.B\tilde{C}_0 \tag{21}$$

$$\tilde{C}_i \stackrel{\mathrm{df}}{=} \tau@b_i.C_i + \tau@\tilde{k}_{\mathrm{ps}}.\tilde{C}_{i+1} + \tau@\tilde{k}_{\mathrm{dps}}.\tilde{C}_{i-1} + b_i.b'.B\tilde{C}_i \tag{22}$$

$$\tilde{C}_6 \stackrel{\mathrm{df}}{=} \tau@b_6.C_6 + \tau@\tilde{k}_{\mathrm{dps}}.\tilde{C}_5 + b_6.b'.B\tilde{C}_6 \tag{23}$$

$$B\tilde{C}_0 \stackrel{\mathrm{df}}{=} \tau@k_0^{\mathrm{Bb}}.(\tilde{C}_0 \mid B \mid B) + \tau@\tilde{k}_{\mathrm{ps}}.B\tilde{C}_1 + \tilde{a}_0.\tilde{a}'.AB\tilde{C}_0 \tag{24}$$

$$B\tilde{C}_i \stackrel{\mathrm{df}}{=} \tau@k_i^{\mathrm{Bb}}.(\tilde{C}_i \mid B \mid B) + \tau@\tilde{k}_{\mathrm{ps}}.B\tilde{C}_{i+1} \tag{25}$$
$$+ \tau@\tilde{k}_{\mathrm{dps}}.B\tilde{C}_{i-1} + \tilde{a}_i.\tilde{a}'.AB\tilde{C}_i$$

$$B\tilde{C}_6 \stackrel{\mathrm{df}}{=} \tau@k_6^{\mathrm{Bb}}.(\tilde{C}_6 \mid B \mid B) + \tau@\tilde{k}_{\mathrm{dps}}.B\tilde{C}_5 + \tilde{a}_6.\tilde{a}'.AB\tilde{C}_6 \tag{26}$$

$$AB\tilde{C}_0 \stackrel{\mathrm{df}}{=} \tau@\tilde{k}_0^{\mathrm{Ab}}.(B\tilde{C}_0 \mid A \mid A) + \tau@\tilde{k}_{\mathrm{ps}}.AB\tilde{C}_1 \tag{27}$$

$$AB\tilde{C}_i \stackrel{\mathrm{df}}{=} \tau@\tilde{k}_i^{\mathrm{Ab}}.(B\tilde{C}_i \mid A \mid A) + \tau@\tilde{k}_{\mathrm{ps}}.AB\tilde{C}_{i+1} + \tau@\tilde{k}_{\mathrm{dps}}.AB\tilde{C}_{i-1} \tag{28}$$

$$AB\tilde{C}_6 \stackrel{\mathrm{df}}{=} \tau@\tilde{k}_6^{\mathrm{Ab}}.(B\tilde{C}_6 \mid A \mid A) + \tau@\tilde{k}_{\mathrm{dps}}.AB\tilde{C}_5 \tag{29}$$

$$A \stackrel{\mathrm{df}}{=} a(x).x.A + \tilde{a}.0 \qquad B \stackrel{\mathrm{df}}{=} b.0 \tag{30}$$

$$P \stackrel{\mathrm{df}}{=} c_A \cdot A \parallel c_B \cdot B \parallel c_C \cdot C_0 \tag{31}$$

(a) Species and process definitions

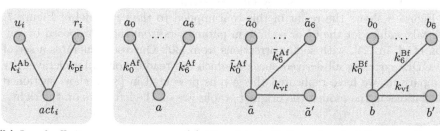

(b) Local affinity networks M_i

(c) The global affinity network Aff

Fig. 7. The $c\pi$ model of the KaiC circadian clock. Parameter i takes values $1\ldots5$ in species definitions and $0\ldots6$ in the affinity networks.

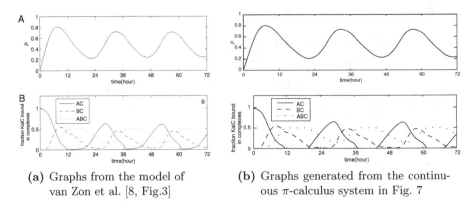

(a) Graphs from the model of van Zon et al. [8, Fig.3]

(b) Graphs generated from the continuous π-calculus system in Fig. 7

Fig. 8. Graphs comparing oscillatory behaviour of the models defined by van Zon et al. [8] and the $c\pi$ terms of §3. The upper graphs show mean phosphorylation level of KaiC over three circadian cycles. The lower ones show the relative amounts of the complexes KaiA-KaiC (active); KaiB-KaiC (inactive); and KaiA-KaiB-KaiC (sequestering KaiA and so inhibiting the phosphorylation of active KaiC).

[8] assumes no differential affinity. The differential affinity of the KaiA-KaiC binding is modelled by differing interaction rates k_i^{Ab} in the collection of local affinity networks M_i (Fig. 7(b)).

Results. There is a close correspondence between the dynamical behaviour of the $c\pi$ model, as generated by the semantics of §2.2, and that of our reference paper [8]. Moreover, we can extract from the species and network declarations of Fig. 7 a set of ODEs that matches those reported in [25, p.14] (up to minor differences due to our alternative modelling of multiple binding).

We have a prototype tool that takes textual descriptions of $c\pi$ systems and applies the semantics of §2.2: exploring the transition state space of species, and then combining these to compute the potential and immediate behaviour of processes. The tool is written in Haskell [26,27], and generates ODEs in a format suitable for numerical analysis by Octave [28,29].

Figure 8 shows the result of this tool applied to the $c\pi$ model of Figure 7. We take values for the 65 or so system parameters from those proposed by van Zon et al. in [25], with some corrections from [30]. Our tool generates a set of 50 ODEs covering all derived species, which are readily solved by Octave. For comparison, we have replicated the graphs presented in [8] to show sustained oscillations in the model: the original graphs are on the left, ours on the right.

4   Discussion

4.1   Alternative Behavioural Semantics

Process algebras offer a distinct level of abstraction compared to more widespread dynamical models of biochemical systems such as ordinary differential equations or Markov chains. A process algebraic description of a system may be translated to more than one such formalism (see e.g. [31]) and so a modeller may choose the most appropriate dynamic paradigm for a given situation. This is true of $c\pi$: aside from the "native" ODE semantics, it is relatively easy to generate Markov chains (by introducing integer quantities of processes instead of real-valued concentrations) and it may also be possible to map the $c\pi$ syntax to other behavioural models (e.g. Petri Nets as in [32]). We see this flexibility as a strength of process algebras like $c\pi$.

4.2 Modelling Evolution

The target application of continuous π is the investigation of Darwinian evolution on the molecular level. At present, we are able to identify two promising concrete applications of $c\pi$ in this context. The first is direct simulation of evolution; the other is analysis of evolutionary robustness.

Evolutionary trajectories. In order to simulate molecular evolution by natural selection, we must be able to express variability, populations and fitness in our process-algebraic framework. While populations of individuals can be modelled simply as collections of processes, the other two concepts use the process-algebraic nature of the model in an essential way.

Variability: We propose addressing qualitative and quantitative variation of pathway topology (connectivity) in two ways. The first is by considering small variations in affinity networks, in order to model changes in the interaction capability of existing active sites. The other is by altering the structure of the species and thus modelling evolution of new sites, domain duplication and similar higher-level discrete events.

It is also possible to consider variation in the initial concentrations of processes and interpret this – particularly in simpler models – as variation in gene expression. It remains to be investigated whether it is biologically sensible to include this type of variation in a model that is otherwise focused on the evolution of network topologies.

Fitness: It is clear that any notion of fitness is problem-dependent and must be defined externally by the modeller. We plan to use a form of quantitative model checking to compute the fitness value. This requires a modeller to formulate a fitness measure in an appropriate logic.

A recently published study [33] uses the process algebra of *Beta-binders* [34] in a similar programme of simulating molecular evolution, which provides some validation of this approach.

Robustness. A further intended application of the calculus is the study of neutrality of biochemical pathways and related concepts such as robustness and evolvability. Since by definition two pathways are neutral with respect to each other if they have the same fitness, the question of determining neutrality can be reduced to that of assessing fitness (see above). We plan to treat it separately, however, as it may be possible to characterize or approximate neutrality without actually computing fitness. For example, we might deem two pathways neutral if they satisfy the same subset of a well-chosen set Φ of sentences in an appropriate logic. Even more desirable would be to characterize neutrality via a suitable behavioural equivalence, requiring no input from the modeller at all. In either case we expected the method to be applicable only to a restricted class of pathways (actual biological entities), whose identification is a challenge in itself.

4.3   Hybrid Modelling

A dynamical system is *hybrid* if its dynamics have both discrete and continuous characteristics. This is a common situation in models for cell biology: for example, consider gene regulation by simple direct negative feedback. A protein is produced at a constant rate (continuous dynamics) until transcription is switched off due to a protein molecule binding to the DNA (discrete event). Continuous π as it is now cannot model this situation because we require every molecular species to be present in some concentration, while the DNA is present as a single copy only. In general, whenever it is impossible (or undesirable) to abstract over gene expression (or at least transcription), we are faced with the need to model a genuinely hybrid situation. In addition, even when the use of a purely continuous approach is conceptually possible, a model may fail to produce characteristic behaviour that depends crucially on stochastic effects. See e.g. [35] for a comparison of the continuous and stochastic approaches.

There are a variety of computational approaches to hybrid modelling, such as hybrid I/O automata [36] and hybrid process algebras [37]. We plan to build on this tradition and extend $c\pi$ with discrete features, in the form of species present as single individuals. Interactions with these species will act as discrete control events on top of the existing continuous semantics, giving truly hybrid process behaviour. We take the *lac* operon molecular system [38] as a suitable target to validate this approach. This is a regulatory network involving protein-DNA interactions which modify the transcription process of several genes; what is more, it is relatively well understood and well known among biologists.

4.4   Refinement of Semantics

At present the potential behaviour ∂P of a process includes a description of every reaction in which it can engage. While this is necessary for compositionality, we believe there is room for improvement in the encoding of this information. Specifically, we need not record for every offered communication precisely what is consumed and what produced; a "net result" is enough (cf. formulae (11) and (13)). Consider for example the process

$$P = c \cdot A, \qquad \text{where } A \stackrel{\text{df}}{=} a.(A \mid B) . \tag{32}$$

In the semantics of §2.2 this communication potential is recorded as $\partial P(A, (;)(A \mid B), a) \neq 0$: in a single communication event over a one A is lost while another is produced along with B. It would be enough just to record that this reaction results in a production of a single new B. It remains to formalize this and extend it to arbitrary concretions.

Finally, we conjecture that $\mathbb{R}^\omega$ is unnecessarily large to serve as $\mathbb{P}$ and that for any process P, its immediate behaviour $\frac{dP}{dt}$ is an element of an ℓ^p space for some fixed p (the space of infinite real sequences with a finite p-norm). Moving from $\mathbb{R}^\omega$ to ℓ^p would allow us to use the rich theory of Banach (and Hilbert, if $p = 2$) spaces to study the properties of the calculus and to approach biological questions about, for example, system trajectories.

References

1. Priami, C., Regev, A., Shapiro, E., Silverman, W.: Application of a stochastic name-passing calculus to representation and simulation of molecular processes. Inf. Proc. Lett. 80 (2001)
2. Regev, A.: Computational Systems Biology: A Calculus for Biochemical Knowledge. PhD thesis, Tel Aviv University (2002)
3. Regev, A., Silverman, W., Shapiro, E.: Representation and simulation of biochemical processes using the pi-calculus process algebra. In: Pacific Symposium on Biocomputing (2001)
4. Kitano, H.: Biological robustness. Nature 5, 826–837 (2004)
5. Wagner, A.: Robustness and Evolvability in Living Systems. Princeton University Press, Princeton (2005)
6. Schuster, P., Fontana, W., Stadler, P., Hofacker, I.: From sequences to shapes and back: A case-study in RNA secondary structures. Proc. Royal Soc. Ser. B 255 (1994)
7. Tomita, J., Nakajima, M., Kondo, T., Iwasaki, H.: No transcription-translation feedback in circadian rhythm of KaiC phosphorylation. Science 307(5707), 251–254 (2005)
8. van Zon, J.S., Lubensky, D.K., Altena, P.R.H., ten Wolde, P.R.: An allosteric model of circadian KaiC phosphorylation. PNAS 104(18), 7420–7425 (2007)
9. Milner, R.: The polyadic π-calculus: A tutorial. Technical Report ECS-LFCS-91-180, LFCS, University of Edinburgh (1991)
10. Milner, R.: Communicating and Mobile Systems: The π Calculus. Cambridge University Press, Cambridge (1999)
11. Parrow, J.: An introduction to the π-calculus. In: Handbook of Process Algebra, pp. 479–543. Elsevier, Amsterdam (2001)
12. Gillespie, D.T.: The chemical Langevin equation. J. Chem. Phys. 113(1), 297–306 (2000)
13. Regev, A., Shapiro, E.: Cellular abstractions: Cells as computations. Nature 419 (2002)
14. Regev, A., Panina, E.M., Silverman, W., Cardelli, L., Shapiro, E.: Bioambients: An abstraction for biological compartments. Theor. Comput. Sci. 325 (2004)
15. Hillston, J.: A Compositional Approach to Performance Modelling. Cambridge University Press, Cambridge (1996)
16. Calder, M., Gilmore, S., Hillston, J.: Modelling the influence of RKIP on the ERK signalling pathway using the stochastic process algebra PEPA. In: Proc. BioConcur. (2004)
17. Heath, J., Kwiatkowska, M., Norman, G., Parker, D., Tymchyshyn, O.: Probabilistic model checking of complex biological pathways. Theor. Comput. Sci. (2007)

18. Calder, M., Duguid, A., Gilmore, S., Hillston, J.: Stronger computational modelling of signalling pathways using both continuous and discrete-state methods. In: Priami, C. (ed.) CMSB 2006. LNCS (LNBI), vol. 4210, pp. 63–77. Springer, Heidelberg (2006)
19. Kitano, H.: Towards system-level understanding of biological systems. In: Foundations of Systems Biology. MIT Press, Cambridge (2001)
20. Stadler, B.M.R., Stadler, P.F., Wagner, G., Fontana, W.: The topology of the possible: Formal spaces underlying patterns of evolutionary change. J. Theor. Biol. 213 (2001)
21. Soyer, O., Salathe, M., Bonhoeffer, S.: Signal transduction networks: Topology, response, and biochemical reactions. J. Theor. Biol. 238 (2006)
22. Plotkin, G.D.: A structural approach to operational semantics. J. Log. Algeb. Progr. 60–61, 17–139 (2004)
23. Ishiura, M., Kutsuna, S., Aoki, S., Iwasaki, H., Andersson, C.R., Tanabe, A., Golden, S.S., Johnson, C.H., Kondo, T.: Expression of a gene cluster kaiABC as a circadian feedback process in cyanobacteria. Science 281(5382), 1519–1523 (1998)
24. Golden, S.S., Johnson, C.H., Kondo, T.: The cyanobacterial circadian system: A clock apart. Current Opinion in Microbiology 1(6), 669–673 (1998)
25. van Zon, J.S., Lubensky, D.K., Altena, P.R.H., ten Wolde, P.R.: An allosteric model of circadian KaiC phosphorylation: Supporting information (2007), http://www.pnas.org/cgi/content/full/0608665104/DC1
26. Haskell, http://www.haskell.org/
27. Peyton Jones, S. (ed.): Haskell 98 Language and Libraries: The Revised Report. Cambridge University Press, Cambridge (April 2003)
28. Octave, http://www.gnu.org/software/octave/
29. Eaton, J.W.: GNU Octave Manual. Network Theory (2002)
30. van Zon, J.S.: A detail of the KaiABC model. Personal communication (2008)
31. Calder, M., Gilmore, S., Hillston, J.: Automatically deriving ODEs from process algebra models of signalling pathways. In: Proc. CMSB (2005)
32. Meyer, R., Khomenko, V., Strazny, T.: A practical approach to verification of mobile systems using net unfoldings. In: Application and Theory of Petri Nets: Proc. ATPN (to appear, 2008)
33. Demate, L., Priami, C., Romanel, A., Soyer, O.: A formal and integrated framework to simulate evolution of biological pathways. In: Calder, M., Gilmore, S. (eds.) CMSB 2007. LNCS (LNBI), vol. 4695, pp. 106–120. Springer, Heidelberg (2007)
34. Priami, C., Quaglia, P.: Beta binders for biological interactions. In: Danos, V., Schachter, V. (eds.) CMSB 2004. LNCS (LNBI), vol. 3082, pp. 20–33. Springer, Heidelberg (2005)
35. Bortolussi, L., Policriti, A.: Connecting process algebras and differential equations for systems biology. In: Process Algebra and Stochastically Timed Activities: Proc. 6th PASTA workshop (2006)
36. Segala, R., Vaandrager, F., Lynch, N.: Hybrid I/O automata. Inf. & Comput. 185(1) (2003)
37. Bergstra, J.A., Middleburg, C.A.: Process algebra for hybrid systems. Theor. Comput. Sci. 335 (2005)
38. Vilar, J.M.G., Guet, C.G., Leibler, S.: Modeling network dynamics: The lac operon, a case study. J. Cell Biol. 161 (2003)

Automatic Complexity Analysis and Model Reduction of Nonlinear Biochemical Systems

Dirk Lebiedz, Dominik Skanda, and Marc Fein

Universität Freiburg, Zentrum für Biosystemanalyse (ZBSA)
dirk.lebiedz@biologie.uni-freiburg.de

Abstract. Kinetic models for biochemical systems often comprise a large amount of coupled differential equations with species concentrations varying on different time scales. In this paper we present and apply two novel methods aimed at automatic complexity and model reduction by numerical algorithms. The first method combines dynamic sensitivity analysis with singular value decomposition. The aim is to determine the minimal dimension of the kinetic model necessary to describe the active dynamics of the system accurately enough within a user-defined error tolerance for particular species concentrations and to determine each species' contribution to the active dynamics. The second method treats the explicit numerical reduction of the model to a lower dimension according to the results of the first method and allows any species combination to be chosen as a parameterization of the reduced model which may either be tabulated in the form of look-up tables or computed in situ during numerical simulations. A reduced representation of a multiple time scale system is particularly beneficial in the context of spatiotemporal simulations which require high computational efforts. Both the complexity analysis and model reduction method operate in a fully automatic and numerically highly efficient way and have been implemented in a software package. The methods are applied to a biochemical example model describing the ERK signaling pathway. With this example, we demonstrate the value of the methods for various applications in systems biology.

1 Introduction

Modern experimental techniques offer various opportunities to reveal nature's secrets to a steadily increasing extend. Coming along with this development, one gets a more and more detailed insight into biochemical reaction mechanisms thus complicating kinetic models tremendously. Similar difficulties with large-scale kinetic models have been encountered, for instance, in the field of combustion in physical chemistry (see e.g. [31]).

In combustion chemistry and biochemical kinetics as well, a broad class of problems can be stated as *ordinary differential equations* (ODE), in particular as well-posed *initial value problems* (IVP)

$$\dot{x}(t) = f(x)$$
$$x(t_0) = x_0 \ . \tag{1}$$

M. Heiner and A.M. Uhrmacher (Eds.): CMSB 2008, LNBI 5307, pp. 123–140, 2008.

With a lot of interacting species being generally involved in these application fields and these species being dynamically coupled on different time scales, the resulting ODEs are highly nonlinear and stiff. For such problems, several methods have been proposed to reduce the model complexity and its dimension. Three main ideas have been pursued in model reduction so far, namely lumping, sensitivity analysis and time scale analysis. The interested reader may refer to [18] for a detailed discussion of lumping techniques and [30, 25, 28] for sensitivity analysis.

Some time scale analysis techniques, for example the Quasi Steady State Approximation (QSSA) [5, 6, 22], which is used to derive the common Michaelis-Menten kinetic equations, are well-known in biology. However, classical QSSA performed "by hand" is not suitable for the application to large systems. To overcome this insufficiency, automatic methods such as Computational Singular Perturbation (CSP) [11, 12] and Intrinsic Low Dimensional Manifold (ILDM) [21, 19, 20] have been developed. The latter has found wide application in the combustion chemistry field.

Most model reduction methods aim at determining a low dimensional manifold which describes long-term dynamics and on which arbitrary system trajectories in phase space condense before approaching their attractor, usually chemical equilibrium. Therefore, this low dimensional manifold approximates the kinetics of the underlying system at the slow time scales assuming fast time scales to be enslaved by the slow ones.

Unfortunately, the widely used ILDM method which is based on eigenvalue analysis of the Jacobian is not that efficient for large scale systems and subject to severe numerical problems if the spectral gap between fast and slow time scales is small. Therefore, different techniques have been suggested [10, 9] and we present a novel and quite general approach based on ideas by Lebiedz et al. [13, 14, 26], which computes an approximation of such a slow low dimensional manifold as a solution of an optimization problem. This method aims at relaxing generalized forces driving chemical reactions as much as possible under given constraints. These constraints are determined by the dimension and the parameterization of the reduced model. Both can be freely chosen.

In the context of model reduction, it is important to first determine the minimal dimension necessary to still describe the system dynamics accurately within a given tolerance. Moreover, it is beneficial to be able to parameterize a low-dimensional approximation of the full model by any chosen species of the full model according to the user's needs. The species chosen for parameterization of the lower-dimensional model are called *reaction progress variables* in the following and it is reasonable to chose those species which significantly contribute to the "active" system dynamics. On this background we present a new method by Lebiedz et al. [14] to analyze the minimal dimension and the contribution of each species to the system dynamics in order to choose this dimension for a reduced model and determine the reaction progress variables needed for explicit model reduction.

To demonstrate the value of a combination of both methods outlined above for applications in systems biology, we apply them exemplarily to a model for the ERK signaling pathway [7, 24] involving 11 biochemical species. We determine the minimal dimension of this system as well as the contribution of each species to the active dynamics of the whole system. After having obtained these results we numerically compute a low dimensional manifold representing the reduced dynamics of the system and compare the manifold with an initial trajectory.

In the following section, we will give a brief overview of the model. In section 3 the complexity analysis method by Lebiedz et al. [14] is presented and afterwards we deal with the model reduction approach by Lebiedz et al. [13, 26]. Finally, we present the numerical results for the example application and discuss them.

2   Model of the Regulatory Influence of RKIP on the ERK Signaling Pathway

In this section, we review the basics of the Ras/Raf/MEK/ERK signaling pathway regulated by RKIP. The ERK signaling pathway is an important subject of research due to the fact that it controls processes like cell differentiation and proliferation. We will briefly present a model of the regulatory influence of RKIP to the ERK signaling pathway introduced in [7] and [24]. The graphical representation of the enzyme kinetic reactions shown in Fig. 1 describes the inhibition of the activation of RAF by RKIP which leads to a "downregulation" of the ERK signaling pathway.

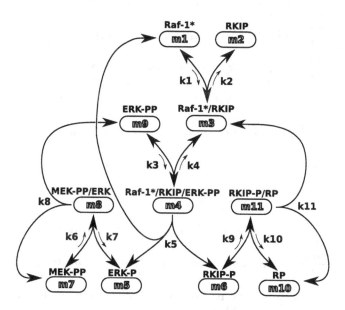

Fig. 1. Graphical representation of the model of the regulatory influence of RKIP to the ERK signaling pathway

Table 1. Parameter values of the ERK signaling pathway

Parameter k_1	k_2	k_3	k_4	k_5	k_6	k_7	k_8	k_9	k_{10}	k_{11}
Value	0.53	0.072	0.625	0.00245	0.0315	0.8	0.0075	0.071	0.92	0.00122	0.87

The concentrations of the different proteins are labeled with m_i ($i = 1, ..., 11$). The corresponding proteins and protein complexes are Raf-1*, RKIP, Raf-1*/RKIP, Raf-1*/RKIP/ERK-PP, ERK-P, RKIP-P, MEK-PP, MEK-PP/ERK, ERK-PP, RP and RKIP-P/RP as indicated in Fig. 1. The suffix -P (-PP) indicates phosphorylated (double phosphorylated) proteins. The forward and backward reactions are indicated by arrows and are labeled by the corresponding rate constants k_i ($i = 1, ..., 11$). The values of these rate constants determined by Cho et al. [7] are shown in Table 2. These values are dimensionless because we are only interested in the complexity reduction and analysis of the model and do not intend to compare the k_i's with each other. The corresponding ODE system by Cho et al. [7] is given by

$$\frac{dm_1}{dt} = -k_1 m_1 m_2 + k_2 m_3 + k_5 m_4 - \kappa m_1 m_5 m_6$$

$$\frac{dm_2}{dt} = -k_1 m_1 m_2 + k_2 m_3 + k_{11} m_{11} - \kappa m_2 m_{10}$$

$$\frac{dm_3}{dt} = k_1 m_1 m_2 - k_2 m_3 - k_3 m_3 m_9 + k_4 m_4$$

$$\frac{dm_4}{dt} = k_3 m_3 m_9 - k_4 m_4 - k_5 m_4 + \kappa m_1 m_5 m_6$$

$$\frac{dm_5}{dt} = k_5 m_4 - k_6 m_5 m_7 + k_7 m_8 - \kappa m_1 m_5 m_6$$

$$\frac{dm_6}{dt} = k_5 m_4 - k_9 m_6 m_{10} + k_{10} m_{11} - \kappa m_1 m_5 m_6 \qquad (2)$$

$$\frac{dm_7}{dt} = -k_6 m_5 m_7 + k_7 m_8 + k_8 m_8 - \kappa m_7 m_9$$

$$\frac{dm_8}{dt} = k_6 m_5 m_7 - k_7 m_8 - k_8 m_8 + \kappa m_7 m_9$$

$$\frac{dm_9}{dt} = -k_3 m_3 m_9 + k_4 m_4 + k_8 m_8 - \kappa m_7 m_9$$

$$\frac{dm_{10}}{dt} = -k_9 m_6 m_{10} + k_{10} m_{11} + k_{11} m_{11} - \kappa m_2 m_{10}$$

$$\frac{dm_{11}}{dt} = -k_9 m_6 m_{10} - k_{10} m_{11} - k_{11} m_{11} + \kappa m_2 m_{10} \ ,$$

where the bold terms are reverse reactions (with a very small backward rate constant κ) which we have added:

$$m_4 \xrightarrow{k_5} m_5 + m_6 + m_1 \qquad\qquad m_4 \underset{\kappa}{\overset{k_5}{\rightleftharpoons}} m_5 + m_6 + m_1$$

$$m_8 \xrightarrow{k_8} m_7 + m_9 \qquad \Longrightarrow \qquad m_8 \underset{\kappa}{\overset{k_8}{\rightleftharpoons}} m_7 + m_9$$

$$m_{11} \xrightarrow{k_{11}} m_3 + m_{10} \qquad\qquad m_{11} \underset{\kappa}{\overset{k_{11}}{\rightleftharpoons}} m_3 + m_{10} \ .$$

Furthermore, we need the mass conservation equations

$$m_1 + m_3 + m_4 = \text{const}$$
$$m_2 + m_3 + m_4 + m_6 + m_{11} = \text{const}$$
$$m_9 + m_4 + m_5 + m_8 = \text{const} \qquad (3)$$
$$m_7 + m_8 = \text{const}$$
$$m_{10} + m_{11} = \text{const}$$

for later purposes.

3   Complexity Analysis: Dynamic Sensitivity Analysis Allowing Orthogonal Decomposition of System Dynamics

The first step for model reduction is the choice of the fixed species (reaction progress variables parameterizing the reduced model). Their choice is in general a priori free. To select a reasonable set of fixed species out of the set of all species it is suitable to perform a complexity analysis in the first place followed by the model reduction.

The goal of our complexity analysis is to determine a reduced dimension for a given error tolerance and the contribution of each species to the active dynamics of the system on a given time scale for a given trajectory in phase space. With this information one can now designate the amount of fixed species or in other words the dimensions of the slow attracting manifold and a suitable choice of reaction progress variables.

In the following we present a novel complexity analysis algorithm by Lebiedz et al. [14]. A similar analysis method has already successfully been applied to large oscillatory chemical systems by Shaik et al. [27]. We use the new algorithm to determine a suitable dimension of the low-dimensional manifold representing the reduced model and a suggestion for the choice of suitable reaction progress variables from the list of species.

The method splits a given trajectory into smaller intervals of equidistant length. On these intervals sensitivity matrix computations are performed corresponding to the propagation of virtual initial value perturbation along the trajectories. By a singular value decomposition of these sensitivity matrices, the extent of contraction/expansion of the perturbation is represented in an orthogonal coordinate system. The aim of the complexity analysis is now to separate *strongly contracting (fast)* directions in phase space from other directions, assume the relaxation of the fast direction via an algebraic equation and compare the solution of the so reduced differential algebraic equation model to the full original differential equation model on the given time interval. Details will be provided in the next section.

Assume now that for the IVP (1) a solution exists

$$x(t) = x(t; t_0, x_0) , \tag{4}$$

which is defined for all $t \geq t_0 \geq 0$.

Moreover, let Δx_0 be an initial perturbation applied at t_0. The deviation from the solution of (4) at a later time point t can be formulated as

$$\Delta x(t) := x(t; t_0, x_0 + \Delta x_0) - x(t; t_0, x_0) . \tag{5}$$

Therefore, linearization yields

$$\Delta x(t) = \frac{x(t; t_0, x_0 + \Delta x_0) - x(t; t_0, x_0)}{\Delta x_0} \cdot \Delta x_0$$

$$\approx \underbrace{\frac{\partial x}{\partial x_0}(t; t_0, x_0)}_{=:W(t,t_0)} \cdot \Delta x_0 , \tag{6}$$

where $W(t, t_0) \in \mathbb{R}^{n \times n}$ is the so called *sensitivity* or *propagation matrix* that propagates the perturbation Δx_0, applied at t_0, along the curve (t_0, x_0).

The *global* aspect lies in the fact that the complete time interval $[0, t_{end}]$ is devided into n sections $[0, T], \ldots, [(n-1)T, nT]$ each of which has length

$$T := \frac{t_{end}}{n} . \tag{7}$$

Let us further assume that on each such section a solution $x(T) = x(T; 0, x_0), \ldots,$ $x(nT) = x(nt; (n-1)T, x((n-1)T))$ exists for the IVP (1). Due to (6) we get the corresponding linearized perturbations $\Delta x(T) \approx W(T, 0) \cdot \Delta x(0), \ldots,$ $\Delta x(nT) \approx W(nT, (n-1)T) \cdot \Delta x((n-1)T)$.

The analysis of the aforementioned propagation matrices $W(iT, (i-1)T), i = 1, \ldots, n$ is the very core of the complexity analysis algorithm because the matrices contain information about the extent of relaxation of perturbations of the species variables on each subinterval and thus allow a separation into relaxing and active dynamical modes.

The *singular value decomposition* offers the opportunity to analyze these matrices and helps to identify strongly contracting modes being enslaved to the remaining ones. It can be stated as follows:

For $W \in \mathbb{R}^{n \times n}$ there exist orthogonal matrices $U = [u_1, \ldots, u_n] \in \mathbb{R}^{n \times n}$ and $V = [v_1, \ldots, v_n] \in \mathbb{R}^{n \times n}$ such that

$$W = U \Sigma V^T , \tag{8}$$

where $\Sigma = \text{diag}(\sigma_1, \ldots, \sigma_n)$ and $\sigma_1 \geq \ldots \geq \sigma_n \geq 0$ are the singular values of W.

Geometrically, (8) means that the column vectors of U correspond with the axes of a n-dimensional ellipse, whose lengths are determined by the singular values $\sigma_1, \ldots, \sigma_n$. The column vectors of V are mapped onto the axes of this *hyperellipse* under the linear transformation W:

$$W v_i = \sigma_i u_i \qquad i = 1, \ldots, n \ .$$

For the sake of better visualization, a sketch of a typical situation in $\mathbb{R}^2$ is depicted in Fig. 2. We advise the readers to consult e.g. [29] for getting a deeper

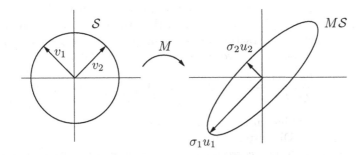

Fig. 2. Singular value decomposition of a 2×2 matrix M. The sphere $\mathcal{S} \in \mathbb{R}^n$ is mapped onto the hyperellipse $M\mathcal{S}$.

understanding of the topic. The crucial advantage of the singular value approach over the analysis of eigenvalues of Jacobian matrices used in [32] is twofold. First, the Jacobian contains only local information in an infinitesimally small time window and in particular if the system is highly nonlinear, this information is of restricted value for the real evolution of the system dynamics in a finite range around the actual system state. Second, since the Jacobian is a non-symmetric matrix in general, a diagonalization by an orthogonal linear transformation is impossible and thus the slow and fast subspaces are not orthogonal to each other. This leads to severe difficulties when computing the contribution of each species to each subspace.

The column vectors of U determine contracting (expanding) directions in phase space and the most contracting direction is characterized by the shortest axis $\sigma_i u_i$ of the hyperellipse and quantified by the corresponding singular value. The general ODE system (1) can be transformed to

$$U^T \dot{x} = U^T f \ , \tag{9}$$

and the *modes* which are defined by $u_i x(t), i = 1, \ldots, n$ can be classified according to their corresponding singular value σ_i: If $\sigma_i < 1$ then mode $u_i x(t)$ is relaxing (perturbations decay in that direction). Conversely, if $\sigma_i > 1$ then the mode is non-relaxing and strongly responds to perturbations. However, if $\sigma_i = 1$ then a constant mode is encountered which maintains perturbations constant.

In this sense, the matrix U can be partitioned into a "relaxing" and an "active" submatrix U_{rel} and U_{act} respectively

$$U = [U_{\text{act}}|U_{\text{rel}}] = [u_1, \ldots, u_r|u_{r+1}, \ldots, u_n] \ . \tag{10}$$

Assuming now the contracting modes to be relaxed and enslaving them to the active modes one can derive a *differential algebraic equation* (DAE) system

$$\begin{aligned} U_{\text{act}}^T \dot{x} &= U_{\text{act}}^T f \\ 0 &= U_{\text{rel}}^T f \ , \end{aligned} \tag{11}$$

with r differential and $n-r$ algebraic equations. In order to determine the number of relaxing modes on each subinterval and thus the dimension of the reduced model, the number of contracting modes assumed to be relaxed is iteratively increased until a user-defined error tolerance TOL is violated

$$\frac{|y_i^*(jT) - y_i(jT)|}{|y_i(jT)|} \leq TOL \ , \quad i = 1, \ldots, r, \quad j = 1, \ldots n \tag{12}$$

where $y_i(jT)^*$ is the differential part of the DAE (11) and $y_i(jT)$ is the solution of the transformed ODE system (9).

The *reduced dimension* r corresponds to the number of active modes and the dynamic behavior of the system is eventually described in an error-controlled way by a model of this reduced dimension. However, according to the linear transformation, the active and relaxing modes correspond to linear combinations of the state variables $x_i(t)$. Hence, in order to get insight into the contribution of each species $x_i(t)$ to the system dynamics, we analyse the subspace spanned by the column vectors of U as suggested in (10): The first r vectors span the active subspace, the following $n - r$ vectors span the subspace representing relaxing processes. According to this, the position of the axes is crucial for determining any contribution of species to the system dynamics separated into active and relaxing ones. The position of the i-th axis with respect to the appropriate subspace is expressed by $p_i^{\text{act}} := \sum_{j=1}^r u_j(i)u_j$ and $p_i^{\text{rel}} := \sum_{j=r+1}^r u_j(i)u_j$, $i = 1, \ldots, n$ respectively with $u_j(i)$ indicating the i-th component of vector u_j.

These projection vectors can be associated with the contribution of each species to the relaxing and active subspace. Hence a relative contribution of each species to the particular subspace (given in %) can be calculated via the formula

$$\begin{aligned} r_i^{\text{act}} &:= \frac{\|p_i^{\text{act}}\|}{\|p_i^{\text{act}}\| + \|p_i^{\text{rel}}\|} \\ r_i^{\text{rel}} &:= \frac{\|p_i^{\text{rel}}\|}{\|p_i^{\text{act}}\| + \|p_i^{\text{rel}}\|} \ , \end{aligned} \tag{13}$$

for each $i = 1, \ldots, n$.

For an application of the model reduction approach presented in the next section, our complexity analysis algorithm is supposed to provide a suitable dimension of the reduced model on a given time interval and to support the choice of appropriate reaction progress variables parameterizing the reduced model. It is reasonable (but not mandatory) to choose species with large contributions to the active subspace here.

4 Model Reduction: Maximal Relaxation of Chemical Forces under Constraints

In this section a novel model reduction method based on optimizing trajectories is applied to the mathematical model (2). In the following we will give an overview of the methodical ideas first introduced by Lebiedz [13] and further refined by Reinhardt et al. [26].

Unlike time scale analysis [23] or lumping techniques [18], the presented method is based on optimizing trajectories in phase space with respect of the relaxation of "chemical forces" along the trajectories. The key idea of this method is to fix the initial concentration of several suitable species and to determine the initial concentrations of the remaining species, a procedure called *species reconstruction*. They are determined such that the resulting trajectory starting from the initial concentrations to the equilibrium is maximally relaxed in terms of "chemical forces" in a suitable sense. This trajectory is then used as a representation of the reduced model for given values of the reaction progress variables parameterizing the reduced model.

These numerical optimized trajectories represent a reduced model in terms of slow attracting manifolds spanned by these trajectories and parameterized by the species with fixed initial concentrations.

This key idea can be realized mathematically by an optimization problem which can generally be formulated as

$$\min_{x_k} \int_0^{t_{\text{end}}} \Phi(x(t))dt \qquad (14a)$$

subject to

$$\frac{dx_k}{dt} = f_k(x) \qquad k = 1, ..., n \qquad (14b)$$

$$x_k(0) = x_k^0 \qquad k \in I_{\text{fixed}} \qquad (14c)$$

$$|x_k(t_{\text{end}}) - x_k^{\text{eq}}| \le \epsilon \qquad k \in I_{\text{fixed}} \qquad (14d)$$

and is subject to conservation relations among the species. x_k are the concentrations of biochemical species, I_{fixed} is the index set that contains the indices of variables with fixed initial values, the so called *reaction progress variables*.

The constraint (14b) includes the dynamics of the biochemical system, i.e. the underlying ODE system (1) in the formulation of the optimization problem. This ensures the consistency of the solution of the optimization problem with the full model.

The exact approach of the equilibrium value x_k^{eq} in a numerical solution of the kinetic equation system would take infinite time because it corresponds to vanishing right-hand sides of the kinetic equations and therefore the dynamics get infinitely slow when approaching the equilibrium point. For this reason the approach to equilibrium is approximated by the mathematical formulation (14) by assuming the system to be close to equilibrium, which means that the deviation of the final value $x_k(t_{end})$ from its equilibrium value x_k^{eq} is assumed to be less or equal to a quantity ϵ. This condition is formulated as the end point constraint for $x_k(t_{end})$ in (14d). The corresponding end time t_{end}, which is a priori unknown, is kept free within the problem formulation and is determined during numerical computations. Alternatively, the time t_{end} can be fixed such that the final state of the system is very close to the chemical equilibrium point, making (14d) redundant.

$\Phi(x(t))$ in (14a) is the objective functional which describes an optimization criterion related to the degree of relaxation of "chemical forces".

Reinhardt et al. [26] state that a suitable criterion $\Phi(x(t))$ should describe the extent of relaxation of "chemical forces" in the evolution of trajectories to equilibrium. This means that the objective functional $\Phi(x(t))$ should be minimal along a trajectory that is as close to equilibrium as allowed by the initial constraints (14d).

Lebiedz [13] considers a generalized concept for the "distance" of a chemical system from its attractor in order to derive a thermodynamic criterion which is related to maximal relaxation of "chemical forces" along phase space trajectories. Reinhardt et al. consider a suitable extension that turned out to be particularly successful when applied to a hydrogen combustion reaction mechanism in order to compute a low-dimensional attracting manifold for model reduction [26].

Under isolated conditions, the attractor of a chemical system is the thermodynamic equilibrium. In Lebiedz' model reduction approach, a special trajectory, called *Minimal Entropy Production Trajectory* (MEPT), which converges towards equilibrium, is calculated such that the sum of affinities of the entropy production rates of single reaction steps is minimized [13].

The entropy production rate is closely related to the concept of chemical affinity which was first introduced by de Donder [8] as the driving force of chemical reactions.

For an elementary reaction step j with the forward and backward reaction rates $\mathcal{R}_{j\rightarrow}$ and $\mathcal{R}_{j\leftarrow}$, the concept of chemical affinity can be related to the concept of entropy production by the following relation [13]:

$$\frac{d_i S_j}{dt} = R \cdot (\mathcal{R}_{j\rightarrow} - \mathcal{R}_{j\leftarrow}) \ln \left(\frac{\mathcal{R}_{j\rightarrow}}{\mathcal{R}_{j\leftarrow}} \right) , \qquad (15)$$

where $d_i S_j/dt$ is the entropy production rate for reaction j and R is the gas constant.

Entropy production rates are additive for several elementary reaction steps. Therefore, the total entropy production rate (the sum of the entropy production rates of all n elementary reaction steps) can be computed for an arbitrary reaction system if kinetic data are available and a detailed elementary reaction step mechanism is known.

In the context of the general optimization problem (14), using entropy production as an optimization functional means:

$$\Phi(x(t)) = \sum_{j=1}^{n} \frac{d_i S_j}{dt} \ . \tag{16}$$

On the basis of the concept of curvature of trajectories in phase space, Reinhardt et al. [26] derive a more fundamentally rooted criterion for the objective functional $\Phi(x(t))$ which is subsequently combined with the entropy production.

This criterion is related to the geometric interpretation of "chemical forces" from a physical point of view. Motivated by this picture the principle of "force = curvature" is transferred to the field of chemical kinetrics and formulated as a corresponding variational principle.

In (bio)chemical systems dissipative forces are active. Slow and fast dynamic modes corresponding to different velocities and thus time scales of chemical reactions result in an anisotropic force relaxation behavior in phase space.

For a chemical system whose dynamics is described by the ODE system

$$\dot{x}(t) = f(x) \ , \tag{17}$$

curvature of the trajectories $x(t)$ as geometrical objects in phase space is considered.

The following relations hold:

$$\ddot{x}(t) = \frac{d^2 x}{dt^2} = \frac{d\dot{x}}{dt} = \frac{d\dot{x}}{dx} \cdot \frac{dx}{dt} = J(\dot{x}(t)) \cdot \dot{x}(t) = J(f(x(t))) \cdot f(x(t)) \tag{18}$$

with $J(f)$ being the Jacobian of the right-hand side of equation (17).

Based on (18), Reinhardt et al. define the curvature of $x(t)$ as the vector norm

$$\|\ddot{x}(t)\| = \|J(f(x(t))) \cdot f(x(t))\| \ . \tag{19}$$

Transferring the fundamental geometric picture of force being equivalent to curvature mentioned above, Reinhardt et al. relate the curvature of trajectories in (17) to the forces driving the chemical system towards equilibrium by subsequent relaxation of dynamical modes. In thermodynamic equilibrium those chemical forces vanish.

As a criterion which characterizes maximal relaxation of chemical forces under given constraints Reinhardt et al. use minimal total ("integrated") curvature of trajectories defined by the objective function

$$\Phi(x(t)) := \|J(f(x)) \cdot f(x)\| \tag{20}$$

in the general optimization problem (14).

Alternatively, the objective function (20) can be interpreted as minimizing the length of a trajectory in suitable Riemannian metrics which expresses distance from equilibrium in a suitable sense.

For any continuously differentiable curve $\gamma(t)$ on a Riemannian manifold, the length L of γ is defined as

$$L(\gamma) := \int_\gamma \sqrt{g_{\gamma(t)}(\dot{\gamma}(t), \dot{\gamma}(t))}dt \ . \tag{21}$$

with $g_{\gamma(t)}$ being a scalar product defined on the tangent space of the curve in each point.

If the Riemannian metrics $g_{\gamma(t)}$ is chosen as

$$g_{\gamma(t)}(f, f) := f^T \underbrace{J^T J}_{\text{positive definite}} f = \|Jf\|^2 \ , \tag{22}$$

then the "length-minimizing" objective functional being equivalent to (20) becomes

$$\min \int_0^{t_{\text{end}}} \sqrt{g_{\gamma(t)}(\dot{x}(t), \dot{x}(t))}dt \ . \tag{23}$$

The solution trajectory of this problem can be interpreted as a geodesic, i.e. a curve which minimizes the length of the path between two points in a possibly curved manifold.

Hence the "distance from equilibrium in a chemical sense" can be formulated here in an explicit mathematical form based on concepts from differential geometry.

To describe the distance of a chemical system from its thermodynamic equilibrium in a very general way, the Riemannian metrics

$$\hat{g}_{\gamma(t)}(f, f) := f^T \underbrace{J^T \cdot A \cdot J}_{\text{positive definite}} f =: \|Jf\|_A^2 \tag{24}$$

can be considered, where A is a positive definite matrix.

As a possible choice for A, Reinhardt et al. [26] propose a diagonal matrix with entries

$$a_{kk} = \sum_{j=1}^n \nu_{kj} \frac{d_i S_j}{dt} \qquad (k = 1, ..., m) \tag{25}$$

which represents an anisotropic "kinetic weighting" of the phase space directions by including the entropy production rate.

Here n is the number of reactions, ν_{kj} are the stoichiometric coefficients describing the degree to which the chemical species k participates in reaction j, and $d_i S_j/dt$ is the entropy production rate of reaction j.

a_{kk} is the sum of the entropy production rates of all elementary reactions in which species k takes part.

A is positive definite since according to the Second Law of Thermodynamics $d_i S_j/dt > 0$ holds for any spontaneous process, and therefore $a_{kk} > 0$ for all $k = 1, ..., n$.

Thus the objective functional $\Phi(x(t))$ can be restated as

$$\Phi(x(t)) = \|Jf\|_A^2 \ . \tag{26}$$

We use this criterion in the biochemical example application presented in this paper and demonstrate its success for the purpose of model reduction.

5   Numerical Results

In the following, we apply a combination of the methods presented in the previous sections to the example model system introduced in section 2. The benefit of combining the two methods lies in the fact that a user, who is interested in model reduction, can choose the dimension of the reduced model and the reaction progress variables parameterizing the reduced model freely in accordance with the problem-determined needs on the basis of the complexity analysis.

For the complexity analysis, we use an implementation of the method by Lebiedz et al. [14] described in section 3. The code is written in C and uses DAESOL [1] in order to compute the sensitivity matrices, an efficient and robust integrator for stiff ODE and DAE systems based on a *backward differentiation formula* (BDF) which additionally implements a numerically stable and accurate differentiation scheme, the *internal numerical differentiation* (IND) [2].

The initial values for the trajectory in phase space, to which the complexity analysis and subsequent model reduction is applied, are:

$$
\begin{array}{llll}
m_1 = 67.95 & m_2 = 0.372 & m_3 = 0.091 & m_4 = 58.25 \\
m_5 = 0.088 & m_6 = 0.228 & m_7 = 66.33 & m_8 = 25.17 \\
m_9 = 176.41 & m_{10} = 160.95 & m_{11} = 2.244 \ .
\end{array}
\tag{27}
$$

They are identical to those chosen by Petrov et al. [24] who accomplish model reduction via QSSA for the same model system. Numerical integration of system (2) is performed over the interval $[0, t_{end}]$ with $t_{end} = 20$ time units. After this time the equilibrium point will be reached. We use equidistant time steps of length $T = 0.001$ time units for the intervals to calculate the sensitivity matrices. Therefore, the trajectory is split into $200{,}000$ pieces on which the reduced dimensions as well as the contributions of the species to the currently active dynamics are computed with an user-defined error tolerance $TOL = 10^{-10}$. In the following we present and discuss the numerical results.

In Fig. 3 the minimal dimension is plotted versus time. The system contains 11 species from which we have to substract the 5 mass conservation equations (3) to obtain the dimension of the space in which the dynamics can take place. That means that the reduced dimension must lie between 6 and 0, where 0 corresponds to the equilibrium state.

In Fig. 4, contributions of selected 4 out of 11 species (given in percent) are depicted which have been calculated via formula (13). Among these species, we illustrate three obviously active ones (m_5, m_8 and m_{11}) and one non-active, i.e. relaxing, (m_6) respectively. The contribution of the inactive species m_6 quickly falls to a constant low level which indicates that this species is relatively unimportant for the active system dynamics. Therefore, if one is free in choice, this species may not be appropriate for use as a reaction progress variable for model reduction in contrast to the three others. The explicit computation of the low dimensional manifold which represents the reduced dynamics is performed via the algorithm described in section 4 which has been implemented within the software package MUSCOD-II [15, 16, 17]. MUSCOD-II has been designed for solving

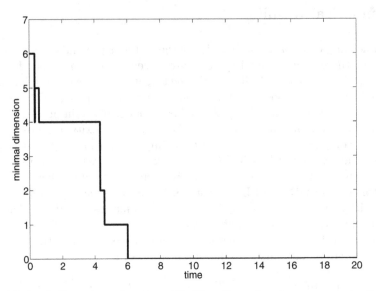

Fig. 3. Minimal dimension of system (2). The dimension decreases gradually when approaching equilibrium.

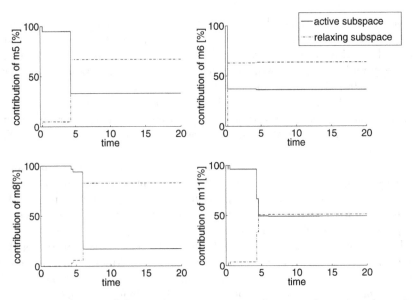

Fig. 4. Contribution of the species m_5, m_6, m_8 and m_{11} to the system dynamics

large scale nonlinear optimization problems and optimal control problems based on a sophisticated multiple shooting approach suggested by Bock [4, 3]. The numerical integrator used along with MUSCOD-II is DEASOL (see above). According to the results obtained by calculating the minimal dimension, the appropriate dimension should be chosen to be 4 because it retains this value for the longest

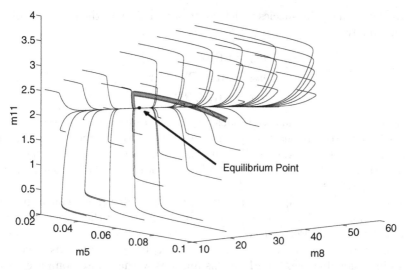

Fig. 5. A 2-dimensional manifold spanned by the reaction progress species m_5 and m_8. The initial trajectory is displayed as light shaded bold line moving towards the equilibrium point.

time until reaching the vicinity of the equilibrium point (compare Fig. 3). Unfortunately, a 4-dimensional manifold cannot be visualized well. Instead we have computed a 2-dimensional one, which also approximates the slow dynamics appropriately (see Fig. 5), however with a lower accuracy in regions far from the equilibrium point.

According to the complexity analysis results, we have chosen two species with large contributions to the active dynamics as reaction progress variables (m_5 and m_8).

For the visualization of the manifold, optimized trajectories spanning this manifold have been calculated by the algorithm described in section 4 with initial values within the following ranges

$$m_5 : 0.022 - 0.088$$
$$m_8 : 10.0 - 55.0 \ .$$

These trajectories may be tabulated and used as a reduced model for further numerical applications. As shown in Fig. 5, the initial trajectory firstly converges towards the computed manifold and thereafter approaches the equilibrium point on the manifold.

6   Conclusion

In this paper we have presented two new numerical methods for complexity analysis as well as model reduction which have been applied to a biochemical system of the influence of RKIP to the ERK signaling pathway. More precisely, we have determined the species which contribute mainly to the dynamics of that system, namely

$$m_5 \mathrel{\widehat{=}} \text{ERK-P}$$
$$m_8 \mathrel{\widehat{=}} \text{MEK-PP/ERK}$$
$$m_{11} \mathrel{\widehat{=}} \text{RKIP/RP} .$$

Petrov et al. [24] state that RKIP possibly plays a minor role in inhibiting the ERK signaling pathway near the quasi steady state of the system. They conclude this from the fact that the equilibrium point does not depend on the initial concentration of RKIP (where the QSSA assumptions are valid) but rather on the Raf-1*/RKIP/ERK-PP complex.

According to the complexity analysis presented above, we can argue that RKIP does not contribute much to the active dynamics of the system thus reinforcing the results by Petrov et al..

Moreover, we have shown that a trajectory of this system is mostly confined to a 4-dimensional space. Based on this analysis we have also computed a low dimensional manifold spanned by optimized trajectories representing a reduced model. We have demonstrated that the initial trajectory quickly converges to the manifold at first and then proceeds to the equilibrium point on the manifold. Our approach is of general value for complexity analysis and model reduction of biochemical systems.

Acknowledgment

We thank H.G. Bock who generously provided the software package MUSCOD-II.

Moreover, we are indebted to V. Reinhardt and J. Kammmerer for implementation and enhancement of the numerical methods used in this paper.

A special thank goes to the Deutsche Forschungsgemeinschaft (DFG), the Sonderforschungsbereich 568 and the FRISYS program as part of the BMBF systems biology initiative FORSYS for financial support.

References

[1] Bauer, I., Finocchi, F., Duschl, W.J., Gail, H.-P., Schlöder, J.P.: Simulation of chemical reactions and dust destruction in protoplanetary accretion discs. Astron. Astrophys. 317, 273–289 (1997)

[2] Bock, H.G.: Numerical treatment of inverse problems in chemical reaction kinetics. In: Ebert, K.H., Deuflhard, P., Jäger, W. (eds.) Modeling of Chemical Reaction Systems. Springer Series in Chemical Physics, vol. 18, pp. 102–125. Springer, Heidelberg (1981)

[3] Bock, H.G.: Randwertproblemmethoden zur Parameteridentifizierung in Systemen nichlinearer Differentialgleichungen. Bonner Mathematische Schriften, vol. 183. University of Bonn, Bonn (1987)

[4] Bock, H.G., Plitt, K.J.: A multiple shooting algorithm for direct solution of optimal control problems. In: Proc. 9th IFAC World Congress Budapest. Pergamon Press, Oxford (1984)

[5] Bodenstein, M.: Eine Theorie der photochemischen Reaktionsgeschwindigkeiten. Z. Phys. Chem. 85, 329–397 (1913)

[6] Chapman, D., Underhill, L.: The interaction of chlorine and hydrogen. The influence of mass. J. Chem. Soc. Trans. 103, 496–508 (1913)

[7] Cho, K.-H., Shin, S.-Y., Kim, H.-W., Wolkenhauer, O., McFerran, B., Kolch, W.: Mathematical modeling of the influence of RKIP on the ERK signaling pathway. In: Priami, C. (ed.) CMSB 2003. LNCS, vol. 2602, pp. 127–141. Springer, Heidelberg (2003)

[8] de Donder, T., van Rysselberghe, P.: Thermodynamic Theory of Affinity: A Book of Principles. Stanford University, Menlo Park (1936)

[9] Gorban, A., Karlin, I.: Invariant Manifolds for Physical and Chemical Kinetics. Springer - Lecture Notes in Physics, vol. 660. Springer, Heidelberg (2005)

[10] Gorban, A., Karlin, I., Zinovyev, A.: Constructive methods of invariant manifolds for kinetic problems. Phys. Rep. 396, 197–403 (2004)

[11] Hadjinicolaou, M., Goussis, D.A.: Asymptotic solutions of stiff PDEs with the CSP method: the reaction diffusion equation. SIAM J. Sci. Comput. 20, 781–810 (1999)

[12] Lam, S.H., Goussis, D.A.: The CSP method for simplifying kinetics. J. Chem. Kinet. 26, 461–486 (1994)

[13] Lebiedz, D.: Computing minimal entropy production trajectories: An approach to model reduction in chemical kinetics. J. Chem. Phys. 120, 6890–6897 (2004)

[14] Lebiedz, D., Kammerer, J., Brandt-Pollmann, U.: Automatic network coupling analysis for dynamical systems based on detailed kinetic models. Phys. Rev. E 72(041911) (2005)

[15] Leineweber, D.B.: Efficient reduced SQP methods for the optimization of chemical processes described by large sparse DAE models. Fortschritt-Berichte VDI Reihe 3, Verfahrenstechnik, vol. 613. VDI-Verlag GmbH, Düsseldorf (1999)

[16] Leineweber, D.B., Schäfer, A., Bock, H.G., Schlöder, J.P.: An efficient multiple shooting based reduced SQP strategy for large-scale dynamic process optimization – part I: Theoretical aspects. Comput. Chem. Engng. 27, 157–166 (2003)

[17] Leineweber, D.B., Schäfer, A., Bock, H.G., Schlöder, J.P.: An efficient multiple shooting based reduced SQP strategy for large-scale dynamic process optimization – part II: Software aspects and applications. Comput. Chem. Engng. 27, 167–174 (2003)

[18] Li, G., Pope, S.B., Rabitz, H.: New approaches to determination of constrained lumping schemes for a reaction system in the whole composition space. Chem. Eng. Sci. 46, 95–111 (1991)

[19] Maas, U.: Coupling of chemical reaction with flow and molecular transport. Appl. Math. 40, 249–266 (1995)

[20] Maas, U.: Efficient calculation of intrinsic low-dimensional manifolds for the simplification of chemical kinetics. Springer – Computing and Visualization in Science 1, 69–81 (1998)

[21] Maas, U., Pope, S.B.: Simplifying chemical kinetics: Intrinsic low-dimensional manifolds in composition space. Combust. Flame 88, 239–264 (1992)

[22] Michaelis, L., Menten, M.L.: Die Kinetik der Invertinwirkung. Biochem. Z. 49, 333–369 (1913)

[23] Okino, M.S., Mavrovouniotis, M.L.: Simplification of mathematical models of chemical systems. Chem. Rev. 98, 391–406 (1998)

[24] Petrov, V., Nikolova, E., Wolkenhauer, O.: Reduction of nonlinear dynamic systems with an application to signal transduction pathways. IET Syst. Biol. 1(1), 2–9 (2007)

[25] Rabitz, H., Kramer, M., Dacol, D.: Sensitivity analysis in chemical kinetics. Annu. Rev. Phys. Chem. 34, 419–461 (1983)

[26] Reinhardt, V., Winckler, M., Lebiedz, D.: Approximation of slow attracting manifolds by trajectory-based optimization approaches. J. Phys. Chem. A 112, 1712–1718 (2008)

[27] Shaik, O.S., Kammerer, J., Górecki, J., Lebiedz, D.: Derivation of a quantitative minimal model for the photosensitive Belousov-Zhabotinsky reaction from a detailed elementary-step mechanism. J. Chem. Phys. 123(234103) (2005)

[28] Tomlin, A.S., Pilling, M.J., Turányi, T., Merkin, J.H., Brindley, J.: Mechanism reduction for the oscillatory oxidation of hydrogen sensitivity and quasi-steady state analyses. Combust. Flame 91, 107–130 (1992)

[29] Trefethen, L.N., Bau, D.: Numerical Linear Algebra. SIAM, Philadelphia (1997)

[30] Turányi, T.: Sensitivity analysis of complex kinetic systems. Tools and applications. J. Math. Chem. 5, 203–248 (1990)

[31] Warnatz, J., Maas, U., Dibble, R.W.: Combustion. Physical and Chemical Fundamentals, Modeling and Simulation, Experiments, Pollutant Formation, 3rd edn. Springer, Heidelberg (2001)

[32] Zobeley, J., Lebiedz, D., Kammerer, J., Ishmurzin, A., Kummer, U.: A new time-dependent complexity reduction method for biological systems. Trans. Comput. Syst. Biol. 1, 90–110 (2005)

Formal Analysis of Abnormal Excitation in Cardiac Tissue

Pei Ye[1], Radu Grosu[1], Scott A. Smolka[1], and Emilia Entcheva[2]

[1] Computer Science Department, Stony Brook University, NY 11794, USA
[2] Biomedical Engineering Department, Stony Brook University, NY 11794, USA

Abstract. We present the Piecewise Linear Approximation Model of Ion Channel contribution (PLAMIC) to cardiac excitation. We use the PLAMIC model to conduct formal analysis of cardiac arrhythmic events, namely Early Afterdepolarizations (EADs). The goal is to quantify (for the first time) the contribution of the overall sodium (Na^+), potassium (K^+) and calcium (Ca^{2+}) currents to the occurrence of EADs during the plateau phase of the cardiac action potential (AP). Our analysis yields exact mathematical criteria for the separation of the parameter space for normal and EAD-producing APs, which is validated by simulations with classical AP models based on complex systems of nonlinear differential equations. Our approach offers a simple formal technique for the prediction of conditions leading to arrhythmias (EADs) from a limited set of experimental measurements, and can be invaluable for devising new anti-arrhythmic strategies.

1 Introduction

Excitable cells are those cells capable of generating and propagating electrical signals without damping. They are essential biological building blocks, determining functionality in the brain, heart, skeletal and smooth muscles.

An *action potential* (AP) is a change in an excitable cell's membrane potential caused by the flow of different ions across the cell membrane. The left panel in Fig. 1 illustrates a normal AP waveform for a guinea-pig heart cell. By convention, a normal AP follows a well defined cycle of "depolarization" (the rising phase), followed by "repolarization" (the falling phase). Furthermore, in qualitative terms, the "repolarization" phase can be divided in "early repolarization", "plateau" and "final repolarization".

Under some pathological conditions leading to a prolonged repolarization phase, the morphology of the AP can be altered by an abnormal secondary depolarization, termed **Early Afterdepolarization** (EAD). By clinical definition [1,2], EADs occur before the completion of repolarization of an AP (as illustrated in the right panel of Fig. 1).

Such cellular-level events can give rise to undesired new excitation waves and can precipitate life-threatening heart activation sequences, e.g. tachyarrhythmias, especially in patients with Long QT syndrome [3,4]. As critical arrhythmia triggers, EADs have been of interest to cardiac researchers for several decades [5].

M. Heiner and A.M. Uhrmacher (Eds.): CMSB 2008, LNBI 5307, pp. 141–155, 2008.

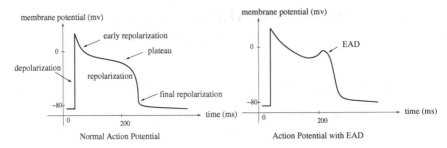

Fig. 1. EAD in cardiac myocyte

Attempts have been made to uncover the ionic mechanisms underlying EADs, so that their occurrence can be predicted as well as effectively treated. Various studies have found that the reactivation of calcium (Ca^{2+}) or sodium (Na^+) channels or abnormally reduced potassium (K^+) current can lead to this phenomenon [6,7,8]. Yet, a unified view of EAD mechanisms along with predictive criteria are lacking.

In this paper, we present the *Piecewise Linear Approximation Model of the Ion Channel contribution* (PLAMIC) as a basis for understanding and analyzing the biochemical mechanisms underlying the formation of EADs during the cardiac action potential. The derivation of the PLAMIC model can be understood as follows. Let V^{Na^+}, $V^{Ca^{2+}}$ and V^{K^+} denote the integral contributions to the AP due to the sodium, calcium and potassium channels, respectively; i.e. the voltages the ionic currents flowing through these channels induce. Further, let V^{NaK} denote the combined sodium and potassium voltage.

A key observation is that during normal and abnormal APs, the behavior of $V^{Ca^{2+}}$ and V^{NaK} corresponds to triangular-like functions of opposite polarity (see Fig. 3). As such, in the PLAMIC model, $V^{Ca^{2+}}$ and V^{NaK} are approximated in a piecewise-linear fashion using two very simple triangular functions, each of which naturally comprises a rising phase and a falling phase. The PLAMIC model also incorporates an AP-morphology-related (exponential) decay function, which can be fitted across different cell types.

A main advantage of the PLAMIC model then is its highly constrained parameter space, essentially limited to the peak voltage values and their occurrence in time of the two triangular functions. The model is therefore amenable to a closed-form, voltage-monotonicity analysis on the AP cycle during repolarization. We in fact show that the absence of a monotonically decreasing AP V ($\frac{dV}{dt} < 0$) during the plateau phase of repolarization is a necessary and sufficient condition for EAD. We furthermore provide specific conditions on the parameter space (involving the relative slopes of the two triangular functions, the relative occurrence of their peaks, and their relative magnitudes) for EAD occurrence.

We also performed an experimental validation of the conditions derived from the above-described formal analysis of the PLAMIC parameter space, assembling a test set of normal and abnormal APs from the widely accepted Luo-Rudy model ventricular cell model [9]. Our results demonstrate that the results of our formal analysis can be used as a valid classifier for EAD prediction.

The organization of the rest of the paper is as follows: Section 2 provides a formal definition of the PLAMIC model. Section 3 conducts a model-based analysis of the conditions under which EADs occur. Section 4 uses computer simulations with the Luo-Rudy model to validate our results. Section 5 offers our concluding remarks and directions for future work.

2   The PLAMIC Model

Mathematical modeling of excitable cells has a long tradition, starting with the first empirically-derived ionic model of the action potential in a giant squid axon proposed by Hodgkin and Huxley in 1952 [10]. Subsequently, more ion channels and complex biophysical processes have been included in these models, although the general mathematical framework for representing the ion-channel contribution has remained essentially the same.

The model we propose adopts an abstraction based on voltage, i.e., it deals with the superposition of the voltages generated by the individual ion channels. We study the occurrence of EADs as a disturbance in the subtle balance between the underlying ion currents using their voltage surrogates.

The advantage of using superposition of the voltages, as opposed to the ionic currents directly, is the integral (smoother) nature of the former in the RC-circuit model that approximates the electrical behavior of the cell membrane. This facilitates the curve-fitting process and allows for simpler mathematical expressions to be employed and further linearized in a piecewise fashion. The result is the *Piecewise Linear Approximation Model of the Ion-Channel contributions* (PLAMIC).

We illustrate the idea of the PLAMIC model starting from a modification of traditional ionic models based on the Hodgkin-Huxley formalism. The main equation used in these ionic models is presented in Eqn. 1.

$$C\dot{V} = -\sum I_i(t) + I_{st}(t) \qquad (1)$$

where $\dot{V}$ is the time derivative of the membrane potential V, C is the equivalent capacitance of the cell membrane, $\sum I_i(t)$ is the sum of all the ion currents flowing in or out of the cell membrane, and $I_{st}(t)$ is the stimulation current.

$\sum I_i(t)$ may incorporate a number of individual currents for different cell types. For example, in the Luo-Rudy model [9] (LRd), a widely accepted ventricular cell model, currents can be grouped by ion species as in Eqn. 2.

$$\sum I_i(t) = I_{Na}(t) + I_K(t) + I_{Ca}(t) \qquad (2)$$

where I_{Na}, I_K, and I_{Ca} are the sodium, potassium and calcium overall ion currents, respectively. The top row of Fig. 2 plots these three components of the LRd model for a normal AP.

Using for each component current the corresponding voltage, Eqns. 1 and 2 can be equivalently rewritten into the following form (Eqn. 3):

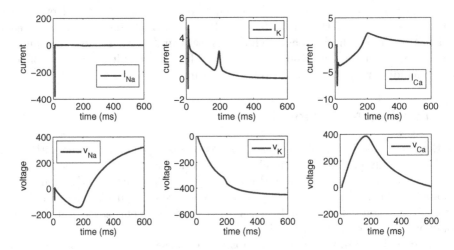

Fig. 2. Individual ionic currents and their corresponding voltages in the LRd model

$$C\dot{V}_{Na} = -I_{Na}(t), C\dot{V}_K = -I_K(t)$$
$$C\dot{V}_{Ca} = -I_{Ca}(t), C\dot{V}_{st} = I_{st}(t) \qquad (3)$$
$$V = (V_{Na}(t) + V_K(t) + V_{Ca}(t)) + V_{st}(t)$$

where V_{Na}, V_K, V_{Ca} and V_{st} are the voltages obtained via integration from I_{Na}, I_K, I_{Ca} and I_{st}, respectively.

The motivation behind Eqn. 3 is to first calculate the voltages from the individual currents and then obtain the overall membrane potential via super-position. Note the much smoother appearance of the voltage curves (bottom row) compared to the "spikey" current curves (top row) in Fig. 2. Furthermore, grouping the sodium and potassium voltages into one combined voltage yields the opposing triangular-like (and thus inherently linearizable) voltage functions depicted in Fig. 3.

The essentially triangular-shaped voltage functions suggests the use of two linear segments (linked together in a triangular form) to approximate the combination voltage due to the sodium and potassium currents (denoted as the NaK voltage), and the individual voltage due to the calcium current alone (the Ca voltage).

2.1   Definition of the PLAMIC Model

Two linear segments, forming a triangle (also known as a Lagrange hat function), are used to represent each of the NaK and Ca voltages. Two of the triangle vertices (beginning and end) are fixed on the time-axis and the triangle shape varies by shift in the free (peak) vertex. The triangular function is shown in Fig. 4 (A). It is essentially a two-piece linear function starting from point $(0,0)$ and ending at $(T_S, 0)$, where T_S is the total simulation time for the generation of an AP.

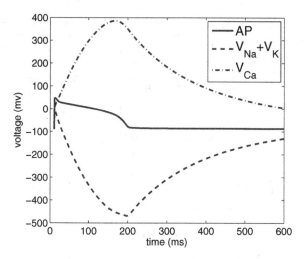

Fig. 3. AP, combined sodium and potassium voltage, and calcium voltage in LRd model

By fixing the simulation time T_S, each function is determined solely by the switching (peak) point (t_{max}, v_{max}). The mathematical definition of the triangular function is given by Eqn. 4, where the superscript $u \in \{Ca, NaK\}$ is used to distinguish the voltage functions corresponding to the different current types.

$$f^u(t) = \begin{cases} \frac{v_{max}^u}{t_{max}^u} t, & t \leq t_{max}^u; \\ \frac{T_S - t}{T_S - t_{max}^u} v_{max}^u, & t_{max}^u < t \leq T_S. \end{cases} \quad (4)$$

The functions for v_{Ca} and v_{NaK} are then defined simply as follows:

$$v_{Ca}(t) = f^{Ca}(t) \quad (5)$$

$$v_{NaK}(t) = f^{NaK}(t) \quad (6)$$

The overall action potential v is the superposition of the two (Eqn. 7).

$$v(t) = v_{Ca}(t) - v_{NaK}(t) + g(t) \quad (7)$$

where $g(t)$ is a decay function related to the AP morphology. It is defined by V_{max}, the absolute difference between the resting potential and the maximum voltage during upstroke, and by D ($D < 0$), an action-potential-duration parameter which can be adjusted across different cell types (Eqn. 8).

$$g(t) = V_{max} e^{Dt} \quad (8)$$

The decay function qualitatively reflects the passive component of the cell-membrane response: an RC circuit will exhibit exponential decay after the upstroke due to capacitor discharge. In the PLAMIC model, this passive decay is used in conjunction with the superposed opposite potentials (NaK and Ca).

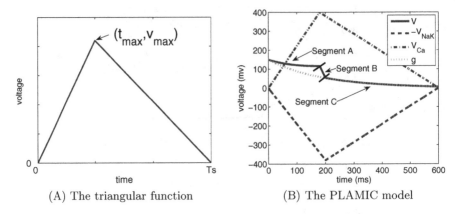

(A) The triangular function   (B) The PLAMIC model

Fig. 4.

Based on the relative magnitude of t_{max}^{Ca} and t_{max}^{NaK} (i.e. which voltage reaches its peak first), the AP equation for v (Eqn. 7) has two alternative formulations. In each case, v is represented as a three-segment function, referred to in the following equations as segments A, B and C, respectively.

First, let $a_1^c = \frac{v_{max}^{Ca}}{t_{max}^{Ca}}$, $a_2^c = \frac{-v_{max}^{Ca}}{T_S - t_{max}^{Ca}}$, $b^c = \frac{v_{max}^{Ca}}{T_S - t_{max}^{Ca}} T_S$, $a_1^k = \frac{v_{max}^{NaK}}{t_{max}^{NaK}}$, $a_2^k = \frac{-v_{max}^{NaK}}{T_S - t_{max}^{NaK}}$, and $b^k = \frac{v_{max}^{NaK}}{T_S - t_{max}^{NaK}} T_S$.

Case I: $t_{max}^{Ca} < t_{max}^{NaK}$

$$v(t) = \begin{cases} a_1^c t - a_1^k t + V_{max} e^{Dt}, & t \le t_{max}^{Ca}, \text{ segment A;} \\ (a_2^c t + b^c) - a_1^k t + V_{max} e^{Dt}, & t_{max}^{Ca} < t \le t_{max}^{NaK}, \text{ segment B;} \\ (a_2^c t + b^c) - (a_2^k t + b^k) + V_{max} e^{Dt}, & t \ge t_{max}^{NaK}, \text{ segment C.} \end{cases}$$

(9)

Case II: $t_{max}^{Ca} \ge t_{max}^{NaK}$

$$v(t) = \begin{cases} a_1^c t - a_1^k t + V_{max} e^{Dt}, & t \le t_{max}^{NaK}, \text{ segment A;} \\ a_1^c t - (a_2^k t + b^k) + V_{max} e^{Dt}, & t_{max}^{NaK} < t \le t_{max}^{Ca}, \text{ segment B;} \\ (a_2^c t + b^c) - (a_2^k t + b^k) + V_{max} e^{Dt}, & t \ge t_{max}^{Ca}, \text{ segment C.} \end{cases}$$

(10)

In Fig. 4 (B), one of the possible implementations of the PLAMIC model (case I) is shown. The overall PLAMIC-abstracted AP is given as a solid line, with its three segments annotated accordingly. We plot $-v_{NaK}$ instead of v_{NaK} to reveal the similarity to the LRd AP parameters shown in Fig. 3.

3   Formal Analysis of the PLAMIC Model

3.1   Monotonicity and EADs

EADs are secondary depolarization phenomena that arise during the repolarization phase; i.e. they disrupt the normal voltage return to rest. Therefore, a monotonicity analysis of the AP is an appropriate test for EADs. For example, it is safe to claim that a monotonically decreasing AP v ($\frac{dv}{dt} < 0$) is a sufficient condition for the absence of EADs. The opposite statement does not always hold, i.e. it is *not* always the case that if AP v is not universal decreasing, there is an EAD. For example, a "notch" in the early repolarization phase is common in many cardiac cells and is not considered an EAD (Fig. 5 (A)). Furthermore, in some cases, the membrane may transiently hyperpolarize; i.e. an undershoot may occur, with the potential lower than the resting potential during final repolarization. This non-monotonic case is also not an EAD (Figure 5 (B)).

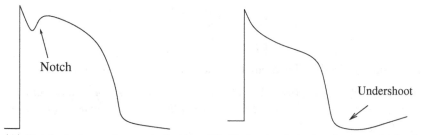

(A) Notch during early repolarization. (B) Undershoot during final repolarization.

Fig. 5. Non-monotonic APs that do not exhibit EADs

If, however, the monotonicity analysis is restricted to the "plateau" phase of the repolarization process, any deviation from monotonic decay will effectively be an EAD. In order to define the plateau phase in the PLAMIC model, let *notch-delay* be the cell-type-specific initial time segment of the repolarization phase during which a notch may occur. The PLAMIC plateau phase is then defined to consist of the suffix of segment A beginning at *notch-delay* followed by segment B. For most physiological choices of $(t^u_{\max}, v^u_{\max})$, $u \in \{\mathrm{Ca}, \mathrm{NaK}\}$, this definition of the plateau phase coincides closely with its physiological counterpart.

Based on the above monotonicity discussion, the following definition will serve as the theoretical basis of our formal analysis of EAD in the PLAMIC model.

Definition 1. *The PLAMIC model contains an EAD if $\dot{v} > 0$ at some point during the plateau phase.*

In Section 3.2, we present a monotonicity analysis of the PLAMIC plateau phase for both Cases I and II, and derive the exact conditions for EAD occurrence. Physiological explanations for these conditions are discussed as well.

3.2   Monotonicity Analysis of the PLAMIC Model

Case I. Case I is the most physiologically feasible scenario in cardiac cells. In simulation data of normal cardiac APs using the LRd model, $t^{\mathrm{Ca}}_{\max} < t^{\mathrm{NaK}}_{\max}$ holds at all times. As the PLAMIC-based voltage is a piecewise-linear function, monotonicity is analyzed on a per-segment basis.

Segment A. The first derivative of v within this segment is given by the following equation:

$$\frac{dv}{dt} = a_1^c - a_1^k + V_{max}De^{Dt} \tag{11}$$

Imposing the condition $\frac{dv}{dt} > 0$ yields:

$$t > \frac{1}{D}\ln(\frac{a_1^k - a_1^c}{V_{max}D}) \tag{12}$$

Further examination of Eqn. 12 shows that the existence of a positive real solution for t requires the following conditions to hold:

$$\begin{cases} a_1^c > a_1^k > 0 \\ 0 < \frac{a_1^k - a_1^c}{V_{max}D} < 1 \\ t < t_{max}^{Ca} \end{cases}$$

which are summarized in Theorem 1 as the major result for case I.

Theorem 1. $(a_1^k < a_1^c < (a_1^k - V_{max}D)) \wedge (t_{max}^{Ca} > \frac{1}{D}\ln(\frac{a_1^k - a_1^c}{V_{max}D}))$ *is a sufficient condition for a case-I occurrence of EAD during the suffix of segment A beginning at notch-delay.*

In Fig. 6, we plot the different possibilities of the relative magnitudes of a_1^k and a_1^c. Table 1 summarizes the relationship between these values and the occurrence of EAD.

Table 1. Summary of conditions for the existence of EAD in segment A

Condition	EAD	No EAD
$a_1^c < a_1^k$		Fig. 6 (A)
$a_1^k < a_1^c < a_1^k - V_{max}D$	Fig. 6 (D)	Fig. 6 (C)
$a_1^k - V_{max}D < a_1^c$	Fig. 6 (B)	

An intuitive physiological explanation of the above result is that the existence of EAD is closely related to the relative speeds of the voltage increase due to different ion currents, represented by a_1^c and a_1^k.

At the beginning of the plateau phase, the AP follows a decreasing trend, which requires the calcium current to have an upper bound ($a_1^c < a_1^k - V_{max}D$); otherwise, the AP curve will be increasing through this segment. Furthermore, for an EAD to form, the balance has to be in favor of the calcium-current contribution ($a_1^k < a_1^c$). The last condition ensures that the calcium current has enough time to accumulate for the formation of an EAD ($t_{max}^{Ca} > \frac{1}{D}\ln(\frac{a_1^k - a_1^c}{V_{max}D})$).

Segment B. As in the analysis for segment A, we first determine the expression for $\frac{dv}{dt}$:

$$\frac{dv}{dt} = a_2^c - a_1^k + V_{max}De^{Dt} \tag{13}$$

Since $\frac{dv}{dt} < 0$ throughout this segment ($a_2^c < 0$, $-a_1^k < 0$ and $V_{max}De^{Dt} < 0$), no EAD is possible in segment B.

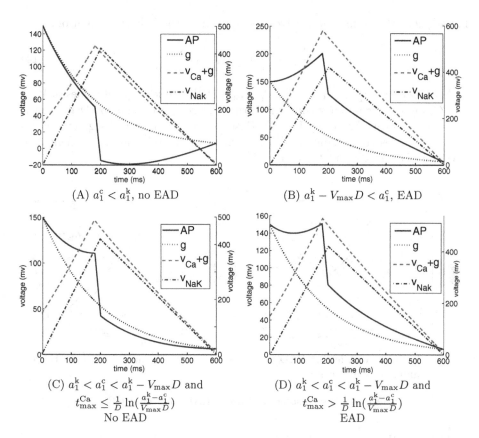

Fig. 6. PLAMIC-based analysis for EAD occurrence in segment A. The AP and decay functions are plotted as solid lines and use the y-axis on the left; v_{NaK} and $v_{\text{Ca}} + decay$ are plotted in dashed lines and use the y-axis on the right. The same conventions apply to Figs. 7 and 8.

Case II. The defining segment-A equation for v is exactly the same as in case I, modulo the replacement of $t_{\max}^{\text{Ca}}$ with $t_{\max}^{\text{NaK}}$ in the time bound for t. Following the case-I analysis for segment A, this observation yields the following condition for the occurrence of EADs:

$$\begin{cases} a_1^c > a_1^k > 0 \\ 0 < \frac{a_1^k - a_1^c}{V_{\max} D} < 1 \\ t < t_{\max}^{\text{NaK}} \end{cases}$$

Similarly, the major result for Case II can be summarized as follows.

Theorem 2. $(a_1^k < a_1^c < (a_1^k - V_{\max}D)) \wedge (t_{\max}^{\text{NaK}} > \frac{1}{D}\ln(\frac{a_1^k - a_1^c}{V_{\max}D}))$ *is a sufficient condition for a case-II occurrence of EAD during the suffix of segment A beginning at notch-delay.*

For segment B, the first derivative of v is given by the following equation.

$$\frac{dv}{dt} = a_1^c - a_2^k + V_{\max} D e^{Dt} \tag{14}$$

As $a_1^c > 0$ and $a_2^k < 0$, and $a_1^c - a_2^k + V_{\max} D e^{Dt} > 0$ for typical values of $V_{\max}$ and C, we observe an increasing AP during this segment. Thus, by Definition 1, segment B always has case-II EAD. Based on whether or not segment A has EAD, two cases are possible: EAD commences in (the tail end of) segment A or it commences in segment B; see Fig. 7.

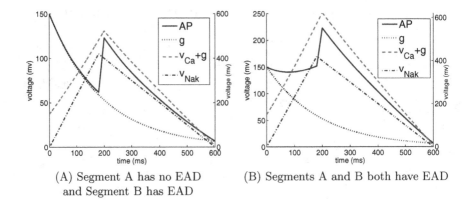

(A) Segment A has no EAD
and Segment B has EAD

(B) Segments A and B both have EAD

Fig. 7. The existence of Case-II EAD for segment B

Although this particular EAD morphology was not observed in the computer simulations we performed with the LRd model, this does not preclude its actual occurrence. Further examination of experimental data is needed to confirm or deny the physiological relevance of this case.

4   Experimental Validation of the PLAMIC Model

In this section, we consider the experimental validation of the PLAMIC model, specifically, the validity of Theorem 1 as an EAD predictor (classification rule) during the plateau phase of the AP cycle. To this end, we applied the protocols presented in [5] to the LRd cardiac-myocyte model to reproduce a number of AP curves with EADs. We also obtained the corresponding voltages for the calcium and the combined sodium and potassium currents using the integration method of Eqn. 3.

For each AP, in order to obtain the PLAMIC model parameters $(t_{\max}^u, v_{\max}^u)$, $u \in \{Ca, NaK\}$, we took the maximum value of $V_{Na} + V_K$ as $v_{\max}^{NaK}$, and the time at which it occurs as $t_{\max}^{NaK}$. Data points $(t_{\max}^{Ca}, v_{\max}^{Ca})$ were obtained in a similar fashion. The constant coefficients in our experiments are defined as $V_{\max} = 150$, offset$=127$ (defined below), and $D = -0.0052$. These values have been

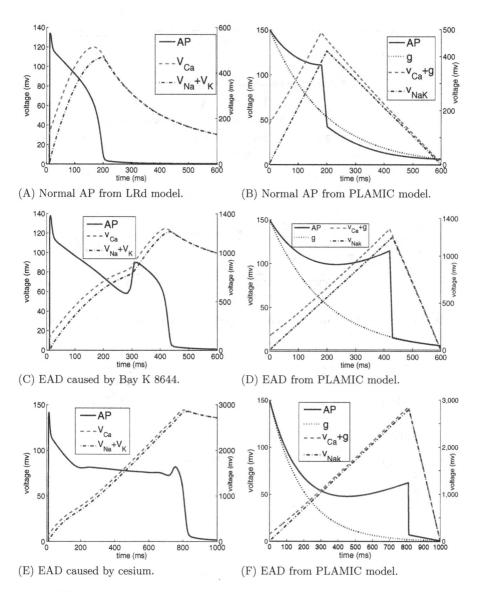

(A) Normal AP from LRd model.

(B) Normal AP from PLAMIC model.

(C) EAD caused by Bay K 8644.

(D) EAD from PLAMIC model.

(E) EAD caused by cesium.

(F) EAD from PLAMIC model.

Fig. 8. Comparison of AP curves from LRd and the PLAMIC model

chosen to match the LRd simulation results, but can be varied to fit different
AP morphologies and cell types.

A side-by-side comparison of the AP curves obtained from the LRd and
PLAMIC models for both normal and EAD-producing APs is illustrated in
Fig. 8. The top-left panel shows a normal AP and an EAD-exhibiting AP, triggered by a calcium-current-enhancing drug, Bay K 8644. The top-right panel
shows the PLAMIC model simulation for the two cases, which uses a piecewise-
linear approximation of the current-inducing voltages obtained from the LRd

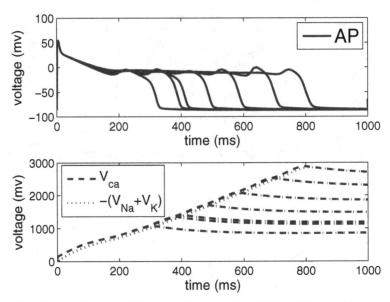

Fig. 9. Simulation of normal AP and APs including EAD with variable timing and severity

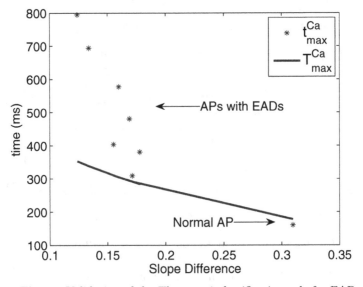

Fig. 10. Validation of the Theorem 1 classification rule for EADs

model. The bottom row shows similar results for the LRd and PLAMIC models for EADs induced by the administration of cesium, resulting in a substantial prolongation of the repolarization phase.

The AP curves generated by the PLAMIC model qualitatively match the LRd curves, with an AP morphology that is more stylized due to the simplicity of the linear functions on which the PLAMIC model is based. Nevertheless, the EAD phenomenon and variations of the repolarization phase are well captured by the much simpler PLAMIC model.

To validate Theorem 1, formulated for the PLAMIC model, we need only focus on case I since the condition $t_{max}^{Ca} < t_{max}^{NaK}$ is always true in the LRd model. We also need to reformulate (the last condition of) Theorem 1 for the following reason. In the LRd model, v_{max}^{Ca}, the maximum value of V_{ca} during one AP cycle, serves as the sole contributor to the positive portion of the voltage. In the PLAMIC model, however, the positive part is composed of the linear function $v_{Ca}(t)$ and the decay $g(t)$. Thus, when calculating the slope a_1^c in the LRd model, it is not accurate to use v_{max}^{Ca} directly. Rather, a "decay" factor given by $V_{max}e^{Dt_{max}^{Ca}}$ should be subtracted from v_{max}^{Ca}.

The reformulation of Theorem 1 is given in Eqn. 15, where $\tilde{a}_1^c$ is the corrected slope and *offset* is a constant used to ensure a non-negative AP value, as in the PLAMIC model.

$$
t_{max}^{Ca} > \frac{1}{D} \ln(\frac{a_1^k - \tilde{a}_1^c}{V_{max}D})
$$
$$
\text{where } \tilde{a}_1^c = \frac{v_{max}^{Ca} + offset - V_{max}e^{Dt_{max}^{Ca}}}{t_{max}^{Ca}} \tag{15}
$$
$$
a_1^k = \frac{v_{max}^{NaK}}{t_{max}^{NaK}}
$$

In order to test the validity of the derived condition for EAD occurrence given by Eqn. 15, we have assembled a test suite of LRd simulation data consisting of one normal AP and seven APs with variable EADs. The simulation results for both normal AP and abnormal APs are presented in Fig.9. The top panel shows the AP curves and the bottom panel shows $-(V_{Na} + V_K)$ and V_{Ca} as defined by Eqn. 3.

Let $T_{max}^{Ca} \equiv \frac{1}{D} \ln(\frac{a_1^k - \tilde{a}_1^c}{V_{max}D})$ be the *threshold value* for the LRd model. That is, according to Def. 1 and Thm. 1, an LRd AP should be EAD-producing if and only if $t_{max}^{Ca} > T_{max}^{Ca}$.[1] Note that since C and V_{max} are fixed for the LRd model, T_{max}^{Ca} is a function of $a_1^c - \tilde{a}_1^k$, the slope difference.

For each AP, we calculate $a_1^c - \tilde{a}_1^k$ using the data points (t_{max}^u, v_{max}^u), $u \in \{Ca, NaK\}$, obtained via numerical simulation from the LRd model, and calculate the threshold time T_{max}^{Ca} derived from our formal analysis. This allows us to then compare the t_{max}^{Ca} values with the T_{max}^{Ca} values. The results of these comparisons are given in Fig. 10, where we plot t_{max}^{Ca} and T_{max}^{Ca} as a function of the slope difference $a_1^c - \tilde{a}_1^k$.

As can be seen in Fig. 10, for all APs with EAD, we have that $t_{max}^{Ca} > T_{max}^{Ca}$. Conversely, for all APs without EAD (only one such AP in our data set), $t_{max}^{Ca} < T_{max}^{Ca}$. Physiologically, these results suggest that the cells generating EADs spend an amount of time greater than the threshold in letting calcium accumulate and thereby dominate the effects of repolarizing potassium in order to produce such abnormal secondary depolarization. Regardless of the underlying physiology, the

---

[1] The other conditions required by Theorem 1 for EAD occurrence, $a_1^k < a_1^c < (a_1^k - V_{max}D)$, are needed to ensure the existence of a positive real solution for t and are not considered here.

results of Fig. 10 demonstrate that Theorem 1 can be used as a valid classifier for EAD prediction, as suggested by the formal analysis.

5   Conclusions

In this paper, we presented the PLAMIC model, a new, simplified model of the action potential in excitable cells. Despite its simplicity and piecewise-linear nature, the PLAMIC model preserves ties to main ionic species and the time course of their contributions to the AP. This allowed us to analyze biological phenomena of clinical importance: early afterdepolarizations (EADs). Unlike the original, highly nonlinear system of equations typically used to model an AP, the PLAMIC model proved amenable to formal analysis.

Specifically, with the PLAMIC model, we were able to explore the parameter space, without having to rely on exhaustive simulations, and to derive basic rules for the conditions under which EADs may occur. Overall, such conditions relate to the subtle balance of different ionic currents during the plateau phase of the repolarization process. While this result is somewhat intuitive and not surprising, to the best of our knowledge, our study is the first to formalize it and to provide quantitative rules for prediction of normal and EAD-containing APs based on the abstracted representation of the contributing ionic currents. We successfully validated the classification rules obtained by formal analysis with the PLAMIC model by computer simulations with widely accepted, detailed nonlinear AP models.

The utility of the PLAMIC model is rooted in its direct links to experimentally measurable parameters, and the relatively easy derivation of the EAD classification rules for a wide range of AP shapes and different cell types and species. Such a prediction tool can be very useful in designing new anti-arrythmic therapies and in confirming the safety of any genetic or pharmacological manipulations of excitable cells that may lead to alterations in the balance of ionic currents.

There are several limitations of the PLAMIC model. First, due to its simplicity, the AP curves are only qualitatively reproduced. Second, as the PLAMIC model studies the *overall contribution* of an ionic current to changes in the AP; details about the components of a current (steady-state behavior, kinetics parameters), which may be important, are lacking. For example, calcium handling constitutes an important aspect of cardiac-cell function, especially with regard to electromechanical coupling. Our model only indirectly reflects the effects of intracellular calcium on the action potential (AP). In particular, with $V_{Ca}(t)$, we have modeled the integral contribution of calcium fluxes to the AP. In the Luo-Rudy model, for example, this term would correspond to the sum of the L-type Ca^{2+} channel (which has a Ca^{2+} sensitive gate), the Na/Ca exchanger and the background Ca^{2+} current. By qualitatively capturing the behavior of the Luo-Rudy model, especially with respect to monotonicity in the post-upstroke AP, we indirectly take into account changes in intracellular Ca^{2+}, although the PLAMIC model lacks parameters directly associated with these changes. While developing the model, our goal was to maintain simplicity so that monotonicity analysis could be performed on its parameter space for EAD-predictive purposes; as such, the PLAMIC model focuses on transmembrane fluxes only.

Future work includes validation of the PLAMIC model using actual experimental data with relevant statistical measures. Furthermore, we will explore the derivation of a more accurate excitable-cell model for EAD prediction, yet one that retains the possibility of formal analysis. Our work in using hybrid automata to model excitable cells [11] is one possible formal framework for this research direction.

References

1. Cranefield, P.F., Aronson, R.S.: Cardiac arrhythmias: the role of triggered activity and other mechanisms. Futura Publishing Company (1988)
2. Fozzard, H.: Afterdepolarizations and triggered activity. Basic Res. Cardiol. 87(Suppl. 2), 105–113 (1992)
3. Hiraoka, M., Sunami, A., Zheng, F., Sawanobori, T.: Multiple ionic mechanisms of early afterdepolarizations in isolated ventricular myocytes from guinea-pig hearts. QT Prolongation and Ventricular Arrhythmias, pp. 33–34 (1992)
4. January, C., Moscucci, A.: Cellular mechanism of early afterdepolarizations. QT Prolongation and Ventricular Arrhythmias, pp. 23–32 (1992)
5. Zeng, J., Rudy, Y.: Early afterdepolarizations in cardiac myocytes: mechanism and rate dependence. Biophysical J. 68, 949–964 (1995)
6. Homma, N., Amran, M., Nagasawa, Y., Hashimoto, K.: Topics on the Na+/Ca2+ exchanger: involvement of Na+/Ca2+ exchange system in cardiac triggered activity. J. Pharmacol. Sci. 102, 17–21 (2006)
7. Clusin, W.: Calcium and cardiac arrhythmias: DADs, EADs, and alternans. Crit. Rev. Clin. Lab. Sci. 40, 337–375 (2003)
8. Charpentier, F., Drouin, E., Gauthier, C., Marec, H.L.: Early after/depolarizations and triggered activity: mechanisms and autonomic regulation. Fundam. Clin. Pharmacol. 7, 39–49 (1993)
9. Luo, C.H., Rudy, Y.: A dymanic model of the cardiac ventricular action potential: I. simulations of ionic currents and concentration changes. Circ. Res. 74, 1071–1096 (1994)
10. Hodgkin, A.L., Huxley, A.F.: A quantitative description of membrane currents and its application to conduction and excitation in nerve. J. Physiol. 117, 500–544 (1952)
11. Ye, P., Entcheva, E., Smolka, S., Grosu, R.: A cycle-linear hybrid-automata model for excitable cells. IET Systems Biology 2, 24–32 (2008)

The Distribution of Mutational Effects on Fitness in a Simple Circadian Clock

Laurence Loewe[1] and Jane Hillston[1,2]

[1] Centre for System Biology at Edinburgh,
The University of Edinburgh, Edinburgh EH9 3JU - Scotland
Laurence.Loewe@ed.ac.uk
[2] Laboratory for Foundations of Computer Science,
The University of Edinburgh, Edinburgh EH8 9AB, Scotland
jeh@inf.ed.ac.uk

Abstract. The distribution of mutational effects on fitness (DME^F) is of fundamental importance for many questions in biology. Previously, wet-lab experiments and population genetic methods have been used to infer the sizes of effects of mutations. Both approaches have important limitations. Here we propose a new framework for estimating the DME^F by constructing fitness correlates in molecular systems biology models. This new framework can complement the other approaches in estimating small effects on fitness. We present a notation for the various DMEs that can be present in a molecular systems biology model. Then we apply this new framework to a simple circadian clock model and estimate various DMEs in that system. Circadian clocks are responsible for the daily rhythms of activity in a wide range of organisms. Mutations in the corresponding genes can have large effects on fitness by changing survival or fecundity. We define potential fitness correlates, describe methods for automatically measuring them from simulations and implement a simple clock using the Gillespie stochastic simulation algorithm within StochKit. We determine what fraction of examined mutations with small effects on the rates of the reactions involved in this system are advantageous or deleterious for emerging features of the system like a fitness correlate, cycle length and cycle amplitude. We find that the DME can depend on the wild type reference used in its construction. Analyzing many models with our new approach will open up a third source of information about the distribution of mutational effects, one of the fundamental quantities that shape life.

1 Introduction

Evolutionary theory has been very successful in predicting the fate of mutations in various settings, assuming that the mutational effect on fitness is known. Determining the actual effects of mutations is difficult. While many biological wet-lab experiments have been conducted with the aim of determining the effects of new mutations, these have been particularly successful for mutations with large effects, as experimental noise obscures small effects [1].

This is unfortunate, as the evolutionary fate of mutations with big effects on fitness is rather simple to understand: many advantageous ones become fixed in the population,

M. Heiner and A.M. Uhrmacher (Eds.): CMSB 2008, LNBI 5307, pp. 156–175, 2008.

so that in some future generation all individuals will have inherited a copy, while deleterious (harmful) ones are removed rather quickly. The most interesting questions are currently posed by mutations of "small", but not "too small" effects on fitness. Here the difference between "small" and "too small" depends on a threshold set by the effective population size, where mutations in the "too small" category are behaving as if they had no effect on fitness and thus exhibit simple neutral dynamics. Since all organisms have a genome with a large number of opportunities to mutate in different ways, it has become custom to summarise these possibilities in the form of a distribution of mutational effects on fitness (DME^F), which associates a frequency of the occurrence of mutational changes with each mutational effect and abstracts the various molecular causes that determine the size of that effect. Various evolutionary theories make varying assumptions about this distribution [1] and their quality as a predictive tool often depends on the underlying DME^F. These theories are important for understanding the evolution of genomic sequences and thus play a crucial role in efforts to interpret the sequence of the human genome [1]. Because of the paramount importance of DME^Fs, recent work in population genetics has started to estimate DME^Fs directly from sequence data [1,2]. Such methods are not limited by the lack of sensitivity seen in wet-lab experiments, as they exploit the sensitivity of the evolutionary process on DME^Fs to infer the location and shape of a given type of DME^F in systems that evolve according to well understood forces. The drawbacks include:

Limited mechanistic details. Current population genetic estimates of DME^Fs from DNA sequence data are descriptions of observations that lack a rigorous underpinning in the form of a mechanistic model of mutational effects. There is little information for distinguishing various types of distributions (e.g. gamma, lognormal), once certain broad criteria are met.

Limited applicability. Each distribution is only a snapshot of a specific DME^F for a specific organism. While comparing such snapshots helps to ascertain common features of DME^Fs, such descriptive results do not help with further explorations.

Sensitive to evolutionary process. All methods that estimate DME^Fs from DNA sequence data require a set of assumptions about the evolutionary process that led to the sample of DNA sequences used for the inference. These assumptions can be difficult to test and may cast doubts on estimates of a DME^F. Since several evolutionary processes can lead to similar features in a set of sequences, it can be challenging to disentangle their effects from those of the underlying DME^F [3].

In this paper we propose a third approach to the study of distributions of mutational effects on fitness, besides the direct experiments and the population genetical methods mentioned above. Our main contribution is to describe the approach and to demonstrate how it works in principle in a simple circadian clock model. We suggest that molecular systems biological models can be used to obtain much of the evolutionary interesting properties of a DME^F for a particular limited model system. Combining the *in silico* experimental techniques of molecular systems biology with knowledge of the study system from the wet-lab experiments allows the following improvements:

More mechanistic details. Molecular systems biology allows the construction of rigorous mechanistic models of biological systems that maintain a close link to biological reality. This reduces errors in estimates that are caused by biologically misleading abstractions. In addition, such computational models allow the further exploration of parameter space at the low cost of simulations as opposed to the often prohibitive costs or difficulties of performing equivalent wet-lab experiments.

Precision. The precise control over every aspect of a model that comes with *in silico* models makes it possible in principle to compute emerging properties with a very high degree of precision. Depending on the stochastic nature and computational complexity of the model, it may still be too costly for some models to achieve the level of precision that some evolutionary questions require. However, we anticipate that many useful models can be analysed without such problems. In addition, advances that reduce the cost of computing and improve the speed of algorithms can be translated into increasingly precise estimates of DME^Fs.

To demonstrate the feasibility of our new approach we deliberately choose a simple model to make it easier to focus on the fundamental challenges that arise from this new perspective. Such challenges include (i) the construction of computable fitness correlates that can be used as surrogates for biological fitness in the wild, (ii) the accuracy with which such fitness correlates need to be (and can be) computed and (iii) fundamental biological questions about distributions of mutational effects. For example, how often will a change of an underlying reaction rate improve or degrade overall functionality? Will the relative size of effects on the emerging properties of the system be larger or smaller than the relative size of mutational effects on reaction rates?

Results demonstrate that our new approach can be used in principle to infer interesting properties of distributions of mutational effects, where details strongly depend on the model under focus. The rest of the paper is structured as follows. In Section 2 we present some background on key ideas from evolutionary biology which we will use in the remainder of the paper. Section 3 outlines our framework for taking a systems biology approach to the study of the distribution of mutational effects. The model we consider in this paper is presented in Section 4 whilst its analysis is described in Section 5. A discussion of the results is given in Section 6 and conclusions in Section 7.

2 Background

Instead of obtaining DME^Fs directly, our basic strategy is to (i) build a mechanistic model of how the phenotype changes depending on lower level changes in reaction rates that are ultimately caused by DNA changes, (ii) define a function that computes fitness from the phenotype and (iii) use random perturbations together with (i) and (ii) to determine the DME^F. While some commonly used models in quantitative genetics also compute a phenotype as an intermediate step towards computing the distribution of mutational effects [4], the approach presented here can include much more mechanistic detail by building on molecular systems biological data. DME^Fs can be used to quantify *robustness*. Understanding robustness [5] is important for drug design [6].

2.1 A Nomenclature of Distributions of Mutational Effects (DMEs)

In this subsection we explain precisely what we mean by *distributions of mutational effects*. This is necessary to avoid confusion when discussing the various distributions. We consider each of the terms in turn (for examples, see Figures 8+9):

Effects. The *effects* are the changes in the emerging high-level systemic property under focus in the investigated system. DME^Y is used to denote a DME of the emergent system property Y. All Y are high-level properties, so a superscript is used.

We denote DMEs that describe the effects on fitness in the wild by DME^F, where the *fitness* can be easily linked to a trait like survival rate or fecundity that can be observed in its natural environment. In the more limited example of our circadian clock DME^L describes variations in the length of a cycle and DME^A variations in the amplitude of the oscillations. *Effects* are changes in *phenotype* properties.

Mutations. The *mutations* are low-level genotype changes that perturb the wild type reference system and cause phenotypic effects to change. At the lowest level *mutations* are DNA changes. In the absence of a mechanistic model for predicting enzymatic reaction rates from DNA, *mutations* can also be introduced as reaction rate changes, as the mechanistic chain of causality that links DNA changes and fitness changes must pass through the corresponding reaction rates at some point. DM_XE is used to denote the genotypic perturbations that are introduced into property X to measure a DME. All X are low-level properties, so a subscript is used.

If *mutations* are a representative sample of naturally occurring DNA changes, we omit X, as this is the most natural and most important DME. In the more limited example of our circadian clock we can only change the reaction rates listed in Table 2. For example $DM_{v_d}E^L$ denotes the *distribution of mutational changes in protein degradation rate v_d that have effects on the length of clock cycles L*.

Distribution sign. One may want to focus only on increases or decreases of the values of a DME. For example, advantageous mutations in the DNA that increase fitness could be analyzed separately from survival compromising mutations that decrease fitness. Here we denote an increase and decrease with the additional letter 'I' and 'D', respectively. If these occur in a high-level emerging property of the system, the letters are superscript, if in low-level mutational changes, the letters are subscript. Specifying nothing is equivalent to 'DI'.

Thus a *distribution of increasing mutational changes in protein production rate k_s that have only decreasing effects on fitness* is denoted by $D_I^D M_{k_s} E^F$.

If we wish to be very general, we simply specify DME. If we want to be more specific, we include the additional information according to the notation introduced above. Since all DMEs describe how the emergent properties of complex systems change in response to changes in lower level components, some generalities may emerge from their study.

2.2 Fitness and Selection Coefficients

Fitness is the highest level function of any biological system. As such it is difficult to define rigorously [7]. Fitness correlates have been used successfully in the study of life-history evolution [8]. We propose that it is possible to define meaningful fitness correlates that are computable from molecular systems biological models. For simplicity, we will assume that W, the absolute fitness in the wild, can be estimated by observing a high level organismic fitness correlate in wet-lab experiments and that this is proportional to the fitness correlate F that we compute *in silico*. Thus we can define:

$$F_M = F_{WT}(1 + s) = \frac{W_M}{W_{WT}},$$

where the subscripts M and WT denote the mutant and the wild type. Here the wild type is considered to be relatively 'mutation free' and s is the selection coefficient, commonly used in population genetics to denote the effects of a mutation on fitness. Using this approach we can compute DMEs for F and any emerging property of our model if we specify an underlying distribution of how reaction rates are affected by DNA changes. Expressing our results as s allows direct comparison with population genetics results.

2.3 Circadian Clocks

Circadian clocks are the internal molecular clocks that govern large parts of the molecular machinery of life. They frequently have a huge impact on the behaviour of organisms. They are responsible for waking us up in the morning and they make us feel tired in the evening. Such clocks are of paramount importance for the vast majority of organisms from Cyanobacteria [9] through fruitflies to humans [10]. Much recent work has focused on elucidating the various molecular components that perform the chemical reactions that oscillate with a daily rhythm. This has resulted in a series of models with increasing numbers of interlocking feedback loops [11]. In this paper we focus on one of the simplest models for a circadian clock that exists. This decision is motivated by a desire to focus the reader's attention on the basic principles of our new approach and on fundamental aspects of observing DMEs in clocks. We also wanted to apply our new approach to simple systems first to collect experience before analyzing complex ones.

3   Evolutionary Systems Biology

Evolutionary genetics and molecular biology have both been very successful in furthering our understanding of the natural world. However, after decades of research some familiar simplifying assumptions are now reaching their limits and evolutionary biologists are getting increasingly interested in the molecular details of their systems. At the same time molecular biologists progressively recognize the merit of quantitative modelling. Growing genomics and systems biology datasets provide a strong motivation for exploring realistic models at the interface (e.g. [12]). Increasingly detailed models of intracellular processes could help understand evolution by deriving DME^Fs *ab initio* by computing fitness correlates. Below we define one possible fitness correlate for circadian clocks. The following procedure can estimate a DME in a particular system:

1. Define a wildtype for use as a fixed 'mutation-free' reference point.
2. Treat each protein like a system in order to link DNA changes to changes in protein function by assuming a DME^{rate}, which denotes a realistic distribution of mutational effects on reaction *rates* for DNA changes within protein-coding and regulatory sequences. If necessary, scale the frequencies of mutations for a given *rate* change by an estimate of the number of base pairs in the DNA that influence this *rate* to reflect the varying mutational target sizes in the system under investigation.
3. Compute enough samples [13] to obtain a $DM_{rate}E^F$, which denotes a distribution of the changes in the fitness correlate that emerges from the lower level distribution(s) of the underlying reaction rate changes.

4. Plot the differences to the mutation free reference as a $DM_{rate}E^F$, on one logscale for decreasing and on another logscale for increasing effects to visualise changes in the frequency of small effects from potential random noise expectations. Compare the number of fitness increases and decreases with the increases and decreases of the underlying reaction rate distributions in order to establish whether the molecular structure within biomolecules, or the network structure of biochemical reaction systems, has a larger influence on the DME of the fitness correlate.

The particular importance of small mutational effects in long-term evolution emphasizes the need for a careful analysis of numerical issues while computing fitness correlates. In plotting DME^Fs, logscales were found to be more helpful than linear scales due to their ability to visualize very small differences.

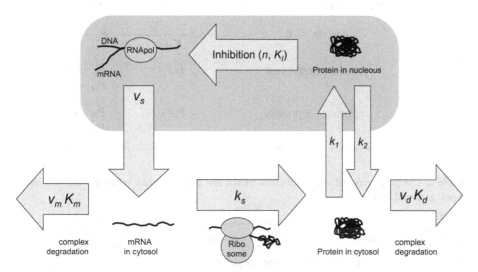

Fig. 1. Overview over the basic negative transcriptional feedback system that implements the simple clock analysed here

4 Model

The very simple model of a circadian clock that we use has been described elsewhere [14] and is closely related to the elementary transcriptional feedback oscillator described by Goodwin [15].

The basic reaction scheme is found in Figure 1. Briefly, the RNA polymerase complex transcribes a gene into mRNA, which is exported into the cytosol, where it accumulates at constant rate v_s. The ribosome translates the mRNA into a protein which accumulates at rate k_s. The protein can migrate between the nucleus and the cytosol, where the rates of transport are k_1 and k_2. Transportation into the nucleus is assumed here to be equivalent to turning the protein into a repressor, so potentially more than one reaction might be subsumed here. If enough copies of the protein have accumulated in the nucleus, they can cooperatively bind to the DNA and thus prohibit the

Table 1. Stochastic simulation implementation of the simple clock model. The kinetics of the chemical reactions shown here is governed by the propensity functions that determine which reaction out of this list will occur next. Once it has occurred, the species counts are adjusted according to the transition entry.

Number	Reaction	Propensity function	Transition
1	gene $\rightarrow$ gene + mRNA	$(v_s\Omega)\dfrac{(K_I\Omega)^n}{(K_I\Omega)^n + P_N^n}$	$M \rightarrow M + 1$
2	mRNA $\rightarrow \emptyset$	$(v_m\Omega)\dfrac{M}{(K_m\Omega) + M}$	$M \rightarrow M - 1$
3	mRNA $\rightarrow$ mRNA + protein	$k_s M$	$P_C \rightarrow P_C + 1$
4	protein $\rightarrow \emptyset$	$(v_d\Omega)\dfrac{P_C}{(K_d\Omega) + P_C}$	$P_C \rightarrow P_C - 1$
5	protein $\rightarrow$ repressor	$k_1 P_C$	$P_C \rightarrow P_C-1$ $P_N \rightarrow P_N+1$
6	repressor $\rightarrow$ protein	$k_2 P_N$	$P_N \rightarrow P_N-1$ $P_C \rightarrow P_C+1$

Table 2. The parameters of our basic clock model and their assumed values for the two 'wild types' explored here

Parameter	Meaning	*Neurospora*	24h-clock
Ω	Size of system	10^5	10^5
n	Degree of Hill-type cooperativity	4	4
K_I	Threshold for Hill-type repression	1	1
v_s	Effective rate of mRNA accumulation in cytosol	1.6	1.6
v_m	Maximal effective turnover of mRNA degradation	0.505	0.505
K_m	Michaelis-Menten constant for mRNA degradation	0.5	0.5
k_s	Effective rate of protein production in cytosol	0.5	0.5
v_d	Maximal effective turnover of protein degradation	1.4	1.4
K_d	Michaelis-Menten constant for mRNA degradation	0.13	0.13
k_1	Effective rate of repressor accumulation in nucleus	0.5	0.4623
k_2	Effective rate of repressor movement out of nucleus	0.6	1.2

binding of the RNA polymerase complex, effectively shutting down the production of mRNA. This cooperative binding is described by kinetics of the Hill type with a given degree of cooperativity, n, and a threshold constant for repression, K_I. To allow transcription to start again, mRNA and the protein are constantly degraded by reactions of the Michaelis-Menten type. Here v_m denotes the maximal effective turnover rate of the mRNA degradation complex (with Michaelis-Menten constant K_m). The corresponding reaction for the protein is described by v_d and K_d.

This model can be described by the following ordinary differential equations (ODEs), where M, P_C and P_N denote the concentrations of mRNA, cyctosolic protein and nuclear repressor, respectively. The change in the concentration of mRNA is given by

$$\frac{dM}{dt} = v_s \frac{K_I^n}{K_I^n + P_N^n} - v_m \frac{M}{K_m + M}, \tag{1}$$

the change in concentration of the cytosolic protein is given by

$$\frac{dP_c}{dt} = k_s M - v_d \frac{P_C}{K_d + P_C} - k_1 P_C + k_2 P_N \tag{2}$$

and the change in the concentration of the repressor form of the protein in the nucleus is given by

$$\frac{dP_N}{dt} = k_1 P_C - k_2 P_N. \tag{3}$$

To translate these ODEs into chemical reaction equations, we followed the scheme described in [16]. To this end all molecular concentrations in the ODEs are turned into actual molecule counts by multiplying them by Ω, the parameter that describes the scale of the system. Table 1 gives the important quantities that were used to compute the propensity functions and the stoichiometry matrix in the stochastic simulations of the system.

Such a model will have a degree of approximation due to the presence of the reaction with Hill kinetics, since it has been shown that a direct application of Gillespie's algorithm to implement Hill's kinetic law can lead to an overestimate of the variance when compared to a more faithful low-level representation of the actual elementary reactions [17]. Mass action and Michaelis-Menten reactions do not suffer from this problem [18,19].

We used two sets of reaction rates as the starting points for our simulations: (i) the original set of parameters that Leloup *et al.* [14] used to describe a simple model of the 22h cycle circadian clock in *Neurospora crassa* and (ii) a modification of their parameter combination which we introduced to approximate a 24h cycle with the same set of reactions. Table 2 summarises the corresponding parameters.

5   Model Analysis

5.1   Simulations

We employed stochastic simulations to measure the emerging features of our model. To allow for flexibility in the analysis and speed of computation, we employed StochKit 1.0 (http://www.engineering.ucsb.edu/ cse/StochKit/), which implements a variety of algorithms that speed up Gillespie's Direct Method algorithm for stochastic simulation under particular sets of circumstances. For example, when large numbers of molecules are in the system, the library can choose to use a tau-leaping algorithm. It then no longer simulates every single reaction but rather estimates the number of reactions that will

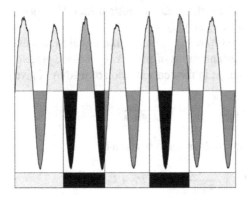

Fig. 2. The four states of a system with external and internal oscillations. It is possible to unambiguously assign one of the four states given in the table on the right to every point on a time course. Assignment to one of the four states is indicated by different shades in the time course on the right. We estimate the threshold from the observations as half the distance between the minima and maxima.

happen in a particular period of time. Our implementation of StochKit automatically switched between adaptive tau leaping [20] and fully detailed stochastic simulations. An overview of the corresponding methods has been presented elsewhere [20,21] and is also included in the StochKit manual.

5.2   Measuring Fitness Correlates

In order to explore the construction and behaviour of fitness correlates, we defined a simple biologically credible measure that we expect to be linked to fitness in many realistic situations. A schematic overview of the core principle is given in Figure 2.

Basically the existence of an external cycle (day or night) and an internal cycle (molecule count high or low) allows the definition of four states that describe all states that such a system can be in. Either it is 'in phase' (which can mean day or night) or it is 'anti-cyclic' (molecule count high, while external light is low and vice versa). If the internal system oscillates with a 24-hour period, then selection will favour mutations that help the cell to organise its patterns of gene activity around that cycle. However, if the internal oscillations are faster or slower, then the benefit of mutations that link particular genes with a particular state of the clock will be very limited, if existent at all: the genes that are in phase today will be out of phase in a few weeks and thus the long-term expectation of such a clock is probably not very different from random noise. From such considerations we can derive two measures of fitness, the cyclical fitness correlate F_C, defined as:

$$F_C = \frac{T_{1D} + T_{0N}}{T_{tot}}, \tag{4}$$

and the anti-cyclical fitness correlate F_A defined as;

$$F_A = \frac{T_{0D} + T_{1N}}{T_{tot}}, \tag{5}$$

where T_{1D}, T_{0N}, T_{0D} and T_{1N} sum over all time when the system is "On" during "Day", "Off" during "Night", "Off" during "Day" and "On" during "Night", respectively. All these quantities scale with the total time that the system has been under observation, T_{tot}. Based on such a definition, these fitness correlates can never be larger than 1. While any of the two measures can become zero under some special circumstances, we argue that more often the minimal value is 0.5, based on the random expectation of the complete absence of an internal cycle. For our 24h-clock we found F_A to be high and F_C to be low. Therefore we report only F_A below.

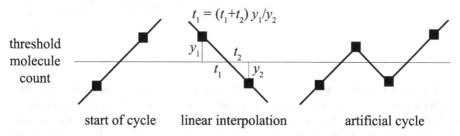

start of cycle   linear interpolation   artificial cycle

Fig. 3. Transitions of the clock. We use a threshold to distinguish the 'on' and 'off' states of the clock and keep track of the interpolated transition times to measure cycle length. In the presence of high levels of stochastic noise, artificial cycles can be generated. This was no problem above $\Omega > 10000$ in our system. Squares denote observations.

5.3 Measuring Cycle Length and Amplitude

To obtain a robust understanding of DMEs in a particular system it is preferable to investigate several of the emerging higher level properties of the system under investigation. In our case we also wanted to explore properties that could be determined more precisely than our present implementation of direct fitness correlates.

We decided to automatically observe the cycle length L and amplitude A, where the amplitude is the difference in molecule counts between the highest and the lowest point of a cycle. To define the beginning of a new cycle requires a threshold between the number of molecules at which the clock is considered to be 'off' and the same count in the 'on' state. We implemented this by using the same threshold required for our fitness measurements. Thus we stored the past state and determined for each current state, whether the threshold had been passed in upwards or downwards direction (Figure 3). If it had, the transition time was interpolated and cycle length was recorded. If it had not, it was checked whether the current value was a new extremum, facilitating the observation of cycle amplitudes. To get a good estimate of the true transition time we computed the intersection of the threshold with the line joining the two closest observations using the law of proportionality (see Figure 3, linear interpolation).

This system worked very well for large values of Ω. However, analyses at smaller values of Ω showed a sudden increase in the corresponding standard error estimates. Further scrutiny revealed that this was due to rare cases, where stochastic fluctuations had temporarily crossed the threshold, bucking the trend for just a moment and thereby triggering what the code considered a new, very short, cycle (Figure 3, right). This

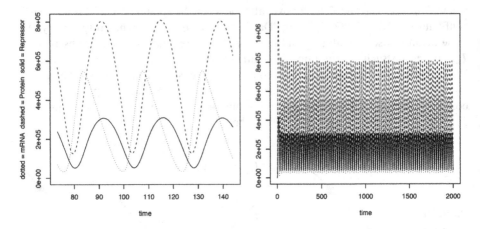

Fig. 4. Oscillations of our 24h-clock at $\Omega = 100000$. As in all our simulations the initial concentration for all the reactants was 1 molecule. To avoid any influence from initial concentrations, we allowed the clock to run for 50 hours before starting another 50 hour period, where merely maxima and minima were recorded to automatically estimate the threshold for fitness, cycle length and amplitude at half way between the two. Before actual observations started after calibration, two more cycles would be discarded, so that most observations would span the time from about 150 – 2000 hours.

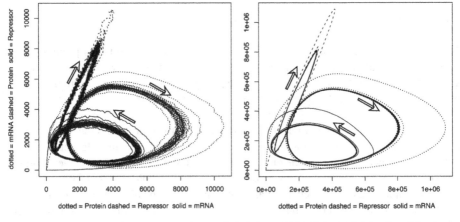

Fig. 5. Limit cycles for the 24h-clock at $\Omega = 1000$ (left) and 100 000 (right). Smaller Ω increase noise even more, but the general presence of oscillations is remarkably robust at this parameter combination, even if $\Omega = 10$.

phenomenon became prevalent at about $\Omega < 6000$ in the parameter combinations that we tested. To remedy this problem, two thresholds will have to be set up in such a way that a cycle is only recognised as such, if it has crossed both thresholds.

5.4   Basic Clock Behaviour

Our clock models do not behave differently from those analysed in the literature. Figure 4 demonstrates the extraordinary regularity and long-term stability of the oscillations at $\Omega = 100000$. If Ω is reduced, noise is increased, as can be seen in the limit cycles of Figure 5. To obtain solid estimates of the stochastic variability of the Neurospora and our 24h-clock, we observed 6380 and 6927 single simulations for 2000 hours (less the calibration period). The resulting distribution of anti-cyclic fitness, cycle length and amplitude can be found in Figures 6 and 7 (see the next section for an explanation of the DME plots in these figures).

5.5   Bootstraps and DME Estimates

Bootstraps. To obtain robust estimates of a DME^F is a statistical challenge. If an underlying $D^D ME^{k_1}$ or $D^I ME^{k_1}$ is assumed to map DNA sequence changes to repressor production rate changes, then one would like to know the effects on the distributions of emerging properties given by $D_{DI}M_{k_1}E^{F_A}$, $D_{DI}M_{k_1}E^L$ and $D_{DI}M_{k_1}E^A$.

Here we propose to use a slight modification of the statistical bootstrap technique to achieve this. Bootstrapping in statistics was introduced to estimate the unknown distribution U of variates that are computed by a known function f from a known distribution D [13]. This is achieved by repeatedly sampling (with replacement) from the known distribution (or dataset). Then the function f is applied to each sample $\mathbf{x}$ to obtain samples from the unknown distribution:

$$U \sim f(\mathbf{x}), \text{ where } x \sim D \tag{6}$$

Thus U can be quantified rigorously if enough samples can be generated. Here we use as D the underlying $D_D M_{k_1}E$ or $D_I M_{k_1}E$ and as U any of the emerging properties (F_A, L, A) distributions specified above. f is specified by our simulation system that implements and observes the circadian clock model. To quantify U, we plot it in the DME plots shown in Figures 8-9. This approach allows us to detect changes in the distribution of emerging features that are caused by differences in the underlying low-level $D_D M_{k_1}E$ or $D_I M_{k_1}E$.

Design of the DME plot. DMEs are notorious for being difficult to visualize due to conflicting requirements. A Biologist would typically want to get an overview of deleterious, neutral and advantageous mutations at the same time, which is simple on a linear scale. However recent results have shown that the DME^F is highly leptokurtic [1,2], implying that most mutations have very small effects and would thus be lost in something that looks like a bar around zero on a linear scale. Thus a logscale seems the most appropriate way to visually convey information about most DMEs. We decided to follow a pragmatic approach that combines the best of both worlds by neglecting parameter ranges that are biologically uninteresting and implemented a corresponding plotting function in R (http://www.r-project.org/). The code first constructs a histogram of bins for decreasing effects that are equally spaced on a logscale. Then it does the same for increasing effects. The focus of the plot is on values within user-defined upper and lower limits of interest, merely checking for the existence of other values. Then

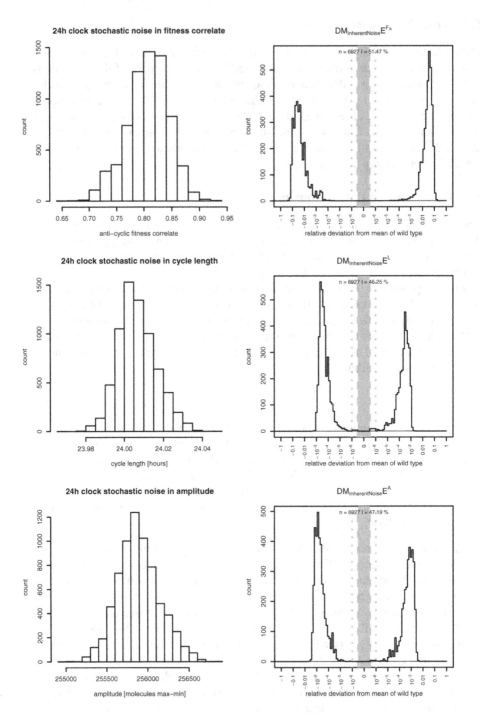

Fig. 6. The stochastic variability of the emerging features of the 24h-clock parameter combination. See Section 5.5 for an explanation of the right part of the figure.

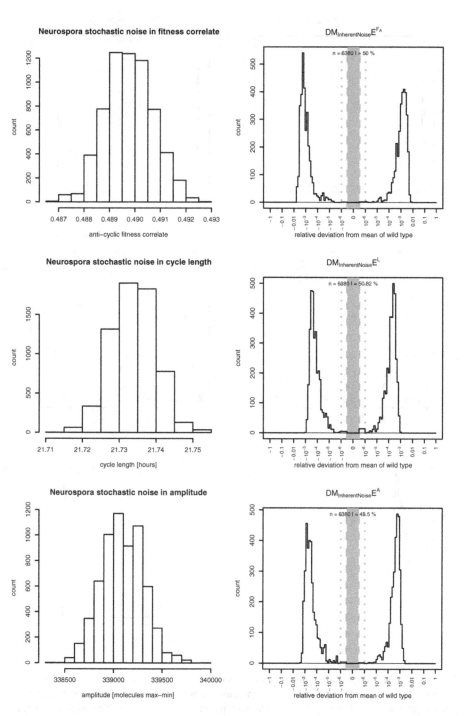

Fig. 7. The stochastic variability of the emerging features of the Neurospora clock parameter combination. See Section 5.5 for an explanation of the right part of the figure.

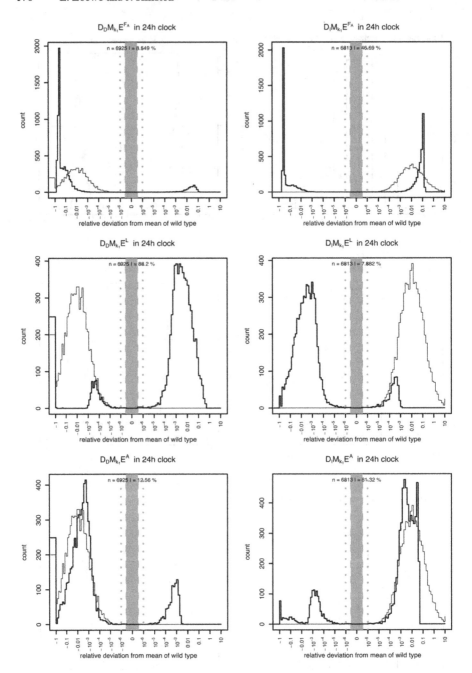

Fig. 8. These DMEs show the effects of assuming a low level lognormal $D_D M_{k_1} E$ and $D_I M_{k_1} E$ as distribution of generated genotypes on the emerging phenotypic features anti-cyclic fitness F_A, cycle length L and amplitude A for the 24h-clock parameter combination. The thick line gives the high level DME, the thin line the low level DME, n the sample size and I the fraction of increasing effects on a high level.

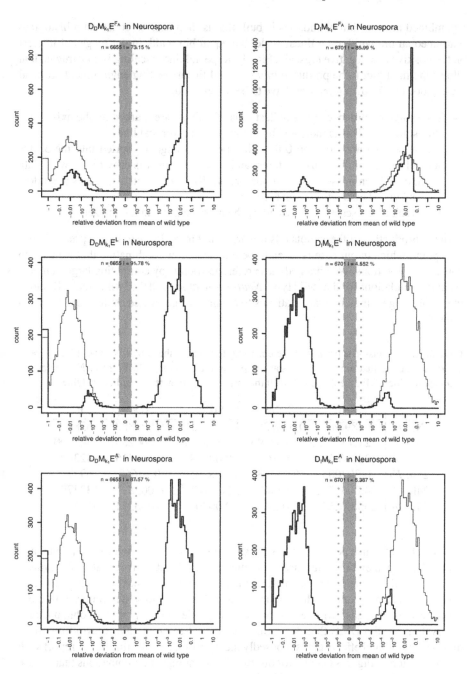

Fig. 9. These DMEs show the effects of assuming a low level lognormal $D_D M_{k_1} E$ and $D_I M_{k_1} E$ as distribution of generated genotypes on the emerging phenotypic features anti-cyclic fitness F_A, cycle length L and amplitude A for the Neurospora parameter combination. The thick line gives the high level DME, the thin line the low level DME, n the sample size and I the fraction of increasing effects on a high level.

a combined array of bin boundaries is built that is then used to construct a histogram with unequal bin width on a linear scale, but equal bin width in the ranges of interest on the positive and negative logscale. Finally, a special linear scaling is constructed that allows the final plot to be produced in a standard linear plotting environment. To read these plots, the following features have to be understood:

- The smallest borders of the smallest bins of interest are marked by the axis labels closest to zero. This is indicated by the grey dotted vertical lines.
- All values that are closer to 0 than the specified range are sorted into the 3 bins defined by the limits of user interest and $\pm10^{-15}$. Thus it is easy to see what data might have been missed. The bin borders of $\pm10^{-15}$ are plotted at values that allow for easy visual distinction from zero.
- The break in the scales is indicated by the massive greying around zero.

In the production of these plots it is of paramount importance to have a precise reference point, which in our case is taken to be the parameter combination that we used to start our explorations. We obtained these reference points by computing large numbers of single simulations for the 'wild type' *Neurospora* clock and the 'wild type' 24h-clock and combining their elementary statistics to obtain the aggregated estimates reported in Table 3.

Table 3. High precision estimates of the emerging features of the two 'wild type' clock parameter combinations that are used as a starting points for exploring DMEs here. N denotes the *Neurospora* clock, 24h, the 24h-clock with the respective parameters specified in Table 2.

	Mean	StDev	StErr	CV	n
N: F_A	0.4897883	0.0009251	1.45×10^{-7}	0.001888	6380
N: L	21.7338705	0.05796	1.07×10^{-7}	0.002667	542300
N: A	339093.583	1836	0.00339	0.005415	542300
24h: F_A	0.807005	0.03768	5.44×10^{-6}	0.04669	6927
24h: L	24.0066962	0.06960	1.30×10^{-7}	0.002899	533379
24h: A	255891.547	1620	0.00304	0.006331	533379

Since the reference points used for constructing DMEs are infinitesimally small and our model system exibits a significant amount of stochasticity, any repeated observation of an identical parameter combination will lead to what looks like many small increasing and decreasing changes. The amount of such stochastic noise determines how close the corresponding peaks will be to zero on the logscale. It is important to obtain a null-observation for the DME that determines its natural stochasticity, to avoid reporting spurious mutational effects that supposedly increase or decrease fitness. We report such an observation in Figures 6 and 7 for our two clock parameter combinations that we use as starting points for estimating real DMEs.

Equipped with such a framework, we can now present an example analysis of a simple system using DMEs. Figure 8 and 9 show the DMEs that result from computing many samples varying the rate k_1 with which the repressor accumulates in the nucleus. We ran four sets of simulations. In one set the rate was decreased by an amount that was sampled from a lognormal distribution, while keeping the smallest resulting rates at zero. The lognormal distribution had location $\mu = 0.1$ and shape $\sigma = 2$ on the log scale. In the other set the rate was increased by an amount that was sampled from the same lognormal distribution. We chose a lognorm distribution, because it had performed well in previous population genetical tests [2]. This analysis was performed for the *Neurospora* and 24h-clock parameter sets to obtain an impression for how different DMEs are when sampled from different points in parameter space.

The results in Figure 8 and 9 show that it depends on the starting point and other parameters, whether a particular parameter change will be advantageous or deleterious. For example decreases in k_1 led to frequent increases in F_A and A for Neurospora, but mostly decreased these emerging properties for the 24h-clock. In other cases the high level effects appear to follow the low level effects rather closely. For example, A in Figure 8 follows the distribution of k_1 so closely that one would expect that at this point the intra-molecular structural effects of the corresponding enzymes have a larger influence on clock amplitude than the larger biochemical reaction network. However, this depends on other properties of the system, as the same is not true in Neurospora. It is not the purpose of this paper to discuss all corresponding DMEs in this simple clock model, but rather to demonstrate that the approach which has been presented is capable of producing the raw data that is needed for more comprehensive analyses.

6   Discussion

We have introduced a new framework for quantifying distributions of mutational effects using molecular systems biological models and presented a compact notation for navigating the complex multi-layered world of DMEs. We have demonstrated how this new approach works in principle and addressed fundamental challenges by estimating $DM_{k_1}E^{F_A}$, $DM_{k_1}E^{L}$ and $DM_{k_1}E^{A}$ in a simple model of a circadian clock. We were able to observe significant changes in the DME that exceeded the noise present in our system. The changes in this simple model show that it is important which parameter combination is used as a starting point for estimating a DME. While some parameter combinations lead to a majority of decreases, others mostly increase fitness. A more comprehensive analysis of this and other models is needed to determine how frequent fitness increasing effects are on a larger scale.

We deliberately did not include entrainment here to focus the reader's attention on our new framework. Our analysis showed that circadian clocks without entrainment will in most cases have a fitness in the wild that will approximate the absence of a circadian clock. Other, non-circadian clocks might still be important, but realistic models of robust circadian clocks in the wild need to include entrainment. This can be done by allowing for time-dependent changes in the reaction rates used in propensity functions.

Future work can improve the accuracy of our estimates by using more computing power. This will help building more realistic models, which is important, as the quality of our DME^F estimates depends on the quality of the molecular systems biological models used. Recent advances in molecular systems biology provide hope for the construction of quality models in an increasing number of model systems. Any such models are likely to be closer to biological reality than most of the extremely abstract and simple models of mutational effects that have been used in population genetics so far. As any DME^F that has been observed by this new approach is specific to a very specific model, one can start comparing many different DME^Fs from many different systems. Such work will show how specific such DMEs are, and how often general features emerge that are robust to much of the underlying complexity.

7 Conclusions

We presented the first comprehensive application of our new framework for estimating distributions of mutational effects. Using the example of a simple circadian clock we demonstrated several fundamental features of this approach. Circadian clocks have been analysed before with systems biology methods [14,16,22,23], but the distribution of mutational effects has not been quantified in these systems before. Many more models need to be analysed in order to determine the general features that DMEs may exhibit.

Acknowledgements. We thank Ozgur Akman for extensive discussions of molecular clocks, Martha Loewe for help with LaTeX, John Welch and three anonymous reviewers for helpful comments on this manuscript and the BBSRC and EPSRC for funding. The Centre for Systems Biology at Edinburgh is a Centre for Integrative Systems Biology (CISB) funded by BBSRC and EPSRC, reference BB/D019621/1.

References

1. Eyre-Walker, A., Keightley, P.D.: The distribution of fitness effects of new mutations. Nat. Rev. Genet. 8, 610–618 (2007)
2. Loewe, L., Charlesworth, B.: Inferring the distribution of mutational effects on fitness in Drosophila. Biology Letters 2, 426–430 (2006)
3. Keightley, P.D., Eyre-Walker, A.: Joint inference of the distribution of fitness effects of deleterious mutations and population demography based on nucleotide polymorphism frequencies. Genetics 177, 2251–2261 (2007)
4. Martin, G., Lenormand, T.: A general multivariate extension of Fisher's geometrical model and the distribution of mutation fitness effects across species. Evolution 60, 893–907 (2006)
5. Kitano, H.: Towards a theory of biological robustness. Mol. Syst. Biol. 3, 137 (2007)
6. Kitano, H.: A robustness-based approach to systems-oriented drug design. Nat. Rev. Drug Disc. 6, 202–210 (2007)
7. Brommer, J.E.: The evolution of fitness in life-history theory. Biol. Rev. Camb. Philos. Soc. 75, 377–404 (2000)
8. Stearns, S.C.: The evolution of life histories. Oxford University Press, Oxford (1992)
9. Rust, M.J., Markson, J.S., Lane, W.S., Fisher, D.S., O'Shea, E.K.: Ordered phosphorylation governs oscillation of a three-protein circadian clock. Science 318, 809–812 (2007)
10. Panda, S., Hogenesch, J.B., Kay, S.A.: Circadian rhythms from flies to human. Nature 417, 329–335 (2002)
11. Brunner, M., Káldi, K.: Interlocked feedback loops of the circadian clock of Neurospora crassa. Mol. Microbiol. 68(2), 255–262 (2008)
12. Gjuvsland, A.B., Plahte, E., Omholt, S.W.: Threshold-dominated regulation hides genetic variation in gene expression networks. BMC Syst. Biol. 1, 57 (2007)
13. Efron, B., Tibshirani, R.D.: An introduction to the bootstrap. Chapman & Hall, New York (1993)

14. Leloup, J.C., Gonze, D., Goldbeter, A.: Limit cycle models for circadian rhythms based on transcriptional regulation in Drosophila and Neurospora. J. Biol. Rhythms 14(6), 433–448 (1999)
15. Goodwin, B.C.: Oscillatory behavior in enzymatic control processes. Adv. Enzyme Regul. 3, 425–438 (1965)
16. Gonze, D., Halloy, J., Goldbeter, A.: Deterministic versus stochastic models for circadian rhythms. J. Biol. Phys. 28, 637–653 (2002)
17. Bundschuh, R., Hayot, F., Jayaprakash, C.: Fluctuations and Slow Variables in Genetic Networks. Biophys. J. 84, 1606–1615 (2003)
18. Arkin, A.P., Rao, C.V.: Stochastic chemical kinetics and the quasi-steady-state assumption: application to the Gillespie algorithm. J. Chem. Phys. 11, 4999–5010 (2003)
19. Cao, Y., Gillespie, D.T., Petzold, L.: Accelerated Stochastic Simulation of the Stiff Enzyme-Substrate Reaction. J. Chem. Phys. 123(14), 144917–144929 (2005)
20. Cao, Y., Gillespie, D.T., Petzold, L.: Adaptive explicit-implicit tau-leaping method with automatic tau selection. J. Chem. Phys. 126, 224101 (2007)
21. Gillespie, D.T.: Stochastic simulation of chemical kinetics. Annu. Rev. Phys. Chem. 58, 35–55 (2007)
22. Bradley, J.T., Thorne, T.: Stochastic Process Algebra models of a Circadian Clock. In: Nicol, D.M., Priami, C., Nielson, H.R., Uhrmacher, A.M. (eds.) Simulation and Verification of Dynamic Systems, Dagstuhl Seminar Proceedings, Dagstuhl, Germany (2006), http://drops.dagstuhl.de/opus/volltexte/2006/705
23. Stenico, M.: Modelling molecular systems with discrete concentration levels in the context of process algebra PEPA: Stochastic and deterministic interpretations. MSc.Thesis, University of Trento (2006)

SED-ML – An XML Format for the Implementation of the MIASE Guidelines

Dagmar Köhn[1] and Nicolas Le Novère[2]

[1] Research Training School dIEM oSiRiS, University of Rostock, Germany
[2] European Bioinformatics Institute, Hinxton, CB10 1SD, UK

Abstract. Share and reuse of biochemical models have become two of the main issues in the field of Computational Systems Biology. There already exist widely-accepted formats to encode the structure of models. However, the problem of describing the simulations to be run using those models has not yet been tackled in a satisfactory way. The community believes that providing detailed information about simulation recipes will highly improve the efficient use of existing models. Accordingly a set of guidelines called the Minimum Information About a Simulation Experiment (MIASE) is currently under development. It covers information about the simulation settings, including information about the models, changes on them, simulation settings applied to the models and output definitions. Here we present the Simulation Experiment Description Markup Language (SED-ML), an XML format that enables the storage and exchange of part of the information required to implement the MIASE guidelines. SED-ML is independent of the formats used to encode the models – as long as they are expressed in XML –, and it is independent of the software tools used to run the simulations. Several test implementations are being developed to benchmark SED-ML on simple cases, and pave the way to a more complete support of MIASE.

1 Introduction

As Systems Biology transforms into one of the main fields in life sciences, the number of available computational models is growing at an ever increasing pace. At the same time, their size and complexity are also increasing. The need to build on existing studies by reusing models therefore becomes more imperative. It is now generally accepted that one needs to be able to exchange the biochemical and mathematical structure of models. Guidelines, such as the *Minimum Information Requested in the Annotation of Models* (MIRIAM [1]), describe the information that needs to be exchanged to properly understand a model; computer formats, such as SBML [2] or CellML [3], allow people to implement those guidelines and exchange models between a large diversity of tools.

However, the computational modeling procedure is not limited to the definition of the model structure. According to the MIRIAM specification, "the model, when instantiated within a suitable simulation environment, must be able to reproduce all relevant results given in the reference description that can readily

M. Heiner and A.M. Uhrmacher (Eds.): CMSB 2008, LNBI 5307, pp. 176–190, 2008.

be simulated" [1]. MIRIAM does not impose to list those relevant results, or to describe how to obtain them. It became nevertheless clear that the description of simulation experiments was mandatory to correctly exchange, re-use and interpret models. This led to the development of the *Minimum Information About a Simulation Experiment* (MIASE). Obtaining a desired numerical result often requires to run complex simulation tasks on original and perturbed models. Furthermore, the same model can provide various results when simulated using different approaches. Well-known examples are systems that exhibit steady-state when simulated with deterministic approaches, and oscillation or multistationarity when simulated with stochastic methods. MIASE addresses exactly these problems by providing a list of mandatory information required for the production – or reproduction – of a given set of simulation results. This information can be split into the following four categories:

Information about the models simulated
MIASE recommends to explicitly define all models used in a simulation by providing a specific name and the source of each model. The use of a model as such is often not sufficient to get a desired simulation result, therefore changes that have to be applied to the model before the simulation must be described in detail. Examples are the assignment of a new value (e. g. constant, initial concentration), or the change of a mathematical expression (e. g. using different enzyme kinetics).

Information about the simulation methods used
Each simulation can be characterized by certain types of simulation procedures to be run (e. g. steady-state, time course) and the simulation algorithms used to perform them. The information has to be sufficiently detailed so that no arbitrary choices have to be made when setting up the simulations.

Information about the tasks performed
Once simulation settings and changes on the models have been defined, the simulation tasks undertaken to complete the simulation experiment need to be specified. Typically, that will involve describing how a simulation procedure has to be applied to a specific model, and in which order.

Information about the outputs produced
It is often necessary to define the transformations that have to be performed on the raw output of the simulation tasks, and how to provide the final results. These results can be numerical or graphical. For instance, a model of a periodic process can provide just time courses showing oscillations; or it can, on the contrary, provide phase diagrams, which are more explicit in describing the relationship between variables. An even more striking example of the necessity for output definitions is the bifurcation diagram.

The adoption of MIASE will be greatly fastened, both on the generation and the reuse sides, if the required information is encoded in a standard format – produced and understood by simulation software. The object model (SED-OM) presented in this paper is a platform independent prototype model encoding MIASE guidelines for simple simulation experiments. We also present an XML based implementation of that model (SED-ML) which is introduced with a detailed example in section 3. Related efforts are compared and discussed in section 4.

2   The Simulation Experiment Description Object Model

The *Simulation Experiment Description Object Model* (SED-OM) is a formal representation of the MIASE guidelines using the Unified Modeling Language (UML [4]). The top-level classes of the SED-OM can be seen in Figure 1 and will be described in more detail in the following section. For clarification, the SED-OM class names are put in brackets using `typewriter` font.

2.1   Information on the Model and Model Changes

A MIASE based simulation description will in many cases make use of more than just one model. That is why all models have to be defined clearly for later reference. In SED-OM, all models involved in the simulation experiment are in a list of models (see Figure 2). Each model (`Model`) has its own unambiguous identifier (`id`). Additionally, it may have a name (`name`) and it may hold information about its encoding (`type`). As most simulation tools support only particular formats, it is strongly recommended to provide the type of model encoding (e. g. CellML). Information about the model format helps simulation tools to decide whether the model can be loaded directly or has to be converted into another format first. SED-OM also requires the source of the model to be defined (`source`). It is not within the scope of SED-OM to store model representations, but to provide a secure way of accessing them. The source should be reliable, meaning it should point to a repository of curated models in order to ensure the correctness and validity of the model.

Models often need to be modified before subjected to a simulation task. Those changes can be direct atomic changes on simple attributes of a model, such as changes on a parameter or on the initial concentration (`ChangeAttribute`). Complex changes, depending on other values, can be described using mathematical expressions (`ChangeMath`). Those expressions are mathML [5] constructs that are made up of parameters defined in a list of parameters (`Parameter`) and variables defined in a list of variables (`Variable`). A variable object holds a reference to an already defined model and targets a certain XML element within that model using XPath. Finally, a general class (`ChangeXML`) allows to replace any piece of XML code by another valid one, including void which amounts to a deletion.

The XML Path Language (XPath [6]) has been chosen to target model elements. Apart from being a natural choice when working with XML files, XPath expressions allow to unambiguously identify any (syntactical) part of a model that can be altered. XPath offers a very convenient way of describing changes on the model independently of the actual model representation format: An XPath expression defines a path through an XML document and points to a particular XML element or XML attribute within the document. The only restriction imposed upon the model by the usage of XPath is that it has to be available in an

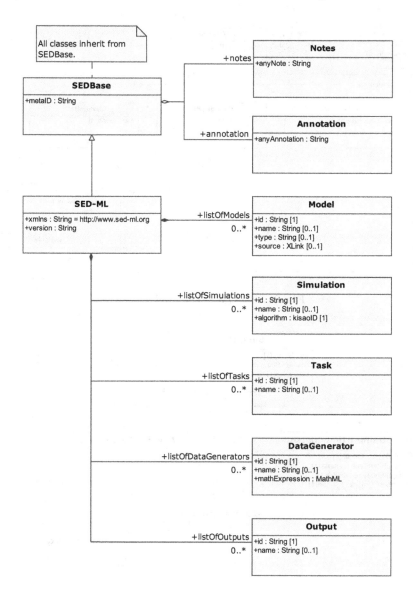

Fig. 1. SED-OM – Top level classes

XML based format. The addressed XML element can be a leaf element, or an element containing a whole mathematical expression. In principle, everything that can be addressed by an XPath expression can be modified. Other solutions that were considered for the definition of changes on a model would have involved the creation of change classes for each supported language format, depending on the current version of the standard and its syntactical naming of the model elements.

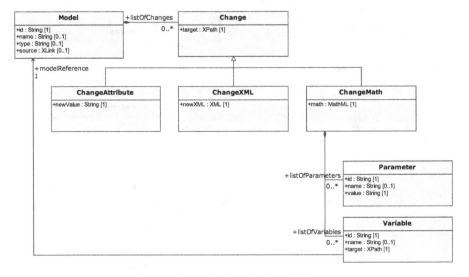

Fig. 2. SED-OM – The Model class

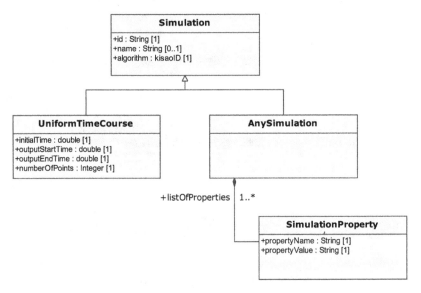

Fig. 3. SED-OM – The Simulation class

2.2 Information on the Simulation Settings

A simulation is typically characterized by the simulation algorithm used, the settings applied to the simulation algorithm, and the simulation type.

Each simulation (`Simulation`, see Figure 3) can be referred to by an identifier (`id`). It might also contain a name (`name`) and a reference to the simulation algorithm used to run the experiment (`algorithm`). This algorithm reference is an identifier corresponding to a KiSAO term. The *Kinetic Simulation Algorithm Ontology* (KiSAO [7]) is an effort to characterize and categorize existing algorithms for the simulation of quantitative models within the field of Systems Biology. Using terms from an ontology rather than agreed-upon strings allows for reasoning. The simplest reasoning procedure is to find that algorithm available from KiSAO which is the closest to the one described in the simulation description, if the latter is not available for the user.

Depending on the chosen simulation algorithm different settings have to be applied. The necessary information which settings that are can be retrieved from the KiSA ontology which will provide the according information about additional settings for each simulation algorithm covered by the ontology. At the current state of development, KiSAO does not allow for extracting the mandatory simulation algorithm settings. As a consequence, the storage of simulation algorithm settings are not yet possible.

Very important is the type of simulation that should be launched. SED-OM defines the different types of simulations as sub-classes of the `Simulation` class. For the time being, `UniformTimeCourse` simulations are supported. The inclusion of further simulation types has been postponed to future versions of SED-OM as the integration of classes with different but overlapping attributes is not trivial. Until then, the `AnySimulation` class functions as a generic place holder for all additional simulation types. Depending on the type of simulation, different additional information has to be provided, such as the initial simulation time for uniform time courses. For the `AnySimulation` class, those simulation properties can explicitly be defined in the `SimulationProperty` class through the name and the value of the property (`propertyName`, `propertyValue`). For particular simulation types derived from the general simulation class, those attributes are already defined in the SED-OM, e. g. `initialTime` in the `UniformTimeCourse` class.

2.3   Information on the Simulation Task

In a simulation experiment, simulation approaches described in a `Simulation` object are combined with specific models described in a `Model` object. In SED-OM, the association between those two objects is supported through the definition of tasks (`Task`, see Figure 4). Each task contains one reference to a model and one reference to a simulation. The task itself can be referenced by its own identifier (`id`) and might have an additional name (`name`).

By providing the opportunity of explicitly linking models to simulations, a redundant definition of models as well as of simulation settings is avoided – a single model can easily be used with several different simulations and vice versa.

2.4   Information on the Output

One important part of SED-OM is the description of a simulation experiment based on particular (changed or unchanged) models. However, just as important is the definition of the results to be produced and the way to provide them (see Figure 5).

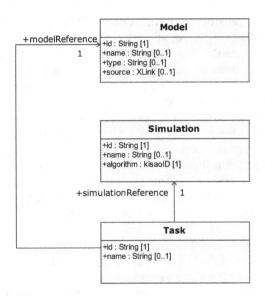

Fig. 4. SED-OM – The Task class

The output class (`Output`) can be referred to by an identifier (`id`) and an optional name (`name`). The SED-OM allows for the definition of different kinds of outputs, which can either be specified as simple data tables (i. e. reports) or as plots. Reports (`Report`) consist of a number of columns (`Column`); a formula defines how to generate the data written in each column (see the `DataGenerator` class further down). In addition, the SED-OM provides structures for two dimensional plots (`Plot2D`) and three dimensional plots (`Plot3D`). A two dimensional plot displays a number of curves (`Curve`) and a three dimensional plot displays a number of surfaces (`Surface`). Curves and surfaces refer to the data to be mapped on the according axes, and precise if the mapping is logarithmic or not. The aim of the output class is to define concisely the procedure leading to a certain output rather than to define *how* it should be presented to the user. Nonetheless, since all classes may have notes attached, it is always possible to store meta data such as information on the output shape or labels for curves.

The formulas used to generate the data are described in the `DataGenerator` class (see Figure 6). All types of output reference an instance of that class. In doing so it does not matter whether the data is calculated for plots or for the columns of a report. One example for such a calculation is the definition of the x-axis of a plot (referenced through the `Curve` class by an `xDataReference`). Each data generator has an identifier (`id`) and might have a name (`name`). A single data generator consists of a list of variables (`Variable`), a list of parameters (`Parameter`) and a mathematical expression (`Math`). A variable definition is a reference to an existing variable in one of the defined models. However, instead of referencing an element in a particular model using the model id, the variable definition refers to the the task that simulates the model. As every task uses only one model and one simulation setting description, this reference is

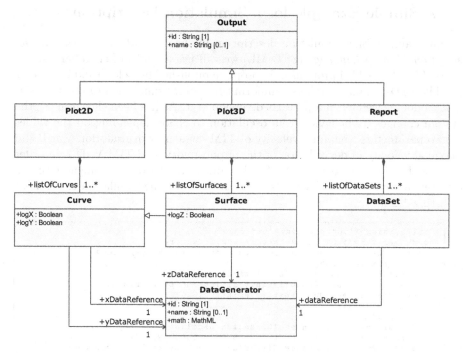

Fig. 5. SED-OM – The Output class

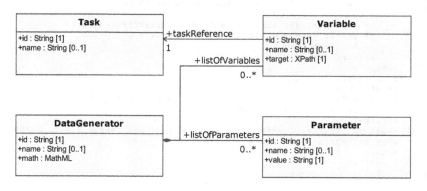

Fig. 6. SED-OM – The DataGenerator class

unambiguous. The variable inside the model is addressed via XPath expressions
for the same reasons justifying the use of XPath in the description of model
changes. Parameters are values introduced additionally to be used in the mathematical post-processing of the variable's values. To facilitate calculations based
on the defined parameters and variables, mathematical expressions (Math) can
be constructed using mathML. The use of a data generator could, for instance,
lead to the following definition of a plot: "Take variable v1 of model m02 and
multiply its values by 2. Use the result as the abcissa x-axis of a 2D plot".

3   A Simple Example for a Simulation Description

As an example for a simulation description in the *Simulation Experiment Description Markup Language* (SED-ML) we will use a model of circadian oscillations of PER and TIM proteins in Drosophila published by Leloup and Goldbeter [8]. The SED-ML file describes a uniform time course simulation run on the original model, as well as on a perturbed version of it. As has been shown in [8], the system changes its behavior from oscillation to chaos depending on the values of two parameters (maximal velocity of TIM messenger degradation V_mT and maximal velocity of degradation of the bi-phosphorylated TIM V_dT). In order to show the difference between both behaviors, a simulation experiment with two different parameter settings is described in the following simulation experiment in listing 1.1.

```xml
 1  <?xml version="1.0" encoding="utf-8"?>
 2  <sedML version="1.0" xmlns="http://www.miase.org/">
 3    <notes>Changing a system from oscillation to chaos</notes>
 4    <listOfSimulations>
 5      <uniformTimeCourse id="simulation1"
 6      algorithm="KiSAO:0000071" initialTime="0" outputStartTime="50"
 7      outputEndTime="1000" numberOfPoints="1000" />
 8    </listOfSimulations>
 9    <listOfModels>
10      <model id="model1" name="Circadian Oscillations" type="SBML"
11      source="urn:miriam:biomodels.db:BIOMD0000000021" />
12      <model id="model2" name="Circadian Chaos" type="SBML" source="model1">
13        <listOfChanges>
14          <changeAttribute target="/sbml/model/listOfParameters/
15          parameter[@id='V_mT']/@value" newValue="0.28">
16          </changeAttribute>
17          <changeAttribute target="/sbml/model/listOfParameters/
18          parameter[@id='V_dT']/@value" newValue="4.8">
19          </changeAttribute>
20        </listOfChanges>
21      </model>
22    </listOfModels>
23    <listOfTasks>
24      <task id="task1" name="Baseline" modelReference="model1"
25      simulationReference="simulation1">
26      </task>
27      <task id="task2" name="Modified parameters" modelReference="model2"
28      simulationReference="simulation1">
29      </task>
30    </listOfTasks>
31    <listOfDataGenerators>
32      <dataGenerator id="time" name="Time">
33        <mathExpression>
34          <math>
35            <apply>
36              <plus />
37              <csymbol encoding="text"
38              definitionURL="http://www.sbml.org/sbml/symbols/time">time
39              </csymbol>
40            </apply>
41          </math>
42        </mathExpression>
43      </dataGenerator>
44      <dataGenerator id="tim1" name="tim mRNA (total)">
45        <listOfVariables>
46          <variable id="v1" taskReference="task1"
47          target="/sbml/model/listOfSpecies/species[@id='Mt']" />
48        </listOfVariables>
```

```
49    <mathExpression>
50      <math>
51        <apply>
52          <plus />
53          <ci>v1</ci>
54        </apply>
55      </math>
56    </mathExpression>
57  </dataGenerator>
58  <dataGenerator id="tim2" name="tim mRNA (changed parameters)">
59    <listOfVariables>
60      <variable id="v2" taskReference="task2"
61        target="/sbml/model/listOfSpecies/species[@id='Mt']" />
62    </listOfVariables>
63    <mathExpression>
64      <math>
65        <apply>
66          <plus />
67          <ci>v2</ci>
68        </apply>
69      </math>
70    </mathExpression>
71  </dataGenerator>
72  </listOfDataGenerators>
73  <listOfOutputs>
74    <plot2D id="plot1" name="tim mRNA with Oscillation and Chaos">
75      <listOfCurves>
76        <curve logX="false" logY="false" xDataReference="time"
77          yDataReference="tim1" />
78        <curve logX="false" logY="false" xDataReference="time"
79          yDataReference="tim2" />
80      </listOfCurves>
81    </plot2D>
82  </listOfOutputs>
83  </sedML>
```

Listing 1.1. Encoding of simulation settings using SED-ML

The original model used for the simulation experiment is model number 21 in BioModels database [9]. This is specified by the **source** attribute of the first model entry in the list of models (l. 11). The second model defined in the SED-ML file references the first one (**source="model1"**, l. 12). Contrary to the first model definition, this XML element contains a sub-element **listOfChanges** (ll. 13-20) that has two **changeAttribute** elements, each defining one change in the XML model representation. Both changes apply new values to existing parameters: The parameter V_mT is adapted in the first change definition (**newValue="0.28"**, l. 15), and the parameter V_dT is adapted in the second change definition (**newValue="4.8"**, l. 18).

In lines four to eight, the simulation settings are stored: The simulation has been characterized as a uniform time course (ll. 5-7) running from timepoint zero to 1000, but starting the output at timepoint 50. The simulation algorithm is specified by a KiSAO id (**KiSAO:0000071**, l. 6) which corresponds to the ontology entry "livermore solver for ordinary differential equations" (LSODE).

After the models have been defined and the simulation settings have been stored, the next step is to combine both of them by creating simulation tasks (**listOfTasks**, ll. 23-30). The first task runs the original model with the (only) simulation setting defined (**model1** with **simulation1**, ll. 24-26). The second task

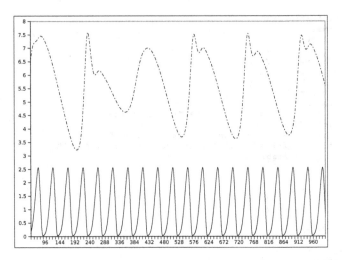

Fig. 7. Simulation result gained from the SED-ML description (created using COPASI 4.2 (Build 22) and Gnumeric Spreadsheet 1.7.11): tim mRNA concentration (line) and tim mRNA concentration with updated parameters V_mT and V_dT (dotted line)

then runs model2 with modified parameters using the same simulation setting (model2 with simulation1, ll. 27-28).

The information given so far is sufficient for the design of valid and repeatable simulation experiments. SED-ML also offers structures for the specification of desired outputs. The example in listing 1.1 creates three different data generator elements: The first data generator (ll. 32-43) is a simple specification of time using the time construct available from SBML. The second data generator element (ll. 44-56) points to the species with identifier Mt which is the total amount of tim mRNA (as can be gained from the SBML model description file) that comes out of the task task1. The third data generator again points to the total amount of tim mRNA, but it is now using the values coming out of task task2 (ll. 57-69). Note that task2 – in contrast to task1 – is performed on the changed model (model2).

The last part of the SED-ML file describes a plot (listOfOutputs, ll. 70-79). It consists of two curves (ll. 75-76). Both curves plot time on their x-axis; however, the first curve plots the total amount of tim mRNA using the original model (yDataReference="tim1"), and the second curve plots the total amount of tim mRNA applying the parameter changes (yDataReference="tim2"). The result of the simulation following the specifications in the SED-ML file is shown in Figure 7.

4   Related Work

The problem of simulation experiment descriptions is not new to Systems Biology and has been addressed by several groups before. Of course, each simulation tool that is capable of storing simulation settings uses its own internal storage format.

For example, COPASI [10] uses an XML based format for encoding the selected simulation algorithm and the task definitions. However, those formats can only be used with a specific simulation tool, and therefore simulation experiment descriptions cannot be exchanged with others.

Standardization communities face the problem of describing simulation experiments as well. One example is the ongoing discussion about the CellML Metadata Specification [11] in the CellML community. The authors mention the need to not only describe a model but also to describe "details of any particular simulation being run". The proposed solution is to extend the CellML meta data concept by additional simulation description concepts. CellML meta data, and thus also the simulation meta data, are specified using the Resource Description Framework (RDF [12]). With help of the CellML Metadata Specification, one or more simulation runs can be associated and described in one model specification. The description covers information about the type of simulation and about the simulation algorithm used (including the name of the linear solver, the specification of the iteration method and the multistep method used). Apart from that, the specification of step size and starting values for the simulation are supported. The CellML Metadata Specification is currently in the state of a discussion draft. Unlike SED-ML, the approach chosen by the CellML community will store simulation specification details inside the model definition and thus be restricted to the use of CellML models. By suggesting to refer to a model rather than being part of it, SED-ML enhances reusability of simulation descriptions and supports the description of simulation experiments using not only a single model, but a number of models – which could even be encoded in different description formats.

A specification to characterize simulation experiments has been proposed in the SBML community as well [13]. Along with the development of SBML Level 3 extensions, the author proposes the description of simulation settings. Although part of Level 3 extensions, the description of simulation runs is suggested to be included inside the SBML model. So as to define simulation runs, the proposal consists of several parts: (1) the definition of changes on the model, such as updates on initial values, model parameters and others; (2) the specification of simulation parameters and the storage of (time,value) pairs to maintain simulation results; and (3) the definition of plots through specification of the axes and the data that should be shown in the output. The proposal is currently available as a DTD [14] draft. Again, the approach as it has been introduced in [13] does not follow the ideas of a simulation description format independent of software tools and model description languages. Additionally, the inclusion of simulation results is proposed. This is not considered to be part of SED-ML, but in our opinion should be covered by other efforts.

5   Discussion and Future Work

In this paper, an approach for the description of simulation experiments has been introduced. The novel idea is to define a set of minimal guidelines detailed enough to unambiguously define a simulation experiment, independent of specific simulation tools and particular model description languages. A first version of a model for the realization of those guidelines has been proposed (SED-OM) and has been encoded using XML (SED-ML). A sample simulation experiment in the SED-ML format has been described in detail. It showed that the SED-OM can be applied to existing models. The use is restricted to simple simulation experiments though.

SED-ML can encode simulation experiments being run with several models, which can even exist in different formats (e. g. comparing simulation results of a CellML model and an SBML model). SED-ML can specify different simulation settings applicable to the same model (e. g. running a model with a stochastic and a deterministic simulation algorithm). Combinations of both are also possible, it is easily conceivable to set up a simulation experiment that results in an output comparing a parameter of a CellML model to a parameter of an SBML model, depending on different simulation algorithms.

However, there are a number of important issues in simulation experiments that are currently not covered by the SED-OM. The description of more complex simulation tasks, e. g. parameter scans, is not yet supported. The difficulty here is to decide how to describe the range of parameter changes that have to be applied to a model. One option is to do that in the `Task` class, another option is to extend the functionality of the `Change` class. Furthermore, the current SED-OM allows to freely combine variables from different tasks in one output – although the combination is depending on integrity restrictions. For example, the output of variables from different simulation settings in one plot is only possible as long as all participating simulations produce the same time points. Another complex task that is not yet supported is the linear execution of simulation experiments, meaning that the result of one simulation is used as the input for another simulation task. For example, the result of a steady state analysis will lead to a model with changed parameters. If that model then should be simulated using a time course simulation, the results of the steady state analysis have to be applied to the original model before. The definition of such sequences is not yet supported by the SED-OM.

For all those reasons, the SED-OM must be further discussed. The use of SED-ML and test implementations in different simulation tools will help enhancing the coverage and robustness of the format.

6   Resources

If you want to contribute to the SED-OM and SED-ML development, or should you have any questions or comments, please contact the authors or visit the website on `http://www.ebi.ac.uk/compneur-srv/sed-ml`. The current SED-OM and sample SED-ML instances can be downloaded from the MIASE project homepage on sourceforge `http://www.sourceforge.net/miase`. For discussions of the MIASE guidelines, please join the mailing list `miase-discuss@lists.sourceforge.net` or visit the web site on `http://www.ebi.ac.uk/compneur-srv/miase`.

Acknowledgements

The authors would like to thank everybody from the community involved in the discussion of the MIASE guidelines. Special thanks goes to all members of the `miase-discuss` mailing list who have been very active in discussing both the MIASE guidelines and the SED-OM format, in particular Frank Bergman (roadRunner), Ion Moraru (VCell), Sven Sahle (COPASI) and Henning Schmidt (SBToolbox). Exchanges with Sven Sahle enriched the discussion part. Part of the work was funded by the Marie Curie program and by the German Research Association (DFG Research Training School "dIEM oSiRiS" 1387/1).

References

1. Le Novère, N., Finney, A., Hucka, M., Bhalla, U.S., Campagne, F., Collado-Vides, J., Crampin, E.J., Halstead, M., Klipp, E., Mendes, P., Nielsen, P., Sauro, H., Shapiro, B., Snoep, J., Spence, H., Wanner, B.: Minimum Information Requested In the Annotation of biochemical Models (MIRIAM). Nature Biotechnology 23(12), 1509–1515 (2005)
2. Hucka, M., Bolouri, H., Finney, A., Sauro, H., Doyle, J., Kitano, H., Arkin, A., Bornstein, B., Bray, D., Cuellar, A., Dronov, S., Ginkel, M., Gor, V., Goryanin, I., Hedley, W., Hodgman, T., Hunter, P., Juty, N., Kasberger, J., Kremling, A., Kummer, U., Le Novère, N., Loew, L., Lucio, D., Mendes, P., Mjolsness, E., Nakayama, Y., Nelson, M., Nielsen, P., Sakurada, T., Schaff, J., Shapiro, B., Shimizu, T., Spence, H., Stelling, J., Takahashi, K., Tomita, M., Wagner, J., Wang, J.: The Systems Biology Markup Language (SBML): A medium for representation and exchange of biochemical network models. Bioinformatics 19, 524–531 (2003)
3. Lloyd, C., Halstead, M., Nielsen, P.: CellML: its future, present and past. Progress in Biophysics & Molecular Biology 85, 433–450 (2004)
4. Object Management Group (OMG): Unified Modeling Language (UML), Version 2.1.2 (2007), http://www.omg.org/spec/UML/2.1.2/
5. Ausbrooks, R., Buswell, S., Carlisle, D., Dalmas, S., Devitt, S., Diaz, A., Froumentin, M., Hunter, R., Ion, P., Kohlhase, M., Miner, R., Poppelier, N., Smith, B., Soiffer, N., Sutor, R., Watt, S.: Mathematical Markup Language (MathML) Version 2.0, 2nd edn. (2003)
6. World Wide Web Consortium (W3C) Recommendation: XML Path Language (XPath) (1999), http://www.w3.org/TR/xpath
7. Köhn, D., Le Novère, N.: The Kinetic Simulation Algorithm Ontology (KiSAO). Website (2007), http://www.ebi.ac.uk/compneur-srv/kisao/
8. Leloup, J., Goldbeter, A.: Chaos and birhythmicity in a model for circadian oscillations of the per and tim proteins in drosophila. Journal of theoretical biology 198(3), 445–459 (1999)
9. Le Novère, N., Bornstein, B., Broicher, A., Courtot, M., Donizelli, M., Dharuri, H., Li, L., Sauro, H., Schilstra, M., Shapiro, B., Snoep, J., Hucka, M.: Biomodels database: a free, centralized database of curated, published, quantitative kinetic models of biochemical and cellular systems. Nucleic Acids Research 34(Database issue) (January 2006)
10. Hoops, S., Sahle, S., Lee, C., Pahle, J., Simus, N., Singhal, M., Xu, L., Mendes, P., Kummer, U.: Copasi a complex pathway simulator. Bioinformatics 22(24), 3067–3074 (2006)

11. Miller, A.: CellML simulation metadata specification – a specification for simulation metadata (2007), http://www.cellml.org/specifications/metadata/simulations
12. World Wide Web Consortium (W3C) Recommendation: Resource Description Framework (RDF) (1997), http://www.w3.org/RDF
13. Kopalov, F.: SBML extensions for level 3: Experiments, simulation, parameters, results and plots (March 2008) (unpublished proposal)
14. World Wide Web Consortium (W3C) Specification: Document Type Definition (DTD) (1998), http://www.w3.org/XML/1998/06/xmlspec-report

On Parallel Stochastic Simulation of Diffusive Systems

Lorenzo Dematté[1,2] and Tommaso Mazza[1]

[1] The Microsoft Research - University of Trento
Centre for Computational and Systems Biology
Piazza Manci, 17, 38100, Povo (TN), Italy
{dematte,mazza}@cosbi.eu
[2] Department of Information Engineering and Computer Science (DISI),
University of Trento

Abstract. The parallel simulation of biochemical reactions is a very interesting problem: biochemical systems are inherently parallel, yet the majority of the algorithms to simulate them, including the well-known and widespread Gillespie SSA, are strictly sequential. Here we investigate, in a general way, how to characterize the simulation of biochemical systems in terms of Discrete Event Simulation. We dissect their inherent parallelism in order both to exploit the work done in this area and to speed-up their simulation. We study the peculiar characteristics of discrete biological simulations in order to select the parallelization technique which provides the greater benefits, as well as to touch its limits. We then focus on reaction-diffusion systems: we design and implement an efficient parallel algorithm for simulating such systems that include both reactions between entities and movements throughout the space.

Keywords: Parallel and distributed simulation, reaction-diffusion systems, Gillespie SSA.

1 Introduction

In computational biology, the interest on multi-processor computing is growing over the years, even if ubiquitous and parallel computing require deep knowledge both on the bio-reality and on the tools in charge of handling and interpreting it. Indeed, the correct parallel computation of whatever problem must take into account four milestones: (i) the best computational splitting policy; (ii) how to handle synchronization among the computational workers, (iii) the more suitable hardware architecture and software packages to use and (iv) the nature of the inherent parallelism.

There are problems naturally parallelizable and others purely serial. According to the case, the additional computing power afforded by new machines can be used to advantage of one or of the other. To enhance the efficiency of Monte Carlo simulations, Single Replication in Parallel (SRIP) and Multiple Replications in Parallel (MRIP) computational paradigms have been widely contemplated in the past and deemed to be appropriate.

M. Heiner and A.M. Uhrmacher (Eds.): CMSB 2008, LNBI 5307, pp. 191–210, 2008.

Single Replication in Parallel. The SRIP approach is based on the decomposition of a stochastic trajectory into logical processes, running on different processors and communicating by means of message passing protocols [1]. For naturally divisible problems, it shows elevated performances in speed-up and scale-up benchmarks. Significant drawbacks originate from the necessity for warranty of synchronism.

Multiple Replications in Parallel. The MRIP method speeds up simulation by launching independent replications on multiple computers and using different random seeds in such a way the processes result approximatively uncorrelated [2]. In contrast to SRIP, MRIP can be easily applicable to any system, independent of the inherent system parallelism. However, the fact that a single replication cannot be executed on a unique processor and that outputs (or pieces of them) almost deterministic are identical when replicated, make the use of MRIP approaches sometimes inappropriate [3]. The MRIP and SRIP approaches are not exclusive, i.e., it is possible to use MRIP and SRIP in the same simulation program.

In biology, whereas the MRIP policy, well understood and investigated for a long time [2], [4], [5], [3], [6], [7], [8], [9], [10], [11], finds straightforward application to real case-studies [12], [13], the SRIP policy has a rather vague characterization. SRIP methods can be further divided into two opposite subcategories which include: (a) methods that exploit *data-parallelism* (or *loop-level parallelism*), namely that exemplify simulation of interacting particles on a finite grid in which individual processors are in charge of simulating the state of each site [14]; (b) methods that exploit *task-parallelism* (or *functional parallelism*), namely that divide the computation of a realization into a set of sub-computations among cooperative processors by computational dependency criteria [1], [15]. To date, the research in distributed-parallel processing has successfully solved many related problems; however, it has not led yet to a portable and efficient tool for distributing stochastic simulation in the field of computational biology. We aim to move the attention of the reader toward our target by going through the theoretical bases and strategic decisions which configure our insight.

In particular, the next section will introduce the Gillespie Stochastic Simulation Algorithm (SSA), the most known and the *de-facto* standard for the simulation of biochemical systems at microscopic and mesoscopic level and a possible extension for simulating bigger systems where spatiality and diffusion are important variables. Next, we will briefly introduce the category of computer-simulation systems known as DES and the work done on these systems in the light of parallel and distributed computing. We will show how the SSA can be reformulated in term of a DES system, and we will show the characteristics the algorithm assumes when it runs in a parallel environment. Section 4 will present how these concepts were used in the designing and implementation of a reaction-diffusion simulator that runs on HPC clusters, and Sections 5 and 6 will close the paper with an example and considerations about future work.

2   The Gillespie SSA

The stochastic approach to chemical kinetics was first employed by Delbruck in the '40s. The basic assumptions of this approach are that a chemical reaction occurs when two (or more) molecules of the right type collide in an appropriate way, and that such collisions are *random* in a system of molecules in thermal equilibrium. Whenever two molecules come into a certain proximity, they can react with some probability: collisions are frequent, but those with the proper orientation and energy, that is the collisions that allow molecules to react together, are infrequent. In [16] and [17], Gillespie introduced the additional assumption that the system is in thermal equilibrium. This assumption means that the considered system is a well-stirred mixture of molecules, where the number of non-reactive collisions is much higher than the number of chemical reactions. It makes possible to state that the molecules are randomly and uniformly distributed at all times. The derived stochastic method becomes computationally lighter than the classical methods in charge of predicting collisions by estimating the collision volume of each particle.

The so called *Stochastic Simulation Algorithm* (SSA) models a general biological system as a set of pairs *(entity type, quantity)* and a set of possible interactions between the entities. In the case of biochemical models, *entities* are molecules and *interactions* are coupled chemical reactions. Therefore, we can reduce the necessary parameters for describing a system to:

– the *entities*, usually referred to as *species*, present in the system $S_1, ..., S_N$;
– the number and type of *interactions*, called *reaction channels*, through which the molecules interact $R_1, ..., R_M$;
– the state vector $\mathbf{X}(t)$ of the system at time t, where $X_i(t)$ is the number of molecules of species S_i present at time t.

The state vector $\mathbf{X}(t)$ is a vector of random variables, that does not take account of the position and velocity of the single molecules. For each reaction channel R_j, a function a_j, called *propensity function* for R_j, is defined as:

$$a_\mu = h_\mu c_\mu, \text{ for } \mu = 1, \ldots, M \qquad (1)$$

such that h_μ is the number of distinct reactant combinations for reaction R_μ and c_μ is a constant depending on physical properties of the reactants. The c_μ constant is usually called *base rate*, or simply *rate* of an action, while the value of the function a_μ is called the *actual rate*.

Gillespie derived a physical correct *Chemical Master Equation* (CME) from the above representation of biochemical interactions. Intuitively, this equation shows the stochastic evolution of the system over time, which is indeed a Markov process. Gillespie also presented in [17] an exact procedure, called *exact stochastic simulation*, to numerically simulate the stochastic time evolution of a biochemical system, thus generating one single trajectory. The procedure is based on the *reaction probability density function* $P(\tau, \mu)$, which specifies the probability that the next reaction is an R_μ reaction and that it occurs at time τ:

$$P(\tau, \mu) = \begin{cases} a_\mu \exp(-a_0 \tau) & \text{if } 0 \leq \tau < \infty \text{ and } \mu = 1, \dots, M \\ 0 & \text{otherwise} \end{cases}$$

where a_μ is the *propensity function* and a_0 is the sum of a_μ, $\mu = 1, \dots, M$.

The *reaction probability density function* is used in a stochastic framework to compute the probability of an action to occur. The way of computing the combinations h_μ and, consequently the *actual rate* a_μ, varies with the different kind of reactions. In the case of first-order reactions, h_μ is equal to the number of entities (the *cardinality*) of the one reactant, while in the case of second-order reactions, h_μ corresponds to the number of all possible interactions that can take place among the reactants.

2.1   Simulation of Reactive-Diffusive Systems

When studying a single localized pathway, the macroscopic description of its kinetics usually suffices. On the other end, many biological processes are not local and, often, they take place in an inhomogeneous medium, the *cytosol*, where spatially localized fluctuations of inorganic catalysts and intracellular diffusion can play an important role. When dealing with such processes, it is mandatory to explicitly consider the cell geometry and, in general, spatial conformations and diffusion processes.

Several algorithms for the simulation of reactive-diffusive systems exist; each of them uses a different abstraction that gives a different level of detail which influences both the accuracy of the simulation and its execution speed. Chemical and biochemical reactions can be simulated in a very precise and detailed way using *molecular dynamics* [18], a form of computer simulation where atoms are allowed to interact under known physics laws. In these simulations, details about the chemical reactions, like formation and bonds breaking between single atoms, are explicitly simulated as well as the position and energy of every atom in the system. These methods have been applied to a wide range of problems of chemical and biological interest, such as chemical reactions in solution and enzymes and solvent effects on electronic excited states.

Other methods, like the one used by Bray et al. in *Smoldyn* [19], operate at a coarser level of detail, where molecules have an identity and an exact position in a continuous space, but no volume, shape or inertia. Moreover, every molecule of interest is represented as an individual point, while those that are not of interest (water, non-reactive molecules, etc.) are not represented. Molecules move at rates specified by a *diffusion coefficient* and diffuse in random directions and distances calculated by means of the *Fick's* second law. Bimolecular reactions take account of the spatial relations; a bimolecular reaction occurs if two reactants approach each other within a *binding radius*, a radius that is different (typically smaller) than the physical radius of the molecules, and that depends on the diffusion coefficients and on the reaction rate constant. Simulated space is continuous; on the other hand, simulated time is discrete as reactions, computation of movements and update of the position are done at fixed time steps.

2.2   Reaction-Diffusion with the Gillespie Method

The Gillespie algorithm, introduced in Sec. 2, allows to simulate chemical reactions in an efficient way. Every collision that leads to a reaction is explicitly simulated, but collisions that do not lead to a reaction are not. The stochastic behaviour of the chemical system is preserved, as molecules are still represented as discrete quantities, but information on a single molecule, and with it any positional information, are lost. Moreover, the assumptions made by Gillespie explicitly rule out diffusion from the system: since the solution is in thermal equilibrium, it is assumed that diffusion is instantaneous so that each molecule has the same probability of reacting with every other molecule in the system. The algorithm works well locally, but cannot be used to represent complex pathways that span over a considerable extension of reactions taking place in an inhomogeneous medium.

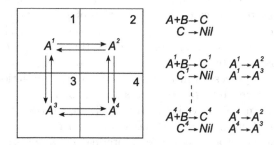

Fig. 1. The extension to the Gillespie SSA proposed by Bernstein. On the left, a discretization of the space into four cells. On the right, the species and the reactions added to the system in order to deal with diffusion.

A proposed extension is the *discretization* of the space by subdivision into logical sub-volumes, often referred to as *cells*. The dimension of a cell is chosen to be small enough for the sub-space to be homogeneous and for the enclosed entities to have almost instantaneous diffusion, so that the assumptions made by the Gillespie algorithm are valid inside a single cell; furthermore, spatial information is added to the system by duplicating every species S. New species with the same characteristics of S and with an index identifying its position on the grid are added to the system $(S_1, S_2, ..., S_n)$; diffusion is represented by first-order reactions among species. This method, proposed by Bernstein [20], is depicted in Fig. 1.

The advantage of this approach is that the algorithm in charge of simulating the reaction-diffusion system does not change; it is possible to add more species to model the molecules in different compartments and add reactions to "diffuse" between adjacent compartments, and then to use the existing tools and algorithms to simulate the modified system.

An efficient implementation for simulating reactive-diffusive systems by using spatial structures is used in the *next subvolume method* (Elf et al. [21]). The underlying theory is the same utilized by Bernstein, as both are based on the exact realizations of the Markov process described by the Reaction Diffusion Master Equation. The algorithm uses three data structures: (i) a *connectivity matrix*, (ii) an *event queue* and (iii) a *configuration matrix*, used to naturally partition reactions into sub-volumes. Instead of mapping movements of entities using different species, the *direct method* [17] is used on each sub-volume to compute the time for the next event, i.e. a chemical reaction or a diffusion event. Then, the *next reaction method* [22] is used to identify the sub-volume where the first event will occur. The event is simulated, then the reaction and diffusion times in the volume (or volumes, in case of diffusion) are updated using the *direct method* again.

3   Discrete Event Simulation (DES)

In DES, the life of a *system* is modelled as a sequence of timed events. With this approach, a system is set up by a collection of *processes* $P = \{p_1, p_2, \ldots\}$ and of *activities* or *events* $E = \{e_1, e_2, \ldots\}$. A *process* is fully characterized by a finite set of *states* $S = \{s, s', \ldots\}$. At any given time, each process has exactly one *active state*. Each state s has a set of *actions* $A_s = \{\alpha_s, \alpha'_s, \ldots\}$ that can be performed when the process is in that state; the aim of an action is to change the current active state. *Activities* or *events* are sets of actions that are executed together to transform the state of the system. Here, we refer to the state of a system z as the collection of all the active states of the processes in the system. A *run* is thus meant as a sequence of interleaved system states and events: $r : z_0|e_0 \rightarrow z_1|e_1 \rightarrow z_2|e_2 \ldots z_{(u-1)}|e_{(u-1)} \rightarrow z_u$. As opposed to continuous simulation, in discrete event simulation, state changes of the simulated system are assumed to happen at discrete points of the virtual time and are thus controlled by uncontinuous functions, resulting in a succession of *events*.

DES can be used to simulate stochastic processes. In a stochastic process, each state is partially but not fully determined by the previous one. Typically, a stochastic process can have one or more deterministic *arguments*[1] and their values range over an index collection of non-deterministic random variables X_i with certain probability distributions. Such functions are equally known as *realisations* or *simple paths*. The view of a stochastic process as an indexed collection of random variables is the most common one. The *events* to be executed are bound to the set of random variables X_i, that determine which event will be executed and when. A simulation executes events in nondecreasing time-stamp order so that the virtual time (the time-stamp of the last executed event) never decreases.

Indeed, the occurrence of an event typically causes four actions: (a) progression of the virtual time to the *timestamp* of the simulated event; (b) changes of the state of the simulated system; (c) scheduling of new events and (d) descheduling of other events. Thus the basic data structure of a DES program consists of: (i) a virtual simulation clock; (ii) a timestamp ordered list of pending events and (iii) the state variables.

---

[1] We consider the *time* as always present among the arguments.

3.1 Parallel and Distributed Discrete Event Simulation (PDES and DDES)

In this summary, we deal with *parallelism at model function level*. In particular, we focus on methods which make intensive use of multiprocessors architectures for DES and which can be classified in between the following two classes: *parallel discrete event simulation* (PDES) and *distributed discrete event simulation* (DDES).

In PDES and DDES, a simulation model is partitioned into regions or domains[2]. Each region is simulated by a so-called *logical process* (LP). Each LP consists of [23]: (i) a spatial region R_i of the simulated system; (ii) a simulation engine SE_i executing the events belonging to the region R_i and (iii) a communication interface, enabling LPs to send messages to and receive messages from other LPs.

LPs are mapped onto distinct processors with (as an assumption) no common memory. Thus, every LP can only access a subset of the state variables $S_i \subset S$, disjoint to the state variables assigned to the other LPs. The simulation engine SE_i of each LP processes two kinds of events: *internal* events which have no direct causal impact on the state variables held in the other LPs and *external* events that can change the state variables in one or more other LPs. If an external event is processed, the LP holding the state variables that are to be changed is informed through a message sent by the LP. The message routing between the LPs is done by a communication system, connecting the LPs. Incoming messages are stored in input queues, one for each sending process.

Due to different virtual time progression within the various LPs, the causality principle is hard to be guaranteed and special considerations have to be made to obtain the same simulation results from DDES as from sequential DES. The two most commonly used synchronization protocols in DDES are: (i) The *conservative* (or Chandy-Misra) synchronization protocol developed by Chandy and Misra [24], [25] and (ii) the *optimistic* (or time warping) simulation protocol based on the virtual time paradigm proposed by Jefferson [26].

3.2 Conservative vs. Optimistic

The basic idea of the conservative protocol is to absolutely avoid the occurrence of causality violations. It is granted by strictly freezing the computation of an event e with virtual time (VT) t_e until when no messages with VT lower than t_e will be received. Under the assumption of FIFO message transport, this is achieved by only simulating an event if its VT is lower than the minimum of the timestamps of all events in all input queues.

An obvious problem arising in conservative simulation is the possibility of deadlocks [27]. Some deadlock resolution schemes have been developed during the last years. Among them, the more interesting are: [24] which avoids deadlock by the use of NULL-messages and [25] which detects and recovers deadlock, in advance. Some optimization protocols are discussed in [28], [29] (Null-messages

---

[2] For the purposes of this paper, only spatial decomposition is considered; however, the concepts illustrated here are also suitable for decompositions into general domains.

approach), in [30] (NULL-messages on request), in [31], [32] (lookahead computation), and in [33], [34] (local deadlock detection).

In contrast to the conservative protocol, there is no blocking mechanism in the optimistic one. An event is simulated even if it is not safe to process. Thus, causality errors are allowed to occur, but are later detected and solved. To guarantee causality, a mechanism called *time warp* or *rollback* has been designed. Time warp is optimistic in the sense that each processor P_0 executes events in timestamp order under the optimistic assumption that causality is not being violated. At any point, however, P_0 may receive a straggler event E, that should have been executed before the last several events already executed by P_0. In this case, it rolls back to a check pointed system state that corresponds to a time-stamp which is a global minimum among all VT (global virtual time) and less than the straggler's time-stamp. Processor P_0 resumes its execution from this point, and P_0 processes E, in the right time-stamp order.

A successful optimistic DDES minimizes the runtime costs of (i) state-saving system state, (ii) rollback, (iii) global virtual time (gvt) computation, and (iv) interprocessor communication.

3.3   Characterization of the Gillespie SSA as a PDES Algorithm

From a computational point of view, a biochemical system designed to be simulated with SSA can be seen as a collection of interacting processes, where each process can be in a different state among a set of discrete states. In this view, *biochemical species* are treated as *processes* that are able to perform a set of actions, changing their state in response to an external or internal action; *reactions* can be codified as *events* that are composed of a number of complementary actions, so that the execution of a reaction results in a simulation event that executes two (in the case of mono-molecular reactions) or more (in the case of bi-molecular reactions) actions in two or more processes (see Fig. 2).

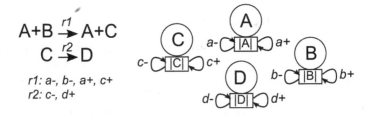

Fig. 2. A set of species and a set of reaction (left) represented as a set of processes and events (right). Each event is composed by two or more actions that modify the state of each process, typically decreasing or increasing the counter for the cardinality of the corresponding species.

So, a *biochemical system* $S = (P, E)$ can be seen as a set of processes $P = \{p_1, ..., p_n\}$, each holding a set of states and a set of actions that can be performed to modify its state, and a set of events $E = \{e_1, ..., e_m\}$, each composed by a set of actions; typically, for every process p there will be two actions $(a+, a-)$ in charge of decreasing and increasing the counter for the cardinality of the corresponding species. However, it is possible and sometimes useful to add additional state variables and corresponding actions.

It is easy to see that, following this computational view, the simulation of a biochemical system with the Gillespie algorithm becomes a DES, where event times are generated by sampling an exponential distribution. The fact that times are generated by an exponential distribution leads to some insights in how this particular DES can be parallelised. In particular, we will show that it is almost never convenient to parallelize biochemical systems by using a conservative approach. In support of our analysis, we shall consider a *dependency graph* between events, defined as the graph of reactions introduced by Gibson and Bruck [22].

Definition 1. *Let Reactants(e) and Products(e) be the sets of reactants and products, respectively, involved in the event e.*

Here, for *reactants* we indicate the processes whose actions decrease the cardinality of their state variable, identified with the name of the process and a '-' suffix. For example, the event $e1$ in Fig. 3 is composed by the actions $\{a-, b-, c+\}$; the actions $a-$ and $b-$ modify the state of A and B, so $Reactants(e1) = \{A, B\}$. *Products* are defined in a similar way as the processes whose actions increase the cardinality of their state variable.

Definition 2. *Let DependsOn(e) be the set of processes whose state change affects the execution time of the event e, and Affects(e) the set of processes whose state changes when an event is executed.*

Following the description of the SSA given in Sec. 2, $Reactants(e) = DependsOn$ (e). Typically, $Affects(e) = Reactants(e) \cup Products(e)$, or better, the set of processes on which the actions in e act. Sometimes, when two actions are complementary (i.e. one cancels the effects of the other), the set can be a little smaller. This is the case of the event $e3$ in Fig. 3, where $e-$ cancels $e+$ and $Affects(e3)$ can be reduced to $\{D, F\}$.

Definition 3 (Dependency graph). *The dependency graph of a biochemical system S is a directed graph $G(V, E)$ in which the set of nodes V corresponds to the set of events and there is a directed edge between each pair of nodes $(V(e1), V(e2))$ if and only if $Affects(e1) \cap DependsOn(e2) \neq \emptyset$*

The dependency graph can be used to show that the dependencies within reactions, united with the times sampled from an exponential distribution, in many cases lead to the need for sequential execution.

Definition 4. *Considering a system S, its dependency graph can be partitioned into a set of strongly connected components. We call the set of processes and events belonging to a strongly connected component of cardinality greater than one a subsystem (see Fig. 4).*

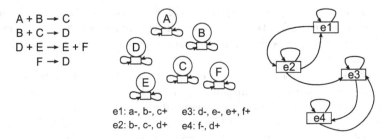

Fig. 3. The dependency graph (right) for a simple biochemical system (left)

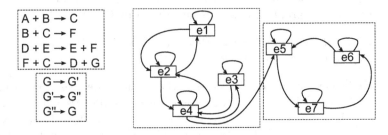

Fig. 4. A biochemical system partitioned into subsystems

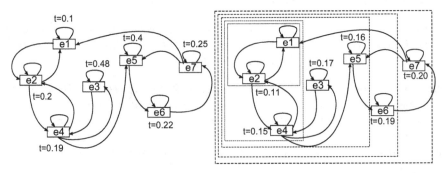

Fig. 5. Due to the exponential distribution used to generate execution times, the execution of an event can lead to the re-computation of the times of all the events in the same subsystem during the successive simulation steps

Proposition 1. *The execution of an event may lead to the need to recompute the next execution time for all the events in a subsystem.*

Whenever an event is executed, the next execution time of the events depending on it, i.e. its neighbours in the dependency graph, must be updated. Since times are exponentially distributed, there is no lower bound that guarantees us that the times we are going to recompute will be higher than a certain threshold.

The events with the new, lower, timestamps will in turn lead to the need for recomputing the time of other events, with the possibility of generating lower timestamps for the events they affect. By definition, in a strongly connected

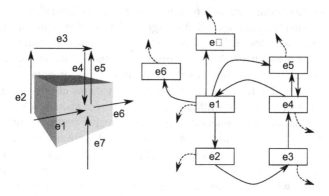

Fig. 6. Due to the diffusion events, a reaction-diffusion system has only one single subsystem

component there exists a path between any two vertexes, so it is possible that the generation of new times ripples and affects all the events in the subsystem (see Fig. 5). From this proposition, we can immediately derive two corollaries:

Corollary 1. *In a subsystem, the absence of causality errors is guaranteed whenever actions are executed in increasing time stamp order (*zero lookahead*).*

Corollary 2. *Since the absence of causality errors is guaranteed only if actions are executed one after the other, a pure conservative approach to PDES -which allows actions to be executed only when they cannot incur in causality errors- has a lookahead of zero, leading to a serialized execution where no speedup is possible.*

In [35], the authors considered several techniques for obtaining the *lookahead* necessary for concurrent execution of events under the conservative approach, such as artificially inserting lookahead into the computation and relaxing ordering constraint. We examined these approaches as well, and we came to the conclusion that using these techniques would lead to unacceptable compromises concerning the accuracy of the simulation.

An alternative approach for having some lookahead even in presence of exponential distributed random numbers is *pre-sampling*. Pre-sampling is a technique proposed by Nicol [36] for computing lookaheads in queueing network simulations with exponential distributed service time, and then used also for federated military simulations by Loper and Fujimoto [37]. At a glance, this technique seems to be applicable even in our domain. It carries a number of problems that makes it infeasible for our simulations. As noted by Nicol and Fujimoto, the service time variation has a strong effect on speedup. Under high variation, very small lookahead values are possible, meaning that lookahead is computed more often, thereby incurring in increased overhead. Furthermore, they also notice that rich interconnections between simulated entities, such as those used for simulating a continuous spatial environment, cause increased uncertainty in future behavior, resulting again in small lookaheads, with poor performances especially when using exponential distributed times.

Fujimoto conclusions that this technique requires (i) fixed sized time intervals, (ii) the same distribution for all messages, (iii) precise timestamps with few random number samples and (iv) knowledge concerning the number of messages produced in the near future [37] convinced us to discard it, as reaction-diffusion biochemical simulations do not meet any of these requirements.

Indeed, it is possible to make two crucial observations about reactive-diffusive simulations: (a) in a reaction-diffusion systems where species are free to diffuse in every direction, the dependency graph for diffusive events is fully connected; thus, the whole system is made of a single big subsystem (see Fig. 6) and (b) many biological systems show a little number of big subsystems; compounds, molecules and enzymes in a cell are reused over and over, forming big interconnected networks with loops. Indeed, regulation and transcription processes are often based on feedback loops, that shows up as connected components (subsystems) of dependency graphs.

In conclusion, the SSA can be characterized as a DES. Of the two main approaches to parallelize DES, the optimistic one is the most promising: as the two corollaries show, a pure conservative approach, united with exponentially distributed times and the particular dependency structure of biochemical systems, is very likely to perform poorly.

4　An Optimistic Reaction-Diffusion Simulator

As a proof of concept, we designed and developed a parallel stochastic reaction-diffusion simulator. While designing the simulator, we kept three goals in mind: (a) *correctness*: the simulator must respect the assumptions underlying the Gillespie method and its extension; (b) *scalableness*: the addition of further processing power must result in an increased execution speed-up; (c) *fastness*: the speed measured after running on a single processor must be comparable with that would be achieved if the simulator was strictly sequential.

4.1　Distributed Simulator Design

The first goal can be met by implementing the extension of the Gillespie method with diffusion events we introduced in Sec. 2.1, and the second and third goals can be fulfilled by using an approach based on PDES with an optimistic scheduling policy, as discussed in Sec. 3.

Notice that these two objectives must be considered together, as they heavily influence each other. Some methods, like the one presented in [38], chose to ignore correctness, violating the assumptions made by Gillespie and the properties stated by Bernstein with the aim to obtain fast parallel execution through volume subdivision. The algorithm, as the authors themselves say, can be useful in some cases, but it is not correct in a general sense. Indeed, when the spatial localization of molecules becomes important for the purposes of the experiment, the algorithm produces incorrect results.

For an effective implementation of the simulation algorithm as a PDES, spatial structures and partition of reactions into sub-volumes must be provided.

Moreover, state information should be maintained in a decentralized way, avoiding to keep global shared state informations, whenever it is possible. Partial local state can be processed and updated concurrently by different processors. The Next Subvolume Method (NSM) and algorithms derived from it employ spatial partitioning into sub-volumes, but they maintain information of execution times in global data structures; therefore they are not immediately adaptable to a parallel environment (even if a distributed version of the algorithm was recently proposed by Jeschke et. al. [35]).

We take a slightly different approach with respect to the NSM; we also divide reactions into sub-volumes (*cells*), but we consider each cell on the two or three-dimensional grid as an almost autonomous entity. We assume that every cell knows and stores its local information: concentrations of species, diffusion and reaction rates, next reaction time, as well as references to its neighbours. In each cell there are some dependency relations, both between species inside the same cell and between those in neighbour cells that can diffuse into it (see Fig. 7a). We have noticed that each cell on the grid can *evolve* (i.e. execute simulation events) independently from the other cells if the executed events do not violate the restrictions imposed by the dependencies. Following the optimistic approach, we let each cell evolve independently, up to a diffusion event occurs. When a neighbour notifies to the current cell a diffusion event with a clock T_{diff} smaller than the current clock T_{act}, reactions with times between T_{diff} and T_{act} are marked as *straggler*. So, we rollback every action executed within T_{diff} and T_{act}, recompute propensities and reaction times and restart the simulation of the events in that cell from time T_{diff}.

In our initial implementation, each cell was mapped onto a logical process (LP). The assignment of logical processes to physical processes may be done either dynamically, possibly by using a load balancing algorithm, or statically, by exploiting spatial locality to reduce communication overhead.

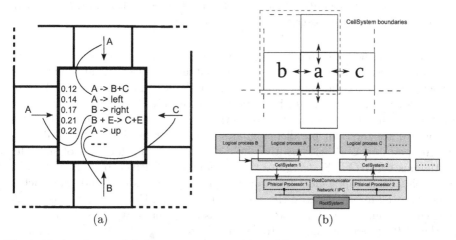

(a)   (b)

Fig. 7. Each cell is modelled as a process in a PDES (a). Cells are grouped into systems in order to reduce communication overhead (b).

4.2   Performance Considerations

The third goal, *fastness*, is not easy to achieve because a lot of practical, real world considerations have to be taken into account. The SSA was designed to run efficiently on hardware of the late '70; and it is indeed very efficient. An efficient implementation of the Gillespie algorithm can process and simulate roughly 10^5 reaction events per second; that is, a simulation loop takes approximately 10000-30000 CPU cycles to execute. Since a simulation loop is so fast, it is really difficult to speed it up by means of a parallel architecture. Execution of diffusion or reaction events on different processors requires synchronization in order to exchange messages. In the best of the hypotheses, processes can run on a single multi-core machine, where communication is done using shared memory and mutexes. According to the literature and to our own tests, even in this case the mere cost of context switching and proceeding the execution on a different thread (roughly 5000 CPU cycles) can easily result comparable to the loop time (see Fig. 8).

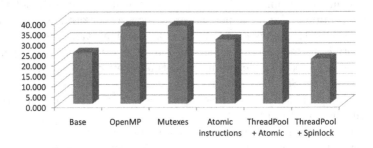

Fig. 8. The execution time of a parallel simulation (running on 2 processors) using various techniques of synchronization and inter-thread communication, compared to a serial simulation (Base). Notice that the overhead for running on multiple threads actually *increases* the execution times in all but the last case, where we used a pool of threads and hand-written assembly code for synchronization.

On a shared memory architecture, the problem is even worse. Even if current HPC architectures can rely upon very low-latency connections and upon very efficient message passing implementations (like the MPI interface we used), communication overheads can vanish any performance gain. For this reason, we chose a coarser granularity, in order to reduce the overhead to the minimum. For this reason, we designed our parallel simulator in a hierarchical way (Fig. 7b): cells are grouped into *Cell Systems*, that hold the partial state for a set of spatially contiguous cells. *Cell Systems* are then grouped and driven by a *Root System* that holds some topological information on the Cell Systems and that caches some essential information on the system global state. Every *System* has a specialized communicator. The *Root System* has a communicator based on MPI to let the *Cell Systems* it manages to communicate with each other across processor and machine boundaries. The *Cell Systems* have a single threaded, shared memory communicator in charge of maximizing the performance and reduce the overhead on a single processor or core. A further layer can be added with the aim to manage groups of *Cell Systems* which execute on different CPUs or cores on the same computation node, i.e. on a machine that shares the same memory and that does not need for network communication or message passing in case of inter-groups interactions.

```
CELLSYSTEM ():
  while  true  do
    NextAction := FastestCell().FastestAction;
    StateChange := Action.Execute();
    History.Add(StateChange);
    UpdateClock(StateChange);
    if  Action.IsDiffusion()
      if  Action.TargetCell ∉ CellSystem.Cells
        RootComm.Notify(StateChange);

    else
        Action.TargetCell.Notify(StateChange);

    if  RootComm.HasNotification
      Event := RootComm.HasNotification;
      switch  Event.Type
        case  ROLLBACK  :
          DoRollback(Event.Time);

        case  DIFFUSION  :
          Event.TargetCell.Notify(
              Event.DiffusionAction);


ROOTSYSTEM ():
  while  true  do
    Timer := StartTimer();
    Event := WaitForEvents(Timer, RootComm);
    switch  Event.WakeReason
      case  TIMERTICK  :
        SendCheckpointCommand(GlobalTime);
        SystemState := RecvCheckpointData();
        DoCheckpoint(SystemState);

      case  COMMUNICATION  :
        switch  Comm.Type
          case  ERROR  :
            BroadcastRollback(COMM.Time);

          case  DIFFUSION  :
            TrgtSystem :=
                LookupSystem(COMM.SourceCell);
            TrgtSystem.ForwardDiffusion(COMM);
        CurrentGlobalTime := Event.UpdateTime();
```

Fig. 9. Pseudo-code for *CellSystem* and *RootSystem*

Cells and *Cell Systems* communicate through a consistent interface, that is *transparent*, and that allows cells to communicate any diffusion information without taking care of the hierarchy. To communicate a diffusion from the cell C_1 to the cell C_2, C_1 sends a message to its *Cell System*; if C_2 is on the same physical processor (i.e. it belongs to the same Cell System), the information is directly propagated. If instead the *Cell System* realizes that C_2 does not belong to the set of cells it manages, it forwards the information up to the next System, until it reaches a System that knows C_2 or until it reaches the *Root System*. In the second case, the information is propagated using inter-thread communication or MPI messages (see the pseudo-code in Fig. 9).

The *Root System* is also responsible for checkpointing the system state, computing the global virtual time and propagate rollbacks, if necessary. In order to minimize the interprocessor communication, each *Cell System* has an incremental state history held in its own memory, as a queue of performed events. During the checkpoint phase, the *Root System* receives the partial state histories from the *Cell Systems*, computes the GVT, and commit all the events up to it. The commit is done by saving the system state to disk in an incremental way or, in alternative, by reconstructing the complete state on the fly. Each *Cell System* can then flush its own history up to the new GVT.

When a *Cell System* receives a straggler event, first it examines its queue and then it marks all the events with time-stamps greater than the straggler's one, performing a very quick rollback. If these events involve other *Cell Systems*, it informs the *Root System*, that take care of forwarding any rollbacks to the other *Cell System(s)*. These examine their partial state and, if necessary, perform rollback on their state history. At this point, the *Root System* informs the *Cell Systems* to resume their computations. Since cells are grouped into *Cell Systems* -which are called LPs in PDES terminology- secondary rollbacks, and thus propagation of rollbacks, are very infrequent.

Care is taken that the two operations of checkpointing and rollback do not interfere with each other by means of a barrier.

5   Example

The simulator we developed accepts an input file that specifies reactions, reaction rates and diffusion coefficients, as well as the initial location of the chemicals. We also developed a visualizer, that is able to read the execution traces produced by the simulator and display them as a 3D rendering of the simulated volume.

We tested our simulator with some models (enzymatic reactions, oscillatory networks, chemotaxis pathway) under realistic conditions: most or all the molecules not attached to membranes have been let to move and, mostly important, the diffusion coefficients have been set always higher or at least comparable to the reaction rates. Such conditions obviously increase the number of messages sent, making harder for our simulator to appropriately scale. However, it is fundamental to provide a realistic model that respects the assumptions we made [20].

```
var predator : rate 100;
var prey : rate 100;

predator + prey -> predator + predator [55];
prey -> prey + prey [15];
predator -> nil [10];
```

Fig. 10. The input file for the 3D Lotka-Volterra model

As an example, we introduce a spatial version of the Lotka-Volterra predator-prey model. This model was chosen because it is simple but realistic; many other algorithms proposed in the same fields use completely artificial scenarios, with diffusion and reaction rates that are unrealistic (especially the ratio between them). Furthermore, a more complex model would not have contributed to the discussion.

This model allowed us to use rates taken from literature. With these using these rates, the model exhibits a different and interesting behaviour when ran in an environment that includes spatial information [39]. The results we obtained (see Fig. 11) are consistent with what we expected and with what is found in the literature [39].

This model allowed us to perform some initial performances estimations, listed in Table 1. We measured the execution time of the serial version of the algorithm, where all the inter-process communications were removed and substituted with direct manipulation of data structures in shared memory, and of the optimistic

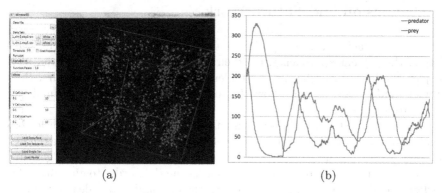

(a)   (b)

Fig. 11. A time-step of the Lotka-Volterra simulations (a) and the variation in cardinality of each species over time (b)

Table 1. Times in second for the execution of $5 \cdot 10^4$ simulation steps, 400 entities, (min/max/avg of five runs), and speedup for the parallel algorithm (*: on a 3D grid)

N cells	Serial	2 Cores		5 Cores		12 Cores	
256	1.5	14.8	0.1x	-	-	-	-
10000	13.1	10.7	1.22x	-	-	-	-
16384	17.4	(12.3/15.4/13.5)	1.29x	(9.1/10.1/9.4)	1.86x	-	-
26896	64.1	(34.7/42.8/38.7)	1.66x	(14.2/17.2/15.1)	4.25x	(7.0/9.4/8.0)	8.06x
32768*	75.8	(42.7/47.3/45.3)	1.67x	(18.4/20.7/19.2)	3.95x	(16.7/17.1/16.9)	4.49x

parallel algorithm. The hardware used for the simulation consists of PCs with AMD Opteron 64-bit CPUs at 2.4GHz, 4GB of Ram, interconnected with a 10Gbps Infiniband connection. We can observe that the overhead is significant when dealing with a small 16x16 2D grid, for a total of 256 node; the overhead starts to be less heavy starting with a 100x100 2D grid. As the grid becomes larger and larger, given a fixed number of subsystems, the diffusion events between different subsystems becomes less frequent. Note that a number of cells in the tens or hundreds of thousands is not unrealistic; for example, data for the last row of Table 1 were obtained for a 32x32x32 3D grid.

These preliminary figures are far from being complete performance measures, but they give an indication of the feasability of our approach. Since we do not have implemented or investigated work subdivision or load balancing techniques at this stage, we expect our method to perform well when each subsystem have to deal with the same amount of work, e.g. when there are not too big differences in the total concentration of elements (or *crowding*), with the same order of magnitude of activity. Note that concentration and activity of different compounds can vary significantly without any problems. Furthermore, using *Cell Systems* as computational units, we expect that our method performs well even in the hard but very common case of diffusion rates higher than reaction rates, as seen in the previous example.

6   Conclusion and Future Work

One of the obstacles on the way of computational systems biology is the scalability of the current approaches, i.e. their ability to deal with bigger and more complex models. With the aim to understand higher level behaviours, these complex models need for both powerful modelling tools and efficient simulation engines to analyse them.

In this paper we tackled the problem of designing a parallel simulator for biochemical systems, based on the theory developed by Gillespie, from both a theoretical and a practical point of view. The design of parallel and distributed algorithms requires indeed both a strong theoretical background, in order to guarantee that the designed algorithm is equivalent to the serial one, and a good deal of practical tricks and experience in order to make it really scalable and efficient.

Here we presented some first steps in this direction; although the results we obtained so far are promising, a lot of work needs to be done. In particular, Jeschke et. al. [35] conducted a parallel research on the same topic, focusing on the analysis of communication costs and on sizing of the window for optimistic execution in a distributed grid environment. It will be interesting to incorporate their studies and analysis of the window size to our framework, to see which are the differences between their grid-based and our HPC based approaches. Other problems we need to face are the analysis of the obtained data, whose dimension grows at an impressive rate when dealing with spatial simulations, load-balancing techniques for workload subdivision and analysis of the rollback mechanisms on different biochemical systems. Finally, we would like to perform an in-depth study of the performances, with different checkpoint frequencies, different number of nodes, different policy of cell allocation between nodes and different state saving strategies.

References

1. Fujimoto, R.M.: Parallel discrete event simulation. Comm. ACM 33(10), 30–53 (1990)
2. Ewing, G.C., McNickle, D., Pawlikowski, L.: Multiple replications in parallel: Distributed generation of data for speeding up quantitative stochastic simulation. In: Proceedings of the 15th Congress of Int. Association for Matemathics and Computer in Simulation, pp. 397–402 (1997)
3. Glynn, P.W., Heidelberger, P.: Analysis of initial transient deletion for parallel steady-state simulations. SIAM J. Scientific Stat. Computing 13(4), 904–922 (1992)
4. Newman, M.E.J., Barkema, G.T.: Monte Carlo Methods in Statistical Physics. Oxford University Press, Oxford (2000)
5. Glynn, P.W., Heidelberger, P.: Analysis of parallel replicated simulations under a completion time constraint. ACM TOMACS 1(1), 3–23 (1991)
6. Glynn, P.W., Heidelberger, P.: Experiments with initial transient deletion for parallel, replicated steady-state simulations. Management Science 38(3), 400–418 (1992)
7. Lin, Y.B.: Parallel independent replicated simulation on a network of workstations. ACM SIGSIM Simulation Digest 24(1), 73–80 (1994)

8. Yau, V.: Automating parallel simulation using parallel time streams. ACM TOMACS 9(2), 171–201 (1999)
9. Hybinette, M., Fujimoto, R.M.: Cloning parallel simulations. ACM TOMACS 11(4), 378–407 (2001)
10. Bononi, L., Bracuto, M., D'Angelo, G., Donatiello, L.: Concurrent replication of parallel and distributed simulation. In: Proceedings of the 19th ACM/IEEE/SCS PADS Workshop, pp. 430–436 (2005)
11. Streltsov, S., Vakili, P.: Parallel replicated simulation of markov chains: implementation and variance reduction. In: Proceedings of the 25th conference on Winter simulation, pp. 430–436 (1993)
12. Tian, T., Burrage, K.: Parallel implementation of stochastic simulation for large-scale cellular processes. In: Proceedings of of Eighth International Conference on High-Performance Computing in Asia-Pacific Region, pp. 621–626 (2005)
13. Burrage, K., Burrage, P.M., Hamilton, N., Tian, T.: Computer-intensive simulations for cellular models. In: Parallel Computing in Bioinformatics and Computational Biology, pp. 79–119 (2006)
14. Schwehm, M.: Parallel stochastic simulation of whole-cell models. In: Proceedings of ICSB, pp. 333–341 (2001)
15. Mazza, T., Guido, R.: Guidelines for parallel simulation of biological reactive systems. In: Proceedings of NETTAB 2008, Bioinformatics Methods for Biomedical Complex System Applications, pp. 83–85 (2008)
16. Gillespie, D.: A general method for numerically simulating the stochastic time evolution of coupled chemical reactions. J. Phys. Chem. 22, 403–434 (1976)
17. Gillespie, D.: Exact stochastic simulation of coupled chemical reactions. Journal of Physical Chemistry 81(25), 2340–2361 (1977)
18. McCammon, J.A., Harvey, S.C.: Dynamics of Proteins and Nucleic Acids. Cambridge University Press, Cambridge
19. Andrews, S.S., Bray, D.: Stochastic simulation of chemical reactions with spatial resolution and single molecule detail. Phys. Biol. (1), 137–151 (2004)
20. Bernstein, D.: Exact stochastic simulation of coupled chemical reactions. PHYSICAL REVIEW E 71 (April 2005)
21. Elf, J., Ehrenberg, M.: Spontaneous separation of bi-stable biochemical systems into spatial domains of opposite phases. Syst. Biol. 1(2) (December 2004)
22. Gibson, M., Bruck, J.: Efficient exact stochastic simulation of chemical systems with many species and many channels. J. Phys. Chem. 104, 1876–1889 (2000)
23. Ferscha, A.: Parallel and Distributed Simulation of Discrete Event Systems. McGraw-Hill, New York (1996)
24. Chandy, K.M., Misra, J.: Distributed simulation: A case study in design and verification of distributed programs. Comm. ACM 24(11), 198–206 (1981)
25. Chandy, K.M., Misra, J.: Asynchronous distributed simulation via a sequence of parallel computations. IEEE Trans. on Software Engineering SE-5(5), 440–452 (1979)
26. Jefferson, D.R.: Virtual time. ACM Transactions on Programming Languages and Computer Systems 7(3), 404–425 (1985)
27. Holt, R.C.: Some deadlock properties of computer systems. ACM Computing Surveys 4(3), 179–196 (1972)
28. Cai, W., Turner, S.J.: An algorithm for distributed discrete-event simulation - the 'carrier null message' approach. In: Proceedings of the SCS Multiconference on Distributed Simulation, vol. 22, pp. 3–8 (1990)
29. Wood, K.R., Turner, S.J.: A generalized carrier-null method for conservative parallel simulation. In: Proceedings of the 8th PADS Workshop, pp. 50–57 (1994)

30. Bain, W.L., Scott, D.S.: An algorithm for time synchronization in distributed discret event simulation. In: Proceedings of the SCS Multiconference on Distributed Simulation, vol. 19, pp. 30–33 (1988)
31. Groselj, B., Tropper, C.: The time-of-next-event algorithm. In: Proceedings of the SCS Multiconference on Distributed Simulation, vol. 19, pp. 25–29 (1988)
32. Cota, B.A., Sargent, R.G.: A framework for automatic lookahead computation in conservative distributed simulations. In: Proceedings of the SCS Multiconference on Distributed Simulation, vol. 22, pp. 56–59 (1990)
33. Prakash, A., Ramamoorthy, C.V.: Hierarchical distributed simulations. In: Proceedings of the 8th International Conference on Distributed Computing Systems, pp. 341–348 (1988)
34. Rukoz, M.: Hierarchical deadlock detection for nested transactions. Distributed Computing 4, 123–129 (1991)
35. Jeschke, M., Ewald, R., Park, A., Fujimoto, R., Uhrmacher, A.: Parallel and distributed spatial simulation of chemical reactions. In: Proceedings of the 22nd ACM/IEEE/SCS PADS Workshop (2008)
36. Nicol, D.M.: Parallel discrete-event simulation of fcfs stochastic queueing networks. SIGPLAN Not. 23(9), 124–137 (1988)
37. Loper, M.L., Fujimoto, R.M.: Pre-sampling as an approach for exploiting temporal uncertainty. In: PADS 2000, pp. 157–164 (2000)
38. Ridwan, A., Krishnan, A., Dhar, P.: A parallel implementation of gillespie's direct method. In: Bubak, M., van Albada, G.D., Sloot, P.M.A., Dongarra, J. (eds.) ICCS 2004. LNCS, vol. 3037, pp. 284–291. Springer, Heidelberg (2004)
39. Schinazi, R.B.: Predator-prey and host-parasite spatial stochastic models. The Annals of Applied Probability 7(1), 1–9 (1997)

Large-Scale Design Space Exploration of SSA

Matthias Jeschke and Roland Ewald

University of Rostock
Institute of Computer Science, Modelling and Simulation Group
Albert-Einstein-Str. 21, 18059 Rostock, Germany

Abstract. Stochastic simulation algorithms (SSA) are popular methods for the simulation of chemical reaction networks, so that various enhancements have been introduced and evaluated over the years. However, neither theoretical analysis nor empirical comparisons of single implementations suffice to capture the general performance of a method. This makes choosing an appropriate algorithm very hard for anyone who is not an expert in the field, especially if the system provides many alternative implementations. We argue that this problem can only be solved by thoroughly exploring the design spaces of such algorithms. This paper presents the results of an empirical study, which subsumes several thousand simulation runs. It aims at exploring the performance of different SSA implementations and comparing them to an approximation via τ-Leaping, while using different event queues and random number generators.

Keywords: Stochastic Simulation Algorithms, Performance Evaluation.

1 Introduction

When simulating biological systems that contain species with low amounts of elements, stochastic effects cannot be ignored. For example, in [1] intracellular viral kinetics are studied with two approaches, one being continuous and deterministic, the other being discrete and stochastic. The results reveal significant differences when the initial quantity of viral genetic material is very small. While the deterministic simulation always shows a spreading of the viral infection, the virus can be degraded before it is able to infect a cell if stochastic fluctuations are considered.

In general, the time evolution of a stochastic system with a discrete set of states $S = \{0, \ldots, K\}, K \in \mathbb{N}$ can be written as a *master equation* (see e.g. [2] for a detailed derivation). Basically, a first order differential equation $dP_k(t)/dt$ is defined for each possible state $k \in S$, which describes the time evolution of the state probability function. The master equation is then the set of coupled differential equations $\{dP_0(t)/dt, \ldots, dP_K(t)/dt\}$. While it is possible to analytically solve the master equation for systems with few states, this problem is usually intractable for larger state spaces. For example, if the state of a chemical system at time t can be described by the vector $\boldsymbol{X}(t) = (X_0(t), \ldots, X_9(t))$, with $X_i(t) \in [0, 9]$ representing the amount of the i-th species, the total number of states (and therefore the number of coupled differential equations that need to be solved) is 10^{10}. Note that when a final simulation time t_f is given, not all states might be reachable during the interval $[t_0, t_f]$ and hence it is possible to restrict the number of states to consider. For example, in [3] subsets of

M. Heiner and A.M. Uhrmacher (Eds.): CMSB 2008, LNBI 5307, pp. 211–230, 2008.
© Springer-Verlag Berlin Heidelberg 2008

the state space are used to efficiently approximate the solution to the master equation at different time points $t_0 < t \leq t_f$. It is shown that this approach can produce more accurate statistics and executes faster compared to performing many SSA realizations when both the time parameter t_f and the model are not too large.

In 1977, Gillespie presented a stochastic simulation algorithm (SSA, also often called Gillespie's algorithm) [4] that allows the generation of *exact* trajectories for such systems. Here, 'exact' refers to the probability of a generated trajectory, which equals its actual probability if the system satisfies certain preconditions (e. g., it has to be in thermal equilibrium). Repeatedly generating trajectories with SSA, i. e., sampling elements of the universe defined by the master equation, allows to exactly approximate it. The evolution of the system can be interpreted as a continuous Markov process with *reaction propensities* representing the transition rates between states. In chemical reaction networks, the reaction propensities rely on the amount and kind of existing particles, so that they need to be re-calculated after each simulation step.

SSA therefore requires considerable computational effort, and since the result of each run is merely one *sample* of the system's behavior, general statements can only be derived from repeated simulation runs with constant model parameters, i. e., simulation replication. Although single trajectories may reveal some interesting stochastic aspects of the given system, result analysis relies on statistics if general statements about the system behavior shall be made, for example in terms of mean values and variances. Unfortunately, gaining statistically significant results may require many hundred simulation replications. This situation is aggravated by additional methods that are usually employed on top of such simulation experiments, such as parameter estimation, sensitivity analysis, or optimization. They require the evaluation of many parameter combinations, each of which requires many replications in turn. To speed up computation, Gillespie's original algorithms have been greatly improved during the last decade (see section 3). However, the improvements are still up to some debate [5].

Another way to gain performance is to approximate the SSA. The τ-Leaping algorithm [6], for example, abandons the idea of single reaction events. Instead, the method carefully chooses time intervals of size τ and determines the number of occurrences for each reaction within the interval, so that it can produce trajectories with fewer iterations. As the method is not exact, it is possible that simulation outcomes are biased. Additionally, the increased performance comes at the expense of a more complex algorithmic description, which includes several parameters to adjust the algorithm's functioning (see section 3.2). Other approximate variants include, e. g., hybrid, multi-scale algorithms [7] or a nested hierarchy of SSA [8].

Apart from different opinions on the merits of each algorithm, their empirical analysis is hampered by factors that are hard to capture: the underlying hardware and operating system, the programming language and compiler, the quality of the implementation, and also the structure of the models used for benchmarking. The abundance of these factors hampers reproducibility and comparability of results and requires a thorough experimental evaluation to make reliable conclusions, since an algorithm's actual performance is often influenced by them [9]. For example, today's CPUs may give algorithms an advantage that are relatively slow in theory, but benefit from caching hierarchies [10]. Therefore, theoretical complexity analysis can be regarded as a first step toward understanding the benefits of an algorithm, but performance evaluation is mandatory for *all* implementations.

We address this need by conducting a large-scale performance evaluation for SSA variants as well as approximations, for which we also assess accuracy and various parameter settings. The paper is structured as follows: Sections 2 and 3 provide some background, i. e., they introduce some related work and survey the most important SSA variants. Sections 4 and 5 describe our benchmark model and experimentation methodology, while sections 6 and 7 analyze the results and conclude the paper.

2 Background and Related Work

We conducted the performance study with JAMES II, a modelling and simulation framework written in Java. It already provides support for several simulation approaches from Computational Biology [11], is extensible via a plug-in mechanism [12], and provides an experimentation layer that explicitly supports simulator performance evaluation. Moreover, performance data can be managed conveniently by a performance database, which facilitates result analysis [13]. In JAMES II, re-usable algorithms and data structures are provided as plugins of a certain *type*. They are managed by a central Registry, which lets every plug-in request other plug-ins for solving subtasks. This allows to reuse existing components and makes it easy to exchange one plug-in for a subtask with another. The resulting algorithms to be executed are tree-like hierarchies of sub-algorithms, where a parent node is relying on its child nodes to solve some parts of the given problem.

For example, Gibson and Bruck pointed out that random number generation and event queue implementation are crucial for the performance of their Next Reaction Method [14]. We could therefore regard their SSA variant as a parent node with two children: one is a generator of pseudo-random numbers, the other is an event queue. The performance of the overall algorithm is defined by such a tree and depends on *all* nodes, i. e., the random number generator, the event queue, and the SSA implementation itself. Gibson and Bruck's approach is usually implemented with a heap as event queue – but is this the best choice, and how important is it? Such questions can only be answered by exploring the design space, e.g. by evaluating to which degree the overall performance depends on each (sub-)algorithm. JAMES II supports such explorations by providing tested implementations of various common utilities for simulator development, including event queues [15] and random number generators (RNGs).

Algorithm performance evaluation has a long tradition in computer science. Seminal work on the problem of comparing algorithms for a given problem has been presented by Rice in 1976 [16]. He introduced a formal framework for the so-called *algorithm selection problem*. It subsumes various sub-problems, but the basic problem is to identify an algorithm $a \in \mathbb{A}$ that performs good on a given problem $p \in \mathbb{P}$. In JAMES II, the set $\mathbb{A}$ of available algorithms consists of all algorithm trees applicable to a given simulation problem. Identifying efficient methods now requires to explore the multidimensional performance space $\mathbb{R}^n$ for each $a \in A$. Rice defines a performance mapping $perf : \mathbb{A} \times \mathbb{P} \to \mathbb{R}^n$ for this, which is then mapped to $\mathbb{R}$ by a norm defined on $\mathbb{R}^n$. As will be discussed later, this norm usually depends on the user and the application at hand. These theoretical foundations are reflected in the performance database for JAMES II, which allows to store simulation problems, performance measurements, and corresponding algorithm trees in a concise and consistent manner [13].

Comparative studies on efficiency are quite common and cover virtually all kinds of algorithms, e. g., for database query resolution, load balancing, or simply sorting. If sufficient data regarding algorithm efficiency is available for a class of problems, these can be combined to form *algorithm portfolios* [17,18]. A portfolio consists of several algorithms to solve the same problem. If it is designed well, there is at least one very efficient algorithm for any problem of a given class. Similar approaches have been developed in the field of Problem Solving Environments, which tackle the algorithm selection problem in a more general manner (e. g., PYTHIA II [19]).

Regarding the class of SSA problems, special attention must be paid to *stiff systems*, which span multiple time scales. Here, the original SSA is not very efficient, as most computing power is spent on executing fast reactions that, when summed up, have only little impact on the system's state. An interesting numerical method for dealing with stiffness was presented in [20]: The state space is partitioned into sets of micro states (referred to as aggregates or macro states), such that slow transitions only occur between aggregates. When the aggregated state space is significantly smaller than the original one, an approximation for the evolution of the system can be given by a continuous time Markov process with states corresponding to the aggregates.

Work concerned with development and improvement of concrete SSA variants will be surveyed in the next section. Note that in the context of this work stiff systems and specific algorithms developed to handle these are not considered and left for future work.

3   SSA Variants

The original stochastic simulation algorithm as introduced by Gillespie simulates every single reaction that occurs inside the system and produces exact trajectories according to the underlying master equation. Variants of this basic algorithm belong to the class of *exact* SSA algorithms [21]. While being accurate, their execution time increases dramatically when simulating stiff systems or systems with a large number of elements. To speed up computation, *approximate* algorithms were introduced that perform "jumps" on the time line, i. e., aggregate reactions that occur in certain time intervals.

The following sub-sections will provide a short overview of the individual algorithm classes, but general terms and variables common to all SSA algorithms shall be introduced at first. The state of the system at time t is represented by the state vector

$$\boldsymbol{X}(t) = [X_0(t), \ldots, X_{N-1}(t)]^T, \; X_j, N \in \mathbb{N},$$

i. e., the vector that holds the number of elements for each of the N species in the model. The species can interact through M reaction channels $\{R_0, \ldots, R_{M-1}\}$ and a stochastic rate constant c_i is assigned to each reaction $R_i, i \in [0, M-1]$. This constant represents information about the physical properties of the reactants as well as the volume and temperature of the modeled system. For each reaction, two state change vectors

$$\boldsymbol{v_i}^R = [v_{i0}^R, \ldots, v_{i(N-1)}^R]^T,$$
$$\boldsymbol{v_i}^P = [v_{i0}^P, \ldots, v_{i(N-1)}^P]^T, \ v_{ij} \in \mathbb{N}$$

are constructed that describe the population change of the reactant and product species, respectively. Note that $v_{ij}^R > 0$ if species j participates as a reactant in reaction R_i and $v_{ij}^P > 0$ if species j is a product of R_i. With $\boldsymbol{X}(t) = \boldsymbol{x}$, the *propensity* $a_i(\boldsymbol{x})$ of a reaction R_i is the product of the stochastic rate constant c_i with the number of distinct reaction pairs, denoted by $H_i(\boldsymbol{x})$. In general, $H_i(\boldsymbol{x})$ can be written as

$$H_i(\boldsymbol{x}) = \prod_{j=0}^{N-1} h_{ij}(\boldsymbol{x}), \text{ with } h_{ij}(\boldsymbol{x}) = \begin{cases} \begin{pmatrix} x_j \\ v_{ij}^R \end{pmatrix} & \text{if } x_j > 0, v_{ij}^R > 0, \\ 0 & \text{if } x_j = 0, v_{ij}^R > 0, \\ 1 & \text{if } v_{ij}^R = 0. \end{cases} \tag{1}$$

For example, in case of a simple second order reaction equation $A + B \rightarrow C$, $H_i(\boldsymbol{x}) = |A(t)| \cdot |B(t)|$, i.e., the product of the number of individual elements of species A and B.

3.1 Exact Variants

Despite the variety of exact SSA implementations and optimizations, it is possible to identify algorithmic sub-routines that are common to all versions. Basically, each variant performs the following steps:

1. initialize state vector $\boldsymbol{X}(0) = \boldsymbol{x_0}$ and global time $t = 0$
2. main loop
   (a) let $\boldsymbol{X}(t) = \boldsymbol{x}$
   (b) $\tau, \mu \leftarrow SelectNextReaction(\boldsymbol{x}), \tau \in \mathbb{R}, \mu \in [0, M-1]$
   (c) update state vector: $\boldsymbol{X}(t + \tau) \leftarrow \boldsymbol{x} - \boldsymbol{v}_\mu^R + \boldsymbol{v}_\mu^P$
   (d) update global time: $t \leftarrow t + \tau$

While state and time update are equal for all algorithms, the procedure for determining the time and the index of the reaction that will occur next ($SelectNextReaction(\boldsymbol{x})$) might differ drastically between the approaches.

Direct and First Reaction Method (DRM, FRM). With the work that introduced the general principle of the stochastic simulation algorithm, Gillespie also provided two implementations of the SSA [4]. The first version is called the *Direct Reaction Method* and calculates successive next event times based on the sum of the individual reaction propensities. What reaction actually occurs is determined by randomly selecting a reaction with a probability that corresponds to its propensity.

The $SelectNextReaction(\boldsymbol{x})$ function for the DRM algorithm can be written as

1. Update the propensities $a_i(\boldsymbol{x})$, $i \in [0, M-1]$
2. Calculate the sum of all reaction propensities

$$a_g(\boldsymbol{x}) \leftarrow \sum_{i=0}^{M-1} a_i(\boldsymbol{x}) \tag{2}$$

3. Draw two random numbers u_1 and u_2 from the uniform distribution $U(0,1)$
4. Calculate τ as a sample from the exponential distribution $Exp(a_g)$

$$\tau \leftarrow \frac{-ln(u_1)}{a_g(\boldsymbol{x})} \tag{3}$$

5. Determine pending reaction with μ being the smallest integer that satisfies

$$\sum_{i=0}^{\mu-1} \frac{a_i(\boldsymbol{x})}{a_g(\boldsymbol{x})} < u_2, \ \mu \in [0, M-1] \tag{4}$$

6. Return τ, μ

In contrast, the second implementation, called *First Reaction Method*, samples event times for each reaction and proceeds with the reaction having the smallest time of next event. For this, the algorithm needs to sample M random numbers per iteration, which can be rather costly compared to the two samples the DRM needs (one for the next event time, the second for determining the reaction to execute). The $SelectNextReaction(\boldsymbol{x})$ function for the FRM as pseudo code:

1. Update the propensities $a_i(\boldsymbol{x})$, $i \in [0, M-1]$
2. Draw M random numbers $\{u_0, \ldots, u_{M-1}\}$ from the uniform distribution $U(0,1)$
3. Determine next event time and pending reaction

$$(\tau, \mu) \leftarrow (\min_{i \in [0, M-1]} \left\{ \frac{-ln(u_i)}{a_i(\boldsymbol{x})} \right\}, \ \operatorname*{argmin}_{i \in [0, M-1]} \left\{ \frac{-ln(u_i)}{a_i(\boldsymbol{x})} \right\}) \tag{5}$$

4. Return τ, μ

Next Reaction Method (NRM). DRM and FRM recalculate the propensities for *all* reactions after a state change. However, the propensity of a reaction is only changed when the amounts of its reactants have been affected by the reaction executed before. It is therefore unnecessary to recalculate propensities for reactions whose reactants were neither reactants nor products of the last reaction. Doing so introduces overhead, particularly for sparse reaction networks, i.e., loosely coupled networks in which every species only participates in a small number of reactions.

Following this line of thought, Gibson and Bruck [14] implemented a variant of the SSA that uses a dependency graph to determine the set of reaction propensities that have to be updated after processing a reaction. Additionally, Gibson and Bruck illustrated how the event times of affected reactions can be reused to calculate new event times by linear interpolation. Hence, it is only necessary to generate a single random number, which is used for generating the executed reaction's time of next event. NRM can be regarded as an extension of Gillespie's First Reaction Method. The dependency graph is constructed during initialization, followed by the calculation of reaction propensities and the first set of next event times for all reactions. The reactions are then enqueued in an event queue structure ordered by increasing event time. The first reaction is dequeued and executed. The $SelectNextReaction(\boldsymbol{x})$ procedure looks as follows:

1. Get the set D_μ of all reactions that depend on the last executed reaction μ from dependency graph and update propensities and event times

$$\forall \nu \in D_\mu : \tau_\nu \leftarrow \left(\frac{a_{\nu,old}(x)}{a_{\nu,new}(x)} \right)(\tau_\nu - t) + t \tag{6}$$

2. Requeue reactions
3. Draw random number u from uniform distribution $U(1,0)$
4. Update propensity and next event time for μ

$$\tau_\mu \leftarrow \frac{-ln(u)}{a_\mu(x)} \tag{7}$$

5. Enqueue μ
6. Dequeue reaction μ_{new} with the smallest next event time τ_{new}
7. Return τ_{new}, μ_{new}

Optimized Direct Method (ODM). In [22], Cao et al. analyzed the computational efficiency of different SSA implementations (DRM, FRM, NRM) and proposed an optimized version of Gillespie's original Direct Reaction Method. Based on a thorough cost analysis, two optimizations have been identified:

Reorder reaction indices: In the original DRM algorithm, the pending reaction is determined by summing up the weighted individual propensities until the sum exceeds the sampled uniform random number. If some reactions fire more frequently than others, e. g., in stiff systems, a reordering of the reaction indices can speed up this search. In the presented approach, reactions are ordered by decreasing firing frequency. Therefore, reactions having a higher propensity are processed first during the summation, making it more likely that the propensity sum already exceeds the threshold after only considering a few reactions (instead of summing up many reactions with a very low individual propensity). Cao et al. suggest to use pre-simulation for finding an optimal order of the indices.

Update propensities only when necessary: The second optimization was inspired by the Next Reaction Method. The ODM also uses a dependency graph to determine the reactions that need to be updated after the state has changed. Furthermore, instead of recalculating the propensity sum in every iteration, the old propensity for an affected reaction is simply subtracted and the new propensity added to the current value of a_g (see equation 2).

3.2   Approximative Variants

τ-Leaping. A step toward an approximative strategy has been made by Gillespie with the introduction of the τ-Leaping algorithm [6]. The basic idea is to allow larger time intervals and approximate the number of firings for each reaction during an interval. Let us assume that τ was selected as a suitable interval, then the state at time $t + \tau$ with $X(t) = x$ can be written as

$$X(t + \tau) = x + \sum_{i=0}^{M-1} K_i(\tau; x, t)v_i, \tag{8}$$

where $K_i(\tau; \boldsymbol{x}, t)$ is a random variable that denotes how often the reaction R_i has occurred in the time interval $[t, t + \tau)$. By defining a condition on the selection of τ, $K_i(\tau; \boldsymbol{x}, t)$ can be approximated by a Poisson distribution $P_i(a_i(\boldsymbol{x}), \tau)$ [23]. This condition is the *leap condition* and it restricts τ to be small enough, so that the sum of reaction propensities is nearly constant during the jump. But as the Poisson distribution is unbounded, it could happen that the number of reaction firings exceeds the amount of reactants at time t, resulting in negative species populations. Different approaches have been developed to prevent this. Tian et al. [24] use a bounded binomial distribution for determining the number of reaction firings. In [25], reactions are classified as either *critical* or *non-critical* at the beginning of each iteration. The class of a reaction depends on the number of available reactants. A next event time is calculated for both reaction classes. If the next reaction is non-critical, then no critical reaction may fire during the leap, otherwise only one critical reaction gets executed. In any case, the number of occurrences for each non-critical reaction is sampled from the Poisson distribution (see equation 8).

Gillespie also argued that τ-Leaping should be abandoned for some iterations if the selected interval τ is smaller than a multiple of $1/a_g(\boldsymbol{x})$, denoted by γ. This can be implemented by falling back to an exact SSA implementation for a predefined number of steps. Summing up, the τ-Leaping algorithm ca be written as follows (this is a shortened version of the algorithm presented in [26]):

1. Initialize initial state vector $\boldsymbol{X}(0) = \boldsymbol{x}_0$, global time $t = 0$, threshold parameter γ and number $h \in \mathbb{N}$ of SSA steps that get executed if leap is smaller than threshold $\gamma/a_g(\boldsymbol{x})$
2. Main loop
    2.1 $\boldsymbol{X}(t) = \boldsymbol{x}$
    2.2 Split set of reactions into critical and non-critical
    2.3 Calculate τ^n as the next event time for non-critical reactions
    2.4 if $\tau^n < \gamma/a_g(\boldsymbol{x})$ then perform h SSA steps and goto (2.1)
    2.5 Calculate τ^c as the next event time for a critical reaction
    2.6 if $\tau^n < \tau^c$
        – $\tau \leftarrow \tau^n$
        – No critical reaction fires during τ
        else
        – $\tau \leftarrow \tau^c$
        – Select critical reaction μ randomly, with point probability $a_\mu^c(\boldsymbol{x})/a_g^c(\boldsymbol{x})$, that fires only once during τ
    2.7 Sample for each non-critical reaction ν its number of firings during τ from $P_\nu^n(a_\nu^n(\boldsymbol{x}), \tau)$
    2.8 Update state and global time

There are a numerous τ-Leaping variants. In [27], Rathinam et al. introduced the implicit τ-Leaping for handling stiff systems. Another version, called K-leaping, was presented in [28] and overcomes the problem that occurs when a Poisson distribution is used and the number of firings for a reaction is so high that the resulting state change violates the leap condition. In this algorithm, a value for a variable K is calculated for each iteration, similar to the calculation of τ in the original τ-Leaping. This K restricts the total number of occurrences for all reactions. Based on this, the value for the leap τ and the number of firings for each individual reaction is determined.

4 Benchmark Models

We used three types of benchmark models for our design space exploration: a *Linear Chain System (LCS)*, a *Totally Independent System (TIS)* (both from [22]) and a *Cyclic Chain System (CCS)*. Every model consisted of N species $\{S_0, \ldots, S_{N-1}\}$.

Linear Chain System (LCS). The N species can undergo $N - 1$ reactions, with the product of reaction R_i participating as reactant for R_{i+1}:

$$R_n : S_n \xrightarrow{c_n} S_{n+1}, \; n \in [0, N - 2]$$

This model describes a loosely coupled system with one reaction affecting at most two propensity values. If the index of the executed reaction R_i is either 0 or $N - 2$, only the value for $a_i(x)$ needs to be updated, otherwise $a_{i+1}(x)$ needs to be adapted in addition. Considering this, the NRM is expected to perform better than the DRM, because the recalculation of $N - 1$ propensities per iteration is necessary for the latter.

Totally Independent System (TIS). In this model, a single reaction changes only the amount of its reactant, hence there are no dependencies between reactions:

$$R_n : S_n \xrightarrow{c_n} \emptyset, \; n \in [0, N - 1]$$

Similarly to the LCS, the DRM needs to update N reaction propensities, although only one propensity is actually affected. Again, the dependency graph of the NRM helps filtering out unnecessary updates.

Cyclic Chain System (CCS). The Cyclic Chain System encompasses the following reaction network structure:

$$R_n : \sum_{i=0}^{k} S_{(n+i) \bmod N} \xrightarrow{c_n} \sum_{i=k+1}^{2k+1} S_{(n+i) \bmod N}, \; n \in [0, N - 1], k \in \left[0, \left\lceil \frac{N-1}{2} \right\rceil\right]$$

It is easy to see that the execution of an arbitrary reaction affects in any case $2k + 2$ propensity values. The value of k determines the coupling of the system, with $k = 0$ representing a loosely coupled and $k = \lceil \frac{N-1}{2} \rceil$ a totally coupled system. With increasing k, the NRM needs to update more propensities and the questions arises if – in case of only few reactions and for some high value for k – the DRM might outperform the NRM. In this scenario, the performance of the DRM should heavily depend on the random number generator, as it always needs to generate an additional sample per iteration, compared to the NRM.

We restricted our study to these three models, as thoroughly evaluating even such a small set of models already requires considerable efforts (see section 5). Using synthetic benchmark models instead of real-world examples offers several advantages in the context of algorithm design space exploration:

- **Comparability:** As two of these models have been introduced by Cao et al. [22], it is possible to compare the runtime performance of the realizations. Large deviations may indicate subtle implementation errors or the dependency on additional, yet undiscovered, factors (such as specific hardware or compiler optimizations).
- **Ease of Implementation:** Due to their simple structure, the synthetic models can be generated automatically and are easily prepared for any SSA simulator. This is important for re-validating our results with other implementations.
- **Scalability:** The synthetic models are built to scale. Their size can be varied by adjusting N, which allows to analyze the efficiency of a simulator when the problem size grows. There are many algorithms that perform very well on small problems, but become more and more inefficient when the problem size is increased. Scalable models help to find the problem size for which an algorithm is most efficient.
- **Parameterization:** Parameters of synthetic models allow to investigate algorithm performance on *classes* of real-world models. For example, adjusting k for CCS controls the degree of interdependency between reactions.
- **Analytical results:** It is usually much easier to derive analytical results for synthetic models, since they exhibit a regular and simple structure. This may guide the developers in case of invalid results and could facilitate result analysis in general.

Nevertheless, the experimentation with models of real systems, such as the Heat Shock Response model also used for benchmarking in [22], is an important aspect, as it reveals where typical 'use cases' are situated in the problem space $\mathbb{P}$ (see section 2). A comparison with curated models from the *BioModels Database* [29] shows that our benchmark models cover a relevant subset of $\mathbb{P}$ with respect to the number of species and reactions, as well as the ratio of those (ratio between reactions and species (benchmark/BioModels): $\approx 1.0/1.52$; maximum species: $600/105$; maximum number of reactions: $600/300$). Note that this only compares fairly basic model properties but not, for example, the distribution of rate constants etc.

5   Experimentation Methodology

5.1   Evaluated Algorithms

JAMES II provides an implementation of DRM, two realizations of NRM, and the τ-Leaping method described in section 3.2. All methods rely on random number generators (RNGs). DRM generates two numbers per iteration, NRM generates one number per iteration, and for τ-Leaping this depends on the dynamics of the model, e. g., additional numbers need to be sampled if the τ gets rejected frequently.

Six JAMES II plugins for random number generation have been used in this study: the default Java RNG, a custom implementation of a linear congruential generator (LCG) with the same parameters as Java's RNG, the Mersenne Twister [30], a recursion-with-carry generator (Marsaglia's mother of all RNGs [31]), ISAAC (a cryptographically secure RNG, [32]), and RANDU, which is a classical LCG that is not of practical relevance any more, due to its strong correlations. Using random numbers for stochastic simulations is not a trivial task, since the pseudo-random numbers may correlate and therefore bias a stochastic simulation [33,34]. This is particularly dangerous when

RNGs are poorly initialized [35]. However, the SSA variants do not rely on high-dimensional tuples of random numbers, so that correlations should be very rare. We therefore focused our RNG-related investigations on runtime performance. Another important aspect of RNGs in the context of stochastic simulation is the size of their seed, as it limits the maximal number of trajectories that can be generated [36]. We initialized all RNGs with seeds of type long.

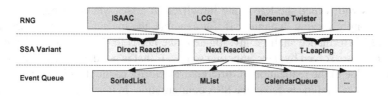

Fig. 1. Combinations of JAMES II algorithms for SSA

The NRM implementations also rely on another data structure, namely an event queue to manage the reactions and their time of next event. JAMES II provides 13 event queue implementations, including a simple sorted list, a heap, the MList (an event queue known for its good performance [37]), and the Calendar Queue [38]. This extends the design space to be explored from a couple of configurations to more than hundred: DRM and τ-Leaping should be executed with every RNG, which results in 12 configurations, but we have two NRM realizations and these need to be tested with all RNGs *and* all event queues. With two NRM variants, six RNGs, and 13 event queues, this results in $2 \cdot 6 \cdot 13 = 156$ additional configurations for NRM evaluation alone, i.e., 168 configurations when including DRM and τ-Leaping. Figure 1 illustrates the combinatorial explosion.

5.2 Performance Measurements

Before investigating the performance of SSA design alternatives, it has to be defined which aspects of their performance are of interest, and how these can be measured. An algorithm's performance has many facets, e.g., accuracy with respect to certain simulation outcomes, execution speed, memory load, network load, energy consumption, stability, and so on.

In principle, the n performance facets of interest can be expressed as an element of $\mathbb{R}^n$ ([16], see discussion in sec. 2). Now, the end user has to weight these aspects with regards to the problem at hand and the desired outcomes, which can be defined as a norm $\| \ \|$ on the performance space. An algorithm $a_1 \in \mathbb{A}$ has a better performance than an algorithm $a_2 \in \mathbb{A}$ for a problem $p \in \mathbb{P}$, iff $\|perf(a_1, p)\| \geq \|perf(a_2, p)\|$ (see sec. 2). This study is focused on two performance measurements: execution speed and accuracy. Weighting both measurements against each other should be left to the user.

Execution speed. Both mean and variance of an algorithm's execution time are relevant, since a large variance hints at its dependence on aspects of the simulation problem that have not been controlled during the experiment. For example, τ-Leaping may exhibit a large variance when tested with a bistable benchmark model, which has one equilibrium state that is disadvantageous for τ-Leaping's runtime performance (e. g., small amounts of species that lead to a frequent rejection of τ^n). The (potentially high) dependency of execution speed on input can be regarded as the risk of using the algorithm [39].

Two additional factors need consideration when measuring execution times: the noise introduced by hardware and operating system, and the potential bias introduced by the benchmark model. The first problem is common to all studies on algorithm performance; it can be resolved by replication. The second problem requires to limit the set of seeds with which the RNGs are initialized. It arises when a benchmark model exhibits strongly varying dynamics. If, for example, the impact of a sub-algorithm on the overall performance shall be investigated, one needs to ensure that *equivalent* input problems are solved by the competing setups. If that is not the case, execution time differences due to the realization of different trajectories cannot be distinguished from execution time differences due to the SSA configuration. In this context, both problems are solved by limiting the number of RNG seeds for execution time experiments and conducting numerous replications. Note that initializing two distinct SSA variants with equivalent RNGs would lead to *different* trajectories, as each variant uses random numbers differently. Hence, a limited set of RNG seeds only facilitates the comparison of sub-algorithms that have been plugged into the same SSA variant and do not change their execution logic.

Accuracy. Additional efforts are required to quantify the accuracy of approximative SSA variants. As generating trajectories with SSA realizations can be regarded as sampling the underlying master equation, the generated trajectories of, e. g., τ-Leaping and DRM should belong to the same universe. Each trajectory can be regarded as an *independent* sample from that universe: we initialized each RNG with a different seed for sampling trajectories, i. e., the RNGs produce independent pseudo-random numbers.

Unfortunately, the data to be tested is intractably large. Consider the Totally Independent System (TIS) with $N = 600$ species, each initially comprised of $X_i = 10000$ particles. Six million reactions have to be computed until the system halts, and their reaction times and indices define the trajectory. Using 6 bytes per reaction, i. e., four bytes for the reaction time and two bytes for the reaction index, a single trajectory still has a size of ≈ 36 Megabyte. Now, this sampling has to be repeated *many* times, which would result in trajectory data of many Gigabytes. To circumvent this problem, we sample the current state of the model for k fixed points in time. This limits the required storage per trajectory to k state vectors, e. g., $3 \cdot 600 = 1800$ integers in case of sampling the aforementioned TIS setup thrice.

Now, we can compare the sampled state vector sets S_1 and S_2, generated by two SSA variants for a fixed point in simulation time. The state vectors are checked species-wise for deviations in their empirical distributions: if, for example, the average amount of species x in S_1 is *significantly* higher than in S_2, both sets might *not* be sampled from the same universe and the approximative variant of the two may therefore be inaccurate with respect to the given species and the given point in time.

This allows to apply the well-known Kolmogorov-Smirnov (KS) test [40, p. 577 pp.] to check if the sampled distributions of two SSA variants deviate significantly for a single species. Note that this method can only falsify. Our null hypothesis is that all amounts of a species in S_1 and S_2 have been sampled from the *same* universe, i. e., their

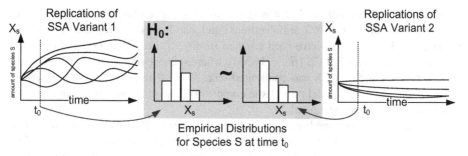

Fig. 2. Comparing the results of two SSA variants. Here for both SSA variants the state is saved during *each* replication at time point t_0, resulting in empirical distributions for each species S. The null hypothesis H_0 states that these distributions do not differ significantly.

distributions do *not* differ significantly. The basic idea is illustrated in figure 2. If the null hypothesis can be rejected by a test, it means that there is a statistically significant difference between the samples of both variants. Still, this species-wise approach does *not* suffice to prove that two sets of *state vectors* are sampled from the same universe, let alone two sets of trajectories. All exact SSA variants should be accurate by their very definition, but it is crucial to measure the accuracy of approximative methods. We chose DRM as a benchmark method and included NRM as another exact variant to control the quality of the tests.

5.3 Experiments

Our experiments are mainly focused on the two performance measurements introduced before: accuracy and execution speed. Accuracy has been investigated for the LCS model with $N = 101$ and only the first species is present with $S_0 = 10000, S_{i>0} = 0$. Its linear structure should amplify and propagate errors that could be hidden and counterbalanced otherwise. The model parameters are equal to those from [22]. DRM, NRM, and τ-Leaping have been used to sample 1000 trajectories each. Three state vectors have been recorded at times $t_i = (5.0, 15.0, 25.0)$.

All three benchmark models have been used for assessing execution speed. As depicted in figure 1, the combinatorial nature of the problem yields 156 configurations of NRM, six configurations of DRM, and nine setups for τ-Leaping: the default configuration and eight altered setups, as the τ-Leaping method from [26] is parameterizable with four parameters. ϵ influences the determination of τ, as the expected propensity change for each reaction is bounded by $\epsilon \cdot a_g$, with $0 < \epsilon \ll 1$. γ adjusts the lower threshold for τ; if τ is below the threshold, the algorithms switches to its SSA fallback and executes h SSA steps (see sec. 3.2). Finally, n_c controls the separation of critical and non-critical reactions. If there are less than n_c elements for any reactant participating in a given reaction, it is regarded as critical. Default values have been suggested in [26] for each of these parameters: $\epsilon = 0.03$, $\gamma = 10$, $h = 100$, and $n_c = 10$. Each of them was altered in both directions to assess their impact on τ-Leaping's execution speed. The lower and higher values were generated by multiplication with 0.1 and 10 respectively.

All in all, this amounts to 171 configurations that are applied to various problem instances (i.e., different RNG initializations) and also have to be replicated. The parameters for LCS and TIS have been adopted from [22], i.e., $N = 600$ and only the first species is present in LCS ($S_0 = 10000$), whereas every species is initialized with 10000 particles in TIS. CCS was configured with $N = 10$ and $k = 3$, so that it represents a well-connected model of realistic complexity. The rate constants have been set to 1.0 throughout all experiments.

The simulation runs have been executed on a Windows XP 64 workstation with two 2.5 GHz QuadCore Xeon Processors and 8 GB of RAM. It achieves a Java Scimark 2.0 [41] result of 482.2 points. Each instance of JAMES II was assigned to a single core, one core was always left to the operating system and the MySQL server hosting the performance database [13]. To reduce additional bias from observation code or database I/O, only state vectors for the experiments concerned with accuracy were stored. Another JAMES II plug-in was used to do so, and the data was written via JDBC into an additional MySQL database for simulation data. The R language was used for statistical result analysis [42].

6   Result Analysis

Performance. Figures 3, 4, and 5 show the results from performance experiments on the three model types defined in section 4. In figure 3, the performance for a variety of simulation configurations is plotted for the LCS and CCS model type. Note that we only plotted the results with different random number generators for the configurations *(DRM;*)*, *(NRMA;TwoList2;*)*, and *(NRMA;TwoList;*)*, since the data only indicated an insignificant dependency between execution speed and RNG implementation for

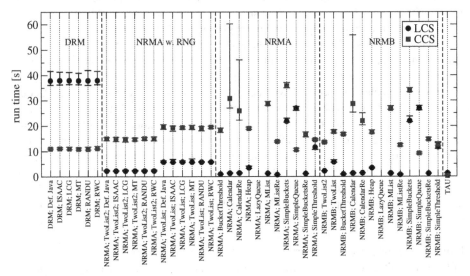

Fig. 3. Performance comparison of DRM, NRM, and τ-Leaping implementations with different random number generators and event queues for the LCS and CCS model

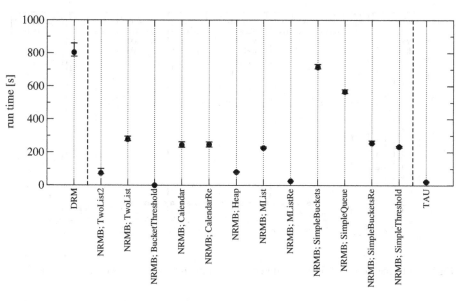

Fig. 4. Performance comparison of DRM, NRM, and τ-Leaping with different event queues for the TIS model

these two models. The default Java RNG has been used in the remaining configurations. These results are somewhat surprising, as the overhead of generating random numbers is expected to have a recognizable impact on SSA performance (e. g., in [14]). This is not the case, which leads us to the conclusion that other factors are more important for efficient SSA realizations.

For example, the execution speed of DRM and NRM differs greatly between LCS and CCS. In case of LCS, NRM outperforms DRM by several orders of magnitude. In contrast DRM computes the CCS model much faster than most NRM configurations. This is due to the different model structures, we suspect that the event queue overhead simply exceeds the overhead of recalculating all propensities for this CCS setup. Moreover, it is apparent that the second NRM implementation (NRMB) is slightly faster than NRMA for all configurations.

Figure 4 shows the performance of the SSA variants for the TIS model. Surprisingly, the configuration *(NRMB;SimpleQueue)*, which was very fast for the CCS model (cf. fig. 3), is much slower in comparison to other NRMB configurations. This can only be explained by differing event queue usage patterns imposed by the models; the performance of NRM is mostly determined by the used event queue and the model at hand. For example, the TIS model contains a lot of reactions with roughly the same propensities. Requeuing a reaction will therefore be very inefficient for the SimpleQueue, as it is based on a sorted list and reactions are likely to be requeued near its end. This is not so problematic for more sophisticated data structures like the MList.

The execution speeds of the τ-Leaping setups described in section 5.3 are shown in figure 5. Except for the outlier maximum value of $\epsilon = 0.3$ configuration, only one parameter setup exhibits interesting behavior, the one with $n_c = 1$, i. e., where nearly

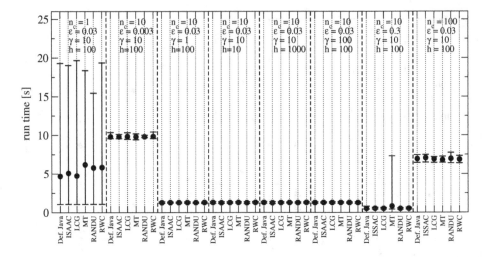

Fig. 5. Performance comparison of different configurations for the τ-Leaping algorithm (LCS)

no reactions are defined as critical. This results in a huge overhead, since τ-Leaping avoids negative species numbers by finding a small enough τ. Searching for this τ requires additional random numbers, which explains the increased impact of RNGs on execution speed.

Accuracy. As can be seen in figures 3 and 4, τ-Leaping outperforms any other SSA variant when it comes to execution speed. However, as table 1 shows, the null hypothesis regarding the equivalence DRM's and τ-Leaping's empirical distributions gets rejected for several state variables.

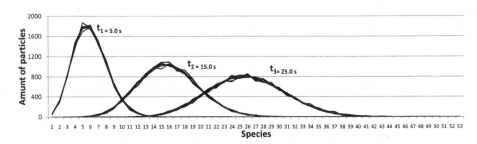

Fig. 6. LCS trajectories for time $t_1 = 5.0$, $t_2 = 15.0$, and $t_3 = 25.0$

Although the null hypothesis for the equivalence of DRM's and NRM's empirical distributions gets rejected as well, this is only the case for variables very far from the mean (see fig. 6). This is not the case for τ-Leaping. Histograms of the empirical distributions show a shift of the mean when compared with DRM's reference distribution. These findings correspond to the results presented in [26].

Summary. Our study revealed several interesting aspects of SSA performance. First of all, our design space exploration showed that a heap is not the most efficient event

Table 1. KS-Test results for NRM and τ-Leaping algorithm at time points $t = 5$ and $t = 25$

	t = 5						t = 25					
	NRM			τ-Leaping				NRM			τ-Leaping		
Species	p	D	Rejected	p	D	Rejected	Species	p	D	Rejected	p	D	Rejected
S_0	0.73	0.03	0	0	0.18	1	S_8	0.86	0.03	0	0.99	0.02	0
S_3	0.14	0.05	0	0.08	0.06	0	S_{14}	0.96	0.02	0	0	0.2	1
S_6	0.92	0.02	0	0	0.23	1	S_{20}	0.31	0.04	0	0.24	0.05	1
S_9	0.25	0.05	0	0	0.27	1	S_{26}	0.16	0.05	0	0	0.18	1
S_{12}	0.45	0.04	0	0	0.31	1	S_{32}	0.64	0.03	0	0	0.12	1
S_{15}	0	0.09	1	0	0.1	1	S_{38}	0.7	0.03	0	0	0.17	1
S_{18}	0	0.47	1	1	0.07	0	S_{44}	0.08	0.06	0	0.01	0.07	1
S_{21}	0.12	0.05	0	1	0	0	S_{50}	0	0.5	1	1	0	0
S_{24}	1	0	0	1	0	0	S_{56}	0	0.1	1	1	0	0
S_{27}	1	0	0	1	0	0	S_{62}	1	0	0	1	0	0

queue implementation for NRM (as was suggested in [14]). In fact, NRM execution speed depends heavily on the model *in combination* with the event queue. This result encourages further research on that dependence. RNG performance, in contrast, seems to only have a negligible impact on execution time for most setups. As expected, τ-Leaping was faster than any other SSA variant, but we proved that there is a statistically significant difference in its empirical distributions. Finally, we have to concede that the actual execution times presented in [22] are for some configurations much faster than ours, particularly for the DRM when applied to the TIS model. On the other hand, this illustrates the need for performance studies on every single simulation system.

7 Conclusions and Outlook

This paper presents the results of exploring the SSA design space on a large scale ($\approx$ 40.000 simulation runs, see section 6). It highlights the difficulties of conducting an unbiased comparison of SSA approaches and shortly sketches how these can be circumvented (sec. 5). We propose a statistical methodology to empirically detect inaccuracies of approximative variants, which shares some basic ideas with experiments presented in [26]. Our method employs a species-wise Kolmogorov-Smirnov test and is applied to several points in time.

Having quantitative information on the performance of SSA variants is often helpful. First of all, it identifies the most efficient implementations, which could be integrated into an algorithm portfolio. Users would not have to choose an algorithm on their own, this could be done by the simulation system instead. Another benefit is the qualitative knowledge gained by analyzing the data, which allows to focus further research efforts to the most promising algorithms and variants, which can then be pushed to their full potential. Finally, the mechanisms for conducting such a large-scale performance study facilitate a quick and unbiased evaluation of new algorithms or benchmark models within a predefined setup, which enables simulator validation against proven and tested implementations.

Besides speeding up the simulation of chemical reaction networks, our results could also help to choose appropriate SSA configurations for related methods, e. g., in multi-scale simulators [43] or simulators for the stochastic π calculus [44]. Future work will address the automated exploration of larger SSA design spaces and the integration of new algorithms, data structures, and benchmark models. Additionally, a more systematic exploration of the τ-Leaping parameter space might provide optimized parameter settings for different model types.

Acknowledgements

We thank Adelinde M. Uhrmacher for her helpful advice and her comments on an earlier version of this work.

References

1. Srivastava, R., You, L., Summers, J., Yin, J.: Stochastic vs. deterministic modeling of intracellular viral kinetics. Journal of Theoretical Biology 218, 309–321 (2002)
2. Gillespie, D.: A rigorous derivation of the chemical master equation. Physica A Statistical Mechanics and its Applications 188, 404–425 (1992)
3. Macnamara, S., Burrage, K., Sidje, R.B.: Multiscale modeling of chemical kinetics via the master equation. Multiscale Modeling & Simulation 6(4), 1146–1168 (2008)
4. Gillespie, D.: Exact Stochastic Simulation of Coupled Chemical Reactions. Journal of Physical Chemistry 81(25) (1977)
5. Sandmann, W.: Simultaneous stochastic simulation of multiple perturbations in biological network models (2007)
6. Gillespie, D.: Approximate accelerated stochastic simulation of chemically reacting systems. The Journal of Chemical Physics 115(4), 1716–1733 (2001)
7. Cao, Y., Gillespie, D.T., Petzold, L.R.: The slow-scale stochastic simulation algorithm. J. Chem. Phys. 122(1) (January 2005)
8. Weinan, E., Di, L., Vanden-Eijnden, E.: Nested stochastic simulation algorithms for chemical kinetic systems with multiple time scales. J. Comput. Phys. 221(1), 158–180 (2007)
9. McGeoch, C.: Experimental algorithmics. Communications of the ACM 50(11), 27–31 (2007)
10. LaMarca, A., Ladner, R.: The influence of caches on the performance of sorting. In: SODA 1997: Proceedings of the eighth annual ACM-SIAM symposium on Discrete algorithms, Philadelphia, PA, USA. Society for Industrial and Applied Mathematics, pp. 370–379 (1997)
11. Uhrmacher, A., Himmelspach, J., Jeschke, M., John, M., Leye, S., Maus, C., Röhl, M., Ewald, R.: One modeling formalism & simulator is not enough! - a perspective for computational biology based on james ii. In: Proceedings of the 1st FSMB Workshop, London. LNCS. Springer, Heidelberg (2008)
12. Himmelspach, J., Uhrmacher, A.: Plug'n simulate. In: Proceedings of the 40th Annual Simulation Symposium, pp. 137–143. IEEE Computer Society, Los Alamitos (2007)
13. Ewald, R., Himmelspach, J., Uhrmacher, A.: An algorithm selection approach for simulation systems. In: Proceedings of the 22nd ACM/IEEE/SCS Workshop on Principles of Advanced and Distributed Simulation (PADS 2008) (2008)
14. Gibson, M., Bruck, J.: Efficient Exact Stochastic Simulation of Chemical Systems with Many Species and Many Channels. J. Chem. Physics 104, 1876–1889 (2000)
15. Himmelspach, J., Uhrmacher, A.: The event queue problem and pdevs. In: Proceedings of the SpringSim 2007, DEVS Integrative M&S Symposium, SCS, pp. 257–264 (2007)

16. Rice, J.: The algorithm selection problem. Advances in Computers 15, 65–118 (1976)
17. Gomes, C., Selman, B.: Algorithm portfolio design: Theory vs. practice. In: Proc. of the 13th Conf. on Uncertainty in Artificial Intelligence (UAI 1997), pp. 190–197. Morgan Kaufmann, San Francisco (1997)
18. Leyton-Brown, K., Nudelman, E., Andrew, G., Mcfadden, J., Shoham, Y.: Boosting as a metaphor for algorithm design. In: Rossi, F. (ed.) CP 2003. LNCS, vol. 2833, pp. 899–903. Springer, Heidelberg (2003)
19. Houstis, E.N., Catlin, A., Rice, J., Verykios, V., Ramakrishnan, N., Houstis, C.: Pythia ii: A knowledge/database system for managing performance data and recommending scientific software. ACM Transactions on Mathematical Software 26(2), 227–253 (2000)
20. Busch, H., Sandmann, W., Wolf, V.: A Numerical Aggregation Algorithm for the Enzyme-Catalyzed Substrate Conversion (2006)
21. Cai, X., Wang, X.: Stochastic modeling and simulation of gene networks - a review of the state-of-the-art research on stochastic simulations. Signal Processing Magazine, IEEE 24(1), 27–36 (2007)
22. Cao, Y., Li, H., Petzold, L.: Efficient formulation of the stochastic simulation algorithm for chemically reacting systems. The Journal of Chemical Physics 121(9), 4059–4067 (2004)
23. Gillespie, D.: The chemical langevin equation. The Journal of Chemical Physics 113(1), 297–306 (2000)
24. Tian, T., Burrage, K.: Binomial leap methods for simulating stochastic chemical kinetics. The Journal of Chemical Physics 121(21), 10356–10364 (2004)
25. Cao, Y., Gillespie, D., Petzold, L.: Avoiding negative populations in explicit Poisson tau-leaping. J. Chem. Phys. 123, 054104 (2005)
26. Cao, Y., Gillespie, D.T., Petzold, L.R.: Efficient step size selection for the tau-leaping simulation method. J. Chem. Phys. 124(4) (January 2006)
27. Rathinam, M., Petzold, L.R., Cao, Y., Gillespie, D.T.: Stiffness in stochastic chemically reacting systems: The implicit tau-leaping method. The Journal of Chemical Physics 119, 12784–12794 (2003)
28. Cai, X., Xu, Z.: K-leap method for accelerating stochastic simulation of coupled chemical reactions. The Journal of Chemical Physics 126, 4102 (2007)
29. EMBL-EBI: Biomodels database, 10 (accessed July 18, 2008), http://www.ebi.ac.uk/biomodels/
30. Matsumoto, M., Nishimura, T.: Mersenne twister: a 623-dimensionally equidistributed uniform pseudo-random number generator. ACM Trans. Model. Comput. Simul. 8(1), 3–30 (1998)
31. Marsaglia, G.: The Marsaglia random number CDROM including the Diehard battery of tests of randomness (1995), http://www.stat.fsu.edu/pub/diehard/
32. Jenkins, B.: ISAAC, a fast cryptographic random number generator (1996), http://www.burtleburtle.net/bob/rand/isaacafa.html
33. Hellekalek, P.: Good random number generators are (not so) easy to find. Math. Comput. Simul. 46(5-6), 485–505 (1998)
34. Grassberger, P.: On correlations in "good" random number generators. Physics Letters A 181(1), 43–46 (1993)
35. Matsumoto, M., Wada, I., Kuramoto, A., Ashihara, H.: Common defects in initialization of pseudorandom number generators. ACM Trans. Model. Comput. Simul. 17(4) (September 2007)
36. Marsaglia, G.: Seeds for random number generators. Commun. ACM 46(5), 90–93 (2003)
37. Goh, R., Thng, I.: Mlist: An efficient pending event set structure for discrete event simulation. International Journal of Simulation - Systems, Science & Technology 4(5-6), 66–77 (2003)
38. Brown, R.: Calendar queues: a fast 0(1) priority queue implementation for the simulation event set problem. Commun. ACM 31(10), 1220–1227 (1988)

39. Huberman, B., Lukose, R., Hogg, T.: An economics approach to hard computational problems. Science 275, 51–54 (1997)
40. Sheskin, D.J.: Handbook of Parametric and Nonparametric Statistical Procedures, 4th edn. Chapman & Hall/CRC, Boca Raton (January 2007)
41. Pozo, R., Miller, B.: Java scimark, http://math.nist.gov/scimark2/
42. R Development Core Team: R: A language and environment for statistical computing. R Foundation for Statistical Computing, Vienna, Austria (2005)
43. Takahashi, K., Kaizu, K., Hu, B., Tomita, M.: A multi-algorithm, multi-timescale method for cell simulation. Bioinformatics 20 (2004)
44. Phillips, A., Cardelli, L.: A correct abstract machine for the stochastic pi-calculus. Transactions on Computational Systems Biology (2005)

Statistical Model Checking in *BioLab*: Applications to the Automated Analysis of T-Cell Receptor Signaling Pathway*

Edmund M. Clarke[1], James R. Faeder[2], Christopher J. Langmead[1],
Leonard A. Harris[2], Sumit Kumar Jha[1], and Axel Legay[1]

[1] Computer Science Department, Carnegie Mellon University, Pittsburgh PA
[2] Department of Computational Biology,
University of Pittsburgh School of Medicine, Pittsburgh PA

Abstract. We present an algorithm, called BioLab, for verifying temporal properties of rule-based models of cellular signalling networks.

BioLab models are encoded in the BioNetGen language, and properties are expressed as formulae in probabilistic bounded linear temporal logic. Temporal logic is a formalism for representing and reasoning about propositions qualified in terms of time. Properties are then verified using sequential hypothesis testing on executions generated using stochastic simulation. BioLab is optimal, in the sense that it generates the minimum number of executions necessary to verify the given property. BioLab also provides guarantees on the probability of it generating Type-I (i.e., false-positive) and Type-II (i.e., false-negative) errors. Moreover, these error bounds are pre-specified by the user. We demonstrate BioLab by verifying stochastic effects and bistability in the dynamics of the T-cell receptor signaling network.

1 Introduction

Computational modeling is an effective means for gaining insights into the dynamics of complex biological systems. However, there are times when the nature of the model itself presents a barrier to such discovery. Models with stochastic dynamics, for example, can be difficult to interpret because they are inherently non-deterministic. In the presence of non-deterministic behavior, it becomes non-trivial to determine whether a behavior observed in a simulation is typical, or

---

* This research was sponsored by the GSRC (University of California) under contract no. SA423679952, National Science Foundation under contracts no. CCF0429120, no. CNS0411152, and no. CCF0541245, Semiconductor Research Corporation under contract no. 2005TJ1366, Air Force (University of Vanderbilt) under contract no. 18727S3, International Collaboration for Advanced Security Technology of the National Science Council, Taiwan, under contract no. 1010717, the Belgian American Educational Foundation, the U.S. Department of Energy Career Award (DE-FG02-05ER25696), a Pittsburgh Life-Sciences Greenhouse Young Pioneer Award, National Institutes of Health grant GM76570 and a B.A.E.F grant.

M. Heiner and A.M. Uhrmacher (Eds.): CMSB 2008, LNBI 5307, pp. 231–250, 2008.

an anomaly. In this paper, we introduce a new tool, called BIOLAB, for *formally* reasoning about the behavior of stochastic dynamic models by integrating techniques from the field of *Model Checking* [8] into the BIONETGEN [12, 13] framework for rule-based modeling. We then use BIOLAB to verify the stochastic bistability of T-cell signalling.

The term "Model Checking" refers to a family of automated techniques for formally verifying properties of complex systems. Since its inception in 1981, the field of Model Checking has made substantial contributions in industrial settings, where it is the preferred method for formal verification of circuit designs. Briefly, the system is first encoded as a model in a formal description language. Next, properties of interest (e.g., absence of deadlock) are expressed as formulae in temporal logic. Temporal logic is a formalism for representing and reasoning about propositions qualified in terms of time. Given a model, $\mathcal{M}$, a set of initial states, S_0, and a property, ϕ, a model checking algorithm automatically determines whether the model satisfies the formula.

Historically, Model Checking has most often been applied to engineered systems, and thus the majority of Model Checking algorithms are designed for such systems. Recently, however, there has been growing interest in the application of Model Checking to biology (e.g.,[5, 6, 19, 21, 22]). Biological systems present new challenges in the context of formal verification. In particular, biological systems tend to give rise to highly parameterized models with stochastic dynamics. Biologists are generally interested in determining whether a given property is (or is not) sensitive to a plausible set of initial conditions and parameter values. Model checking algorithms targeting biological applications must therefore apply to stochastic, multi-parameter models.

BIOLAB models stochastic biochemical systems using the BIONETGEN modeling language. The set of initial states (i.e., S_0) comprise a user-specified set of initial conditions and parameter values. Properties are expressed in probabilistic bounded linear temporal logic. BIOLAB then statistically verifies the property using sequential hypothesis testing on executions sampled from the model. These samples are generated using variants of Gillespie's algorithm [15, 11, 32], which ensures that the executions are drawn from the "correct" underlying probability distribution. This, combined with the use of sequential hypothesis testing provides several guarantees. First, BIOLAB can bound the probability of Type-I (i.e., false-positive) and Type-II (i.e., false-negative) errors, with regard to the predictions it makes. These error bounds are specified by the user. Second, BIOLAB is optimal in the sense that it generates the minimum number of executions necessary to determine whether a given property is satisfied. The number of required executions varies depending on the behavior of the model and is determined dynamically, as the program is running.

The contributions of this paper are as follows: (i) Our method is the first application of *statistical* Model Checking to rule-based modeling of biochemical systems. (ii) Our algorithm provides guarantees in terms of optimality, as well as bounds on the probability of generating Type-I and Type-II errors. (iii) We verify that a stochastic model of T-cell receptor signaling exhibits behaviors that are qualitatively different from those seen in an ordinary differential equation model of the same system [23]. In particular we verify that stochastic effects induce switching between two stable steady states of the system.

2   BioNetGen

Proteins in cellular regulatory systems, because of their multicomponent composition, can interact in a combinatorial number of ways to generate myriad protein complexes, which are highly dynamic [17]. Protein-protein interactions and other types of interactions that occur in biochemical systems can be modeled by formulating rules for each type of chemical transformation mediated by the interactions [18]. The rules can be viewed as definitions of reaction classes and used as generators of reactions, which describe the transformations of molecules in the system possessing particular properties. The assumption underlying this modeling approach, which is consistent with the modularity of regulatory proteins [27], is that interactions are governed by local context that can be captured in simple rules. Rules can be used to generate reaction networks that account comprehensively for the consequences of protein-protein interactions. Examples of rule-based models of specific systems can be found in [16, 3, 1, 26].

BIONETGEN is a software package that provides tools and a language for rule-based modeling of biochemical systems [2, 13]. A formal description of the language and underlying graph theory is provided in [4]. BIONETGEN is similar to the κ-calculus, which has also been developed as a language for rule-based modeling of biochemical systems [9]. Other tools for rule-based modeling are reviewed in [18].

The syntax and semantics of BIONETGEN have been thoroughly described in [13]. Briefly, a BIONETGEN model is comprised of six basic elements that are defined in separate blocks in the input file: *parameters, molecule types, seed species, reaction rules, observables,* and *actions.* Molecules are the basic building blocks of a BIONETGEN model, and are used to represent proteins and other structured biological molecules, such as metabolites, genes, or lipids. The optional *molecule types* block is used to defined the composition and allowable states of molecules. Molecules may contain components, which represent the functional elements of molecules, and may bind other components, either in the same molecule or another molecule. Components may be associated with state variables, which take on a finite set of possible values that may represent conformational or chemical states of a component, e.g., tyrosine phosphorylation. An example of a molecule type declaration is

```
TCR(ab,ITAM~U~P~PP,lck,shp)
```

which is used to define the structure of the T cell receptor in the model presented in Sec. 5.2. The name of the molecule type is given first, followed by a comma-separated list of its components in parenthesis. Any declared component may participate in a bond. In addition, the allowed values of the state variable associated with a component are indicated with $\sim$ followed by a name. In the above example, a molecule of type TCR has four components, three of which (ab, lck, and shp) may be used only for binding and one of which (ITAM) has an associated state variable that takes on the values U, P, or PP—representing the unphosphorylated, phosphorylated, and doubly phosphorylated forms respectively.

The *seed species* block defines the molecules and molecular complexes that are initially present in the system with an optional quantifier. Depending on the semantics used in the simulation of the model (see below) the value of the quantifier may be either continuous or restricted to discrete values. For example, the line

```
Lck(tcr,Y~U,S~U)     LCK
```

in the seed species block specifies that the initial amount of the species comprised of a molecule of Lck with both its Y (tyrosine residue 394) and S (serine residue 453) components in the U (unphosphorylated) state is given by the parameter LCK, which is defined in the *parameters* block. Only species with a non-zero initial amount as declared in the *seed species* block are present in the system at the beginning of simulation.

The *reaction rules* block contains rules that define how molecules in the system can interact. A rule is comprised, in order of appearance, of a set of reactant patterns, a transformation arrow, a set of product patterns, and a rate law. A pattern is a set of molecules that select a set of species through a mapping operation [4]. The match of a molecule in a pattern to a molecule in a particular species depends only on the components that are specified in the pattern (which may include wildcards), so that one pattern may select many different species. Three basic types of operations are carried out by the rules in the T cell model: binding (unbinding) of two molecules through a specified pair of components and changing the state variable of a component. An example of a binding rule is

```
TCR(ab,shp)+pMHC(p~ag) -> TCR(ab!1,shp).pMHC(p~ag!1) b1
```

which specifies that *any* TCR molecule containing unbound ab and shp components may bind through its ab component to a p component in the ag state of a pMHC molecule. In this example, the first reactant pattern, TCR(ab,shp), matches any TCR-containing species with free ab and shp components, independent of the state of the remaining two components. The + operator separates two reactant patterns that must map to distinct species. The transformation arrow may be either unidirectional (->), as in the above rule, or bidirectional (<->), indicating that the rule is to be applied in both the forward and reverse directions (i.e., switching the reactant and product patterns). The product patterns define the configuration of the selected reactant molecules following the application of the rule. Here, the ab component of TCR is bound to the p component of pMHC by the addition of an edge labeled 1, as indicated by the two bond labels (!1) in the products. The parameter b1 specifies the rate constant to be used in determining the rate of the reaction, which is computed as a product of the rate constant and each of the reactant amounts.

The *observables* block contains definitions of model outputs, which are defined as sums over the amounts of species matched by a set of patterns. The output of the TCR model is the level of doubly-phosphorylated ERK, which is specified by the following line in the BNGL file

```
Molecules ppERK ERK(S~PP)
```

where the first item is a keyword defining the type of observable, the second item is the name of the observable, and the final item is a list of patterns that determines the matching species.

The *actions* block specifies the operations that are to be carried out either to generate or simulate a network. As we now discuss, the choice of operations to perform also defines the semantics under which the model elements are interpreted. BIONETGEN uses three basic methods to simulate the time course of observables for a rule-based network: generate-first (GF), on-the-fly (OTF), and network-free (NF). These methods are described in detail in [13]; we provide a brief overview in this section. In GF, rules are iteratively applied to the initial set of seed species until all reachable species and reactions are generated or some other stopping criterion is satisfied. The resulting network can be simulated either by solving a set of ODE's for the average concentration of each species in the system under the influence of the mass action reactions (GF-ODE) or by Gillespie's stochastic simulation algorithm (SSA) [15] to sample the exact solution to the chemical master equations governing the species probabilities (GF-SSA). Both methods generate traces[1] of the species concentrations as a function of time, but the GF-ODE algorithm is deterministic for a given initial state and set of system parameters, whereas each simulation run of GF-SSA from a given initial state represents a stochastic process and may generate a different trace. Like GF-SSA, OTF uses the Gillespie algorithm to generate traces but only generates species and reactions that are reachable within a small number of specified time steps [24, 11]. OTF was originally proposed as a way to maintain computational efficiency for large reaction networks, but is not practical for rule-based models that include oligomerization or attempt a comprehensive description of reaction networks [18, 10]. The NF method [10, 32] avoids explicit generation of species and reactions by simulating molecules as agents and has been shown to have per event cost that is independent of the number of possible species or reactions [10]. NF also relies on the SSA to sample reaction events that govern the evolution of the molecular agents. Because species are not explicitly tracked, the NF method generates traces over observables rather than individual species. This restriction is not an issue for applications to biology because the concentrations of individual species are typically not observable in biological experiments.

3   Model Checking for Stochastic Systems

The following section introduces the concept of *statistical model checking*. We assume the reader is familiar with basic concepts in probability theory.

---

[1] The term "trace" is equivalent to the term "execution". From now, we will use "trace" when we want to emphasize that we are talking about a BIONETGEN Model.

3.1   The Problem

We use $Pr(E)$ to denote the probability of the event E to occur. We consider a system $\mathcal{M}$ whose executions (sequences of states of the system) are *observable* and a property ϕ that is defined as a set of executions. We assume that one can decide whether an execution trace of $\mathcal{M}$ satisfies ϕ, i.e. whether the execution belongs to ϕ. In this paper, the *probabilistic model checking problem* consists in deciding whether the executions of $\mathcal{M}$ satisfy ϕ with a probability greater than or equal to a given threshold θ. The latter is denoted by $\mathcal{M} \models Pr_{\geq \theta}(\phi)$. This statement only makes sense if one can define a probability space on the executions of the system as well as on the set of executions that do satisfy ϕ.

The probabilistic model checking problem can be solved with a *probabilistic model checking algorithm*. Such an algorithm is *numerical* in the sense that it computes the exact probability for the system to satisfy ϕ and then compares it with the value of θ. Successful probabilistic model checking algorithms [7, 20]) have been proposed for various classes of systems, including (continuous time) Markov chains and Markov Decision Processes. The drawback with those approaches is that they compute the probability for all the executions of the system, which may not scale up for systems of large size.

Another way to solve the probabilistic model checking problem is to use a *statistical model checking algorithm*. In the rest of this section, we recap the statistical model checking algorithmic scheme proposed by Younes in [33].

3.2   Statistical Approach

The approach in [33, 29] is based on hypothesis testing. The idea is to check the property ϕ on a sample set of simulations and to decide whether the system satisfies $Pr_{\geq \theta}(\phi)$ based on the number of executions for which ϕ holds compared to the total number of executions in the sample set. With such an approach, we do not need to consider all the executions of the system. To determine whether $\mathcal{M}$ satisfies ϕ with a probability $p \geq \theta$, we can test the hypothesis $H : p \geq \theta$ against $K : p < \theta$. A test-based solution does not guarantee a correct result but it is possible to bound the probability of making an error. The *strength* (α, β) of a test is determined by two parameters, α and β, such that the probability of accepting K (respectively, H) when H (respectively, K) holds, called a Type-I error (respectively, a Type-II error) is less or equal to α (respectively, β).

A test has *ideal performance* if the probability of the Type-I error (respectively, Type-II error) is exactly α (respectively, β). However, these requirements make it impossible to ensure a low probability for both types of errors simultaneously (see [33] for details). A solution to this problem is to relax the test by working with an *indifference region* (p_1, p_0) with $p_0 \geq p_1$ ($p_0 - p_1$ is the *size of the region*). In this context, we test the hypothesis $H_0 : p \geq p_0$ against $H_1 : p \leq p_1$ instead of H against K. If both the values of p and θ are between p_1 and p_0 (the indifference region), then we say that the probability is sufficiently close to θ so that we are indifferent with respect to which of the two hypotheses K or H is accepted.

3.3   An Algorithmic Scheme

Younes developed a procedure to test $H_0 : p \geq p_0$ against $H_1 : p \leq p_1$ that is based on the *sequential probability ratio test* proposed by Wald[31]. The approach is briefly described below.

Let B_i be a discrete random variable with a Bernoulli distribution. Such a variable can only take 2 values 0 and 1 with $Pr[B_i = 1] = p$ and $Pr[B_i = 0] = 1 - p$. In our context, each variable B_i is associated with one simulation of the system. The outcome for B_i, denoted b_i, is 1 if the simulation satisfies ϕ and 0 otherwise. In the sequential probability ratio test, one has to choose two values A and B, with $A > B$. These two values should be chosen to ensure that the strength of the test is respected. Let m be the number of observations that have been made so far. The test is based on the following quotient:

$$\frac{p_{1m}}{p_{0m}} = \prod_{i=1}^{m} \frac{Pr(B_i = b_i \mid p = p_1)}{Pr(B_i = b_i \mid p = p_0)} = \frac{p_1^{d_m}(1 - p_1)^{m-d_m}}{p_0^{d_m}(1 - p_0)^{m-d_m}}, \quad (1)$$

where $d_m = \sum_{i=1}^{m} b_i$. The idea behind the test is to accept H_0 if $\frac{p_{1m}}{p_{0m}} \geq A$, and H_1 if $\frac{p_{1m}}{p_{0m}} \leq B$. An algorithm for sequential ratio testing consists of computing $\frac{p_{1m}}{p_{0m}}$ for successive values of m until either H_0 or H_1 is satisfied. This has the advantage of minimizing the number of simulations. In each step i, the algorithm has to check the property on a single execution of the system, which is handled with a new Bernoulli variable B_i whose realization is b_i. In his thesis[33], Younes proposed a logarithmic based algorithm (Algorithm 2.3 page 27) SPRT that given p_0, p_1, α and β implements the sequential ratio testing procedure. Computing ideal values A_{id} and B_{id} for A and B in order to make sure that we are working with a test of strength (α, β) is a laborious procedure (see Section 3.4 of [31]). In his seminal paper[31], Wald showed that if one defines $A_{id} \geq A = \frac{(1-\beta)}{\alpha}$ and $B_{id} \leq B = \frac{\beta}{(1-\alpha)}$, then we obtain a new test whose strength is (α', β'), but such that $\alpha' + \beta' \leq \alpha + \beta$, meaning that either $\alpha' \leq \alpha$ or $\beta' \leq \beta$. In practice, we often find that both inequalities hold.

The SPRT algorithm can be extended to handle Boolean combinations of probabilistic properties as well as much more complicated probabilistic Model checking problems than the one considered in this paper[33].

Statistical Model Checker. The SPRT algorithm can be implemented in order to solve the probabilistic model checking problem for a specific class of systems and a specific class of properties. For this, we have to implement :

- *A simulator* that is able to simulate the system and produce observable executions without necessarily constructing its entire state-space.
- *An execution verifier* that is a procedure to decide whether an execution satisfies a given property.

In section 5, we propose BIOLAB, which is an implementation of the SPRT algorithm for models encoded and simulated using BIONETGEN.

4   Statistical Model Checking for CTMCs

A BIONETGEN model can be interpreted as Continuous-time Markov Chain (CTMC), which may be simulated using the stochastic simulation methods described in Sec. 2. In this section, we review CTMCs and then introduce the probabilistic bounded linear temporal logic, which will is used in BIOLAB to define properties over CTMC and thus over BIONETGEN models.

4.1   Continuous-Time Markov Chains

Let $\mathbb{R}$ (resp. $\mathbb{N}$) denote the set of real (resp. natural) numbers and let $\mathbb{R}_{\geq 0}$ and $\mathbb{R}_{>0}$ denote the set of non-negative and strictly-positive real numbers, respectively. $\mathbb{N}$ is the set of natural numbers, and $\mathbb{N}_{\geq 0}$ is the set of strictly positive natural numbers. We now recall the definition of *Structured Continuous-time Markov Chains*.

Definition 1. *A Structured Continuous-time Markov Chain is a tuple* $\mathcal{M} = (S, S_0, R, SV, V)$, *where*

- S *is a finite set of states;*
- $s_0 \in S$ *is the initial state;*
- $R : S \times S \to \mathbb{R}_{\geq 0}$ *is the rate matrix.*
- SV *is a finite set of state variables defined over* $\mathbb{R}_{\geq 0}$. *These variables represent the concentration of each molecular species in the model.*
- $V : S \times SV \to \mathbb{R}_{\geq 0}$ *is a value assignment function providing the value of* $x \in SV$ *in state* s.

Let $\mathcal{M} = (S, S_0, R)$ be a structured continuous-time Markov chain. Let $t \in \mathbb{R}_{>0}$ and $s_1, s_2 \in S$, the probability to go from s_1 to s_2 within t time unit is defined as follows

$$P(s_1, s_2, t) = \frac{R(s_1, s_2)}{T(s_1)}(1 - e^{T(s_1)t}), \tag{2}$$

where $T(s_1) = \sum_{s' \in S} R(s, s')$.

An *execution*, also called *trace*, of $\mathcal{M}$ is a possibly infinite sequence $\sigma = (s_0, t_0) (s_1, t_1)(s_2, t_2) \ldots$ such that for each $i \geq 0$, (1) $p(s_i, s_{i+1}, t_i) > 0$, and (2) $t_i \in \mathbb{R}_{>0}$. Given (s_i, t_i), t_i is the time that is spent in state s_i. Given s_i, $\sum_{j<i} t_j$ is the number of time units spent before reaching s_i. We use $\sigma(i)$ (with $i \geq 0$) and σ^i to reference the i−th state of the execution and the suffix of the execution starting from the pair (s_i, t_i), respectively. Given a set $S' \in S$, we will use Path(S') to denote the set of all the executions whose initial states are in S'.

4.2   Probabilistic Bounded Linear Temporal Logic

BIOLAB is intended to be used as a tool for verifying properties of executions of CTMCs. Users specify properties of interest in the probabilistic bounded linear temporal logic. We now give the syntax and the semantics of bounded linear temporal logic (BLTL).

Let SV be a set of nonnegative real variables and $\sim \; \in \{\geq, \leq, = \}$. A Boolean predicate over SV is a constraint of the form $x \sim v$, where $x \in SV$ and $v \in \mathbb{R}_{\geq 0}$. A BLTL property is built on a finite set of Boolean predicates over SV using Boolean connectives and temporal operators. The syntax of the logic is given by the following grammar :

$$\phi ::= x \sim v \mid (\phi_1 \vee \phi_2) \mid (\phi_1 \wedge \phi_2) \mid \neg \phi \mid \mathbf{X}(\phi) \mid (\phi_1 \mathbf{U^t} \phi_2) \mid (\phi_1 \widetilde{\mathbf{U}}^t \phi_2) \mid \mathbf{D_t}(\phi).$$

The operators $\neg$, $\vee$, and $\wedge$ are the normal *propositional logic operators*, which are read "not", "or", and "and", respectively. The operators $\mathbf{X}$, $\mathbf{U^t}$, $\widetilde{\mathbf{U}}$, and $\mathbf{D}$ are the *temporal operators*. The operator $\mathbf{X}$ is read "next", and corresponds to the notion of "in the next state". The operator $\mathbf{U^t}$ is read "until t time units have passed", and requires that its first argument be true *until* its second argument is true, which is required to happen within t time units. The operator $\widetilde{\mathbf{U}}$ is read "release", and requires that its second argument is true during the first t time units unless this obligation has been released by its first argument becoming true. The operator $\mathbf{D_t}$ is read "dwell", and requires that each time the argument becomes true, it is falsified within t time units.

Two additional temporal operators are in very common use. The first of them is $\mathbf{F^t}$, where $\mathbf{F}$ is read "eventually". The eventually operator requires that its argument becomes true within t units of time. Formally, we have $\mathbf{F^t}\psi = \mathbf{True}\,\mathbf{U^t}\,\psi$. The second operator is $\mathbf{G^t}$, where $\mathbf{G}$ is read "always". This operator requires that its argument stays true during at least t units of time. Formally, we have $\mathbf{G^t}\psi = \mathbf{False}\,\widetilde{\mathbf{U}}^t\,\psi$.

The semantics of BLTL was informally described above. We now present its formal semantics. The fact that the execution $\sigma = (s_0, t_0)(s_1, t_1), \ldots$ satisfies the BLTL property ϕ is denoted by $\sigma \models \phi$. We have the following:

- $\sigma \models x \sim v$ if and only if $V(\sigma(0), x) \sim v$;
- $\sigma \models \phi_1 \vee \phi_2$ if and only if $\sigma \models \phi_1$ or $\sigma \models \phi_2$;
- $\sigma \models \phi_1 \wedge \phi_2$ if and only if $\sigma \models \phi_1$ and $\sigma \models \phi_2$;
- $\sigma \models \neg \phi$ if and only if $\sigma \not\models \phi$.
- $\sigma \models \mathbf{X}\phi$ if and only if $\sigma^1 \models \phi$.
- $\sigma \models \phi_1 \mathbf{U^t} \phi_2$ if and only if there exists $i \in \mathbb{N}$ such that (1) $\sigma^i \models \phi_2$ and (2) $\sum_{j<i} t_j \leq t$, and for each $0 \leq j < i$ $\sigma^j \models \phi_1$.
- $\sigma \models \phi_1 \widetilde{\mathbf{U}}^t \phi_2$ if and only if for each i such that $\sigma^i \not\models \phi_2$ and $\sum_{m<i} t_m \leq t$, there exists $0 \leq j < i$ such that $\sigma^j \models \phi_1$.
- $\sigma \models \mathbf{D_t}(\phi)$ if and only if for each state $\sigma(i)$ such that $\sigma(i) \models \phi$, there exists $j > i$ such that $\sigma(j) \not\models \phi$ and $\sum_{m=i}^{m=j-1} t_m \leq t$.

Remark 1. It should be noted that we can decide whether an infinite execution satisfies a BLTL property by observing one of its finite prefixes.

We assume that properties of Structured Continuous-time Markov Chains are specified with *Probabilistic Bounded Linear Temporal Logic* (BTL).

Definition 2. *A BTL property is a property of the form* $\psi = Pr_{\geq \theta}(\phi)$, *where* ϕ *is a BLTL property.*

We say that the Continuous-time Markov Chain $\mathcal{M}$ satisfies ψ, denoted by $\mathcal{M} \models \psi$, if and only if the probability for an execution of $\mathcal{M}$ to satisfy ϕ is greater than θ. The problem is well-defined since, as it is shown with the following theorem, one can always assign a unique probability measure to the set of executions that satisfy an BLTL property.

Theorem 1. *Let $\mathcal{M}$ be a Continuous-time Markov Chain and ϕ be a BLTL formula. One can always associate a unique probability measure to the set of executions of $\mathcal{M}$ that satisfy ϕ.*

5   Statistical Model Checking of a T Cell Model

5.1   The BIOLAB Algorithm

The BIOLAB algorithm is a statistical model checker that implements the SPRT algorithm introduced in Section 3.3 for checking BTL properties against BIONET-GEN models. BIOLAB uses the BIONETGEN simulation engine described in Section 2 to generate traces by randomly simulating biological models, and then uses a *Bounded Linear Temporal Logic trace verifier* to validate the generated traces against the BLTL part of the given BTL property. Depending on the result of the validation of the generated tracess, the BIOLAB tool decides whether the BTL formula is satisfied/falsified or if more samples are needed in order to make this decision. The structure of the BIOLAB algorithm is outlined in Figure 4.2. The BIONETGEN simulator is used to generate stochastic traces and the trace verifier verifies each of them against the BLTL property. Our *trace verifier* is based on the translation from BLTL to alternating automata [30, 14]. The statistical model checker continues to simulate the BIONETGEN model until a decision about the property has been made.

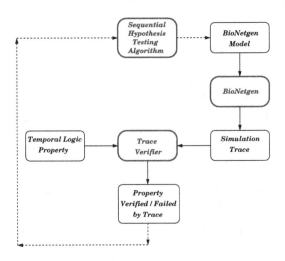

Fig. 1. Architecture of BIOLAB

5.2   The T Cell Receptor Model

T lymphocytes, also known as T cells, play a central role in the immune system by detecting foreign substances, known as antigens, and coordinating the immune response. T cells detect the presence of antigen through surface receptors, called T cell receptors (TCRs), which bind to specific polypeptide fragments that are displayed on the surface of neighboring cells by a protein called the major histocompatibility complex (MHC). Variable regions of the immunoglobulin chains that comprise the TCR give rise to a broad range of TCR binding specificities. Individual T cells (or clonal populations derived from the same precursor) express a unique form of TCR. Processes of positive and negative selection during maturation of T cells in the thymus select T cells possessing TCRs with a weak but nonzero affinity for binding MHC molecules carrying peptides derived from host proteins. High-affinity binding between TCR and peptide-MHC (pMHC) complexes induces a cascade of biochemical events that leads to activation of the T cell and initiation of an immune response. To be effective in detecting antigens while avoiding autoimmunity, T cells must generate strong responses to the presence of minute quantities of antigen—as low as a few peptide fragments per antigen-presenting cell—while not responding to the large quantities of endogenous (host) pMHC expressed on all cells. The T cell appears to maintain this delicate balance between sensitivity and selectivity through a combination of mechanisms that include kinetic proofreading, which discriminates against pMHC-receptor interactions that are too short, positive feedback, which amplifies the response and makes it more switch-like, and negative feedback, which acts in concert with kinetic proofreading to dampen responses to weak stimulation and with positive feedback to enhance the stability of the inactive state.

A computational model incorporating all three of these mechanisms has recently been developed by Lipniacki et al. [23], and serves as the basis for the experiments we conduct here using BioLAB. This model extends previous simplified models of kinetic proofreading [25] and feedback regulation [28] by incorporating mechanistic detail about the involvement of specific signaling molecules. A schematic illustration of the model is presented in Fig. 5.2. Binding of pMHC to the TCR initiates a series of binding and phosphorylation events at the receptor that can lead either to activation or inhibition of the receptor depending on the strength of the stimulus, which is indicated along the kinetic proofreading axis. The rectangular box in the figure represents the TCR complex, which requires three components to make its passage to the activated form. These components are pMHC (P), doubly phosphorylated receptor (T_{pp}), and singly-phosphorylated LCK (L_p). In its active form, the LCK kinase can phosphorylate SHP (S) to produce S_p, which acts as a negative feedback by reversing TCR activation events and blocking TCR activation. LCK also acts through a series of intermediate layers to activate the MAP kinase ERK, a potent activator of transcription, whose active form (E_{pp}) is taken as the final readout of T cell activation. As shown in the figure, activated ERK also provides positive feedback by blocking the activity of S_p.

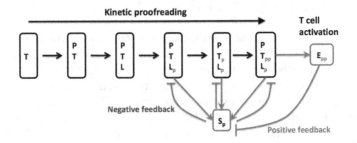

Fig. 2. Overview of the TCR signaling model of Ref. [23]. Lines terminated by flat heads indicate an inhibitory interaction.

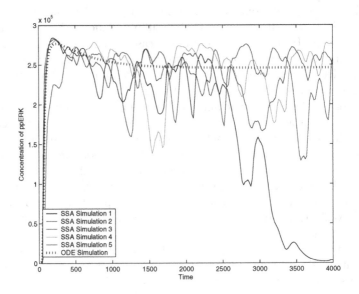

Fig. 3. Traces from deterministic (ODE) and stochastic simulation of the TCR signaling model. $N_1 : 100$, $N_2 : 3000$.

This model captures three important properties of T cell activation, which are sensitivity to small numbers of pMHC with high binding affinity, high selectivity between pMHCs of different affinity, and antagonism, the inhibition of response by pMHC of intermediate affinity. Because only small numbers of high-affinity pMHC ligands are displayed on cell surfaces, stochastic effects have a major influence on the dynamics both of the model and of the initiation of signaling through the TCR. The model also exhibits bistable ERK responses over a broad range of pMHC number and binding affinity. This bistable regime has the interesting property that stochastic trajectories may exhibit completely different dynamics from the deterministic trajectory from the same initial state, and even the average behavior of stochastic trajectories may differ qualitatively from the deterministic behavior (see Fig. 7B of [23] for an example). This divergence between the stochastic and deterministic dynamics was the motivation for using this model of TCR as the basis for the current study, which aims to show that formal verification methods can be useful for the characterization of rule-based biochemical models.

The TCR model has been encoded in the BIONETGEN language (available at `http://bionetgen.org/index.php/Tcr_tomek`) and serves as the basis for the current experiments. The BIONETGEN model is comprised of seven molecule types and 30 rules, which generate a biochemical network of 37 species and 97 reactions. The main output of the model is fraction of ERK that is doubly phosphorylated, denoted by the variable f, which is taken as a measure of T cell activation. For $f < 0.10$ the cell is considered inactive, for $f > 0.5$ the cell is considered active. The response is observed to be switch-like with respect to stimulation strength, measured by the number of agonist (high affinity) pMHC per cell, given by N_1 (see Fig. 2 of [23]). The system also exhibits bistability with respect to f over a wide range of N_1 values (see Fig. 7A of [23]). As shown in Fig. 5.2, under many input conditions traces from stochastic simulations may sample both stable steady states and thus diverge from deterministic traces starting from the same initial conditions, which sample only a single steady state.

5.3   Experiments

We performed several *in-silico* BIOLAB experiments on the T Cell Receptor model. Each of our experiments was performed on a cluster of 40 3GHz computational nodes communicating using the Message Passing Interface.

Property 1. In our first experiment, we were interested in the truth of the hypothesis that the fraction f of doubly phosphorylated ERK stays below a given threshold value with a given probability during the first 300 seconds of simulation. We verified the following property with various values of the probability p and the threshold value γ.

$$Pr_{\geq \theta} \left(\mathbf{G}^{300} \left(ppERK \ / \ totalERK < \gamma \right) \right)$$

The first model we analyzed started with 100 molecules of agonist pMHC ($N_1 = 100$) while antagonist pMHC was absent ($N_2 = 0$). We also set the dissociation constant of agonist pMHC as $1/20$ per second. The results of our experiment are shown in Table 1.

In our second experimental setup, our system started with 100 molecules of agonist pMHC while there were 3000 molecules of antagonist pMHC. We also set the dissociation constant of agonist pMHC as $1/20$ per second and that of antagonist pMHC as $1/3$. The results of our experiment are shown in Table 2.

In the third experiment, there were 100 molecules of agonist pMHC and 1000 molecules of antagonist pMHC. We set the dissociation constant of agonist pMHC as $1/20$ per second and that of antagonist pMHC as $1/3$. We summarize the results in Table 3.

The fourth experiment started with 100 molecules of agonist pMHC and 300 molecules of antagonist pMHC. We used the same dissociation constants as in previous experiment. The results are presented in Table 4.

Table 1. $N_1 : 100$, $N_2 : 0$, Type-I and Type II error : 0.001

Sl. No.	p_1	p_0	γ	Result	Total Number of Samples	Number of Successful Samples	Time
1	0.90	0.95	0.1	No	40	0	232.73
2	0.90	0.95	0.5	No	40	0	221.38
3	0.90	0.95	0.7	No	40	2	221.41
4	0.90	0.95	0.9	No	40	2	233.45
5	0.90	0.95	0.95	Yes	240	236	1162.87

Table 2. $N_1 : 100$, $N_2 : 3000$, Type-I and Type II error : 0.001

Sl. No.	p_1	p_0	γ	Result	Total Number of Samples	Number of Successful Samples	Time
1	0.90	0.95	0.1	No	40	24	54.25
2	0.90	0.95	0.5	Yes	120	97	168.22
3	0.90	0.95	0.7	Yes	240	237	320.30
4	0.90	0.95	0.9	Yes	200	199	263.87
4	0.90	0.95	0.95	Yes	400	385	533.39

Table 3. $N_1 : 100$, $N_2 : 1000$, Type-I and Type II error : 0.001

Sl. No.	p_1	p_0	γ	Result	Total Number of Samples	Number of Successful Samples	Time
1	0.90	0.95	0.1	No	40	4	41.56
2	0.90	0.95	0.5	No	40	14	67.40
3	0.90	0.95	0.7	Yes	200	199	317.90
4	0.90	0.95	0.9	Yes	200	200	278.08
5	0.90	0.95	0.95	Yes	480	459	777.17

The fraction of phosphorylated ERK in the first and the fourth experiments exceeded 0.9 within the first 300 seconds with at least 90% probability. This phenomenon was not observed in the second and the third experiments.

Property 2. In our second experiment, we were interested in the truth of the hypothesis that the system can go from the inactive state to the active state. We verified the following property with various values of the probability p.

$$Pr_{\geq\theta}(\mathbf{F}^{300}(ppERK/totalERK{<}0.1 \wedge \mathbf{F}^{300} (ppERK/totalERK{>}0.5)))$$

Our first model started with 100 molecules of agonist pMHC (with dissociation constant 1/20 per second) while antagonist pMHC was assumed to be absent in the initial state. The results are presented in Table 5.

Table 4. N_1 : 100, N_2 : 300 , Type-I and Type II error : 0.001

Sl. No.	p_1	p_0	γ	Result	Total Number of Samples	Number of Successful Samples	Time
1	0.90	0.95	0.1	No	40	0	96.02
2	0.90	0.95	0.5	No	40	4	108.65
3	0.90	0.95	0.7	No	40	13	89.89
4	0.90	0.95	0.9	No	160	130	322.50
5	0.90	0.95	0.95	Yes	320	312	866.65

Table 5. N_1 : 100, N_2 : 0 , Type-I and Type II error : 0.001

Sl.	p_1	p_0	Result	Total Number of Samples	Number of Successful Samples	Time
1	0.90	0.95	Yes	160	160	412.25
2	0.70	0.75	Yes	120	120	309.58
3	0.50	0.55	Yes	80	80	214.74
4	0.20	0.25	Yes	40	40	88.32
5	0.10	0.15	Yes	40	40	98.84

We note that the number of samples needed to decide the property depends both upon the fraction of samples that satisfied the property and the probability with which we want the property to be satisfied.

In our second experimental setup, our system started with 100 molecules of agonist pMHC while there were 3000 molecules of antagonist pMHC. We also set the dissociation constant of agonist pMHC as 1/20 per second and that of antagonist pMHC as 1/3. We present the results in Table 6.

In the third experiment, there were 100 molecules of agonist pMHC and 1000 molecules of antagonist pMHC. We also set the dissociation constant of agonist pMHC as 1/20 per second and that of antagonist pMHC as 1/3. The results are illustrated in Table 7.

Table 6. N_1 : 100, N_2 : 3000 , Type-I and Type II error : 0.001

Sl.	p_1	p_0	Result	Total Number of Samples	Number of Successful Samples	Time
1	0.90	0.95	No	40	0	24.92
2	0.70	0.75	No	40	0	27.05
3	0.50	0.55	No	80	0	52.19
4	0.20	0.25	No	120	0	86.30
5	0.10	0.15	No	160	0	108.25

Table 7. $N_1 : 100$, $N_2 : 1000$, Type-I and Type II error : 0.001

Sl.	p_1	p_0	Result	Total Number of Samples	Number of Successful Samples	Time
1	0.90	0.95	No	40	27	35.16
2	0.70	0.75	Yes	40	34	34.34
3	0.50	0.55	Yes	120	109	111.30
4	0.20	0.25	Yes	40	37	44.57
5	0.10	0.15	Yes	40	36	45.54

The fourth experiment started with 100 molecules of agonist pMHC and 300 molecules of antagonist pMHC. We used the same dissociation constants as in previous experiment. The outcome of the experiments are shown in Table 8.

Table 8. $N_1 : 100$, $N_2 : 300$, Type-I and Type II error : 0.001

Sl.	p_1	p_0	Result	Total Number of Samples	Number of Successful Samples	Time
1	0.90	0.95	Yes	160	160	346.92
2	0.70	0.75	Yes	120	116	226.08
3	0.50	0.55	Yes	80	80	168.87
4	0.20	0.25	Yes	40	40	81.10
5	0.10	0.15	Yes	40	40	73.11

The second model showed a qualitative difference in behavior from the other three models while quantitative differences in behavior can be seen among all the four models. We verified our hypothesis that the stochastic model of the T Cell Receptor pathway can go from the inactive to the active state with a non-zero probability.

Property 3. In our third set of experiments, we were interested in the truth of the hypothesis that the system can go from the active state to the inactive state. We verified the following property with various values of the probability p.

$$Pr_{\geq\theta}(\mathbf{F^{300}}(ppERK/totalERK{>}0.5\wedge \ \mathbf{F^{300}} \ (ppERK/totalERK{<}0.1)))$$

Our model started with 100 molecules of agonist pMHC (with dissociation constant 1/20 per second) while there was no antagonist pMHC. The results of our experiments are illustrated in Table 9.

Our second model started with 100 molecules of agonist pMHC (with dissociation constant 1/20 per second) while there were 1000 antagonist pMHC (with dissociation constant 1/3 per second). The results of our experiments are illustrated in Table 10.

Table 9. $N_1 : 100$, $N_2 : 0$, Type-I and Type II error : 0.001

Sl.	p_1	p_0	Result	Total Number of Samples	Number of Successful Samples	Time
1	0.90	0.95	No	40	0	107.25
2	0.70	0.75	No	40	0	106.95
3	0.50	0.55	No	80	0	218.42
4	0.20	0.25	No	120	0	168.98
5	0.10	0.15	No	160	0	330.80

Table 10. $N_1 : 100$, $N_2 : 1000$, Type-I and Type II error : 0.001

Sl.	p_1	p_0	Result	Total Number of Samples	Number of Successful Samples	Time
1	0.90	0.95	No	120	79	57.97
2	0.70	0.75	No	280	160	114.62
3	0.50	0.55	No	160	51	66.04
4	0.20	0.25	Yes	120	73	50.06
5	0.10	0.15	Yes	40	21	19.53

Property 4. In our fourth set of experiments, we were interested in asking the question if the system spent more than a certain threshold of time in a given state before leaving that state. We verified the following property with various values of the probability p.

$$Pr_{\geq p} (\mathbf{D_{100}} (ppERK / totalERK > 0.5))$$

The model we analyzed started with 100 molecules of agonist pMHC (with dissociation constant 1/20 per second) while antagonist pMHC was absent. The results of our analysis are presented in Table 11.

Table 11. $N_1 : 100$, $N_2 : 0$, Type-I and Type II error : 0.0001

Sl.	p_1	p_0	Result	Total Number of Samples	Number of Successful Samples	Time
1	0.90	0.95	Yes	160	160	216.21
2	0.70	0.75	Yes	120	120	160.32
3	0.50	0.55	Yes	80	80	109.11
4	0.20	0.25	Yes	40	40	54.33

6   Discussion and Conclusion

In this paper, we have introduced an algorithm, called BIOLAB, for formally verifying properties of stochastic models of biochemical processes. BIOLAB represents the first application of Statistical Model Checking to a rule-based model of signaling, which is specified here using the BIONETGEN modeling framework. BIOLAB is (i) an optimal trace-based method for Statistical Model Checking, which generates the minimum number of traces necessary to verify a property and (ii) BIOLAB provides user-specified bounds on Type-I and Type-II errors.

We demonstrated BIOLAB on a recently-developed BIONETGEN model of the T-cell receptor signaling pathway [23] with two stable states. We verified that both steady states are reachable on a single stochastic trajectory, whereas only a single steady state is reached on a deterministic ODE-based trajectory starting from the same initial conditions. Moreover, we verified that the system will alternate between these two states with high probability. These findings are relevant for understanding the TCR signaling pathway, which, under physiological conditions, must generate a robust response to a handful of stimulatory input molecules.

There are a number of areas for future research in BIOLAB. First, the T-cell receptor signaling model has a number of parameters. We verified properties of the pathway over a range of possible parameter values. In some contexts, it may be preferable to first (re)estimate parameter values for a given model. This can be accomplished by using standard parameter estimation techniques from the fields of Statistics and Machine Learning. One might even incorporate Model Checking into the parameter estimation phase by formally verifying that the parameter estimates reproduce known data, with high probability. Second, our method is presently limited to probabilistic bounded linear temporal logic formulas; we do not allow nested operators. This restriction can be relaxed through the use of different Model Checking algorithms. Pursuit of these two goals is ongoing.

References

1. Barua, D., Faeder, J.R., Haugh, J.M.: Structure-based kinetic models of modular signaling protein function: Focus on Shp2. Biophys. J. 92, 2290–2300 (2007)
2. Blinov, M.L., Faeder, J.R., Goldstein, B., Hlavacek, W.S.: BioNetGen: software for rule-based modeling of signal transduction based on the interactions of molecular domains. Bioinformatics 20(17), 3289–3291 (2004)
3. Blinov, M.L., Faeder, J.R., Goldstein, B., Hlavacek, W.S.: A network model of early events in epidermal growth factor receptor signaling that accounts for combinatorial complexity. BioSyst. 83, 136–151 (2006)
4. Blinov, M.L., Yang, J., Faeder, J.R., Hlavacek, W.S.: Graph theory for rule-based modeling of biochemical networks. In: Priami, C., Ingólfsdóttir, A., Mishra, B., Riis Nielson, H. (eds.) Transactions on Computational Systems Biology VII. LNCS (LNBI), vol. 4230, pp. 89–106. Springer, Heidelberg (2006)

5. Calzone, L., Chabrier-Rivier, N., Fages, F., Soliman, S.: Machine learning biochemical networks from temporal logic properties. In: Priami, C., Plotkin, G. (eds.) Transactions on Computational Systems Biology VI. LNCS (LNBI), vol. 4220, pp. 68–94. Springer, Heidelberg (2006)

6. Chabrier, N., Fages, F.: Symbolic Model Checking of Biochemical Networks. In: Proc. 1st Internl. Workshop on Computational Methods in Systems Biology, pp. 149–162 (2003)

7. Ciesinski, F., Größer, M.: On probabilistic computation tree logic. In: Baier, C., Haverkort, B.R., Hermanns, H., Katoen, J.-P., Siegle, M. (eds.) Validation of Stochastic Systems. LNCS, vol. 2925, pp. 147–188. Springer, Heidelberg (2004)

8. Clarke, E., Grumberg, O., Peled, D.A.: Model Checking. MIT Press, Cambridge (1999)

9. Danos, V., Feret, J., Fontana, W., Harmer, R., Krivine, J.: Rule-based modelling of cellular signalling. In: Caires, L., Vasconcelos, V.T. (eds.) CONCUR 2007. LNCS, vol. 4703, pp. 17–41. Springer, Heidelberg (2007)

10. Danos, V., Feret, J., Fontana, W., Krivine, J.: Scalable simulation of cellular signalling networks. In: Shao, Z. (ed.) APLAS 2007. LNCS, vol. 4807, pp. 139–157. Springer, Heidelberg (2007)

11. Faeder, J.R., Blinov, M.L., Goldstein, B., Hlavacek, W.S.: Rule-based modeling of biochemical networks. Complexity 10, 22–41 (2005)

12. Faeder, J.R., Blinov, M.L., Hlavacek, W.S.: Graphical rule-based representation of signal-transduction networks. In: SAC 2005: Proceedings of the 2005 ACM symposium on Applied computing, pp. 133–140. ACM, New York (2005)

13. Faeder, J.R., Blinov, M.L., Hlavacek, W.S.: Rule-based modeling of biochemical systems with BioNetGen. In: Maly, I.V. (ed.) Systems Biology. Methods in Molecular Biology. Humana Press, Totowa (2008)

14. Finkbeiner, B., Sipma, H.: Checking Finite Traces Using Alternating Automata. Formal Methods in System Design 24(2), 101–127 (2004)

15. Gillespie, D.T.: A general method for numerically simulating the stochastic time evolution of coupled chemical reactions. J. Comp. Phys. 22, 403–434 (1976)

16. Goldstein, B., Faeder, J.R., Hlavacek, W.S.: Mathematical and computational models of immune-receptor signaling. Nat. Rev. Immunol. 4, 445–456 (2004)

17. Hlavacek, W.S., Faeder, J.R., Blinov, M.L., Perelson, A.S., Goldstein, B.: The complexity of complexes in signal transduction. Biotechnol. Bioeng. 84, 783–794 (2003)

18. Hlavacek, W.S., Faeder, J.R., Blinov, M.L., Posner, R.G., Hucka, M., Fontana, W.: Rules for modeling signal-transduction systems. Science STKE 6 (2006)

19. Kwiatkowska, M., Norman, G., Parker, D., Tymchyshyn, O., Heath, J., Gaffney, E.: Simulation and verification for computational modelling of signalling pathways. In: WSC 2006: Proceedings of the 38th conference on Winter simulation, pp. 1666–1674 (2006)

20. Kwiatkowska, M.Z., Norman, G., Parker, D.: Prism 2.0: A tool for probabilistic model checking. In: QEST, pp. 322–323. IEEE, Los Alamitos (2004)

21. Langmead, C., Jha, S.K.: Predicting protein folding kinetics via model checking. In: The 7th Workshop on Algorithms in Bioinformatics. Lecture Notes in Bioinformatics, pp. 252–264 (2007)

22. Langmead, C., Jha, S.K.: Symbolic approaches to finding control strategies in boolean networks. In: Proceedings of The Sixth Asia-Pacific Bioinformatics Conference, pp. 307–319 (2008)

23. Lipniacki, T., Hat, B., Faeder, J.R., Hlavacek, W.S.: Stochastic effects and bistability in T cell receptor signaling. J. Theor. Biol. (in press, 2008)

250 E.M. Clarke et al.

24. Lok, L., Brent, R.: Automatic generation of cellular networks with Moleculizer 1.0. Nat. Biotechnol. 23, 131–136 (2005)
25. McKeithan, T.: Kinetic proofreading in T-cell receptor signal transduction. Proc. Natl. Acad. Sci. 92(11), 5042–5046 (1995)
26. Mu, F., Williams, R.F., Unkefer, C.J., Unkefer, P.J., Faeder, J.R., Hlavacek, W.S.: Carbon fate maps for metabolic reactions. Bioinformatics 23, 3193–3199 (2007)
27. Pawson, T., Nash, P.: Assembly of cell regulatory systems through protein interaction domains. Science 300(5618), 445–452 (2003)
28. Rabinowitz, J.D., Beeson, C., Lyonsdagger, D.S., Davisdagger, M.M., McConnell, H.M.: Kinetic discrimination in T-cell activation. Proc. Natl. Acad. Sci. 93(4), 1401–1405 (1996)
29. Sen, K., Viswanathan, M., Agha, G.: Statistical model checking of black-box probabilistic systems. In: Alur, R., Peled, D.A. (eds.) CAV 2004. LNCS, vol. 3114, pp. 202–215. Springer, Heidelberg (2004)
30. Vardi, M.: Alternating automata and program verification. Computer Science Today, 471–485 (1995)
31. Wald, A.: Sequential tests of statistical hypotheses. Annals of Mathematical Statistics 16(2), 117–186 (1945)
32. Yang, J., Monine, M.I., Faeder, J.R., Hlavacek, W.S.: Kinetic Monte Carlo method for rule-based modeling of biochemical networks (2007) arXiv:0712.3773
33. Younes, H.L.S.: Verification and Planning for Stochastic Processes with Asynchronous Events. PhD thesis, Carnegie Mellon (2005)

On a Continuous Degree of Satisfaction of Temporal Logic Formulae with Applications to Systems Biology

Aurélien Rizk, Grégory Batt, François Fages, and Sylvain Soliman

Projet Contraintes, INRIA Rocquencourt,
BP105, 78153 Le Chesnay Cedex, France
Firstname.Lastname@inria.fr
http://contraintes.inria.fr

Abstract. Finding mathematical models satisfying a specification built from the formalization of biological experiments, is a common task of the modeller that techniques like model-checking help solving, in the qualitative but also in the quantitative case. In this article we propose to go one step further by defining a continuous degree of satisfaction of a temporal logic formula with constraints. We show how such a satisfaction measure can be used as a fitness function with state-of-the-art search methods in order to find biochemical kinetic parameter values satisfying a set of biological properties formalized in temporal logic. We also show how it can be used to define a measure of robustness of a biological model with respect to some specification. These methods are evaluated on models of the cell cycle and of the MAPK signalling cascade.

1 Introduction

Temporal logics [1,2] have proven useful as specification languages for describing the behavior of a broad variety of systems ranging from electronic circuits to software programs, and more recently biological systems in either boolean [3,4,5], discrete [6], stochastic [7,8] or continuous [9,10,4,11] settings.

Because temporal logics allow us to express both qualitative (e.g. some protein is eventually produced) and quantitative (e.g. a concentration exceeds 10) information about time and systems variables, they provide a powerful specification language in comparison with the essentially qualitative properties considered in dynamical systems theory (e.g. multistability, existence of oscillations) or with the exact quantitative properties considered in optimization theory (e.g. curve fitting). In particular, these logics are well suited to the increasingly quantitative, yet incomplete, uncertain and imprecise information now accumulated in the field of quantitative systems biology.

This use of temporal logics relies on a logical paradigm for systems biology [12] which consists in making the following identifications:

$$biological\ model = transition\ system$$
$$biological\ properties = temporal\ logic\ formulae$$
$$biological\ validation = model\text{-}checking$$

M. Heiner and A.M. Uhrmacher (Eds.): CMSB 2008, LNBI 5307, pp. 251–268, 2008.

In this paradigm, temporal logics have been used in many applications, either as query languages of large interaction maps such as Kohn's map of the cell cycle [5,13] or gene regulatory networks [11], or as specification languages of biological properties known or inferred [14] from experiments, and used for validating models, discriminating between models and proposing new biological experiments [6], finding parameter values [9], or estimating robustness [15]. An important limitation of this approach is however due to the logical nature of temporal logic specifications and their boolean interpretation. A yes/no answer to a temporal logic query does not provide indeed any information on how far we are from satisfaction, nor how to guide the search to satisfy a formula. A measure of how close a model is to satisfy a property is needed.

In this paper, we define a continuous violation degree that quantifies how far from satisfaction an LTL formula is in a given model. In order to accommodate the various kinds of quantitative models defined by either ordinary or stochastic differential equation systems [16,17], rule-based languages like SBML [18] or BIOCHAM [19,20], hybrid Petri nets [21,22], stochastic process calculi [23,24], etc..., we represent the behavior of the system simply by numerical traces [14,25,9,10], so our method is rather general. This notion of violation degree is then used for two applications in systems biology: the search of kinetic parameter values in a model, and the quantitative estimation of the robustness of a model by adapting the general framework of Kitano [26] to our temporal logic setting.

Section 2 presents the quantifier free fragment of first-order linear time logic with constraints over the reals, QFLTL($\mathbb{R}$), studied in [14] and used in this paper. Section 3 defines a real-valued degree of satisfaction of an LTL formula using a variable abstraction mechanism which replaces real valued constants in LTL formulae by QFLTL($\mathbb{R}$) variables, and using an aggregation function which composes the distances between the validity domain of these variables and the corresponding constants.

Section 4 shows how such a continuous degree of satisfaction of an LTL formula can be used as a fitness function in local search methods for searching kinetic parameter values in order to satisfy a temporal logic specification. We describe a gradient based method and use the state-of-the-art Covariance Matrix Adaptation Evolution Strategy (CMA-ES) [27] to evaluate the method on models of the budding yeast cell cycle with 8 parameters and of the MAPK signaling cascade with 30 parameters and 7 unknown initial conditions.

In section 5 we propose a definition of a robustness degree of a property w.r.t. a set of model perturbations weighted by probabilities. This definition is inspired by the abstract definition of robustness proposed by Kitano for systems biology [26]. We develop it here in our temporal logic setting and illustrate its relevance by applying it to the previous model of the cell cycle.

2   Preliminaries on Linear Time Logic with Constraints over the Reals

2.1   LTL($\mathbb{R}$)

The *Linear Time Logic* LTL is a temporal logic [2] that extends classical logic with modal operators for qualifying when a formula is true in an infinite sequence of timed states, named a *trace*. The temporal operators are X ("next", for at the next time point), F ("finally", for at some time point in the future), G ("globally", for at all time points in the future), U ("until"), and W (" weak until"). These operators enjoy some simple duality properties, $\neg X\phi = X\neg\phi$, $\neg F\phi = G\neg\phi$, $\neg G\phi = F\neg\phi$, $\neg(\psi\ U\ \phi) = (\neg\phi\ W\ \neg\psi)$, $\neg(\psi\ W\ \phi) = (\neg\psi\ U\ \neg\phi)$. We have $F\phi = true\ U\ \phi$, $G\phi = \phi\ W\ false$.

A version of LTL with constraints over the reals, named LTL($\mathbb{R}$), has been proposed in [10,9] to express temporal properties about molecular concentrations. The atomic formulae of LTL($\mathbb{R}$) are formed with inequality relations and arithmetic operators over the real values of molecular concentrations and of their derivatives. The precise syntax of LTL($\mathbb{R}$) is given in Table 1. As negations and implications can be eliminated by propagating the negations down to the atomic constraints in the formula, we will assume in the following that all LTL($\mathbb{R}$) formulae are in negation free normal form.

Table 1. Syntax of LTL($\mathbb{R}$) formulae

Formula ::= Atom | Formula $\wedge$ Formula | Formula $\vee$ Formula
      | Formula $\Rightarrow$ Formula | $\neg$ Formula
      | X Formula | F Formula | G Formula
      | Formula U Formula | Formula W Formula
Atom ::=   Value Op Value
Op ::= $<$ | $>$ | $\leq$ | $\geq$
Value ::=   float | [molecule] | d[molecule]/dt | Time
      | Value + Value | Value - Value | - Value | Value $\times$ Value
      | Value / Value | Value ^ Value

For instance, F([A]>10) expresses that the concentration of A eventually gets above the threshold value 10. G([A]+[B]<[C]) expresses that the concentration of C is always greater than the sum of the concentrations of A and B. Oscillation properties, abbreviated as oscil(M,K), are defined as a change of sign of the derivative of M at least K times:
F((d[M]/dt > 0) $\wedge$ F((d[M]/dt < 0) $\wedge$ F((d[M]/dt > 0)...)))

LTL($\mathbb{R}$) formulae are interpreted over infinite traces of the form

$$(< t_0, \boldsymbol{x}_0, d\boldsymbol{x}_0/dt >, < t_1, \boldsymbol{x}_1, d\boldsymbol{x}_1/dt >, ...)$$

which give at discrete time points t_i, the concentration values $\boldsymbol{x}_i$ of the molecules, and the values of their first derivatives $d\boldsymbol{x}_i/dt$. Whereas LTL($\mathbb{R}$) formulae are interpreted over infinite traces, the ones we consider are always finite. For instance, in a model described by a system of ordinary differential equations (ODE), and under the hypothesis that the initial state is completely defined, numerical integration methods (such as Runge-Kutta or Rosenbrock method for stiff systems) provide a finite simulation trace. To extend it to an infinite trace, we adopted the solution of adding a loop on the last state, with the assumption that the finite time horizon considered for the numerical integration is sufficiently large to check the properties at hand.

Table 2. Inductive definition of the truth value of an LTL($\mathbb{R}$) formula in a trace π

$s \models \alpha$ iff α is a propositional formula and α is true in the state s,
$\pi \models \phi$ iff $s \models \phi$ where s is the first state of π,
$\pi \models X\psi$ iff $\pi^1 \models \psi$,
$\pi \models \psi \ U \ \psi'$ iff there exists $k \geq 0$ s.t. $\pi^k \models \psi'$ and $\pi^j \models \psi$ for all $0 \leq j < k$.
$\pi \models \psi \ W \ \psi'$ iff either for all $k \geq 0$, $\pi^k \models \psi$.
or there exists $k \geq 0$ s.t. $\pi^k \models \psi \wedge \psi'$ and for all $0 \leq j < k$, $\pi^j \models \psi$.
$\pi \models \neg\psi$ iff $\pi \not\models \psi$,
$\pi \models \psi \wedge \psi'$ iff $\pi \models \psi$ and $\pi \models \psi'$,
$\pi \models \psi \vee \psi'$ iff $\pi \models \psi$ or $\pi \models \psi'$,
$\pi \models \psi \Rightarrow \psi'$ iff $\pi \models \psi'$ or $\pi \not\models \psi$,

It is worth noticing that the semantics of the "next" operator refers to the next time point on the trace and that in adaptive step size integration methods of ODE systems, the step size $t_{i+1} - t_i$ is not constant but determined through an estimation of the error made by the discretization.

Formally, the truth value of an LTL($\mathbb{R}$) formula in a trace π is given in Table 2. These truth values can be computed on traces by model-checking [9].

2.2 QFLTL($\mathbb{R}$)

In [14], the quantifier free fragment of the first-order extension of LTL($\mathbb{R}$), named QFLTL($\mathbb{R}$), has been considered for the purpose of analyzing numerical data time series in temporal logic and computing automatically LTL($\mathbb{R}$) specifications from experimental traces. Syntactically, QFLTL($\mathbb{R}$) adds variables to atomic expressions with the following grammar:

$$\text{Atom} ::= \ \text{Value Op Value} \mid \text{Value Op Variable}$$

For instance, the QFLTL($\mathbb{R}$) formula $G([A] < v)$ expresses the constraint that v is greater than the maximum concentration of A. The restriction that a variable can only appear in the right-hand side of a comparison is motivated by computability results.

As usual, the semantics of a QFLTL($\mathbb{R}$) formula containing variables is defined by its ground instances which are LTL($\mathbb{R}$) formulae. Given a trace π and a QFLTL($\mathbb{R}$) formula $\phi(\boldsymbol{x})$ over a vector $\boldsymbol{x}$ of v real-valued variables, the *constraint satisfaction problem*, $\exists \boldsymbol{x} \in \mathbb{R}^v \ (\phi(\boldsymbol{x}))$, is the problem of determining the valuations $\boldsymbol{v}$ of the variables for which the formula ϕ is true. In other words, we look for the domain of validity $\mathcal{D}_\phi \subset \mathbb{R}^v$ such that $\pi \models \forall \boldsymbol{v} \in \mathcal{D}_\phi \ (\phi(\boldsymbol{v}))$.

In [14], an LTL($\mathbb{R}$) *model-checking algorithm* has been generalized to a QFLTL($\mathbb{R}$) *constraint solving algorithm* which computes the exact domain of validity $\mathcal{D}_\phi$ for any QFLTL($\mathbb{R}$) formula ϕ, in time $O((nf)^{2v})$ where v is the number of variables in ϕ, f the size of the formula and n the length of the trace. This algorithm is at the heart of the methods presented in the following sections.

3 Continuous Satisfaction Degree of LTL($\mathbb{R}$) Formulae

In order to evaluate numerically the adequateness of a model w.r.t. a temporal logic specification, we introduce a continuous violation degree relating a trace of the model to the given constraint LTL formula. When the model satisfies its specification the degree will be null, and the farther the traces from the expected behavior, the biggest the violation degree.

3.1 Variable Abstraction

Our definition of the violation degree of an LTL($\mathbb{R}$) formula relies on an abstraction of the constants occurring in the formula by variables. Starting from an LTL specification ϕ of the expected behavior of a system, we transform it into a QFLTL formula ϕ^* by mapping the constants (i.e. real numbers corresponding to concentration thresholds, amplitudes, etc.) $c_1, \ldots, c_n$ appearing in ϕ, to distinct variables $x_1, \ldots, x_n$. It is worth noting that ϕ^* is a QFLTL formula that can also be seen as a function over $\mathbb{R}^n$ associating a closed LTL formula to an instantiation of its variables.

Definition 1. *Given an LTL($\mathbb{R}$) formula ϕ and a QFLTL abstraction ϕ^*, the objective, noted $var(\phi)$, is the single point in the variable space $\mathbb{R}^n$ of ϕ^*, with x_i equal to c_i for all $1 \leq i \leq n$.*

Example 1. Consider the LTL formula $\phi = F([A] > 20)$ indicating that from experiments it was observed that after some time the concentration of compound A becomes greater than 20. We get $\phi^* = F([A] > x)$ as a QFLTL formula and $\mathbb{R}$ as variable space. We have $var(\phi) = 20$.

Because of the syntactical restriction imposed on the occurrences of variables in the right-hand sides of the inequalities in QFLTL($\mathbb{R}$) formulae, the transformation from ϕ to ϕ^* cannot always be done automatically. However for polynomial expressions over concentrations and derivatives, one can apply the following transformation on atomic expressions:

$$(e_1 \ Op \ e_2)^* \ = \ e \ Op \ x$$

where Op is an inequality operator, $e_1 - e_2$ is a polynomial in the concentrations and derivatives with term c of degree 0, $e = e_1 - e_2 - c$ and x is a new variable introduced for the term $-c$.

More generally, ϕ^* will be a QFLTL formula given with a variable space $\mathbb{R}^n$ that may include variables defined from other ϕ^* variables with linear inequalities, allowing some rescaling between variables if necessary. The objective $var(\phi)$ will be defined explicitly through an instantiation of those variables, i.e. a point in the variable space.

Example 2. Consider the QFLTL formula $\phi^* = F([A] \geq v) \wedge F([A] \leq w)$, let us define the amplitude variable $amp = v - w$ and use it as the only variable for our variable space $\mathbb{R}$. We can set as objective that the amplitude of variation of the compound A is at least 10 with $var(\phi) = 10$.

3.2   Quantitative Satisfaction

Given a QFLTL formula ϕ^* and a numerical trace T, the QFLTL($\mathbb{R}$) constraint solving algorithm of [14] computes the exact *domain of validity* for ϕ^* on T, as the domain of the variables $D_{\phi^*}(T) \subset \mathbb{R}^n$.

Definition 2. *The* violation degree *of a numerical trace T to an LTL formula ϕ, noted $vd(T, \phi)$ is the Euclidean distance between $D_{\phi^*}(T)$ and $var(\phi)$, i.e.* $min_{v \in D_{\phi^*}(T)} d(v, var(\phi))$.

Example 3. In the example 1 and given a mathematical model of our system, let us suppose that the QFLTL constraint solving algorithm applied to ϕ^* on simulation trace T computes $D_{\phi^*}(T) =]-\infty, 15]$ as domain for variable x. Since $var(\phi) = 20$ we get $vd(T, \phi) = 5$, i.e. the violation degree is 5 since the compound reaches a maximum of 15 whereas the formula expresses that the threshold 20 be reached.

For the specification of example 2, suppose that the constraint solving computes the domains of v and w: $D(v) =]-\infty, 15]$ and $D(w) = [10, +\infty[$. For this formula ϕ, the maximum value of $D(v)$ represents the maximum value of $[A]$ and the minimum value of $D(w)$ its minimum value in the trace. The domain for variable amp is $D_{\phi^*}(T) =]-\infty, 5]$ since we know that $amp = v - w$, and thus, since $var(\phi) = 10$, we obtain $vd(T, \phi) = 5$, i.e. the amplitude of the curve is 5 whereas we wanted it to be 10.

Note that if T is such that ϕ is satisfied then $vd(T, \phi) = 0$ since $var(\phi) \in D_{\phi^*}(T)$. However when ϕ is not valid on T, the violation degree vd provides a quantitative measurement of its degree of non satisfaction. The use of this measure is illustrated in the following sections to improve parameter search for biological models and to define a quantitative notion of robustness of a system w.r.t. a temporal logic formula.

4   Kinetic Parameter Search Using Violation Degree

The violation degree provides a measure of how far a given numerical trace is from an LTL specification. It is thus quite natural to use this measure to guide the search when trying to satisfy such a formula by replacing the scanning of parameter values described in [9] by a much more efficient local search method which makes evolve parameter sets by exploring a neighborhood of the current parameter set and by choosing the one which minimizes the violation measure.

4.1   Principle

Let us consider an LTL formula ϕ, an SBML/BIOCHAM reaction model with initial conditions and known parameter values, a set of unknown parameters to explore and for each of those an interval of search. We consider the problem of finding a set of values of the unknown parameters such that the violation degree of the corresponding trace T obtained by numerical simulation is $vd(T, \phi) = 0$.

A generic optimization algorithm for parameter search can be described as follows:

Algorithm 1 (generic parameter search method)

1. *Set the current point in the parameter space to a random point belonging to the provided search box, compute a numerical simulation with trace T and the corresponding violation degree $vd(T, \phi)$;*
2. *if $vd = 0$ jump to 5.*
3. *for each point in a defined neighborhood of the current point, compute a trace and its violation degree;*
4. *based on the violation degrees of the neighbors, determine the next point of the iteration, set the current point to this point, update current vd and go to 2.*
5. *Return the current point in the parameter space.*

This procedure can be interrupted after a given number of steps, returning the best parameter set (minimizing the violation degree). It can also be restarted with a new initial point (step 1) several times in order to diversify the search.

A naive method would be to define as neighborhood of the current parameter state the parameter sets obtained by modifying one parameter by values $\pm \delta$; and to choose as next parameter set the best neighbor.

More efficient instances of this algorithm can be obtained however, by combining state-of-the-art nonlinear optimization methods with the computation of our violation degree used as a blackbox fitness function. In the following sections, we use the Covariance Matrix Adaptation Evolution Strategy (CMA-ES) of Hansen and Ostermeier [27]. This method uses a probabilistic definition of the neighborhood, and stores information in a covariance matrix in order to replace the approximate gradient and Hessian of a quasi-Newton method by an evolutionary algorithm.

4.2   Evaluation on Cell Cycle Models

In this section we present the application of the parameter search method outlined above to the budding yeast cell cycle model of [28]. This model displays how proteins cdc2 and cyclin interact to form the heterodimer Cdc2-Cyclin~{p1,p2} known as maturation promoting factor (MPF) and playing a key role in the control of mitotic cycles. The reaction rules of the model are the following:

```
MA(k1)         for                      _ => Cyclin.
MA(k3)         for   Cyclin + Cdc2~{p1} => Cdc2-Cyclin~{p1,p2}
MA(k4p)        for Cdc2-Cyclin~{p1,p2} => Cdc2-Cyclin~{p1}
AUTOCAT(k4)    for Cdc2-Cyclin~{p1,p2} => Cdc2-Cyclin~{p1}
MA(k6)         for Cdc2-Cyclin~{p1}    => Cyclin~{p1} + Cdc2
MA(k7)         for          Cyclin~{p1} => _
MA(k8)         for                 Cdc2 => Cdc2~{p1}
MA(k9)         for          Cdc2~{p1} => Cdc2
```

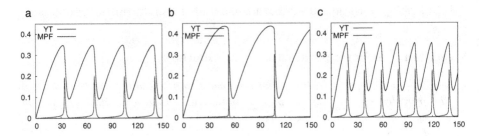

Fig. 1. Dynamical behavior of the cell cycle model. The plots represent total cyclin (YT) and maturation promoting factor (MPF). (a) Oscillatory behavior obtained with parameter values k_{Tyson}. (b) Higher MPF peaks obtained with k_{Tyson}^{*} (solution of problem S1). (c) Shorter oscillations period obtained with k_4^{*} (solution of problem S4).

MA(k) denotes Mass Action law kinetics with parameter k while ~{p1} and ~{p1,p2} denote phosphorylated forms of a molecule. The rate of reaction 4 is described by:

AUTOCAT(k4)= k4*[Cdc2-Cyclin~{p1,p2}]*[Cdc2-Cyclin~{p1}]^2.

We use as reference point k_{Tyson} the values of the kinetic parameters determined in [28]. The simulation for k_{Tyson} of the system of ODEs extracted from these rules, given in appendix, is displayed in Figure 1. The total amount of cyclin presents oscillations of period 35 while MPF exhibits activity peaks with same period.

Using the optimization method CMA-ES together with our violation degree as a parameter search method we wonder whether it is possible to find values of the kinetic parameters corresponding to higher MPF peaks or oscillations with higher amplitudes or shorter periods.

Search Problem S1: Higher MPF Peaks (2 Parameters Unknown) Two parameters, k4 and k6, have been found in [28] to play a particular role for the existence of oscillations. Depending on their values the system exhibits either a steady state behavior or limit cycle oscillations. We wonder whether it is possible to obtain higher MPF peaks by changing values of k4 and k6 only, all other parameters remaining at the value k_{Tyson} chosen in [28]. More precisely, we want to reach at least MPF peaks of 0.3, the maximum amount of MPF for k_{Tyson} being 0.19.

Therefore we define the LTL specification : $\phi_1 = F([MPF] > 0.3)$ with the corresponding QFLTL formula being :

$$\phi_1^{*} = F([MPF] > max)$$

The variable space associated to ϕ_1^{*} is $\mathbb{R}$ and corresponds to the sole variable max. The objective is $var(\phi) = 0.3$, i.e the target peak value of MPF is 0.3. We have been able to find valid parameter values, denoted k_{Tyson}^{*}, satisfying $vd(T, \phi_1^{*}) = 0$ where T is the corresponding simulated trace (see Figure 1b). k_{Tyson}^{*} is given in Table 3.

To illustrate the path followed during the search from k_2 to k_2^* we computed the violation degree landscape in the k4, k6 parameter space. The resulting landscape is displayed in Figure 2. Note that as all constants of the formula have been abstracted by variables, the violation degree can only be finite. In particular when no oscillations are present in the trace amp will be equal to 0, thus leading to a violation degree of 0.19. Regions where the violation degree is 0.19 correspond to regions of steady state behavior whereas regions with a violation degree between 0 and 0.19 correspond to regions of oscillations.

Under mild assumptions Tyson determined linear equations defining a region in the k4, k6 plane where oscillations occurs, also represented in Figure 2. Our results are fully consistent with his analytical analysis, and provide more information on the amplitude of oscillation w.r;t. parameters k4 and k6.

Search Problem S3 and S4 : Amplitude and Period of Oscillations (All 8 Parameters Unknown)

To illustrate the scalability of the method we carry out two parameter searches on all 8 parameters of the model. The first one (problem S3) is the same query as above with formula ϕ_2^* but with all parameters unknown. The second one is a more complex query used to find shorter oscillation periods of Cdc2 :

$$\phi_3^* = F(\quad d([Cdc2])/dt < 0 \wedge X(d([Cdc2])/dt > 0 \wedge Time = t1$$
$$\wedge X(F(d([Cdc2])/dt > 0 \wedge X(d([Cdc2])/dt < 0 \wedge Time = t2))$$

To specify that the target period is 20, we use the variable space $\mathbb{R}$ corresponding to the variable $per = t2 - t1$ with target $var(\phi) = 20$. Search problem S3 starts from parameter values k_3 satisfying the constraints on their order of magnitude given in [28]. k_3 does not give rise to oscillations. Search problem S4 starts from $k_4 = k_{Tyson}$. In both cases parameter values are found satisfying the query (in 30 s for S3 and 350 s for S4). Results are given in Table 3.

Table 3. Resulting parameter values for search problems S1, S2, S3 and S4

	S1		S2		S3		S4	
	Initial values	Result	Initial values	Result	Initial values	Result	Initial values	Result
$vd(T, \phi)$	0.11	0	0.04	0	0.19	0	15.1	4.90e-4
Parameters	k_{tyson}	k_{tyson}^*	k_2	k_2^*	k_3	k_3^*	k_4	k_4^*
k1	1.50e-2	1.50e-2	1.50e-2	1.50e-2	1.00e-2	**1.14e-2**	1.50e-2	**2.41e2**
k3	2.00e2	2.00e2	2.00e2	2.00e2	1.00e2	**1.13e2**	2.00e2	**2.83e2**
k4p	1.80e-2	1.80e-2	1.80e-2	1.80e-2	1.00e-2	**8.77e-3**	1.80e-2	**2.24e-2**
k4	1.80e2	**8.99e2**	2.00e1	**1.94e2**	1.00e2	**1.82e2**	1.80e2	**2.28e2**
k6	1.00	**3.23**	0.25	**1.41**	1.00	**4.17e-1**	1	**1.13**
k7	0.60	0.60	0.60	0.60	1.00	**1.37**	0.60	**5.99e-1**
k8	1.00e2	1.00e2	1.00e2	1.00e2	1.00e3	**8.99e2**	1.00e2	**1.42e2**
k9	1.00e2	1.00e2	1.00e2	1.00e2	1.00e2	**8.44e1**	1.00e2	**6.94e1**

4.3   Evaluation on MAPK Signal Transduction Model

The MAPK signal transduction model [29] is used to test the scalability of the parameter search method on a larger model. This model, made of a cascade of phosphorylation reactions, consists of the following rules :

```
(MA(k1), MA(k2)) for RAF + RAFK <=> RAF-RAFK.
(MA(k3),MA(k4)) for RAF~{p1} + RAFPH <=> RAF~{p1}-RAFPH.
(MA(k5),MA(k6)) for MEK~$P + RAF~{p1} <=> MEK~$P-RAF~{p1}
    where p2 not in $P.
(MA(k7),MA(k8)) for MEKPH + MEK~{p1}~$P <=> MEK~{p1}~$P-MEKPH.
(MA(k9),MA(k10)) for MAPK~$P + MEK~{p1,p2} <=> MAPK~$P-MEK~{p1,p2}
    where p2 not in $P.
(MA(k11),MA(k12)) for MAPKPH + MAPK~{p1}~$P <=> MAPK~{p1}~$P-MAPKPH.
MA(k13) for RAF-RAFK => RAFK + RAF~{p1}.
MA(k14) for RAF~{p1}-RAFPH => RAF + RAFPH.
MA(k15) for MEK~{p1}-RAF~{p1} => MEK~{p1,p2} + RAF~{p1}.
MA(k16) for MEK-RAF~{p1} => MEK~{p1} + RAF~{p1}.
MA(k17) for MEK~{p1}-MEKPH => MEK + MEKPH.
MA(k18) for MEK~{p1,p2}-MEKPH => MEK~{p1} + MEKPH.
MA(k19) for MAPK-MEK~{p1,p2} => MAPK~{p1} + MEK~{p1,p2}.
MA(k20) for MAPK~{p1}-MEK~{p1,p2} => MAPK~{p1,p2} + MEK~{p1,p2}.
MA(k21) for MAPK~{p1}-MAPKPH => MAPK + MAPKPH.
MA(k22) for MAPK~{p1,p2}-MAPKPH => MAPK~{p1} + MAPKPH.
```

We denote by k_{MAPK} the set of kinetic parameter values used as reference for this model.

Search Problem S5 : Curve Fitting at Specific Time Points (22 Parameters Unknown)

In this example, we investigate the use of our parameter search method as a curve fitting tool at specific time points, on 22 parameter values. In order to express the classical distance between two curves at time points 30 and 60 for instance, we use the following pattern of formulae :

$$\phi_4^* = G(\ Time = 30 \rightarrow [MEK - RAF^\sim\{p1\}] = u$$
$$\wedge Time = 60 \rightarrow [MEK - RAF^\sim\{p1\}] = v)$$

The parameter space of this formula is $\mathbb{R}^2$ is defined by the two variables u and v. We set target $var(\phi)$ to the target values of $[MEK - RAF^\sim\{p1\}]$ at time 30 and 60. Note that this formula can be extended to any number of time points and molecules in order to perform a complete curve fitting, if it is relevant.

This pattern of formulae can be used to search the values of all the 22 parameters of the model to fit the concentration $[MEK - RAF^\sim\{p1\}]$ at six time points. The objective values for these time points are the values of the original model, obtained by simulation with the original parameters k_{MAPK}. The initial values for the search are some random altered values $k_{MAPK_{alt}}$. Numerical simulations obtained with k_{MAPK}, $k_{MAPK_{alt}}$ and the resulting parameter values are given in Figure 3. It took 290 s to obtain the result. This shows that the search method scales up well with the dimension of the parameter space, in comparison with the parameter scanning method which has an exponential time complexity in the number of parameters. Here, the computation time is more dependent on the type of problem (formula used and initial values of the parameters) and on the landscape of the violation degrees than on the number of parameters.

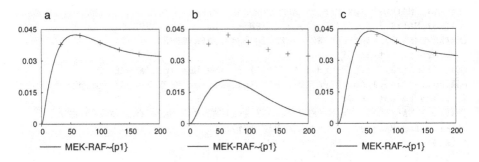

Fig. 3. Dynamical behavior of the MAPK model. The curves display $[MEK - RAF^{\sim}\{p1\}]$. (a) Reference curve obtained with k_{MAPK} (b) Simulated curve obtained with altered parameter values $k_{MAPK_{alt}}$. Points are the reference values taken from curve (a). (c) Simulated curve obtained after curve fitting (solution of problem S5).

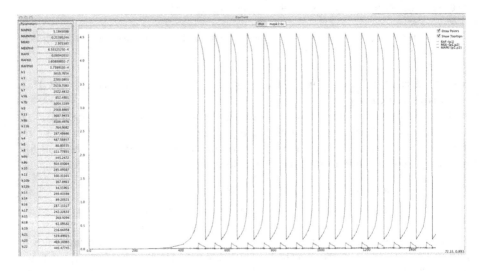

Fig. 4. Oscillations of MAPK found with CMA-ES in BIOCHAM

Search Problem S6: Find Oscillations (30 Kinetic Parameters and 7 Initial Conditions Unknown)

In [30], oscillations have been found in the MAPK cascade model of [29] although this model does not contain any negative feedback reaction. This does not contradict Thomas' necessary condition for sustained oscillations as such a purely directional cascade does contain negative feedback in its influence graph as shown in [31] and analyzed in [32]. However, to know whether these negative circuits in the influence graph are functional, one needs to search for kinetic parameter values and initial conditions that exhibit sustained oscillations.

Just by defining the following formula:

$$\phi_5^* = F(\ [MAPKp1p2] > max \wedge F([MAPKp1p2] < min))$$

using the variable space $\mathbb{R}$ for the single variable $amp = max - min$, and by asking that the amplitude be at least 0.5, setting $var(\phi) = 0.5$, parameter values leading to sustained oscillations, such as the ones depicted in Figure 4, were found in a few minutes.

5 Quantitative Robustness Analysis

5.1 Principle

We have seen in the previous section that our notion of violation degree allows us to use optimization techniques to efficiently guide parameter search given temporal logic properties. Here, we show that the notion of violation degree also allows us to define in a mathematically precise way a degree of robustness of a systems behavior described in temporal logic w.r.t. a set of perturbations, and estimate it computationally. This robustness degree is defined as the inverse of the average violation degree of the property of interest over all admissible perturbations, possibly weighted by their probabilities. This definition is an adaptation of the general definition given by Kitano [26] to our temporal logic setting. Formally, using the notations introduced in previous sections, we set:

Definition 3. *Let P be a set of perturbations, $prob(p)$ be the probability of perturbation p, $T(p)$ be the timed trace of the system under perturbation $p \in P$. The robustness degree $R_{\phi,P}$ of a property ϕ with respect to P is the real value*

$$R_{\phi,P} = \left(\int_{p \in P} vd(T(p), \phi) prob(p) dp \right)^{-1}$$

If the set of perturbations is finite (eg, gene knock outs), the robustness degree is simply the inverse of a finite weighted sum and can be exactly computed. If the set of perturbations is infinite, the robustness degree can be estimated by computing the violation degree between the behavior of the perturbed system $T(p)$ and the specification ϕ for sufficiently many perturbations.

5.2 Evaluation on Cell Cycle Model

Using the same cell cycle model as in section 4.2, we compare the robustness of oscillation properties with regard to perturbations of parameter values k4 and k6 for different points in the parameter space.

We consider that parameter values for k4 and k6 are normally distributed around their reference value with coefficient of variation equal to 0.2. We also enforce that $k4 \geq 0 \wedge k6 \geq 0$. We examine the robustness of the property expressed by ϕ_2^*, that is, MPF oscillations are of amplitude at least 0.19.

The robustness degree of this property is compared for three different values of k4 and k6. These three points in the parameter space of k4 and k6 are indicated by the three points k_A, k_B and k_C in Figure 2. In all cases, the estimation of the robustness degree is done by computing the mean value of the violation degree for 500 samples.

The estimated degree of robustness for parameters k_A, k_B and k_C are respectively 133, 12.9 and 13.5. This is consistent with the location of points k_A, k_B and k_C. Perturbations around point k_A have high probabilities of staying in the region satisfying the specification whereas perturbations around point k_B have high probabilities of moving the system to the region with no oscillation. k_C is more robust than k_B even though, as opposed to k_B, its violation degree is non null. This can be explained by the abrupt transition between oscillating and non oscillating regions near k_B compared to the smoother transition near k_C.

The robustness degree can be estimated for perturbations on any number of parameters. For instance, by computing a robustness estimate for perturbations on all parameters, with coefficient of variation 0.2 for specification ϕ_2^* and parameter values k_{Tyson} and k_3, the estimated robustness degrees for k_{Tyson} and k_3 are 20.7 and 27.1 respectively. This indicates that the oscillations are more robust to variations of the parameters values for k_3 than for the parameters given in the original model of Tyson.

6   Related Work

Probabilistic temporal logics and probabilistic model checking have been used in systems biology [33], e.g. for an analysis of a probabilistic model of the MAP kinase signaling cascade. However these techniques provide information on the probability that a given property is exactly satisfied. They thus provide no quantitative information on unsatisfied formulae and cannot be compared to the satisfaction degree presented in this paper.

More closely related to our continuous satisfaction degree are the linear metrics for quantitative transition systems defined in [34]. These metrics apply to traces and can be characterized by quantitative LTL formulae. LTL formulae are interpreted on the $[0, 1]$ interval. However, no implementation is proposed, and the applicability of this approach to solving optimization and robustness problems is not discussed.

To the best of our knowledge, the most closely related approach is the one proposed by Fainekos and Pappas [35], where a satisfaction degree for temporal logic specifications is defined. Although the two approaches share many similarities, a significant difference is that in [35] the satisfaction degree corresponds to a distance between a trace and the set of traces satisfying a formula, whereas in our case the violation degree corresponds to a distance between a formula and the set of formulae satisfied by the trace. An advantage of the satisfaction degree is that it offers a rather intuitive interpretation, since it corresponds to the minimal perturbation of the trace that can change the truth value of the specification. However, the dimension of the space of traces is in general considerably higher than the dimension of the space of formulae. In the first case, traces are represented in a space of dimension $X^{|\tau|}$ where X the state space and $|\tau|$ the lenght of the trace. In the second case, formulae are represented in a space whose dimension equals the number of variables appearing in the QFLTL formula, typically corresponding to the number of numerical constants appearing in the original LTL formula. Because the computation of satisfaction or violation degree involves set operations, the dimensionality of the corresponding spaces may strongly affect the practical applicability of these methods. Note however that these approaches, handling sets of traces [36,37], and our approach, handling sets of formulae, are *a priori* compatible, and that their combination might combine their benefits.

Concerning robustness, in [38], Chaves and colleagues propose a quantitative measure of robustness corresponding to the volume of the set of valid parameters in the parameter space. This measure thus reflects the proportion of parameters that satisfy exactly the property, as opposed to our measure that represents how close to satisfying the property the system is for various parameters. These two measures provide complementary information on robustness. In [15], robustness is similarly defined with respect to temporal logic specifications. However, it has a Boolean interpretation, since a property is defined as robustly satisfied by an ODE system if it is satisfied by the system for all possible perturbations. As stated earlier, obtaining a quantitative measure of robustness is more informative for many practical problems.

7   Conclusion

We have defined a continuous measure of satisfaction of an LTL($\mathbb{R}$) formula in a numerical trace and shown that it can be computed using the QFLTL($\mathbb{R}$) constraint solving algorithm of [14]. This measure is more informative that the Boolean interpretation of the formulae and can be used in many situations in systems biology to reason about numerical traces.

This measure can be used as a fitness function in state-of-the-art optimization tools to efficiently guide the search of kinetic parameter values in biochemical reaction models in order to satisfy a set of properties formalized in LTL($\mathbb{R}$).

It can similarly be used to estimate the robustness of a model w.r.t. temporal logic specifications, in accordance to Kitano's notion of robustness for systems biology.

The generalization of model-checking to temporal logic constraint solving which is at the basis of the computation of this satisfaction measure thus seems to open new research avenues for the use of temporal logics in systems biology.

Acknowledgements

This work is partly supported by EU FP7 STREP project TEMPO and by the INRA Agrobi project INSIGHT.

References

1. Emerson, E.: Temporal and modal logic. In: van Leeuwen, J. (ed.) Handbook of Theoretical Computer Science. Formal Models and Sematics, vol. B, pp. 995–1072. MIT Press, Cambridge (1990)
2. Clarke, E.M., Grumberg, O., Peled, D.A.: Model Checking. MIT Press, Cambridge (1999)
3. Eker, S., Knapp, M., Laderoute, K., Lincoln, P., Meseguer, J., Sönmez, M.K.: Pathway logic: Symbolic analysis of biological signaling. In: Proceedings of the seventh Pacific Symposium on Biocomputing, pp. 400–412 (2002)
4. Chabrier, N., Fages, F.: Symbolic model checking of biochemical networks. In: Priami, C. (ed.) CMSB 2003. LNCS, vol. 2602, pp. 149–162. Springer, Heidelberg (2003)
5. Chabrier-Rivier, N., Chiaverini, M., Danos, V., Fages, F., Schächter, V.: Modeling and querying biochemical interaction networks. Theoretical Computer Science 325, 25–44 (2004)
6. Bernot, G., Comet, J.P., Richard, A., Guespin, J.: A fruitful application of formal methods to biological regulatory networks: Extending thomas' asynchronous logical approach with temporal logic. Journal of Theoretical Biology 229, 339–347 (2004)
7. Calder, M., Vyshemirsky, V., Gilbert, D., Orton, R.: Analysis of signalling pathways using the continuous time markow chains. In: Priami, C., Plotkin, G. (eds.) Transactions on Computational Systems Biology VI. LNCS (LNBI), vol. 4220, pp. 44–67. Springer, Heidelberg (2006) (CMSB 2005 Special Issue)
8. Heath, J., Kwiatkowska, M., Norman, G., Parker, D., Tymchyshyn, O.: Probabilistic model checking of complex biological pathways. In: Priami, C. (ed.) CMSB 2006. LNCS (LNBI), vol. 4210, pp. 32–47. Springer, Heidelberg (2006)
9. Calzone, L., Chabrier-Rivier, N., Fages, F., Soliman, S.: Machine learning biochemical networks from temporal logic properties. In: Priami, C., Plotkin, G. (eds.) Transactions on Computational Systems Biology VI. LNCS (LNBI), vol. 4220, pp. 68–94. Springer, Heidelberg (2006) (CMSB 2005 Special Issue)
10. Antoniotti, M., Policriti, A., Ugel, N., Mishra, B.: Model building and model checking for biochemical processes. Cell Biochemistry and Biophysics 38, 271–286 (2003)
11. Batt, G., Ropers, D., de Jong, H., Geiselmann, J., Mateescu, R., Page, M., Schneider, D.: Validation of qualitative models of genetic regulatory networks by model checking: Analysis of the nutritional stress response in Escherichia coli. Bioinformatics 21, i19–i28 (2005)
12. Fages, F.: Temporal logic constraints in the biochemical abstract machine BIOCHAM (invited talk). In: Hill, P.M. (ed.) LOPSTR 2005. LNCS, vol. 3901, pp. 1–5. Springer, Heidelberg (2006)
13. Kohn, K.W.: Molecular interaction map of the mammalian cell cycle control and DNA repair systems. Molecular Biology of the Cell 10, 2703–2734 (1999)
14. Fages, F., Rizk, A.: On temporal logic constraint solving for the analysis of numerical data time series. Theoretical Computer Science (2008) (CMSB 2007 special issue)
15. Batt, G., Yordanov, B., Weiss, R., Belta, C.: Robustness analysis and tuning of synthetic gene networks. Bioinformatics 23, 2415–2422 (2007)
16. Segel, L.A.: Modeling dynamic phenomena in molecular and cellular biology. Cambridge University Press, Cambridge (1984)
17. Szallasi, Z., Stelling, J., Periwal, V. (eds.): System Modeling in Cellular Biology: From Concepts to Nuts and Bolts. MIT Press, Cambridge (2006)

18. Hucka, M., et al.: The systems biology markup language (SBML): A medium for representation and exchange of biochemical network models. Bioinformatics 19, 524–531 (2003)
19. Fages, F., Soliman, S., Chabrier-Rivier, N.: Modelling and querying interaction networks in the biochemical abstract machine BIOCHAM. Journal of Biological Physics and Chemistry 4, 64–73 (2004)
20. Calzone, L., Fages, F., Soliman, S.: BIOCHAM: An environment for modeling biological systems and formalizing experimental knowledge. BioInformatics 22, 1805–1807 (2006)
21. Gilbert, D., Heiner, M., Lehrack, S.: A unifying framework for modelling and analysing biochemical pathways using petri nets. In: Calder, M., Gilmore, S. (eds.) CMSB 2007. LNCS (LNBI), vol. 4695, pp. 200–216. Springer, Heidelberg (2007)
22. Matsuno, H., Doi, A., Nagasaki, M., Miyano, S.: Hybrid petri net representation of gene regulatory network. In: Proceedings of the 5th Pacific Symposium on Biocomputing, pp. 338–349 (2000)
23. Priami, C., Regev, A., Silverman, W., Shapiro, E.: Application of a stochastic name passing calculus to representation and simulation of molecular processes. Information Processing Letters 80, 25–31 (2001)
24. Phillips, A., Cardelli, L.: A correct abstract machine for the stochastic pi-calculus. Transactions on Computational Systems Biology, Special issue of BioConcur 2004 (to appear, 2004)
25. Nickovic, D., Maler, O.: Amt: a property-based monitoring tool for analog systems. In: Raskin, J.-F., Thiagarajan, P.S. (eds.) FORMATS 2007. LNCS, vol. 4763, pp. 304–319. Springer, Heidelberg (2007)
26. Kitano, H.: Towards a theory of biological robustness. Molecular Systems Biology 3, 137 (2007)
27. Hansen, N., Ostermeier, A.: Completely derandomized self-adaptation in evolution strategies. Evolutionary Computation 9, 159–195 (2001)
28. Chen, K.C., Csikász-Nagy, A., Györffy, B., Val, J., Novàk, B., Tyson, J.J.: Kinetic analysis of a molecular model of the budding yeast cell cycle. Molecular Biology of the Cell 11, 369–391 (2000)
29. Levchenko, A., Bruck, J., Sternberg, P.W.: Scaffold proteins biphasically affect the levels of mitogen-activated protein kinase signaling and reduce its threshold properties. PNAS 97, 5818–5823 (2000)
30. Qiao, L., Nachbar, R.B., Kevrekidis, I.G., Shvartsman, S.Y.: Bistability and oscillations in the huang-ferrell model of mapk signaling. PLoS Computational Biology 3, 1819–1826 (2007)
31. Fages, F., Soliman, S.: From reaction models to influence graphs and back: a theorem. In: Fisher, J. (ed.) FMSB 2008. LNCS (LNBI), vol. 5054. Springer, Heidelberg (2008)
32. Ventura, A.C., Sepulchre, J.A., Merajver, S.D.: A hidden feedback in signaling cascades is revealed. PLoS Computational Biology (to appear, 2008)
33. Kwiatkowska, M., Norman, G., Parker, D.: Using probabilistic model checking in systems biology. SIGMETRICS Performance Evaluation Review 35, 14–21 (2008)
34. de Alfaro, L., Faella, M., Stoelinga, M.: Linear and branching metrics for quantitative transition systems. In: Díaz, J., Karhumäki, J., Lepistö, A., Sannella, D. (eds.) ICALP 2004. LNCS, vol. 3142, pp. 97–109. Springer, Heidelberg (2004)

35. Fainekos, G., Pappas, G.: Robustness of temporal logic specifications. In: Havelund, K., Núñez, M., Roşu, G., Wolff, B. (eds.) FATES 2006 and RV 2006. LNCS, vol. 4262, pp. 178–192. Springer, Heidelberg (2006)
36. Fainekos, G., Pappas, G.: Robust sampling for MITL specifications. In: Raskin, J.-F., Thiagarajan, P.S. (eds.) FORMATS 2007. LNCS, vol. 4763, pp. 147–162. Springer, Heidelberg (2007)
37. Dang, T., Donze, A., Maler, O., Shalev, N.: Sensitive state space exploration (submitted, 2008), http://www-verimag.imag.fr/~maler/
38. Chaves, M., Sontag, E., Sengupta, A.: Shape, size and robustness: feasible regions in the parameter space of biochemical networks (submitted, 2008) arXiv:0710.4269v1

A Model Checking Approach to the Parameter Estimation of Biochemical Pathways

Robin Donaldson and David Gilbert

Bioinformatics Research Centre, University of Glasgow
Glasgow G12 8QQ, Scotland, UK
{radonald,drg}@brc.dcs.gla.ac.uk

Abstract. Model checking has historically been an important tool to verify models of a wide variety of systems. Typically a model has to exhibit certain properties to be classed 'acceptable'. In this work we use model checking in a new setting; parameter estimation. We characterise the desired behaviour of a model in a temporal logic property and alter the model to make it conform to the property (determined through model checking). We have implemented a computational system called MC2(GA) which pairs a model checker with a genetic algorithm. To drive parameter estimation, the fitness of set of parameters in a model is the inverse of the distance between its actual behaviour and the desired behaviour. The model checker used is the simulation-based Monte Carlo Model Checker for Probabilistic Linear-time Temporal Logic with numerical constraints, MC2(PLTLc). Numerical constraints as well as the overall probability of the behaviour expressed in temporal logic are used to minimise the behavioural distance. We define the theory underlying our parameter estimation approach in both the stochastic and continuous worlds. We apply our approach to biochemical systems and present an illustrative example where we estimate the kinetic rate constants in a continuous model of a signalling pathway.

1 Introduction

Modelling biochemical systems is a key activity in Systems Biology [1], for example in the area of signal transduction pathways [2]. Models can be used to increase the understanding of a biochemical network in terms of the interactions between the components (the topology), or their dynamic behaviour. The representation of such systems can range from the informal, for example pathway diagrams, to the formal, which include qualitative and quantitative descriptions – the latter being stochastic or continuous and requiring kinetic information including reaction rates and concentrations/mass of components [3]. Formal models can permit both simulation of behaviour as well as the analysis of behavioural properties.

One important issue is how models are obtained, a process that usually involves fitting to some trusted data. Model fitting can involve identification of alternative topologies [4], choice of types of kinetic laws and formulae, and estimation of kinetic rate constants and initial concentration/mass values [5]. This

M. Heiner and A.M. Uhrmacher (Eds.): CMSB 2008, LNBI 5307, pp. 269–287, 2008.

is a challenging task in the biochemical field, especially due to the lack of reliable quantitative data.

A biologist or biochemist will often be unsure about exact values of biochemical species over time due to the nature of the wet-lab experimental technology, and will describe behaviour in a semi-quantitative manner. For example, "the concentration of the protein peaks within 2 to 5 minutes and then falls to less than 50% of the peak value within 60 minutes". A significant challenge is how to automatically build a model which conforms to semi-quantitative behaviour. Temporal logic is well-suited to formally represent such semi-quantiative descriptions.

In this paper we report on work to use model checking to drive the estimation of parameter values in biochemical models. We use a probabilistic temporal logic to describe desired behavioural properties and the MC2(PLTLc) model checker [6] to compute how closely the behaviour of a model conforms to the desired behaviour. We use a genetic algorithm to explore model space in order to generate a set of models which exhibit some desired behaviour. We have defined a novel extension of PLTL temporal logic [7] in order to permit a fine grained distance function suitable for use in our model exploration approach. This enables us to operate over both continuous as well as stochastic models.

Given a model with a fixed topology and ranges of parameter values to be explored, we can use a genetic algorithm to explore model space and generate values of kinetic rate constants and initial concentration/masses for which the model exhibits the desired behaviour. We illustrate this approach by considering a continuous model of the well-known MAPK signalling pathway stimulated by EGF [8], and derive values for kinetic rate constants such that the behaviour conforms to that under NGF stimulation. In doing so we confirm the results of [8] which showed that the desired results could be achieved by varying only one parameter, V28, 40-fold however in our approach we perform multi-parameteric fitting and show that the same desired behaviour can be achieved by varying a set of kinetic parameters with V28 only requiring a 16-fold increase.

Our approach contributes to the field of systems biology in terms of model construction from desired behaviour as well as to the field of synthetic biology in terms of system design and construction from desired behaviour properties [9]. This is the first step in a general approach to automatically constructing models based on a formal description of desired behaviour of a model.

This paper is organised as follows. The following section outlines the theory of our approach in both the stochastic and continuous worlds. The next section describes our computational system for parameter estimation, MC2(GA). We next present a case study where we estimate the parameters of a model of a signalling pathway. In doing so, we attempt to answer an important biochemical question concerning this pathway – what are the underlying model differences explaining the cell reactions to different signals. We conclude with a summary of our approach and propose further research ideas.

2 Theory

This section sets out the theory behind our computational system for parameter estimation. First we explain the syntax and semantics of the PLTLc temporal logic. Next, we describe how the desired behaviour of a model can be characterised using PLTLc. Then we define probabilistic domains – the relationship between the values of a free variable in a PLTLc property and the overall probability of the behaviour – and show how they can be helpful in characterising the desired behaviour. Finally, we explain how to build a distance metric of the distance between the model's behaviour and the desired behaviour.

2.1 PLTLc Syntax

Linear-time Temporal Logic (LTL) [10] is the fragment of full Computational Tree Logic (CTL*) [11] without path quantifiers, implicitly quantifying universally over all paths. LTL has been introduced in a probabilistic setting in [7], and extended by numerical constraints over real value variables in [12]. PLTLc combines both extensions, complemented by the filter construct as used in Probabilistic Computational Tree Logic (PCTL) [13] and Continuous Stochastic Logic (CSL) [14]. We start with the LTL with numerical constraints (LTLc) syntax:

$$\phi ::= X\phi \mid G\phi \mid F\phi \mid \phi U\phi \mid \phi R\phi \mid \phi \vee \phi \mid \phi \wedge \phi \mid \neg\phi \mid \phi \rightarrow \phi \mid$$
$$value = value \mid value \neq value \mid value > value \mid value \geq value \mid$$
$$value < value \mid value \leq value \mid true \mid false$$

Numerical constraints over free variables are defined in this logic through the inclusion of free variables denoted by $fVariable$ in the definition of $value$ below – the symbol $ differentiates a free variable from a regular variable. Regular variables are read-only values which form the behaviour of the model, whereas free variables are instantiated during the model checking process to the range of values for which the temporal logic property holds. In our current, implementation free variables are defined to have integer domains initialised to $[0 \rightarrow \infty)$ and describe protein concentrations, numbers of molecules and time. Constraints over free variables, which involve equality/inequality and relational operators, restrict the domain of the free variable, such that with $X \in [0 \rightarrow \infty)$, $X > 5$ sets X to be $[6 \rightarrow \infty)$. If there is a constraint over free variables involving real numbers, then the real numbers are cast to integers. Notice also that disjunction, conjunction, negation and implication of constraints over free variables are allowed. Finally, the values considered in this logic are integers and real numbers, and the four basic arithmetic operations over these values:

$$value ::= value + value \mid value - value \mid value * value \mid value/value \mid$$
$$\$fVariable \mid Variable \mid function \mid Int \mid Real$$

where Int is any integer number and $Real$ is any real number. In our biochemical pathway analysis we define $Variable$ to be the time dependant value of the concentration of any biochemical species in the model, either integers for molecules/levels or real numbers for concentrations, and we define a special variable called $time$ to stand for the values of state time. State time values are the simulation time points such that we can, for example, express properties relative to simulation time. This is especially useful for expressing a property before or after some event, such as introducing a drug into a cell. We provide the ability to define any *function* returning a real or integer value, and in our current system we have chosen to implement the two functions, *max* and *d*. The function *max* operates over all the values of a species to return the maximum of the species' value in the simulation run, thus the peak of a species can be expressed; $Protein = max(Protein)$. We also define a function d which returns the derivative of the concentration of the species at each time point, thus increasing and decreasing species value can be expressed; $d(Protein) > 0$ and $d(Protein) < 0$ respectively.

PLTLc enhances LTLc by the inclusion of a probability operator and filter construct, and the probabilistic interpretation of the domains for the free variables. The top-level definition of PLTLc is:

$$\psi ::= \mathbf{P}_{\unlhd x}[\,\phi\,] \mid \mathbf{P}_{\unlhd x}[\,\phi\{SP\}\,]$$

where ϕ is an LTLc expression. SP is a State Proposition defined to be ϕ without any temporal operator (X, G, F, U, R), and containing *no* free variables without a loss of expressivity. Note that the square and curly brackets are part of PLTLc. Given that $\unlhd \in \{>, \geq, <, \leq\}$, $P_{\unlhd x}$ is any inequality comparison of the probability of the property holding true, for example $P_{\geq 0.5}$. We also permit the expression $P_{=?}$ returning the value of the probability of the property holding true. We disallow equality testing of the probability, $P_{=x}$ because of the representation of real values and the semantics of their equality.

We define filters similar to those used in PCTL and CSL. This permits specifications to refer to the state or states that the property is checked from, rather than default to the initial state. Hence, for a property of the form $\phi\{SP\}$, ϕ is checked from the first state that SP is satisfied.

2.2   PLTLc Semantics

We introduce the semantics of PLTLc in an informal manner to cater to a wide audience. The formal semantics of PLTLc are described in full in [6].

The semantics of PLTLc is defined over a finite set of finite paths through the system's state space – in our case, stochastic or deterministic simulations, or time series data recorded in wet lab experiments.

First, let a path π be a finite sequence of states describing the behaviour of a biochemical system, $\pi = s_0, s_1, ..., s_n$ $(n < \infty)$ and π^i be the subsequence of π starting from state s_i, $i \leq n$, thus $\pi^i = s_i, s_{i+1}, ..., s_n$. Each path in the set of paths can be evaluated to a boolean value as to whether ϕ or $\phi\{SP\}$ holds. When all paths are evaluated, the number of true values in the set over the size of the set yields the overall probability of the PLTLc property. Hence for a stochastic model, where the set of paths is typically > 1, the probability is in the range $[0 \rightarrow 1]$ and calculated through Monte Carlo approximation, whereas a continuous model which contains a single path has a probability of either 0 or 1.

Finally, the two PLTLc functions we have chosen to implement, *max(variable)* and *d(variable)* are defined as follows. *max(variable)* calculates the first state s_{max} in the finite path π for which the value of *variable* is maximal and returns this value. $d(variable)$ calculates for each state s_i in the finite path π the derivative of the value of *variable* between state s_i and s_{i+1}. In the case of the final state in the finite path s_n which contains no next state, the derivative is equal to the derivative of the previous state s_{n-1}.

2.3   Characterising Biochemical Species' Behaviour

The behaviour of biochemical species can be described with PLTLc using four distinct descriptive approaches, with increasing specificity; qualitative, semi-qualitative, semi-quantitative and quantitative. Qualitative uses derivatives of biochemical species concentrations/mass and temporal operators to describe the general trend of the behaviour. Semi-qualitative extends qualitative with relative concentrations. Semi-quantitative extends semi-qualitative with absolute time values. Finally, quantitative extends semi-quantitative with absolute concentration values. For example, transient activation of a biochemical species called Protein can be expressed in these approaches:

qualitative: Protein rises then falls

$\quad$ **P**$_{=?}$ [d(Protein) > 0 U (G(d(Protein) < 0))]

semi-qualitative: Protein rises then falls to less than 50% of peak concentration

$\quad$ **P**$_{=?}$ [(d(Protein) > 0) U (G(d(Protein) < 0) $\wedge$
$\quad\quad$ F([Protein] $< 0.5 * max$[Protein]))]

semi-quantitative: Protein rises then falls to less than 50% of peak concentration at 60 minutes

$\quad$ **P**$_{=?}$ [(d(Protein) > 0) U (G(d(Protein) < 0) $\wedge$
$\quad\quad$ F(time $= 60$ $\wedge$ Protein $< 0.5 * max$(Protein)))]

quantitative: Protein rises then falls to less than $100\mu Mol$ at 60 minutes
$\quad$ **P**$_{=?}$ [(d(Protein) > 0) U (G(d(Protein) < 0) $\wedge$
$\quad\quad$ F(time $= 60$ $\wedge$ Protein < 100))]

In our case study we find that the desired behaviour of the model is most suited to semi-quantitative PLTLc. In fact, the informal explanation of results from biochemical experiments bears a striking similarity to semi-quantitative PLTLc.

2.4   Probabilistic Domains

Each path in the set of paths is also evaluated to a domain of validity, $D_{\phi\ or\ \phi\{SP\}} \subset \mathbb{N}^n$ for n free variables in the PLTLc property, $\$fVar_1, \$fVar_2, ...\$fVar_n$. The domain of validity is defined such that for all valuations v of the n free variables, where $v \in D_{\phi\ or\ \phi\{SP\}}$, the property ϕ or $\phi\{SP\}$ as appropriate holds true for the path. Thus each path has an associated domain of validity, with paths resulting in a boolean value of true having a non-empty domain of validity, i.e. for these paths there must be valuations of the variables for which the property holds.

After the set of domains of validity is evaluated from the set of paths, a probabilistic domain for each of the n free variables in the PLTLc property is calculated. A probabilistic domain associates with each integer value in the domain the probability of the property holding true for that value. If the PLTLc property evaluates to a probability p, then the maximum possible probability of any value in the probabilistic domains is p, such that a property with 0 probability has probabilistic domains with 0 probability for all values. The probabilistic domain of free variable $\$fVar_i$ is calculated by iterating through each integer value I in the probabilistic domain. A count is performed on the set of domains of validity for the number of domains of validity which contain at least one valuation v with $v(\$fVar_i) = I$. This number over the size of the set is the probability of the value I in the probabilistic domain of $\$fVar_i$.

In the case that the system is described by a stochastic model, the probabilistic domains are calculated through Monte Carlo approximation – the number of occurrences of a value for a free variable in each domain of validity in the set over the size of the set. In the case of a continuous model where the size of the set is 1, the probabilistic domain contains probabilities 0 and 1 and can equally be represented by a probabilistic domain or a regular domain.

The semi-quantitative property from the previous section can be enhanced with free variables:

semi-quantitative with free variables. Protein rises then falls to less than 50% of peak concentration at 60 minutes

$$\mathbf{P}_{=?} \left[\left(d(\text{Protein}) > 0 \right) U \left(\text{Protein} \geq \$PeakConc \wedge G(d(\text{Protein}) < 0) \right.\right.$$
$$\left.\left. \wedge F(\text{ time } = 60 \wedge \text{Protein} < 0.5 * max(\text{Protein})) \right) \right]$$

where the probabilistic domain of $\$PeakConc$ associates with each value in the domain the probability of a peak of at least that value.

2.5   Distance Metrics

The distance between a model's behaviour and the desired behaviour can be calculated using a distance metric. We define a metric for the distance, with respect to some property ψ, from the behaviour of the model M to the desired behaviour M_{des}. The distance metric, written $d_\psi(M, M_{des})$, should satisfy the metric properties:

$d(x,y) > 0$ for $x \neq y$ w.r.t. ψ, $d(x,y) = 0$ for $x = y$ w.r.t. ψ

$d(x,y) = d(y,x)$ for all x, y , $d(x,y) \leq d(x,z) + d(y,z)$ for all x, y, z

The metric is domain and behaviour specific, and can be based on the probability of the property or the probabilistic domains.

Perhaps the simplest definition of the metric is the square difference between the model's probability of exhibiting some behavioural property ψ, $P(\psi)$ and desired probability $P_{des}(\psi)$. For example, we may want the property ψ to always hold in which case $P_{des}(\psi)$ is 1. The distance function is then written:

$d_\psi(M, M_{des}) = |P(\psi) - P_{des}(\psi)|^2$

This approach works well in the stochastic world where the model exhibits many behaviours and the probability of the property is in the range $[0 \rightarrow 1]$. However, in the continuous world there is a single behaviour and the probability is either 0 or 1, thus the metric is too coarse grained to be used in a search algorithm in the continuous world. To be useful in the search algorithm, the distance metric should return a value which indicates whether altering the current model has caused its behaviour to be closer to the desired behaviour, therefore providing a gradient for the search algorithm to ascend.

Definitions of the distance metric over probabilistic domains of free variables can result in finer grained distance values, crucial for distance metrics in the continuous world. For a free variable $\$X$ in a property ψ, we can compare the probabilistic domain in the model $\$X$ with the desired probabilistic domain $\$X_{des}$. To do so, we use the residual sum of squares function, RSS:

$$RSS(\$X, \$X_{des}, m, n) = \sum_{i=m}^{n} |\$X(i) - \$X_{des}(i)|^2$$

where m to n is some sub-section of the domain being assessed. Hence, we could desire that a free variable describing the peak concentration value in transient behaviour $\$PeakConc$ of a continuous model is at least $50\mu Mol$. It is then simple to set up a desired probabilistic domain $\$PeakConc_{des}$ with probability 1 for values 0 to 50. A call to $RSS(\$Peakconc, \$PeakConc_{des}, 0, 50)$ would then return a value of how close the current model is to having its peak concentration value at least value $50\mu Mol$.

We can implement a distance metric using the RSS function for any number of free variables we define in our PLTLc property. In the case that we wish to optimise more than one probabilistic domain, we normalise the RSS values between 0 and 1:

$$d_\psi(M, M_{des}) = \frac{RSS(\$X, \$X_{des}, m, n)}{(n - m)} + \frac{RSS(\$Y, \$Y_{des}, u, v)}{(v - u)} + \dots$$

3   Computational System

We implemented a computational system called the Monte Carlo Model Checker with a Genetic Algorithm, MC2(GA). The purpose of this computational system is to estimate the parameters of a model to make it exhibit desired behavioural properties. A genetic algorithm is used to move models through parameter space to minimise their distance to the desired behaviour, checked using a model checker.

A genetic algorithm [15] operates over a *population* of *individuals*, each of which is represented by their *chromosome* containing one or more *genes*. The individuals have an associated fitness based on how "good" their genes are. A selection of individuals from the current population is performed which will be used to create the next generation. Genetic operations on the chromosomes of these selected individuals (reproduction, crossover and mutation) is used to build a next generation with (hopefully) improved overall fitness.

Each model in our MC2(GA) system has a fixed structure and is represented by a chromosome, which is a set of kinetic rate constant values to be estimated (the model's genes) within predefined ranges. The chromosome could equally include initial concentrations/masses.

In the initial generation, a population of models is created by assigning to each model random values within the ranges for the kinetic rate constants. The number of models in the population should be proportional to the size of the parameter space being explored. Each model in the population is evaluated to a fitness value related to the distance of its behaviour to the desired behaviour, hence a model with a smaller distance to the desired behaviour has a higher fitness. This is achieved by formalising the desired behaviour in temporal logic and the novel use of a model checker to calculate the distance of the model to the behaviour. Our approach is to vary models' kinetic rate constant values in order to maximise their fitness values.

After the initial generation which builds a population of models with their related fitness values, a subset of the population is selected to survive. Roulette-wheel selection is used where models in the population are chosen to survive probabilistically, with fitter models having a higher probability of survival. This is done to keep a small number of less fit models in the population such that we do not converge on a solution too early. If the computational system is exploring a high dimensional parameter space, it is important to maintain good coverage of this large space. The population for the next generation is created from the selected models by performing genetic operations on these models' chromosomes representing the kinetic rate constant values. A chromosome may be duplicated (*reproduction*), a section between two chromosomes may be swapped (*crossover*) or a section of one chromosome may be randomly altered (*mutation*) within preferred constraints. The models in the new population are evaluated to their fitness values and then go on to form the next generation.

The best and average fitness value of a model in the population should increase over successive generations of this algorithm. There are stochastic elements to the algorithm however, including random mutation and probabilistic selection of models for the next generation. Hence, it is not always the case that there will be a continual increase in best or average fitness value, though the general trend should increase. Various stopping conditions in this algorithm can be used– we choose to stop after the best fitness of a model in the population has not changed significantly after 10 generations or after a maximum of 100 generations has elapsed.

A population of models with their respective fitness values is returned upon termination of the genetic algorithm. This is quite a powerful result of parameter estimation. By the very nature of a genetic algorithm, we get a set of candidate solutions, which may be representative of more than one general solution type. Hence, with the semi-quantitative description of the desired behaviour, we can get many models, possibly grouped into distinct sub-populations, which exhibit the desired behaviour.

Although any search algorithm which uses a fitness function could be used in this approach, we have chosen a genetic algorithm because it avoids being lost in local minima, which is likely in high dimensional parameter spaces. A genetic algorithm avoids this by maintaining a population of candidate solutions and probabilistically keeping some low fitness solutions in the population between generations.

Our computational system, MC2(GA), currently operates over continuous models only. The desired behaviour of a model is expressed in the PLTLc temporal logic. Models are evaluated to a fitness value through interfaces to the continuous simulator, BioNessie Lite [16] and the Monte Carlo Model Checker for Probabilistic LTLc properties MC2(PLTLc) [17]. The model is simulated for a predefined amount of time and the simulation output is checked for the desired behaviour using MC2(PLTLc). A numerical value for the fitness of the model based on the result of model checking can be computed using the probability of the behavioural property and the probabilistic domains of free variables in the property. The fitness function using the probabilistic domains of free variables has been implemented using the theory described in Section 2.5. We employ the Java Genetic Algorithms Package (JGAP) [18] to move our population of models through parameter space in order to maximise their fitness.

4   Case Study: MAPK Pathway

We illustrate our technique to parameter estimation with a continuous model of the MAPK pathway.

4.1   Biochemical Motivation

The EGF signal transduction pathway conveys Epidermal Growth Factor signals from the cell membrane to the nucleus via the MAP Kinase cascade [2]. The model of the pathway in PC12 cells written in Systems Biology Markup Language (SBML) [19] is the subject of [8]. The same core MAPK cascade can also be stimulated by Nerve Growth Factor (NGF). The reaction of the cell to EGF stimulation is cell proliferation, however the response to NGF is cell differentiation. The active ligand-bound receptor acts as a kinase for the Shc protein. The active Shc and GS complex (ShcGS) binds with the inactive RasGDP complex which enables Son of sevenless homologue protein (SOS) to convert RasGDP to its active RasGTP form. Ras-GTP acts as a kinase to phosphorylate Raf, which phosphorylates MAPK/ERK Kinase (MEK), which in turn phosphorylates Extracellular signal Regulated Kinase (ERK). Feedback regulation of the pathway is through ShcGS dissociation catalysed by phosphorylated ERK. The EGF signal transduction pathway produces transient Ras, MEK and ERK activation whereas NGF stimulation produces sustained activation. The underlying differences of the models describing EGF and NGF stimulation is of key interest to biochemists.

The work in [8], referred to from now on as the original paper, attempted to discover the quantitative differences in initial concentrations and kinetic rate constants between models of these pathways with fixed topology. The authors varied the initial concentrations and kinetic rate constants within biochemically sensible ranges. Simulation was performed with the model using each parameter value in the range and the output was manually inspected for sustained Ras, MEK and ERK activation. A result of this work was the finding that a 40-fold increase in the kinetic rate constant of SOS dephosphorylation can change the behaviour of the model from transient activation to sustained activation.

We suggest that this analysis could be improved by constructing a formal definition of the desired behaviour in temporal logic, and using model checking of the desired behaviour to replace the manual inspection of the simulation outputs. This facilitates the automation detection of a model which exhibits the desired behaviour. We employ this in our computational system, MC2(GA), to vary many kinetic rate constants in the model in parallel to estimate a parameter set of the NGF signal transduction pathway.

4.2   Characterising the Desired Pathway Behaviour

The behaviour of sustained Ras, MEK and ERK activation arising from NGF stimulation observed in wet-lab experiment was described in rather informal statements in the original paper [8].

"The level of RasGTP rapidly reaches a maximum of up to 20% of total Ras within 2 min [then] the level of RasGTP is sustained at around 8% of total Ras."

Similar statements were made about sustained MEK and ERK activation. We have formalised these statements using semi-quantitative PLTLc such that a model could be automatically checked for these behaviours using the MC2(PLTLc) model checker. We formalised these statements in a way to account for biological error by relaxing the constraints, for example that the stable level of RasGTP is 8% to between 5% and 10%:

sustained Ras. Active Ras peaks within 2 minutes to a maximum of 20% of total Ras and is stable between 5% and 10% from at least 15 minutes

$$\mathbf{P}_{=?} \left[\, (\, d(\text{active Ras}) \, > \, 0 \,) \, \wedge \, (\, d(\text{active Ras}) \, > \, 0 \,) \, U \, (\, \text{time} \, \leq \, 2 \, \wedge \right.$$
active Ras $\geq$ 0.15∗total Ras $\wedge$ active Ras $\leq$ 0.2∗total Ras $\wedge$
$d(\text{active Ras}) \, < \, 0 \, \wedge \, (\, d(\text{active Ras}) \, < \, 0 \, \wedge \, \text{time} \, < \, 15 \,) \, U \, (\, G($
(active Ras) $\geq$ 0.05∗total Ras $\wedge$ active Ras $\leq$ 0.10∗total Ras $) \,) \,) \,]$

where the protein RasGTP is found in isolation and in two complexes, thus active Ras = $RasGTP + Ras\_Raf + Ras\_GAP$ and total Ras = $RasGTP + Ras\_Raf + Ras\_GAP + RasGDP + Ras\_ShcGS$.

sustained MEK. Active MEK peaks within 2 to 5 minutes and is stable between 40% and 50% of peak value from at least 15 minutes

$$\mathbf{P}_{=?} \left[\, (\, d(\text{MEKPP}) \, > \, 0 \,) \wedge (\, d(\text{MEKPP}) \, > \, 0 \,) \, U \, (\, \text{time} \, \geq \, 2 \, \wedge \, \text{time} \, \leq \, 5 \, \wedge \right.$$
$d(\text{MEKPP}) \, < \, 0 \, \wedge \, (\, d(\text{MEKPP}) \, < \, 0 \, \wedge \, \text{time} \, < \, 15) \, U \, (\, G($
MEKPP $\geq$ 0.40∗max(MEKPP) $\wedge$ MEKPP $\leq$ 0.50∗max(MEKPP) $) \,) \,) \,]$

sustained ERK. Active ERK peaks within 2 to 5 minutes and is stable between 85% and 100% of peak value from at least 15 minutes

$$\mathbf{P}_{=?} \left[\, (\, d(\text{ERKPP}) \, > \, 0 \,) \wedge (\, d(\text{ERKPP}) \, > \, 0 \,) \, U \, (\, \text{time} \, \geq \, 2 \, \wedge \, \text{time} \, \leq \, 5 \, \wedge \right.$$
$d(\text{ERKPP}) \, < \, 0 \, \wedge \, (\, d(\text{ERKPP}) \, < \, 0 \, \wedge \, \text{time} \, < \, 15 \,) \, U \, (\, G($
ERKPP $\geq$ 0.85 ∗ max(ERKPP) $) \,) \,) \,]$

4.3   Identification of Critical Parameters

The work reported in the original paper [8] varies parameters individually in the model and notes the effect on sustained Ras, MEK and ERK activation by manual inspection of simulation output. We performed a similar analysis in an automated fashion, which made it easy to count the number of parameter values in a particular parameter range that gave our desired behaviour. Hence, rather than a simple yes/no answer, we were able to quantify the significance of a particular parameter regarding a particular behaviour. We used this feature to identify a set of critical parameters to vary in our MC2(GA) system. A further benefit of using an automated approach to detect desired output rather than manual inspection is that we can explore many possible behaviours, generated for example by varying one parameter within a large range.

In the absence of biochemical knowledge of acceptable ranges of kinetic rate constants, we varied each kinetic rate constant in the range ± 2 orders of magnitude from their original value. We simulated the continuous model for 60 minutes using 1,000 parameter values linearly spaced in the range. This produced a set of 1,000 simulation outputs, and we checked each one for the behaviour of sustained Ras, MEK and ERK activation expressed in PLTLc. We then computed the fraction of simulation outputs which satisfy the behavioural property over the number of simulation outputs. We call this fraction the parameter's *significance value*, such that a higher value represents parameter more likely to exhibit the desired behaviour. Each kinetic rate constant in the model was varied individually to produce their significance value.

Any kinetic rate constant with at least one non-zero significance value for sustained Ras, MEK or ERK is called a *critical parameter*. The identified critical parameters are listed in Table 1 along with their respective significance values. Although we ignored initial concentration parameters as they had little effect on sustaining activation in the original analysis, our approach can analyse initial concentrations in the same manner as kinetic rate constants.

From the significance values it is clear that although sustained Ras and ERK activation is quite possible, sustained MEK activation is more difficult to achieve. In fact, when varying parameters individually it was only possible to achieve this using the kinetic rate constant for SOS dephosphorylation, V_28. This was the solution found in the original paper [8] and we note that it has the highest sum of the three significance values. We suspected that other parameter sets could produce our desired behaviour and thus we varied several parameters in parallel using our computational system.

4.4   Genetic Algorithm

We first implemented a fitness function for use in MC2(GA) to describe how close a model is to sustained activation. The descriptions of sustained Ras, MEK and ERK activation given earlier were not particularly helpful in the continuous setting due to the probability being simply 0 or 1. A fitness function based on a description which includes free variables allows greater expressivity using the probabilistic domains. Hence, we have rewritten these descriptions of sustained behaviours using free variables:

Table 1. The identified critical kinetic rate constant parameters in the model with their significance values with respect to sustained Ras, MEK or ERK

parameter	sustained Ras	sustained MEK	sustained ERK
V_20	0.01	0.0	0.001
V_24	0.076	0.0	0.0
V_25	0.023	0.0	0.001
V_27	0.614	0.0	0.0
V_28	0.478	0.151	0.679
k1_14	0.0	0.0	0.778
k1_16	0.0	0.0	0.001
k1_18	0.001	0.0	0.807
k2_18	0.191	0.0	0.0
Km_20	0.001	0.0	0.797
kcat_21	0.001	0.0	0.688
kcat_23	0.001	0.0	0.186
Km_23	0.121	0.0	0.0
Km_25	0.001	0.0	0.157
kcat_26	0.0	0.0	0.001
Km_26	0.0	0.0	0.005

sustained Ras with free variables. Active Ras peaks within 2 minutes to a maximum of 20% of total Ras and is stable between any value in $RasTail1$ and any value in $RasTail2$ from at least 15 minutes

$P_{=?}$ [(d(active Ras) > 0) $\wedge$ (d(active Ras) > 0) U (time $\leq$ 2 $\wedge$ active Ras $\geq$ 0.15*total Ras $\wedge$ active Ras $\leq$ 0.2*total Ras $\wedge$ d(active Ras) < 0 $\wedge$ (d(active Ras) < 0 $\wedge$ time < 15) U (G(active Ras $\geq$ $RasTail1$ $\wedge$ active Ras $\leq$ $RasTail2$)))]

sustained MEK with free variables. Active MEK peaks within 2 to 5 minutes with a peak greater than concentration 20,000 and is stable between any value in $MekppTail1$ and any value in $MekppTail2$ from at least 15 minutes

$P_{=?}$ [(d(MEKPP) > 0) $\wedge$ (d(MEKPP) > 0) U (time $\geq$ 2 $\wedge$ time $\leq$ 5 $\wedge$ MEKPP > 20000 $\wedge$ d(MEKPP) < 0 $\wedge$ (d(MEKPP) < 0 $\wedge$ time < 15) U (G(MEKPP $\geq$ $MekppTail1$ $\wedge$ MEKPP $\leq$ $MekppTail2$)))]

sustained ERK with free variables. Active ERK peaks within 2 to 5 minutes with a peak greater than concentration 350,000 and is stable above any value in $ErkppTail$ from at least 15 minutes

$P_{=?}$ [(d(ERKPP) > 0) $\wedge$ (d(ERKPP) > 0) U (time $\geq$ 2 $\wedge$ time $\leq$ 5 $\wedge$ ERKPP > 350000 $\wedge$ d(ERKPP) < 0 $\wedge$ (d(ERKPP) < 0 $\wedge$ time < 15) U (G(ERKPP $\geq$ $ErkppTail$)))]

The property ψ in our case study was the conjunction of the sustained Ras, MEK and ERK properties expressed with free variables above. To explain the addition of free variables in these properties we consider the active ERK shown in Figure 4.4. This figure illustrates the relationship of a continuous simulation output of ERKPP to the probabilistic domains of the free variable $ErkppTail$ and the desired probabilistic domain $ErkppTail_{des}$. The values in the probabilistic domain $ErkppTail$ with probability 1 are in the range of values from 0 to the tail height. Our desired behaviour of the property is that it peaks within 2 minutes to a concentration of greater than 350,000 (defined in the PLTLc property) and that the tail of the peak remains within 85% of the peak height. We observe a tail characterised by $ErkppTail$ and now characterise our desired tail of at least 85% of the peak height, $ErkppTail_{des}$. This is done by setting values from 0 to 85% of the peak height value in the probabilistic domain of $ErkppTail_{des}$ to probability 1. Hence if the RSS between the model's tail and the desired tail is 0, then the model's tail is within 85% of it's own peak height (our desired property) and if the model's tail falls to concentration 0, then the distance is value 85% of the peak height:

$$erkppDistance = RSS(\$ErkppTail, \$ErkppTail_{des}, 0, max(\$ErkppTail_{des}))$$

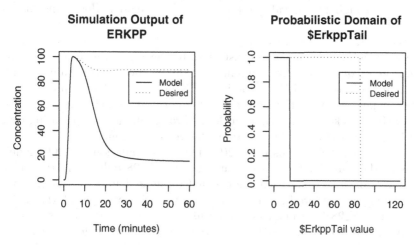

Fig. 1. Continuous simulation output of active ERK (left) with the probabilistic domain of free variables $ErkppTail$ (middle) and the desired probabilistic domain $ErkppTail_{des}$ (right). The probabilistic domain of $ErkppTail$ contains probability 1 for all values from 0 up to the tail height and the desired probabilistic domain $ErkppTail_{des}$ contains probability 1 for all values from 0 up to 85% of the peak height.

Next, recalling that sustained active MEK is defined to be stable between 40% and 50% of the peak, we set the desired probabilistic domain of $MekppTail1$ to have probability 1 for values 0 to 40% of the peak value and the desired probabilistic domain of $MekppTail2$ to have probability 1 for values 50% to 100% of the peak value. The distance of the active MEK behaviour was then:

$$mekppDistance = RSS(\$MekppTail1, \$MekppTail1_{des}, 0, max(\$MekppTail1_{des}))$$
$$+RSS(\$MekppTail2, \$MekppTail2_{des}, min(\$\$MekppTail2_{des}),$$
$$max(\$\$MekppTail2_{des}))$$

Finally, recalling that sustained active Ras is defined to be stable between 5% and 10% of total Ras, we set the desired probabilistic domain of $RasTail1$ to have probability 1 for values 0 to 5% of total Ras and the desired probabilistic domain of $RasTail2$ to have probability 1 for values 10% to 100% of total Ras. The distance of the active Ras behaviour was then:

$rasDistance = RSS(\$RasTail1, \$RasTail1_{des}, 0, max(\$RasTail1_{des})) +$
$RSS(\$rasTail2, \$rasTail2_{des}, min(\$rasTail2_{des}), max(\$rasTail2_{des}))$

The overall metric describing the distance between a model and the desired behaviour can now be defined by averaging the sum of the normalised individual distances, $erkppDistance$, $mekppDistance$ and $rasDistance$, and ranges from 0 (identical) to 1:

$$d_\psi(M, M_{des}) = (\tfrac{erkppDistance}{85\%ERKPeak} + \tfrac{mekppDistance}{90\%MEKPeak} + \tfrac{rasDistance}{95\%TotalRas})/3$$

Note that if the model does not satisfy the PLTLc property – i.e. it does not peak or does not peak within the concentration or time constraints – then all values in the probabilistic domains of the free variables are set to probability 0, thus the distance is 1.

The definition of a model's fitness has opposite semantics from a distance metric – i.e. a high fitness value represents a good model. Hence the fitness function is:

$fitness(M, M_{des}) = 1 - d_\psi(M, M_{des})$

Finally, we added quantitative constraints to the properties to exclude the behaviour being observed with insignificantly small concentration values by requiring that both active ERK and active MEK be greater than approximately half of their peak value in the EGF signalling pathway model. The concentration of RasGTP is kept to a significant level through its definition stating that it must be between 15% and 20% of total Ras concentration.

As a proof of concept, we used MC2(GA) with a population of 2,000 models to estimate the value of the kinetic rate constant V_28 only using the range of ± 2 orders of magnitude from the original value as defined in the previous section. We hoped to reproduce the result from the original paper [8] that a 40-fold increase of V_28 (to value 3,000 $molecule^{-1}minute^{-1}$) produces sustained activation of Ras, MEK and ERK. The fitness of the best model immediately converged to value 1 and the genetic algorithm stopped 10 generations after the convergence, which can be seen in Figure 4. The V_28 value against respective fitness of the models in the final population can be seen in Figure 2. There is a wide range of V_28 values which produce the desired model behaviour – the V_28 value proposed in the original paper of 3000 $molecule^{-1}minute^{-1}$ falls within the range of models with a fitness value 1, which suggests that our genetic algorithm and behavioural properties are correct.

We then applied MC2(GA) to find novel parameter sets which exhibit the desired behaviour. We estimated the values of the 16 critical parameters identified in the previous section, listed in Table 1. We also applied MC2(GA) to the critical parameters without V_28, to assess whether V_28 is crucial to achieving sustained activation. The result of the convergence of these runs are presented in Figure 4. It can be seen from this figure that if the critical parameters are estimated with V_28, then the convergence is quicker and the best model returned was fitter. The best model returned when estimating the critical parameters had

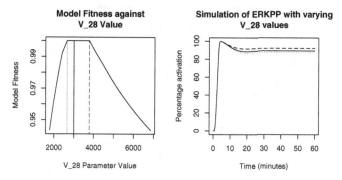

Fig. 2. The population of models from the genetic algorithm (left) and simulation output of three selected models (right). The model proposed in the original paper (solid) has fitness value 1. Also, the model with lowest (dotted) and highest (dashed) value of V_28 whilst maintaining fitness value 1 is shown.

fitness value 1, whereas with V_28 removed the best model returned had a fitness value approximately 0.93.

Figure 3 shows the output of one of the best model returned when estimating the critical parameters with and without V_28. Both behaviours showed good similarity (visually and in terms of fitness value) to the behaviour of the NGF signalling pathway outlined in the original paper. We also found that we can achieve a model with fitness value 1 through a 16-fold increase of V_28, compared with the original paper's 40-fold increase, if we also vary the other critical parameters.

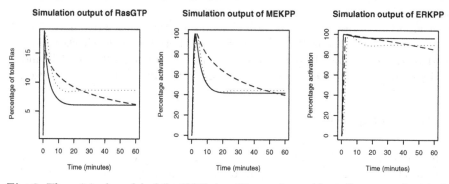

Fig. 3. The original model of the NGF signalling pathway (dotted) compared with the best model returned when varying the critical parameters (solid) and when varying the critical parameters without V_28 (dashed). The best model returned when varying the critical parameters only required a 16-fold increase in V_28 to achieve fitness value 1.

Finally we tested how MC2(GA) copes with a high dimensional parameter space by estimating the values of all 65 kinetic rate constants in the model. Again, these parameters were varied in the range ± 2 orders of magnitude from

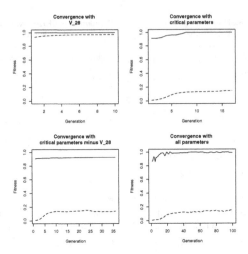

Fig. 4. From top-left to bottom right, the convergence of varying; V_28 only, the critical parameters, the critical parameters without V_28 and all parameters. The fitness of the best model in the population (solid) is shown as well as the average fitness of the models (dashed).

their original values. As shown in Figure 4 there is no convergence after the maximum of 100 generations. However, the best model returned had fitness value approximately 0.99. This is a strong result that even with such a high dimensional parameter space, MC2(GA) still found a viable solution.

To give an idea of the computational expense of our system, a single generation of the genetic algorithm took around ten minutes. Each generation contains on the order of 2000 simulation and model checking operations, thus the evaluation of each model's fitness value took around 300ms. Overall, with a population of M models and N generations, the number of calls to both the model checker and the simulator is $O(N * M)$. In our case study which contained 2,000 models and a maximum of 100 generations, we had around $2 \cdot 10^5$ calls to the model checker.

5   Related Work

Manual parameter estimation of a model can be performed, especially using insight into the real-life system. The authors in [8] manually altered a model of the MAPK pathway to change the behaviour from EGF stimulation to NGF stimulation. They focus their estimation to parameters which have been identified through biochemical experiments as possible targets to explain this difference. This work was accomplished using computational tools, such as a continuous simulator, however it relied heavily on manual inspection of simulation outputs. As such, the degree to which the model could be varied was limited to the amount of manual inspection possible. Hence, it was infeasible for the authors to vary the parameters within a large range or vary parameters in parallel.

Automated parameter estimation approaches employ a function which returns a value of how close model's behaviour is to some desired behaviour. A search algorithm can be used to alter the model to minimise the behavioural distance. Representing the desired behaviour using target data derived from experiments on the real-life system being modelled is an obvious choice. Approaches to parameter estimation using target data are studied and reviewed in [5]. Many functions can compute the difference of a model's output to the target data, such as Maximum Likelihood, Bayes and Weighted Least Squares.

However, the target data of models of biochemical systems results from wet-lab experiments. The data produced from such experiments is typically noisy and sparse with relative concentrations [20]. Any function calculating the difference of the output of a biochemical system model to the target data should account for this. [20] used weights on time-series data derived from wet-lab experiments to account for noise. Furthermore, a Bayesian approach to estimating the difference in model output and target data is the subject of [4].

Although the literature contains approaches to parameter estimation approaches using target data, we found little evidence of implementing parameter estimation with a target behaviour expressed in temporal logic. The closest we have found is [21] which specifies the expected behaviour of a continuous model of a biochemical pathway in the LTL temporal logic. The parameters in the model are varied in some range until a satisfying parameter is found and returned. However, this approach works only for the continuous world and the authors note the computational expense of parameter scans of multiple variables in parallel. They express desire for a "multi-valued measure of satisfaction" of a temporal property rather than the boolean result of LTL checking. This would facilitate the use of LTL as the target function in a search algorithm, allowing many parameters to be varied in parallel.

6   Conclusion

We have shown how we have used probabilistic temporal logic descriptions of biochemical pathway behaviours as the basis for a model checking approach to the parameter estimation of biochemical pathways. This is the first step in a general methodology for behaviour driven model construction.

A key aim of our approach is to be able to operate in both the stochastic and continuous worlds. The PLTLc temporal logic and its simulation-based model checker operate in both these worlds. It is especially important that the model checker is simulation-based as the use of current analytical model checker would be computationally infeasible due to the well-known state space explosion. This is further exacerbated when used within a search algorithm which will typically have many calls to the model checker $- 2 \cdot 10^5$ in our case study. Furthermore, a novel aspect of the numerical constraints in PLTLc is that they can be applied in both the stochastic and continuous worlds; this is crucial to the calculation of the distance of a model's behaviour to the desired behaviour.

We have demonstrated our approach through a case study of a continuous model of the well-known MAPK signalling pathway. This model has previously been manually explored in [8]. They identify a single parameter which when modified by a 40-fold increase produced the desired behaviour. Having first characterising the desired behaviour in PLTLc, we then automatically identified a set of critical parameters. We then used our MC2(GA) system to estimate the values of the critical parameters and discovered novel kinetic rate constant parameter sets which produce the desired behaviour. These include parameter sets which do not require varying the parameter identified in [8]. Finally we showed that the computational system can operate with high dimensional parameter spaces by estimating the values of all 65 kinetic rate constants in the model.

The case study presented in this paper is of a continuous model of a biochemical pathway. However, the theory underlying this approach has been described for both the stochastic and continuous worlds. We are now working on applying this analysis to a stochastic model. Furthermore, we are currently able to estimate the kinetic rate constant and initial concentration/mass values in the model, however we could define theory to vary the model topology (adding, removing or altering reactions). We then could answer questions such as what are the topologies which give rise to particular behaviours of interest. Finally, PLTLc is rather unfriendly to a biologist who is not well versed in temporal logic. We are now developing a user-friendly interface for biologists to describe behaviours using PLTLc.

The computational system, MC2(GA), together with the case study results are available at: www.brc.dcs.gla.ac.uk/software/mc2/mc2ga_bf.

Acknowledgements

This research was supported by the SIMAP project which is funded by the European Commission framework 6 STREP programme.

References

1. Wolkenhauer, O.: Systems Biology: the Reincarnation of Systems Theory Applied in Biology? Briefings in Bioinformatics 2(3), 258–270 (2001)
2. Kolch, W., Calder, M., Gilbert, D.: When kinases meet mathematics: the systems biology of MAPK signalling. FEBS Lett. 579, 1891–1895 (2005)
3. Gilbert, D., Heiner, M., Lehrack, S.: A unifying framework for modelling and analysing biochemical pathways using Petri nets. In: Calder, M., Gilmore, S. (eds.) CMSB 2007. LNCS (LNBI), vol. 4695, pp. 200–216. Springer, Heidelberg (2007)
4. Vyshemirsky, V., Girolami, M.: Bayesian Ranking of Biochemical System Models. Bioinformatics 24(6), 833–839 (2008)
5. Maria, G.: A Review of Algorithms and Trends in Kinetic Model Identification for Chemical and Biochemical Systems. Chem. Biochem. Eng. Q. 18(3), 195–222 (2004)
6. Donaldson, R., Gilbert, D.: A Monte Carlo Model Checker for Probabilistic LTL with Numerical Constraints. Technical report, University of Glasgow, Department of Computing Science (2008)

7. Baier, C.: On Algorithmic Verification Methods for Probabilistic Systems. Habilitation thesis, University of Mannheim (1998)
8. Brightman, F., Fell, D.: Differential feedback regulation of the mapk cascade underlies the quantitative differences in egf and ngf signalling in pc12 cells. FEBS Lett. 482, 169–174 (2000)
9. Gilbert, D., Heiner, M., Rosser, S., Fulton, R., Gu, X., Trybilo, M.: A Case Study in Model-driven Synthetic Biology. In: Biologically Inspired Cooperative Computing: BICC 2008. IFIP, vol. 268, pp. 163–175. Springer, Heidelberg (2008)
10. Pnueli, A.: The Temporal Semantics of Concurrent Programs. Theor. Comput. Sci. 13, 45–60 (1981)
11. Clarke, E.M., Grumberg, O., Peled, D.A.: Model Checking. MIT Press, Cambridge (1999) (third printing, 2001)
12. Fages, F., Rizk, A.: On the Analysis of Numerical Data Time Series in Temporal Logic. In: Calder, M., Gilmore, S. (eds.) CMSB 2007. LNCS (LNBI), vol. 4695, pp. 48–63. Springer, Heidelberg (2007)
13. Hansson, H., Jonsson, B.: A Logic for Reasoning about Time and Reliability. Formal Aspects of Computing 6(5), 512–535 (1994)
14. Aziz, A., Sanwal, K., Singhal, V., Brayton, R.K.: Verifying Continuous-Time Markov Chains. In: Alur, R., Henzinger, T.A. (eds.) CAV 1996. LNCS, vol. 1102, pp. 269–276. Springer, Heidelberg (1996)
15. Goldberg, D.E.: Genetic Algorithms in Search, Optimization and Machine Learning. Addison-Wesley Longman Publishing Co., Inc., Boston (1989)
16. BioNessie: A Biochemical Pathway Simulation and Analysis Tool. University of Glasgow, http://www.bionessie.org
17. MC2(PLTLc) Website: MC2(PLTLc) - PLTLc model checker. University of Glasgow (2008), http://www.brc.dcs.gla.ac.uk/software/mc2/
18. Meffert, K., et al.: JGAP - Java Genetic Algorithms and Genetic Programming Package (2008), http://jgap.sf.net/
19. Hucka, M., Finney, A., Sauro, H.M., Bolouri, H., Doyle, J.C., Kitano, H., et al.: The systems biology markup language (SBML): A medium for representation and exchange of biochemical network models. J. Bioinformatics 19, 524–531 (2003)
20. Fujarewicz, K., Kimmel, M., Lipniacki, T., Świerniak, A.: Parameter estimation for models of cell signaling pathways based on semi-quantitative data. In: BioMed 2006: Proceedings of the 24th IASTED international conference on Biomedical engineering, Anaheim, CA, USA, pp. 306–310. ACTA Press (2006)
21. Calzone, L., Chabrier-Rivier, N., Fages, F., Soliman, S.: Machine Learning Biochemical Networks from Temporal Logic Properties. In: Priami, C., Plotkin, G. (eds.) Transactions on Computational Systems Biology VI. LNCS (LNBI), vol. 4220, pp. 68–94. Springer, Heidelberg (2006)

Compositional Definitions of Minimal Flows in Petri Nets

Michael Pedersen

LFCS, School of Informatics, University of Edinburgh

Abstract. This paper gives algebraic definitions for obtaining the minimal transition and place flows of a modular Petri net from the minimal transition and place flows of its components. The notion of modularity employed is based on place sharing. It is shown that transition and place flows are *not* dual in a modular sense under place sharing alone, but that the duality arises when also considering transition sharing. As an application, the modular definitions are used to give compositional definitions of transition and place flows of models in a subset of the Calculus of Biochemical Systems.

Keywords: Petri nets, minimal flows, minimal invariants, modularity, the Calculus of Biochemical Systems.

1 Introduction

Since their introduction in the early sixties, Petri nets have been used to model concurrent systems in a wide variety of fields [1]. They have long been recognised as a suitable modelling formalism for systems biology; see e.g. [2,3,4] for basic ideas and surveys, [5,6,7,8,9,10] for applications to metabolic pathways, [11,12,13,14,10] for applications to signalling pathways and [15,16] for applications to gene regulatory networks. Applications in model-driven synthetic biology are also starting to emerge [17]. Petri nets are appealing because of their intuitive visual representation as bipartite graphs over *places* and *transitions*, which corresponds well to that of informal biological pathway diagrams: places represent species, transitions represent reactions, and weighted edges represent stoichiometries (see Figure 1 for an example). In addition to their visual appeal, they have a formally defined structure and behaviour and are supported by a large body of simulation and analysis techniques.

High-level extensions of Petri nets enable complex models to be expressed more concisely and at higher levels of abstraction, e.g. as in coloured Petri nets [18]. A key feature of many such extensions is the notion of *modularity*, meaning that a complex model can be composed from modules representing its parts. This is a clear advantage from a modelling point of view as demonstrated in e.g. [19,20] for the yeast pheromone pathway. But modularity can also give rise to modular analysis, potentially reducing computational complexity dramatically, enabling parallel computation, and allowing analysis results to be reused in different contexts.

M. Heiner and A.M. Uhrmacher (Eds.): CMSB 2008, LNBI 5307, pp. 288–307, 2008.

This paper investigates modular analysis in the specific case of Petri net *flows* (also known as *invariants*). Intuitively, a *transition flow* (or T-flow) is a vector representing reaction counts which, when the reactions occur together, have no net effect on species populations. They hence correspond to a notion of cyclic pathways. A *place flow* (or P-flow) is a vector representing species weights for which the weighted sum of species populations is always constant. They hence correspond to chemical conservation relations. More precisely, T and P-flows are natural-number solutions to the equations $Wx = 0$ and $xW = 0$, respectively, where W is the flow matrix of a Petri net which corresponds to the stoichiometry matrix of a biological reaction network. These equations generally have infinitely many solutions, but one can always find finite sets of *minimal* flows which can be combined to generate all other flows. Algorithms for obtaining minimal flows are computationally expensive and the exact algorithmic performance is difficult to estimate [21].

Flow analysis has proven an important tool in biological model validation: the modeller should be able to give biological justification to each minimal flow, otherwise it is likely that the model is incorrect for the intended purpose [13, 10]. Flows are also closely related to the notion of *elementary modes* [22] from metabolic pathway analysis.

The main contributions of this paper are the algebraic definitions of minimal T and P-flows of a Petri net given the minimal T and P-flows of its components (Sections 4 and 5, respectively). We employ a notion of modularity where two modules are composed by merging their shared places/species, also known as *place fusion*. As a second contribution we show that, perhaps contrary to expectation, P and T-flows are *not* dual in a modular sense under place sharing alone, but that the duality arises when also considering composition based on shared transitions/reactions (also known as *transition fusion*, Section 3). Finally, as an application we use our modular definitions to derive compositional definitions of minimal flows in a subset of the Calculus of Biochemical Systems (CBS) [23,19] in Section 6.

Previous efforts have been made towards modular definitions of P-flows in particular, and related work will be discussed in Section 7 before concluding. Detailed proofs of all results can be found in [24] and selected proofs are given in Appendix A. Although the work in this paper is motivated by biological applications, the results are equally applicable outside of biology and flows play an important role in the general analysis of Petri nets; for example, P-flows can be used for determining boundedness of a net, and T-flows can be used for investigating liveness [25].

2 Preliminaries

2.1 Petri Nets

We start with the formal definition of Petri nets.

Definition 1 (Petri net). *A Petri net $\mathcal{P}$ is a tuple $(S, T, W^{\text{in}}, W^{\text{out}})$ where*

– S *is a finite set of* places.
– T *is a finite set of* transitions.
– $W^{\text{in}} : S \times T \to \mathbb{N}$ *is the* flow-in *function*.
– $W^{\text{out}} : S \times T \to \mathbb{N}$ *is the* flow-out *function*.

Define also the derived flow *function* $W(s,t) \overset{\triangle}{=} W^{\text{out}}(s,t) - W^{\text{in}}(s,t)$.

Given a Petri net $\mathcal{P}$, we shall often write $S_{\mathcal{P}}$ for the places of $\mathcal{P}$ and similarly for the transitions and flow functions. As a running example we shall consider simple Petri net models of the foundations of life itself, namely photosynthesis and respiration.

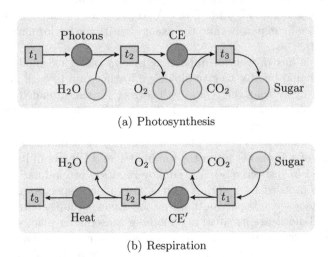

(a) Photosynthesis

(b) Respiration

Fig. 1. Two Petri net models of respectively photosynthesis and respiration, with a distinct shading used for shared places

Example 1. Photosynthesis is the process by which plants produce sugar and oxygen from water, carbon dioxide and sun light (photons). This is modelled by the Petri net in Figure 1(a); places are circles, transitions are squares and arcs (all with the weight 1 omitted) represent the flow function. The first transition provides an unlimited amount of photons. The second converts photons and water into chemical energy (CE) and oxygen, and the third converts chemical energy and carbon dioxide into sugar.

Respiration is the converse process by which e.g. humans use oxygen to break down sugar while producing carbon dioxide and water. This is modelled by the Petri net in Figure 1(b). The first transition breaks down sugar into carbon dioxide and chemical energy CE′, distinct from the chemical energy used in photosynthesis. The second transition utilises this chemical energy and oxygen to make e.g. muscles move, and in the process producing water and heat; the heat is finally removed. Note that both models are strongly simplified and not chemically correct.

If for a Petri net $\mathcal{P}$ we assume some arbitrary but fixed strict total orderings $\prec_s \subsetneq S_{\mathcal{P}} \times S_{\mathcal{P}}$ on places and $\prec_t \subsetneq T_{\mathcal{P}} \times T_{\mathcal{P}}$ on transitions, we can write $S_{\mathcal{P}} = (s_1, \dots, s_m)$ and $T_{\mathcal{P}} = (t_1, \dots, t_n)$ and view the flow functions of $\mathcal{P}$ as $m \times n$ matrices thus:

$$(W_{\mathcal{P}}^{\text{in}})_{i,j} \overset{\Delta}{=} W_{\mathcal{P}}^{\text{in}}(s_i, t_j) \quad (W_{\mathcal{P}}^{\text{out}})_{i,j} \overset{\Delta}{=} W_{\mathcal{P}}^{\text{out}}(s_i, t_j) \quad W_{\mathcal{P}} \overset{\Delta}{=} W_{\mathcal{P}}^{\text{out}} - W_{\mathcal{P}}^{\text{in}}$$

This will allow us to take advantage of matrix operations. Row $(W_{\mathcal{P}}^{\text{in}})_{(i,\cdot)}$ represents the number of tokens *consumed* from place s_i by the respective transitions, and row $(W_{\mathcal{P}}^{\text{out}})_{(i,\cdot)}$ represents the number of tokens *produced* in place s_i by the respective transitions. Row $(W_{\mathcal{P}})_{(i,\cdot)}$ then represents the net effect of transitions on place s_i. The behaviour of Petri nets is defined in the following. Here, and throughout the paper, we use $(\cdot)^{\mathbf{T}}$ to denote vector/matrix transposition.

Definition 2 (Behaviour). *Let* $\mathcal{P} = (S, T, W^{\text{in}}, W^{\text{out}})$ *be a Petri net. Let* $\mathcal{M}(\mathcal{P}) \overset{\Delta}{=} \mathbb{N}^{|S|}$ *be the set of markings of* $\mathcal{P}$. *Then the* transition relation $\to \subseteq \mathcal{M}(\mathcal{P}) \times (\mathbb{N}^{|T|})^{\mathbf{T}} \times \mathcal{M}(\mathcal{P})$ *is defined as follows:* $M \overset{x}{\to} M'$ *if*

1. $M \geq W^{\text{in}}x$
2. $M' = M + W^{\text{out}}x - W^{\text{in}}x = M + Wx$

So in order for a transition count vector x to fire, the marking M must contain enough tokens in each place to supply the inputs of all transitions in x. The marking M' results from removing the tokens consumed by x and adding the tokens produced. The marking of a Petri net then evolves from an initial marking by playing this "token game". But since we consider structural properties only, we shall generally not be concerned with initial markings.

2.2 Petri Net Flows

Transition flows represent transitions which, after they fire, have no net effect on any markings of a Petri net. Place flows represent weights for which the weighted sum of places is constant in any marking reachable from the initial marking. Hence flows give rise to invariance relations. Here is the formal definition:

Definition 3 (T and P-flows). *Let* $\mathcal{P} = (S, T, W^{\text{in}}, W^{\text{out}})$ *be a Petri net. Define*

$$\text{TF}(\mathcal{P}) = \text{TF}(W) \overset{\Delta}{=} \{x \in (\mathbb{N}^{|T|})^{\mathbf{T}} \mid Wx = 0 \land x \neq 0\}$$

$$\text{PF}(\mathcal{P}) = \text{PF}(W) \overset{\Delta}{=} \{y \in \mathbb{N}^{|S|} \mid yW = 0 \land y \neq 0\}$$

The elements of $\text{TF}(\mathcal{P})$ *and* $\text{PF}(\mathcal{P})$ *are called* transition flows *(or T-flows) and* place flow *(or P-flows), respectively.*

Observe that T and P-flows are dual in the following sense:

$$x \in \text{TF}(\mathcal{P}) \Leftrightarrow Wx = 0 \Leftrightarrow x^{\mathbf{T}}W^{\mathbf{T}} = 0 \Leftrightarrow x^{\mathbf{T}} \in \text{PF}(\mathcal{P}^{\mathcal{D}})$$

where the Petri net duality operator $(\cdot)^{\mathcal{D}}$ swaps around the places and transitions in a Petri net and reverses arcs (see [1] for details).

A Petri net generally has infinitely many flows. But it is possible to obtain a finite set of *minimal flows* which can be combined to form all other flows. In the following we shall consider the structure of flows irrespective of whether they are T or P flows. We hence use $F(\mathcal{P})$ and $MF(\mathcal{P})$ to denote the set of either type of flows and minimal flows of $\mathcal{P}$, respectively.

Definition 4 (Support). *The support of a vector $x \in \mathbb{N}^*$, denoted by $\sup(x)$, is the set of indices of non-zero entries in x:* $\sup(x) \triangleq \{i \mid x_i \neq 0\}$.

Definition 5 (Minimal flows). *A flow $x \in F(\mathcal{P})$ is* minimal *if*

1. *x is* canonical, *i.e. the greatest common divisor of non-zero entries of x, written $\gcd(x)$, is 1 and*
2. *x has* minimal support, *i.e. there are no other flows $x' \in F(\mathcal{P})$ with $\sup(x') \subsetneq \sup(x)$.*

We denote by $MTF(\mathcal{P})$ (or $MTF(W_{\mathcal{P}})$) and $MPF(\mathcal{P})$ (or $MPF(W_{\mathcal{P}})$) the sets of minimal T and P-flows of $\mathcal{P}$, respectively.

Example 2. Let us find the minimal flows for the photosynthesis and respiration Petri nets introduced in Example 1. We will do so informally without writing out the flow matrices and full vectors, and instead simply listing the places and transitions which have non-zero entries in the flows.

There are three minimal place flows in the photosynthesis Petri net determined by the places (H_2O, O_2), (CO_2, Sugar) and (CE, H_2O, Sugar). Symmetrically, there are three minimal place flows in the respiration Petri net determined by the places (H_2O, O_2), (CO_2, Sugar) and $(H_2O, \text{Sugar}, CE')$. However, neither Petri net has any transition flows.

The following two theorems are adapted from [21]. They state that $MTF(\mathcal{P})$ and $MPF(\mathcal{P})$ are well-defined, and that any flow can be generated from minimal flows by natural-number linear combinations followed by a division.

Theorem 1. *$MTF(\mathcal{P})$ and $MPF(\mathcal{P})$ are finite and unique.*

Theorem 2. *For any flow $x \in F(\mathcal{P})$ there are $a, \alpha_1, \ldots, \alpha_k \in \mathbb{N}$ and minimal flows $x_1, \ldots, x_k \in MF(\mathcal{P})$ s.t. $x = \frac{1}{a}(\alpha_1 x_1 + \cdots + \alpha_k x_k)$*

We shall also need the following theorem, adapted from [26], which states that any two flows with the same minimal support are multiples of each other.

Theorem 3. *Let $x, y \in F(\mathcal{P})$. If they both have the same minimal support, i.e. there are no other flows $z \in F(\mathcal{P})$ with $\sup(z) \subsetneq \sup(x) = \sup(y)$, then there is $n \in \mathbb{N}$ s.t. either $x = ny$ or $y = nx$.*

Given a set of flows we shall need to filter out the non-minimal ones as in the following definition.

Definition 6 (Minimisation). *Let X be a set of flows. Define minimisation thus:*

$$\min(X) \triangleq \{\frac{x}{\gcd(x)} \mid x \in X \wedge \forall x' \in X.\sup(x') \not\subseteq \sup(x)\}$$

There is a less common definition of minimality which dispenses with the notion of support and defines a flow to be minimal if it cannot be written as the sum of two other flows. This yields a unique set of minimal flows which contains, possibly strictly, the set of minimal flows defined above [21]. The results in this paper are proven valid for both definitions of minimality, and the details can be found in [24].

3   Composition of Petri Nets

In this section we consider how, given two Petri nets $\mathcal{P}_1$ and $\mathcal{P}_2$, these can be composed to form a *parallel* Petri net $\mathcal{P}_1 \mid_p \mathcal{P}_2$. In order for this to be interesting, $\mathcal{P}_1$ and $\mathcal{P}_2$ must have some means of interacting. Two common such means are via *shared places* and *shared transitions*. We will focus on shared places since this appears to be the natural interpretation in the context of chemical reactions. We also see that P and T-flows are *not* dual in a modular sense when considering place sharing alone, but that the duality arises when also considering transition sharing. This allows our results for place sharing to be easily adapted to transition sharing.

3.1   Composition Based on Place Sharing

Following [23] we let the shared places of two Petri nets be determined by syntactic equality of place names rather than introducing explicit place fusion sets. So if two Petri nets $\mathcal{P}_1$ and $\mathcal{P}_2$ both have a place named s, this will be merged to a single place in the composite net $\mathcal{P}_1 \mid_p \mathcal{P}_2$. When modelling large systems one may wish to identify places with different syntactic names, but this can be handled at a higher level of abstraction as in e.g. the Language for Biochemical Systems [19].

  In order to ensure that there is no transition sharing, the parallel composition operation will implicitly rename the transitions of parallel component. This is achieved in a notationally convenient manner by assuming that transitions are strings over the binary alphabet, and prefixing 0 to the transitions of $\mathcal{P}_1$ and 1 to the transitions of $\mathcal{P}_2$.

Definition 7. *Let $\mathcal{P}_1 = (S_1, T_1, W_1^{\text{in}}, W_1^{\text{out}})$ and $\mathcal{P}_2 = (S_2, T_2, W_2^{\text{in}}, W_2^{\text{out}})$ be two petri nets with $T_1, T_2 \subsetneq \{0,1\}^*$. Then define $\mathcal{P}_1 \mid_p \mathcal{P}_2 \triangleq (S, T, W^{\text{in}}, W^{\text{out}})$ where $S \triangleq S_1 \cup S_2$, $T \triangleq \{0t \mid t \in T_1\} \cup \{1t \mid t \in T_2\}$ and for $\text{io} \in \{\text{in}, \text{out}\}$, $b \in \{0,1\}$ we define $W^{\text{io}} : S \times T \to \mathbb{N}$ as follows:*

$$W^{\text{io}}(s, bt) \triangleq \begin{cases} W_1^{\text{io}}(s, t) & \text{if } s \in S_1 \wedge b = 0 \\ W_2^{\text{io}}(s, t) & \text{if } s \in S_2 \wedge b = 1 \\ 0 & \text{otherwise} \end{cases}$$

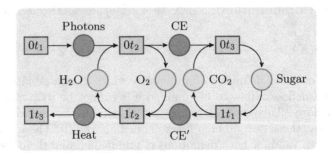

Fig. 2. Photosynthesis and respiration combined into a single Petri net by merging shared places

Example 3. Let $\mathcal{P}_1$ and $\mathcal{P}_2$ be the Petri nets shown in Figure 1. Then $\mathcal{P} = \mathcal{P}_1 \mid_p \mathcal{P}_2$ is shown in Figure 2.

Let us consider the structure of the full flow matrix W arising from the composition of $\mathcal{P}_1$ and $\mathcal{P}_2$ with flow matrices W_1 and W_2. For notational convenience we will assume that all shared places are ordered *after* the non-shared places in $\mathcal{P}_1$, and *before* the non-shared places in $\mathcal{P}_2$. More precisely we assume for $\Delta S = S_{\mathcal{P}_1} \cap S_{\mathcal{P}_2}$ and all $s_1 \in S_{\mathcal{P}_1} \setminus \Delta S$, $s \in \Delta S$ and $s_2 \in S_{\mathcal{P}_2} \setminus \Delta S$ that $s_1 \prec_s s$ and $s \prec_s s_2$. In the running example we could for example order the places as $Photons, CE, H_2O, O_2, CO_2, Sugar, CE', Heat$. Then W_1, W_2 and W can be partitioned as follows where, for $i \in \{1, 2\}$, W_i^s consists of the rows from W_i which represent shared places, and W_i^- are the remaining rows for non-shared places.

$$W_1 = \begin{bmatrix} W_1^- \\ W_1^s \end{bmatrix}, \quad W_2 = \begin{bmatrix} W_2^s \\ W_2^- \end{bmatrix}, \quad W = \begin{bmatrix} W_1^- & 0 \\ W_1^s & W_2^s \\ 0 & W_2^- \end{bmatrix}$$

When considering parallel compositions in the remainder of the paper, we shall write W_1^-, W_1^s, W_2^-, W_2^s and W with the above meaning in mind. We shall furthermore write W_1^+ and W_2^+ to denote respectively the left and right partition of W, i.e. the extensions of W_1 and W_2 with 0-entries for non-shared places from the parallel counterpart. W^- will denote W without the rows $W_1^s W_2^s$ for shared places.

3.2 Modular Duality: Composition Based on Transition Sharing

We have seen that T-flows and P-flows are duals. A natural question then arises of whether this duality holds in the *modular* sense that $\mathrm{PF}(\mathcal{P}_1 \mid_p \mathcal{P}_2) = \mathrm{TF}((\mathcal{P}_1 \mid_p \mathcal{P}_2)^{\mathcal{D}}) = \mathrm{TF}(\mathcal{P}_1^{\mathcal{D}} \mid_p \mathcal{P}_2^{\mathcal{D}})$. The answer is *no*. To see why, let us assume that $T_{\mathcal{P}_1} \cap T_{\mathcal{P}_2} = \emptyset$ and write out the flow matrices $W^{\mathbf{T}}$ of $(\mathcal{P}_1 \mid_p \mathcal{P}_2)^{\mathcal{D}}$ and W' of $(\mathcal{P}_1^{\mathcal{D}} \mid_p \mathcal{P}_2^{\mathcal{D}})$:

$$W^{\mathbf{T}} = \begin{bmatrix} W_1^{-\mathbf{T}} & W_1^{s\mathbf{T}} & 0 \\ 0 & W_2^{s\mathbf{T}} & W_2^{-\mathbf{T}} \end{bmatrix} \quad W' = \begin{bmatrix} W_1^{-\mathbf{T}} & W_1^{s\mathbf{T}} & 0 & 0 \\ 0 & 0 & W_2^{s\mathbf{T}} & W_2^{-\mathbf{T}} \end{bmatrix}$$

The two matrices do not generally have the same dimensions because the dual nets $\mathcal{P}_1^{\mathcal{D}}$ and $\mathcal{P}_2^{\mathcal{D}}$ share transitions rather than places. Hence the modular duality suggested above clearly does not hold in general.

However, we can define the *transition-based* composition operation where transitions (rather than places) of a parallel net are merged based on name equality, and where places are strings over the binary alphabet for the sake of convenient renaming:

Definition 8. *Let* $\mathcal{P}_1 = (S_1, T_1, W_1^{\text{in}}, W_1^{\text{out}})$ *and* $\mathcal{P}_2 = (S_2, T_2, W_2^{\text{in}}, W_2^{\text{out}})$ *be two Petri nets with* $S_1, S_2 \subsetneq \{0,1\}^*$. *Then define* $\mathcal{P}_1 \mid_t \mathcal{P}_2 \triangleq (S, T, W^{\text{in}}, W^{\text{out}})$ *where* $S \triangleq \{0s \mid s \in S_1\} \cup \{1s \mid s \in S_2\}$, $T \triangleq T_1 \cup T_2$, *and for* io $\in \{\text{in}, \text{out}\}$, $b \in \{0,1\}$ *we define* $W^{\text{io}} : S \times T \to \mathbb{N}$ *as follows:*

$$
W^{\text{io}}(bs, t) \triangleq \begin{cases} W_1^{\text{io}}(s, t) & \text{if } t \in T_1 \wedge b = 0 \\ W_2^{\text{io}}(s, t) & \text{if } t \in T_2 \wedge b = 1 \\ 0 & \text{otherwise} \end{cases}
$$

Then the *P-flows* of a parallel net under *transition sharing* are the same as the *T-flows* of the parallel dual nets under *place sharing*, and symmetrically for T-flows under transition sharing:

Theorem 4. *Let* $\mathcal{P}_1$ *and* $\mathcal{P}_2$ *be Petri nets. Then*

1. $\text{TF}(\mathcal{P}_1 \mid_t \mathcal{P}_2) = \text{PF}(\mathcal{P}_1^{\mathcal{D}} \mid_p \mathcal{P}_2^{\mathcal{D}})$.
2. $\text{PF}(\mathcal{P}_1 \mid_t \mathcal{P}_2) = \text{TF}(\mathcal{P}_1^{\mathcal{D}} \mid_p \mathcal{P}_2^{\mathcal{D}})$.

The proof relies on a partitioning of flow matrices similar to the partitioning in the previous subsection, but for shared transitions rather than places. It follows from Theorem 4 that the results for modular flows under place sharing, to be given in the following sections, can be easily adapted to (dual) modular flows under transition sharing.

4   Minimal Transition Flows

We start with an example of how T-flows arise through parallel composition.

Example 4. As noted in Example 2, neither of the photosynthesis and respiration Petri nets has any T-flows. But observe that the composite net in Figure 2 *does* have a single T-flow determined by the transitions $(0t_1, 0t_2, 0t_3, 1t_1, 1t_2, 1t_3)$. How did this flow arise from the parallel composition? To answer this, we need to look at *potential* T-flows of the two nets rather than the *actual* T-flows of which there are none. The potential T-flows are the ones arising from restricting individual components to private places only, i.e. by disregarding the shared places. If we do so, the photosynthesis Petri net has a single T-flow determined by $(0t_1, 0t_2, 0t_3)$, and the respiration net has a single T-flow determined by $(1t_1, 1t_2, 1t_3)$. The T-flow in the parallel net is composed from these two, because the transitions from the two nets operating on shared places cancel each other out.

The general case is slightly more complicated, because there may be many potential T-flows of each parallel component. These T-flows can then be combined by natural-number linear combinations in such a way that the resulting flow has no net effect on shared places. The weights of this natural-number linear combination must be minimal in some sense in order for there to be any hope of minimality of the composite flow in the composite net. A formal definition is given below, where we use the conventions on flow matrix partitioning introduced in Section 3.1; by $[MTF(W_i^-)]$ we mean the matrix consisting of the column vectors in $MTF(W_i^-)$ in some arbitrary order.

Definition 9. Let $\mathcal{P}_1$ and $\mathcal{P}_2$ be Petri nets and let $X_1 = [MTF(W_1^-)]$, $X_2 = [MTF(W_2^-)]$ and W^s be given. Define the following:

1. $X \triangleq \begin{bmatrix} X_1 & 0 \\ 0 & X_2 \end{bmatrix}$

2. $C \triangleq W^s X$.
3. $Z \triangleq \{X\alpha \mid \alpha \in MTF(C)\}$.

We then define $MTF^{Par}(X_1, X_2, W^s) \triangleq \min(Z)$.

To elaborate on this definition, let $m_1 = |T_{\mathcal{P}_1}|$, $m_2 = |T_{\mathcal{P}_2}|$, $n_1 = |MTF(W_1^-)|$ and $n_2 = |MTF(W_2^-)|$. Then X_1 and X_2 are $m_1 \times n_1$ and $m_2 \times n_2$ matrices with the transition flows of respectively $\mathcal{P}_1$ and $\mathcal{P}_2$ *without their shared places*, i.e. of W_1^- and W_2^-. Also,

1. X is an $(m_1 + m_2) \times (n_1 + n_2)$ matrix with columns representing minimal T-flows of W^-.
2. C is an $(|S_{\mathcal{P}_1} \cap S_{\mathcal{P}_2}|) \times (n_1 + n_2)$ matrix with each column c_i representing the effect of the corresponding minimal T-flow x_i on the shared places.
3. Z is a set of linear combinations of the minimal T-flow-columns in X. These linear combinations are chosen in such a way that they have no net effect on the shared species. Note that the set Z is well-defined because $MTF(C)$ is finite and unique by Theorem 1.

Remarks regarding the use of minimisation are made towards the end of the section. The following results state that Definition 9 is sound and complete. Soundness is split into two lemmas, the first of which is needed to prove completeness.

Lemma 1 (Soundness part 1). Let Z be as given in Definition 9. Then

1. $Z \subsetneq TF(\mathcal{P}_1 \mid_p \mathcal{P}_2)$.
2. $\min(Z) \subsetneq TF(\mathcal{P}_1 \mid_p \mathcal{P}_2)$.

The proof uses the definition of C to show that any $X\alpha \in Z$ is a T-flow of W^s. Since X consists of minimal T-flows of W^-, $X\alpha$ is also a T-flow of W^-. Together these give that $X\alpha$ is a T-flow of W and hence of $\mathcal{P}_1 \mid_p \mathcal{P}_2$.

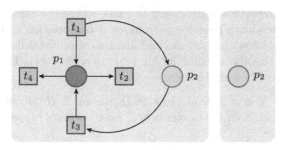

Fig. 3. Two Petri nets illustrating how Definition 9 can give rise to non-minimal flows in Z

Lemma 2 (Completeness). *Let $\mathcal{P}_1$, $\mathcal{P}_2$, X_1, X_2 and W^s be as given in Definition 9. Then* $\mathrm{MTF}(\mathcal{P}_1 \mid_p \mathcal{P}_2) \subseteq \mathrm{MTF}^{\mathrm{Par}}(X_1, X_2, W^s)$.

The proof starts by showing that any $x \in \mathrm{MTF}(\mathcal{P}_1 \mid_p \mathcal{P}_2)$ can be written $x = \frac{1}{a}X\alpha$ where $\alpha \in \mathrm{TF}(C)$ and $a \in \mathbb{N}$ (uses Theorem 2 and the definition of C). Using Euclid's lemma and that x is canonical, we show that a canonical α can in fact be chosen. We then use Theorem 3 and minimality of x to show that any of the minimal-support α which generate x as above is in fact also minimal in C. We arrive at $x \in Z$. To conclude that also $x \in \min(Z)$, we use that any $x' \in Z$ with smaller support than x would also be in $\mathrm{TF}(\mathcal{P}_1 \mid_p \mathcal{P}_2)$ (Lemma 1), hence contradicting minimality of x in $\mathcal{P}_1 \mid_p \mathcal{P}_2$.

Lemma 3 (Soundness part 2). *Let $\mathcal{P}_1$, $\mathcal{P}_2$, X_1, X_2 and W^s be as given in Definition 9. Then* $\mathrm{MTF}^{\mathrm{Par}}(X_1, X_2, W^s) \subseteq \mathrm{MTF}(\mathcal{P}_1 \mid_p \mathcal{P}_2)$

The proof carries on from Lemma 1. To show that the elements of $\min(Z)$ are in fact minimal in $\mathcal{P}_1 \mid_p \mathcal{P}_2$, we use that all minimal-support (although not necessarily canonical) flows are represented in Z by completeness (Lemma 2).

Together the two previous lemmas prove our main T-flow theorem:

Theorem 5 (Soundness and completeness). *Let $\mathcal{P}_1$, $\mathcal{P}_2$, X_1, X_2 and W^s be as given in Definition 9. Then* $\mathrm{MTF}^{\mathrm{Par}}(X_1, X_2, W^s) = \mathrm{MTF}(\mathcal{P}_1 \mid_p \mathcal{P}_2)$

The size of matrices X and C may be reduced by removing columns which have all 0-entries in C; these columns are also flows in the composite net and can be included directly.

The flows in Z may not be minimal, which is why the minimisation function must be applied as a last step. This is illustrated by the following example.

Example 5. Figure 4 shows two Petri nets: the left, $\mathcal{P}_1$, has two places of which one is shared with the right, $\mathcal{P}_2$, consisting of just a single place. The restriction of $\mathcal{P}_1$ to the place p_1 (corresponding to W^-) has four minimal T-flows represented by $x_1 = (t_1, t_2)$, $x_2 = (t_2, t_3)$, $x_3 = (t_3, t_4)$ and $x_4 = (t_1, t_4)$. The "minimal" combinations of these which preserve the flow for the shared place p_2 (corresponding to the minimal flows of C) are $x_1 + x_2 = (t_1, 2 \cdot t_2, t_3)$, $x_1 + x_3 = (t_1, t_2, t_3, t_4)$ and $x_2 + x_4 = (t_1, t_2, t_3, t_4)$. But the latter two flows are not minimal because they strictly contain the support of the first.

Minimisation is however *not* necessary in cases where the minimal flows in X are linearly independent. Then we get unique decomposition in the sense that any flow can be written uniquely as combinations of minimal flows (linear independence fails in the above example, for $x_1 + x_3 = x_2 + x_4$). This can be used in the proof of the following theorem:

Theorem 6. *Let X and Z be as given in Definition 9. If the columns of X are linearly independent, then the elements of Z have minimal support (but still may not be canonical).*

5 Minimal Place Flows

As for T-flows we start by looking at an example of how P-flows in a composite net arise from P-flows in parallel components.

Example 6. In example 2 we listed the three minimal P-flows for each of the two Petri nets in Figure 1. These included $x = (\text{CE}, \text{H}_2\text{O}, \text{Sugar})$ from the first net and $y = (\text{H}_2\text{O}, \text{Sugar}, \text{CE}')$ from the second net. Neither is a flow in the composite net shown in Figure 2 because of interference from the additional transitions. For example, $0t_1$ consumes tokens from Sugar and produces tokens in CE', and this violates the first flow.

However, because x and y have identical weights for their shared places (namely $1 \cdot \text{H}_2\text{O}$ and $1 \cdot \text{Sugar}$), we can "join" them to obtain a new minimal flow $x \frown y = (\text{CE}, \text{H}_2\text{O}, \text{Sugar}, \text{CE}')$ for the composite net.

Here is the formal definition of P-flow joins where again we assume the partitioning of flow matrices given in Section 3.1.

Definition 10 (Flow joins). *Let $\mathcal{P}_1$, $\mathcal{P}_2$ be Petri nets and let $x \in \text{PF}(\mathcal{P}_1)$, $y \in \text{PF}(\mathcal{P}_2)$. Write $x = (x^- x^s)$, $y = (y^s y^-)$ where, for $\Delta S = S_{\mathcal{P}_1} \cap S_{\mathcal{P}_2}$, x^- represents places $S_{\mathcal{P}_1} \setminus \Delta S$, x^s and y^s represent places ΔS, and y^- represents places $S_{\mathcal{P}_2} \setminus \Delta S$. If $x^s = y^s$ we say that x and y are consistent and define their join $x \frown y \overset{\Delta}{=} (x^- x^s y^-)$.*

The join of consistent flows from two parallel nets is a flow in the composite net:

Lemma 4. *Let $\mathcal{P}_1$ and $\mathcal{P}_2$ be Petri nets and let $x \in \text{PF}(\mathcal{P}_1)$, $y \in \text{PF}(\mathcal{P}_2)$. If x and y are consistent then $x \frown y \in \text{PF}(\mathcal{P}_1 \mid_p \mathcal{P}_2)$.*

Conversely, any P-flow z of a parallel composition $\mathcal{P}_1 \mid_p \mathcal{P}_2$ is the join of a P-flow from $\mathcal{P}_1$ and a P-flow from $\mathcal{P}_2$.

Lemma 5. *Let $\mathcal{P}_1$ and $\mathcal{P}_2$ be Petri nets and let $z \in \text{PF}(\mathcal{P}_1 \mid_p \mathcal{P}_2)$. Then there are $x \in \text{PF}(\mathcal{P}_1)$ and $y \in \text{PF}(\mathcal{P}_2)$ s.t. $z = x \frown y$.*

In contrast to Example 6, it is generally not sufficient to join only *minimal* consistent flows. Rather we must obtain two linear combinations of minimal flows from the respective nets in such a way that they become consistent, and then join them to form a flow of the composite net. As for modular T-flows, the weights used in this linear combination must be minimal in some sense in order for there to be any hope of minimality for the resulting join.

The general modular definition of P-flows is given below. Similarly to the definition for T-flows, $[\text{MPF}(\mathcal{P})]$ is a matrix with rows from $\text{MPF}(\mathcal{P})$ in any order.

Definition 11. *Let $\mathcal{P}_1$ and $\mathcal{P}_2$ be Petri nets and let $X = [\text{MPF}(\mathcal{P}_1)]$ and $Y = [\text{MPF}(\mathcal{P}_2)]$ be given. Let X^s and Y^s be the sub-matrices of X and Y containing only columns for shared places $s \in S_{\mathcal{P}_1} \cap S_{\mathcal{P}_2}$. Define the following:*

1. $C \overset{\Delta}{=} \begin{bmatrix} X^s \\ -Y^s \end{bmatrix}$.

2. $Z \overset{\Delta}{=} \{\alpha X \frown \beta Y \mid (\alpha\beta) \in \text{MPF}(C)\}$.

We then define $\text{MPF}^{\text{Par}}(X, Y) \overset{\Delta}{=} \min(Z)$.

To elaborate on this definition, let $\Delta S = S_{\mathcal{P}_1} \cap S_{\mathcal{P}_2}$, $m = |\Delta S|$, $n_1 = |\text{MPF}(\mathcal{P}_1)|$, $n_2 = |\text{MPF}(\mathcal{P}_2)|$. Then

1. C is an $(n_1 + n_2) \times m$ matrix with the first n_1 rows representing minimal P-flows of $\mathcal{P}_1$ and the last n_2 rows representing *negated* minimal P-flows from $\mathcal{P}_2$, but restricted to the m shared places only.
2. Z contains the joins of consistent linear combinations of flows from the two nets. The weights for this linear combination are chosen exactly so that the resulting flows have the same weights for shared places. Note that the set Z is well-defined because $\text{MTF}(C)$ is finite and unique by Theorem 1.

As for T-flows we have soundness and completeness results, and soundness is split into two separate lemmas.

Lemma 6 (Soundness part 1). *Let Z be as given in Definition 11. Then*

1. $Z \subsetneq \text{PF}(\mathcal{P}_1 \mid_p \mathcal{P}_2)$.
2. $\min(Z) \subsetneq \text{PF}(\mathcal{P}_1 \mid_p \mathcal{P}_2)$.

The proof uses Lemma 4 and the definition of C.

Lemma 7 (Completeness). *Let $\mathcal{P}_1, \mathcal{P}_2, X$ and Y be as given in Definition 11. Then* $\text{MPF}(\mathcal{P}_1 \mid_p \mathcal{P}_2) \subseteq \text{MPF}^{\text{Par}}(X, Y)$

The proof first uses Lemma 5 to write any $z \in \text{MPF}(\mathcal{P}_1 \mid_p \mathcal{P}_2)$ as $z = x \frown y$ for some $x \in \text{PF}(\mathcal{P}_1)$ and $y \in \text{PF}(\mathcal{P}_2)$. Then the main challenge is to show that there is some d and $(\alpha\beta) \in \text{MPF}(C)$ s.t. $dx = \alpha X$ and $dy = \beta Y$ (for then we can conclude that $dz \in Z$). First the existence of such $(\alpha\beta) \in \text{PF}(C)$ is shown using Theorem 2, the definition of C and the fact that dx and dy are consistent. Minimality of $(\alpha\beta)$ uses an idea similar to the proof of Lemma 2 (completeness for T-flows).

Lemma 8 (Soundness part 2). *Let $\mathcal{P}_1, \mathcal{P}_2, X$ and Y be as given in Definition 11. Then* $\text{MPF}^{\text{Par}}(X, Y) \subseteq \text{MPF}(\mathcal{P}_1 \mid_p \mathcal{P}_2)$

The proof is similar to that of Lemma 3 (soundness for T-flows). Together the last two lemmas prove our main theorem on modular P-flows:

Theorem 7 (Soundness and completeness). *Let* $\mathcal{P}_1, \mathcal{P}_2, X$ *and* Y *be as given in Definition 11. Then* $\mathrm{MPF}^{\mathrm{Par}}(X, Y) = \mathrm{MPF}(\mathcal{P}_1 \mid_{\mathrm{p}} \mathcal{P}_2)$

The matrix C in Definition 11 can be reduced in size by removing rows with all 0 entries. Because these do not involve shared places, they are also flows of the composite net and can be included directly.

As for the modular definition of T-flows, minimisation is not necessary in cases where the minimal flows in the rows of C are linearly independent:

Theorem 8. *Let* C *and* Z *be as given in Definition 11. If the rows of* C *are linearly independent, then the elements of* Z *have minimal support (but still may not be canonical).*

6 Compositional Definitions of Minimal Flows

The modular definitions of flows given in the previous sections are not compositional for two reasons: 1) the flows of parallel components are given explicitly and not defined inductively. This is because there is no inductive structure on Petri nets and flows per se. 2) In the case of T-flows, the definition of $\mathrm{MTF}^{\mathrm{Par}}$ requires more than just the flows of parallel components to be given. It requires the flows of the components *without shared places* (which are super-sets of the flows of the components) and the flow matrix for shared places.

In response to the first problem, we consider a simple calculus of Petri nets, $\mathcal{CP}$, which is a subset of the Calculus of Biochemical Systems (CBS) [23, 19]. In response to the second problem, we define a compositional T-flow function which returns not T-flows, but *a*) the flow function of the net together with *b*) a *function* mapping shared places to T-flows of the restricted net without these places.

6.1 $\mathcal{CP}$: A Calculus of Petri Nets

In the following, some fixed set of places S and transitions $T \triangleq \{0, 1\}^*$ are assumed.

Definition 12. *The language* $\mathcal{CP}$ *is the set of programs generated by the following grammar, where* $n_i, n'_j \in \mathbb{N}$ *and* $s_i, s'_j \in S$:

$$P ::= \sum n_i \cdot s_i \to \sum n'_j \cdot s'_j \mid P_1 | P_2$$

Intuitively, the program $\sum n_i \cdot s_i \to \sum n'_j \cdot s'_j$ represents a single Petri net transition (reaction) with input places (reactants) $\{s_i\}$, output places (products) $\{s'_j\}$ and flow functions (stoichiometries) given by the n_i and n'_j. In the biological setting, programs thus correspond directly to reactions taking place in parallel.

Example 7. Below are two programs representing respectively photosynthesis and respiration. All stoichiometries are 1 and have hence been omitted. Note the resemblance with chemical reactions. Also note that there are no reactants in the first reaction of photosynthesis, and no products in the last reaction of respiration.

$$P_1 \triangleq \; \to \text{Photons} \mid \text{Photons} + \text{H}_2\text{O} \to \text{CE} + \text{O}_2 \mid \text{CE} + \text{CO}_2 \to \text{Sugar}$$

$$P_2 \triangleq \text{Sugar} \to \text{CE}' + \text{CO}_2 \mid \text{CE}' + \text{O}_2 \to \text{Heat} + \text{H}_2\text{O} \mid \text{Heat} \to$$

The parallel composition $P_1 \mid P_2$ then represents combined photosynthesis and respiration.

The denotational semantics for $\mathcal{CP}$ is given in terms of the set $\mathcal{PN}$ of Petri nets.

Definition 13. *Define* $[\![\cdot]\!] : \mathcal{CP} \to \mathcal{PN}$ *inductively on programs as follows.*

- **Basis:** $[\![\sum n_i \cdot s_i \to \sum n_j' \cdot s_j']\!] \triangleq (\{s_i\} \cup \{s_j'\}, \{\epsilon\}, W^{\text{in}}, W^{\text{out}})$
  where $W^{\text{in}}(s_i, \epsilon) \triangleq n_i$ *and* $W^{\text{out}}(s_j', \epsilon) \triangleq n_j'$.
- **Step:** $[\![P_1 | P_2]\!] \triangleq [\![P_1]\!] \mid_{\text{p}} [\![P_2]\!]$

In the base case, ϵ denotes the empty string over the binary alphabet. Note the compositional nature of the denotation function.

Example 8. Let P_1 and P_2 be as defined in Example 7. Then $[\![P_1]\!]$ and $[\![P_2]\!]$ are given by the Petri nets shown in Figure 1 (modulo transition naming), and $[\![P_1 \mid P_2]\!]$ is given by the composite Petri net shown in Figure 2 (modulo transition naming).

6.2   Flows in $\mathcal{CP}$

We are now in a position to give compositional definitions of minimal flows for $\mathcal{CP}$ programs. To do so, we first define $\mathcal{P} \setminus \Delta S$ to be the Petri net $\mathcal{P}$ without the places ΔS, and similarly $P \setminus \Delta S$ is the program P without the places ΔS. The power set of a set X is denoted by 2^X, and the domain of a function f is denoted by $\text{dom}(f)$.

Definition 14. *Let* $\mathcal{W} = S \times T \hookrightarrow_{\text{fin}} \mathbb{N}$ *be the set of (partial and finite) flow functions. Define the parameterised minimal T-flows,* $\zeta : \mathcal{CP} \to \mathcal{W} \times (2^S \to 2^{(\mathbb{N}^*)})$, *inductively as follows:*

- **Basis:** $\zeta(\sum n_i \cdot s_i \to \sum n_j' \cdot s_j') \triangleq (W_{\mathcal{P}}, h)$
  where $\mathcal{P} \triangleq [\![\sum n_i \cdot s_i \to \sum n_j' \cdot s_j']\!]$ *and* $h(\Delta S) \triangleq \text{MTF}(\mathcal{P} \setminus \Delta S)$.
- **Step:** $\zeta(P_1 | P_2) \triangleq (W, h)$
  where
  - $\zeta(P_1) = (W_1, h_1)$ *and* $\zeta(P_2) = (W_2, h_2)$
  - W *is composed from* W_1 *and* W_2 *as defined in Section 3.*
  - $h(\Delta S) \triangleq \text{MTF}^{\text{Par}}(X_1, X_2, W^{\text{ss}})$.
  - $X_1 \triangleq h_1(\Delta S')$, $X_2 \triangleq h_2(\Delta S')$.

- $\Delta S' \overset{\Delta}{=} \Delta S \cup \Delta S''$.
- $\Delta S'' \overset{\Delta}{=} \{s \in S \mid \exists t_1, t_2 \in T.(s, t_1) \in \mathrm{dom}(W_1) \wedge (s, t_2) \in \mathrm{dom}(W_2)\}$
  (i.e. the places shared between $\mathcal{P}_1$ and $\mathcal{P}_2$).
- W^{ss} *is the sub-matrix of W containing rows for shared places $\Delta S'' \setminus \Delta S$.*

In the inductive step observe how X_1, X_2 and W^{ss} are obtained purely from the results of recursively invoking ζ. Hence the definition is in fact compositional. But compositionality comes at a high price: the return value of ζ "encapsulates" both the flow matrix of the entire net and the flows arising from any restriction of places – all of this information is needed for the composition.

Definition 15. *Define the compositional minimal P-flows, $\xi : \mathcal{CP} \to 2^{(\mathbb{N}^*)}$, inductively as follows:*

- *Basis:* $\xi(\sum n_i \cdot s_i \to \sum n'_j \cdot s'_j) \overset{\Delta}{=} \mathrm{MPF}([\![\sum n_i \cdot s_i \to \sum n'_j \cdot s'_j]\!])$.
- *Step:* $\xi(P_1 \dagger P_2) \overset{\Delta}{=} \mathrm{MPF}^{\mathrm{Par}}(\xi(P_1), \xi(P_2))$.

Note how we again obtained a compositional definition, albeit in a somewhat simpler manner than for T-flows. This illustrates how modular T and P-flows are intricately different and non-dual because more information is needed in the compositional definition of T-flows.

The following theorem says that the compositional definitions given above work as expected. The proof is by induction on the structure of programs, using Theorems 5 and 7

Theorem 9. *Let P be an $\mathcal{CP}$ program and ΔS a set of places. Then*

1. $\zeta(P) = (W_\mathcal{P}, h)$
   where $\mathcal{P} = [\![P]\!]$ and $h(\Delta S) = \mathrm{MTF}(\mathcal{P} \setminus \Delta S)$.
2. $\xi(P) = \mathrm{MPF}([\![P]\!])$.

It follows as a special case of 1) that $h(\emptyset) = \mathrm{MTF}([\![P]\!])$.

7   Related Work and Conclusion

7.1   Related Work

The idea that consistent P-flows from two components can be joined to form a P-flow in the composite net (Lemma 4) is not new. Neither is the converse that any place flow in a composite net is a join of place flows from the parallel component (Lemma 5). These results have been stated previously in some form in [27, 28, 29, 30, 31].

In [29] an algorithm is given for directly computing the minimal P-flows of a "well-formed net" resulting from a place fusion operation, based on the minimal P-flows of the net before fusion. But no proof of correctness is given. In [28] a method similar to Definition 11 is proposed for generative sets of P-flows rather than minimal P-flows. Such a method is also presented for "functional subnets" in [31], which in addition considers how to obtain the modules in the first place. However, in neither case is it clear to us how completeness follows from the proofs given, i.e. that the method in fact results in generative families of P-flows. In contrast, we give a full proof of *minimality* of the resulting P-flows (which is stronger than generativity).

Modular definitions of T-flows have received somewhat less attention than P-flows in the literature. To our knowledge, the only existing explicit work on modular T-flows is [29] (for well-formed nets), but this only shows an *example* of how new T-flows can arise after a place fusion. No general definition is given. In [30], a characterisation of *P-flows* arising from a composition of modules is given based on *both* place sharing *and* transition sharing. The duality elucidated in Theorem 4 suggests that a dual characterisation can be given for *T-flows* under place and transition sharing. Nevertheless, the characterisation does not result in methods for finding minimal or generative sets of flows and hence is of little practical use. It also considers flows in $\mathbb{Z}$ rather than in $\mathbb{N}$ as is more common (and harder).

7.2 Conclusion

As the primary contribution, this paper has presented algebraic definitions for obtaining the minimal P and T-flows of parallel Petri nets given the minimal P and T-flows of its components (with some additional information in the case of T-flows). These definitions have then been proven correct. Although the idea used for minimal P-flows is not new, no complete proof has to our knowledge been given before.

We have also shown modular dualities between T/P-flows under place sharing and T/P-flows under transition sharing. This allows our results for place sharing to be easily adapted to transition sharing. On the other hand, we have seen that T and P-flows are *not* dual in the modular sense under place sharing alone, and hence the existing work on modular P-flows under place sharing does not carry over to T-flows.

As an application we have shown how our modular definitions of T and P-flows can be used to define T and P-flows in a compositional manner for a subset of CBS. This has turned out to be harder for T-flows than for P-flows, thus further illustrating the intricate difference between the two when considering place sharing alone.

Future work may consider the computational complexity of calculating minimal flows using our compositional definitions and investigate if improvements can be made over existing algorithms as in e.g. [28]. The inherent compositionality can also be exploited by distributed computation. On the more theoretical side, a potential next step is to extend the results to higher-level coloured Petri nets. These allow species modifications and complexes to be represented compactly and form a semantical foundation of the full CBS.

Acknowledgements

The author would like to thank Gordon Plotkin and Monika Heiner for useful discussions. This work was supported by Microsoft Research through its European PhD Scholarship Programme.

References

1. Murata, T.: Petri nets: properties, analysis and applications. Proc. IEEE 77(4), 541–580 (1989)
2. Goss, P.J.E., Peccoud, J.: Quantitative modeling of stochastic systems in molecular biology by using stochastic Petri nets. PNAS 95(12), 6750–6755 (1998)
3. Peleg, M., et al.: Using Petri net tools to study properties and dynamics of biological systems. Journal of the American Medical Informatics Association 12(2), 181–199 (2005)
4. Hardy, S., Robillard, P.N.: Modeling and simulation of molecular biology systems using Petri nets: Modeling goals of various approaches. J. Bioinformatics and Computational Biology 2(4), 619–638 (2004)
5. Reddy, V.N., et al.: Petri net representation in metabolic pathways. In: Proc. Int. Conf. Intell. Syst. Mol. Biol., pp. 328–336 (1993)
6. Zevedei-Oancea, Schuster, S.: Topological analysis of metabolic networks based on Petri net theory. Silico. Biol. 3, 323–345 (2003)
7. Heiner, M., et al.: Analysis and simulation of steady states in metabolic pathways with Petri nets. In: Jensen, K. (ed.) Workshop and Tutorial on Practical Use of Coloured Petri Nets and the CPN Tools, pp. 15–34 (2001)
8. Genrich, H., et al.: Executable Petri net models for the analysis of metabolic pathways. J. STTT 3(4), 394–404 (2001)
9. Voss, K., et al.: Steady state analysis of metabolic pathways using Petri nets. Silico. Biol. 3 (2003)
10. Heiner, M., Koch, I.: Petri net based model validation in systems biology. In: Cortadella, J., Reisig, W. (eds.) ICATPN 2004. LNCS, vol. 3099, pp. 216–237. Springer, Heidelberg (2004)
11. Sackmann, A., et al.: Application of Petri net based analysis techniques to signal transduction pathways. BMC Bioinformatics 7(482) (2006)
12. Lee, D.Y., et al.: Colored Petri net modeling and simulation of signal transduction pathways. Metab. Eng. 8(2), 112–122 (2005)
13. Heiner, M., et al.: Petri nets for systems and synthetic biology. In: Bernardo, M., Degano, P., Zavattaro, G. (eds.) SFM 2008. LNCS, vol. 5016, pp. 215–264. Springer, Heidelberg (2008)
14. Taubner, C., et al.: Modelling and simulation of the TLR4 pathway with coloured Petri nets. In: Proc. Annual International Conference of the IEEE Engineering in Medicine and Biology Society, Engineering in Medicine and Biology Society, pp. 2009–2012 (2006)
15. Matsuno, H., et al.: Hybrid Petri net representation of gene regulatory network. In: Pacific Symposium on Biocomputing, vol. 5, pp. 341–352 (2000)
16. Steggles, L.J., et al.: Qualitatively modelling and analysing genetic regulatory networks: a Petri net approach. Bioinformatics 23(3), 336–343 (2007)
17. Gilbert, D., et al.: A case study in model-driven synthetic biology. In: Biologically-inspired cooperative computing. IFIP International Federation for Information Processing, vol. 268, pp. 163–175. Springer, Heidelberg (2008)

18. Jensen, K.: Coloured Petri Nets: Basic Concepts, Analysis Methods and Practical Use, vol. 1. Springer, Heidelberg (1992)

19. Pedersen, M., Plotkin, G.: A language for biochemical systems. In: Heiner, M., Uhrmacher, A.M. (eds.) Proc. CMSB. LNCS. Springer, Heidelberg (2008)

20. Kofahl, B., Klipp, E.: Modelling the dynamics of the yeast pheromone pathway. Yeast 21(10), 831–850 (2004)

21. Krückeberg, F., Jaxy, M.: Mathematical methods for calculating invariants in Petri nets. In: Advances in Petri Nets 1987, covers the 7th European Workshop on Applications and Theory of Petri Nets, London, UK, pp. 104–131. Springer, Heidelberg (1987)

22. Schuster, S., et al.: A general definition of metabolic pathways useful for systematic organization and analysis of complex metabolic networks. Nature Biotechnology 18, 326–332 (2000)

23. Plotkin, G.: A calculus of biochemical systems (in preparation)

24. Pedersen, M.: Compositional definitions of minimal flows in Petri nets. Technical report, University of Edinburgh (2008), http://www.inf.ed.ac.uk/publications/report/1269.html

25. Reisig, W.: Petri nets. EATCS Monograps on Theoretical Computer Science. Springer, Heidelberg (1982)

26. Memmi, G., Roucairol, G.: Linear algebra in net theory. In: Proc. Advanced Course on General Net Theory of Processes and Systems, pp. 213–223. Springer, Heidelberg (1980)

27. Jensen, K.: Coloured Petri Pets: Basic Concepts, Analysis Methods and Practical Use, vol. 2. Springer, Heidelberg (1995)

28. Bourjij, A., et al.: A decentralized approach for computing invariants in large scale and interconnected Petri nets. In: Proc. IEEE International Conference on Systems, Man, and Cybernetics, vol. 2, pp. 1741–1746 (1997)

29. Isabel, C.R.M.: Compositional construction and analysis of Petri net systems. PhD thesis, School of Informatics, University of Edinburgh (1998)

30. Christensen, S., Petrucci, L.: Modular analysis of Petri nets. The Computer Journal 43(3), 224–242 (2000)

31. Zaitsev, D.A.: Decomposition-based calculation of Petri net invariants. In: Cortadella, Yakovlev (eds.) Proc. Workshop on Token based Computing (ToBaCo), Satellite Event of the 25-th International conference on application and theory of Petri nets, pp. 79–83 (2004)

A   Selected Proofs

A.1   Modular T-Flows

Proof (Lemma 2). Take any $x \in \text{MTF}(\mathcal{P}_1 \mid_\text{p} \mathcal{P}_2)$. Then $Wx = 0$, so also $W^s x = 0$ and $W^- x = 0$. Hence $x \in \text{TF}(W^s)$ and $x \in \text{TF}(W^-)$. Observe that X consists exactly of the minimal T-flows of W^-. Therefore, by Theorem 2 there are $\alpha \in \mathbb{N}^{|col(X)|^\mathsf{T}}$ and $a \in \mathbb{N}$ s.t. $x = \frac{1}{a}X\alpha$, i.e. $xa = X\alpha$. There may generally be multiple such α, so pick one which is canonical and has *minimal decomposition-support* in the sense that no other choices have smaller support. Such a canonical choice *is* indeed possible because it is always the case that $\gcd(\alpha)$ divides a. To see this, let $c = \gcd(\alpha)$; then there is a canonical α' s.t. $ax = Xc\alpha' = cX\alpha'$.

Also $\frac{a}{d}x = \frac{c}{d}X\alpha'$ where $d = \gcd(a,c)$. Since x has natural number entries, $\frac{a}{d}$ divides all entries in $\frac{c}{d}X\alpha'$. It follows from Euclid's lemma and $\gcd(\frac{a}{d}, \frac{c}{d}) = 1$ that $\frac{a}{d}$ divides all entries in $X\alpha'$. Canonicity of x then forces $c = d$, and hence $d = \gcd(\alpha)$ divides a as claimed.

We now show that α is a T-flow of C, i.e. that $C\alpha = 0$. The following steps rely on the fact that matrix multiplication is associative:

$$C\alpha = (W^s X)\alpha = W^s(X\alpha) = W^s(xa) = (W^s x)a = 0a = 0$$

Next we show that α is a *minimal* T-flow of C. It is canonical per assumption. To get that α has minimal support, we show that any T-flow α' of C with $\sup(\alpha') \subsetneq \sup(\alpha)$ will also generate x, contrary to our choice of α being the smallest decomposition-support for which this holds. Note here the subtle distinction between minimality of α wrt. decomposition of x and wrt. flows of C; the former holds per assumption, and we will now prove the latter.

So, we have $\sup(\alpha') \subsetneq \sup(\alpha)$ and $C\alpha' = 0$. Then $0 = C\alpha' = (W^s X)\alpha' = W^s(X\alpha')$, so $x' = X\alpha'$ is a T-flow of W^s. Any linear combination of T-flows is also a T-flow, so x' is also a T-flow of W^-. Together these give $x' \in \mathrm{TF}(W)$. Now since $\sup(\alpha') \subsetneq \sup(\alpha)$ it must also hold that $\sup(x') = \sup(X\alpha') \subseteq \sup(X\alpha) = \sup(x)$. Since x has minimal-support, it must be the case that $\sup(x') = \sup(x)$. By Theorem 3, either $x = nx'$ or $x' = nx$ for some $n \in \mathbb{N}$. But x is canonical, so $x' = nx$ i.e. $x = \frac{1}{n}x' = \frac{1}{n}X\alpha'$. This contradicts our original choice of α to be a minimal-support decomposition of x.

We conclude that $\alpha \in \mathrm{MTF}(C)$ and hence $xa = X\alpha \in Z$. Per assumption x is minimal, so there is no other minimal flow $x'' \in \mathrm{MTF}(\mathcal{P}_1 \mid_p \mathcal{P}_2) \supset Z$ (the inclusion is By Lemma 1) with $\sup(x'') \subsetneq \sup(x)$. Hence $x = \frac{xa}{a} \in \min(Z) = \mathrm{MTF}^{\mathrm{Par}}(X_1, X_2, W^s)$.

A.2   Modular P-Flows

Proof (Lemma 7). Take any $z \in \mathrm{MPF}(\mathcal{P}_1 \mid_p \mathcal{P}_2)$. By Lemma 5 there are restrictions $x \in \mathrm{PF}(\mathcal{P}_1)$ and $y \in \mathrm{PF}(\mathcal{P}_2)$ of z s.t. $z = x \frown y$. **Claim:** there are $(\alpha\beta) \in \mathrm{MPF}(C)$ and $d \in \mathbb{N}$ such that

$$dx = \alpha X \text{ and } dy = \beta Y$$

Then $dz = dx \frown dy = \alpha X \frown \beta Y \in Z$. Per assumption z is minimal so there is no other flow $z' \in \mathrm{PF}(\mathcal{P}_1 \mid_p \mathcal{P}_2) \supset Z$ (Lemma 6) s.t. $\sup(z') \subsetneq \sup(z) = \sup(dz)$. Hence $z = \frac{dz}{d} \in \min(Z) = \mathrm{MPF}(X_1, X_2, W^-)$, so we are done.

Proof of claim. By Theorem 2 there are $a, b \in \mathbb{N}$, $\alpha'' \in \mathbb{N}^{|row(X)|}$ and $\beta'' \in \mathbb{N}^{|row(Y)|}$ s.t.

$$ax = \alpha'' X \quad \text{and} \quad by = \beta'' Y \quad \Leftrightarrow$$
$$abx = \alpha'' b X \quad \text{and} \quad aby = \beta'' a Y$$

There may generally be many such (α'', β''), so pick one which has *minimal decomposition-support* in the sense that there are no other choices with smaller support satisfying the above equations.

Now let $c = \gcd(a, b)$, $d = \frac{ab}{c}$, $\alpha = \alpha'' \frac{b}{c}$ and $\beta = \beta'' \frac{a}{c}$. Continuing with the equations from above we then get

$$dx = \alpha X \quad \text{and} \quad dy = \beta Y$$

We know that x and y are consistent, i.e. $x^s = y^s$ where x^s and y^s are the restrictions of x and y to the shared places $S_{\mathcal{P}_1} \cap S_{\mathcal{P}_2}$. Hence also $dx^s = dy^s$. So

$$\alpha X^s = dx^s = dy^s = \beta Y^s \quad \Leftrightarrow$$
$$\alpha X^s - \beta Y^s = 0 \quad \Leftrightarrow$$
$$(\alpha\beta)C = 0$$

It follows that $(\alpha\beta) \in \mathrm{PF}(C)$. We may assume that $(\alpha\beta)$ is canonical, for if it is not, it is always possible to divide through by $\gcd(\alpha\beta)$ since this always divides d. To se why this is the case, let $c = \gcd(\alpha\beta)$. Then there are α' and β' s.t. $dx = c\alpha'X$ and $dy = c\beta'Y$. Now let $e = \gcd(c, d)$ and write $\frac{d}{e}x = \frac{c}{e}\alpha'X$ and $\frac{d}{e}y = \frac{c}{e}\beta'Y$. Since x and y have entries in $\mathbb{N}$, $\frac{d}{e}$ divides all entries in both $\frac{c}{e}\alpha'X$ and $\frac{c}{e}\beta'Y$. From $e = \gcd(c, d)$ and a standard result from number theory we get that $\frac{d}{e}$ divides all entries in $\alpha'X$ and $\beta'Y$. Therefore $\frac{c}{e}$ divides all entries in both x and y, and hence also in $x \frown y = z$. Canonicity of z then forces $c = e$, so c divides d as claimed.

To see that $(\alpha\beta)$ has minimal support in C, suppose towards a contradiction that there is $(\alpha'\beta') \in \mathrm{PF}(C)$ with $\sup(\alpha'\beta') \subsetneq \sup(\alpha\beta) = \sup(\alpha''\beta'')$. From the definition of C it follows that $x' = \alpha'X$ and $y' = \beta'Y$ are consistent, i.e. $x'^s = \alpha'X^s = \beta'Y^s = y'^s$. They are also place flows of $\mathcal{P}_1$ and $\mathcal{P}_2$ respectively. Lemma 4 then gives that $z' = x' \frown y' \in \mathrm{PF}(\mathcal{P}_1 \mid_p \mathcal{P}_2)$. We know that $\sup(z') \subseteq \sup(z)$, but we cannot have $\sup(z') \subsetneq \sup(c)$ since z is minimal. Hence $\sup(z') = \sup(z)$. By Theorem 3, there is some $n \in \mathbb{N}$ s.t.

$$nz = z' = x' \frown y' = \alpha'X \frown \beta'Y$$

But we also know that $nz = n(x \frown y) = nx \frown ny$. Hence

$$nx = \alpha'X \quad \text{and} \quad ny = \beta'Y$$

Per assumption either $\sup(\alpha') \subsetneq \sup(\alpha'')$ or $\sup(\beta') \subsetneq \sup(\beta'')$. This contradicts our original choice of α'' or β'' to have minimal decomposition-support.

On Inner and Outer Descriptions of the Steady-State Flux Cone of a Metabolic Network

Abdelhalim Larhlimi and Alexander Bockmayr

DFG-Research Center Matheon, FB Mathematik und Informatik, Freie Universität
Berlin, Arnimallee 6, 14195 Berlin, Germany
larhlimi@mi.fu-berlin.de,bockmayr@mi.fu-berlin.de

Abstract. Constraint-based approaches have proved successful in analyzing complex metabolic networks. They restrict the range of all possible behaviors that a metabolic system can display under governing
constraints. The set of all possible flux distributions over a metabolic
network at steady state defines a polyhedral cone, the steady-state flux
cone. This cone can be analyzed using an inner description based on sets
of generating vectors such as elementary flux modes or extreme pathways. Another possibility is the use of an outer description based on sets
of non-negativity constraints. In this paper, we study the relationship
between inner and outer descriptions of the cone. We give a generic procedure to show how inner descriptions can be computed from the outer
one. Then we use this procedure to explain why, for large-scale metabolic
networks, the size of the inner descriptions may be several orders of magnitude larger than that of the outer description.

1 Introduction

The huge amount of biological data that is now available has allowed the reconstruction of an increasing number of genome-scale metabolic networks. However,
this information is not sufficient to determine quantitatively the metabolic phe-
notypes that are expressed by metabolic systems under different environmental
conditions. Intuitive reasoning for predicting and analyzing metabolic pheno-
types can be inadequate, often giving incomplete or incorrect predictions. In
this respect, rigorous mathematical and computational methods are strongly
required to investigate the principles of metabolic behaviors.

Although *kinetic modeling* [1] is most appropriate for fully characterizing
metabolic systems, this approach has been hampered by the lack of detailed kinetic information. Moreover, the results that could be drawn from kinetic models
are strongly sensitive to the definition of both kinetic functions and parameters.
In the view of the limits of kinetic modeling, growing attention is being paid
to *constraint-based modeling* [2,3,4]. Rather than attempting to predict exactly
what a metabolic system does, constraint-based methods narrow the range of all
possible behaviors this system can display under physicochemical constraints.
In the present paper, we are specifically concerned with analyzing metabolic
networks subject to the *stoichiometric* and *thermodynamic constraints*. These

M. Heiner and A.M. Uhrmacher (Eds.): CMSB 2008, LNBI 5307, pp. 308–327, 2008.

constraints define the steady-state flux cone which contains all possible flux distributions in a metabolic network subsisting at steady state.

Since the flux cone in general contains infinitely many possible steady-state flux distributions, it is interesting to find out which of these feasible flux distributions are actually displayed by the metabolic network under consideration. Traditionally, *optimization-based approaches* [5,6,7] have been used to search for single optimal behaviors. These methods, which assume that metabolic networks behave optimally driven by an objective, have proved successful in analyzing complex metabolic networks. However, the results are sensitive to the definition of the optimality criterion, which need not be unique. A recent study has shown that a microorganism could use different optimization criteria depending on the environmental conditions [8]. The exploration of all suitable objective functions is still a difficult task. Furthermore, an optimal solution with respect to a suitable objective function need not be unique. Optimization-based techniques often return an arbitrary chosen flux distribution from the set of all optimal solutions. In analogy with the flux cone, the set of all optimal flux distributions is often an infinite convex set and requires an adequate description. A recursive mixed-integer linear programming algorithm has been developed to find all alternative optima [9]. This algorithm was, however, applied only to small networks. Finally, optimization-based approaches consider only optimal states, which form a restricted subset of all possible behaviors of the living system. An interesting alternative to optimization-based modeling is to describe all possible steady-state flux distributions in the metabolic network using *inner* and *outer descriptions* of the flux cone [10,11,12,13].

The purpose of this work is to study the relationship between inner and outer descriptions of the flux cone. We first characterize the outcome of the network reconfiguration in terms of the outer description of the reconfigured cone. The reconfiguration leads to an increase in the size of the description and changes in the reversibility type of reactions. Then we give a generic procedure to show how inner descriptions can be computed from the outer one. We use this procedure to explain why, for large-scale metabolic networks, the size of the inner descriptions may be several orders of magnitude larger than that of the outer description. The organization of this paper is as follows. We start in Sect. 2 with some basic facts about polyhedral cones. In Sect. 3 we recall the definition of the steady-state flux cone. Sect. 4 and 5 give an overview about inner and outer descriptions of the flux cone. In Sect. 6, we analyze the impact of the network reconfiguration. Finally, in Sect. 7, we give a generic procedure to compute inner descriptions from the outer one.

2   Polyhedral Cones

We start with some basic facts about polyhedral cones (see e.g. [14]). A nonempty subset $C \subseteq \mathbb{R}^n$ is called a *(convex) cone* if $\lambda x + \mu y \in C$, whenever $x, y \in C$ and $\lambda, \mu \geq 0$. A cone C is *polyhedral*, if $C = \{x \in \mathbb{R}^n \mid Ax \geq 0\}$, for some real matrix $A \in \mathbb{R}^{m \times n}$. If this is the case, lin.space$(C) = \{x \in \mathbb{R}^n \mid Ax = 0\}$ is called the *lineality space* of C. A cone C is *finitely generated* if there exist $g^1, \ldots, g^s \in \mathbb{R}^n$ such that $C = \text{cone}\{g^1, \ldots, g^s\} \overset{\text{def}}{=} \{\lambda_1 g^1 + \ldots + \lambda_s g^s \mid \lambda_1, \ldots, \lambda_s \geq 0\}$. A fundamental theorem of Farkas-Minkowski-Weyl (see e.g. [14], p. 87) asserts that a convex cone is polyhedral if and only if it is finitely generated. For the rest of this paper, we will consider only polyhedral cones.

An inequality $a^T x \geq 0, a \in \mathbb{R}^n \setminus \{0\}$, is *valid* for C if $C \subseteq \{x \in \mathbb{R}^n \mid a^T x \geq 0\}$. The set $F = C \cap \{x \in \mathbb{R}^n \mid a^T x = 0\}$ is then called a *face* of C. The *dimension* of F is defined as the dimension of the linear subspace generated by F. Any non-zero element $r \in C$ is called a *ray* of C. Two rays r and r' are equivalent, written $r \cong r'$, if $r = \lambda r'$, for some $\lambda > 0$. A ray r is *extreme* if there do not exist rays $r', r'' \in C, r' \ncong r''$, such that $r = r' + r''$.

Pointed cones. A polyhedral cone C is called *pointed* if lin.space$(C) = \{0\}$. Any pointed cone C has a canonical representation

$$C = \text{cone}\{r^1, \ldots, r^s\}, \tag{1}$$

where $r^1, \ldots, r^s$ are the (distinct) extreme rays of C. This representation, which is used in the inner descriptions of the steady-state flux cone [15,16], is minimal and unique up to multiplication by positive scalars.

Non-pointed cones. If C is not pointed, there is no longer such a unique minimal representation. Let t be the dimension of the lineality space of C. Instead of the extreme rays, we consider now the *minimal proper faces* $G^1, \ldots, G^s$ of C, which are the faces of C of dimension $t + 1$. Each G^i can be represented by a row vector a_i^T and a submatrix A_i' of A, with rank $\begin{pmatrix} A_i' \\ a_i^T \end{pmatrix} = n - t$, such that [14]

$$G^i = \{x \in C \mid a_i^T x \geq 0, A_i' x = 0\}, \tag{2}$$

and

$$\text{lin.space}(C) = \{x \in C \mid a_i^T x = 0, A_i' x = 0\}.$$

If we select for each $i = 1, \ldots, s$ a vector $g^i \in G^i \setminus \text{lin.space}(C)$, and vectors $b^0, \ldots, b^t \in \text{lin.space}(C)$ such that $\text{lin.space}(C) = \text{cone}\{b^0, \ldots, b^t\}$, we get

$$C = \text{cone}\{g^1, \ldots, g^s, b^0, \ldots, b^t\}. \tag{3}$$

For each minimal proper face $G^i, i = 1, \ldots, s$, we get

$$G^i = \text{cone}\{g^i\} + \text{lin.space}(C) = \{\lambda g^i + w \mid \lambda \geq 0, w \in \text{lin.space}(C)\} \tag{4}$$

For additional information, we refer to [14].

(3) generalizes (1), but this representation is no longer unique. We may choose an arbitrary base of lin.space(C), and arbitrary rays g^i in $G^i \setminus \text{lin.space}(C)$. However, it follows from (2) that G^i can also be characterized using constraints $a_i^T x \geq 0$, where a_i^T is a row vector from the matrix A that defines the cone. In the context of metabolic network analysis, this leads to a minimal and unique outer description of the steady-state flux cone, based on sets of non-negativity constraints [10].

3   Steady-State Flux Cone

When modeling a metabolic system, we distinguish between *external* and *internal* metabolites [1]. External metabolites are the sources and sinks of the network. For the internal metabolites, we assume that there is no accumulation or depletion at steady state. In general, the classification of a metabolite as external or internal depends on the purpose of the model. It should be noted that this classification has an impact on the algorithmic complexity of analyzing the network [17].

It is also common to distinguish between *internal* and *boundary* reactions in a metabolic network [18]. An internal reaction has the property that its substrates and products each contain at least one internal metabolite. On the other hand, all substrates consumed or all products formed by a boundary reaction are external. Accordingly, a boundary reaction, which is also called an *exchange reaction*, allows the transport of materials across the system boundary, and thus provides a connection between the metabolic system and its environment.

In the context of metabolic pathway analysis, metabolic systems are assumed to operate at steady state so that the rate of production and the rate of consumption of each internal metabolite must be equal. In addition, the flux through each irreversible reaction must be non-negative. Mathematically, the stoichiometric and thermodynamic constraints that have to hold in a metabolic network at steady state can be expressed as follows [12]:

$$Sv = 0, \ v_i \geq 0, \ \text{for all } i \in \mathit{Irr}, \tag{5}$$

where S is the $m \times n$ stoichiometric matrix, with m internal metabolites (rows) and n reactions (columns), and $v \in \mathbb{R}^n$ is the *flux vector*. $\mathit{Irr} \subseteq \{1, \ldots, n\}$ denotes the set of *irreversible* reactions, and $\mathit{Rev} = \{1, \ldots, n\} \setminus \mathit{Irr}$ is the set of *reversible* reactions.

The set of all solutions of the constraint system (5), which corresponds to the set of all possible flux distributions over the network at steady state, defines a polyhedral cone,

$$C = \{v \in \mathbb{R}^n \mid Sv = 0, \ v_i \geq 0, \ \text{for all } i \in \mathit{Irr}\} \tag{6}$$

which is called the *flux cone* [15,19]. Already in [15], we can find the distinction between inner and outer descriptions of this cone, which are called there internal and external representations. The external representation gives a test for determining whether a given flux vector belongs to the cone, while the internal representation allows one to construct flux vectors from a set of generators.

4   Inner Descriptions of the Flux Cone

An inner description of the flux cone allows for describing the infinite flux cone C (defined in equation (6)) by means of a finite set of generating vectors. A key distinction to be made is whether the flux cone is pointed or not. By definition, the flux cone is pointed if its lineality space

$$\text{lin.space}(C) = \{v \in C \mid v_i = 0, \text{for all } i \in Irr\} \tag{7}$$

is reduced to the origin, i.e., no steady-state flux distribution involves only reversible reactions. In particular, if all reactions are irreversible, i.e., $Irr = \{1, \ldots, n\}$, then $\text{lin.space}(C) = \{0\}$ and so the flux cone is pointed. In this case, the flux cone is generated by a unique (up to multiplication by positive scalars) and minimal set of flux vectors that correspond to its extreme rays.

In the presence of reversible reactions, the situation is more involved. Indeed, if some reactions are reversible, the flux cone may be non-pointed and thus no longer has a unique and minimal representation by its extreme rays. To deal with this situation, some approaches propose to reconfigure the metabolic network in order to render the flux cone pointed [15,16]. For this, one considers a subset $SR \subseteq Rev$ of reversible reactions and splits each reversible reaction $j \in SR$ into a forward and a backward reaction, which both are constrained to be irreversible. Let $s = |SR|$ and $SR = \{j_1, \ldots, j_s\}$. For convenience, the stoichiometric matrix $S' \in \mathbb{R}^{m \times (n+s)}$ of the reconfigured network can be written as follows:

$$S'_{*j} = S_{*j} \qquad \text{for all } j \in \{1, \ldots, n\},$$
$$S'_{*(n+k)} = -S_{*j_k} \quad \text{for all } k \in \{1, \ldots, s\}.$$

The set of irreversible reactions in the reconfigured network is given by

$$Irr' = Irr \cup SR \cup \{n+1, \ldots, n+s\}.$$

The *reconfigured flux cone* C', which contains all possible steady-state flux distributions in the reconfigured network, is defined by

$$C' = \{v \in \mathbb{R}^{n+s} \mid S'v = 0, \ v_i \geq 0, \text{for all } i \in Irr'\}. \tag{8}$$

As a result of this reconfiguration, for a well-chosen set SR of split reactions, the reconfigured flux cone C' is pointed and can be described by a unique and minimal set of extreme rays. This reconfiguration has, however, undesirable consequences. On the one hand, the number of variables and constraints increases by s resp. $2s$. This renders more complex the constraint system that defines the reconfigured flux cone. On the other hand, a significant number of rays in the reconfigured cone are extreme for the only reason that the split reversible reactions have been decomposed into forward and backward reactions. In the initial cone, these extreme rays are conically dependent. Accordingly, the number of extreme rays increases by this reconfiguration, which limits the practical applicability of this strategy.

Three main approaches have been proposed to analyze metabolic networks using inner descriptions of the flux cone [15,16,18]. They all determine flux distributions corresponding to a convex basis of the flux cone, but use a different set of reactions that have to be split [13]. If the latter includes all reversible reactions, the reconfigured flux cone is pointed and generated by its extreme rays called *extremal currents* [15]. If only internal reversible reactions are split, the reconfigured flux cone is again pointed and the extreme rays are termed

Table 1. Inner descriptions of the flux cone, with the set of split reversible reactions SR, the characteristics of the reconfigured flux cone C' and of the three inner descriptions. Contrary to elementary modes, the sets of extreme pathways and extremal currents correspond to the extreme rays of their corresponding reconfigured flux cone and so are minimal. However, possibly many of these generating vectors could be in the interior of the original flux cone.

Inner description ID	Split reactions SR	Characteristics of the reconfigured flux cone C'			Charact. of ID	
		Dimension	Number of constraints	lin.space(C')	Unique	Minimal
Extreme pathways	Rev_{int}	$n + \lvert Rev_{int} \rvert$	$m + \lvert Irr \rvert + 2\lvert Rev_{int} \rvert$	$\{0\}$	$\checkmark$	yes
Extremal currents	Rev	$n + \lvert Rev \rvert$	$m + \lvert Irr \rvert + 2\lvert Rev \rvert$	$\{0\}$	$\checkmark$	yes
Elementary modes	$\emptyset$	n	$m + \lvert Irr \rvert$	lin.space(C)	$\checkmark$	no

extreme pathways [16]. Note that if all boundary reactions are irreversible, both concepts are identical. We should also mention that the extremal current and the extreme pathway approach require a reconfiguration of the network even if the initial cone is pointed. Also in this case, the set of extreme rays of the reconfigured cone is much larger than that of the initial cone.

Schuster and Hilgetag [18,20] have proposed a description of the flux cone without any reconfiguration, using *elementary modes (EMs)*. An elementary mode corresponds to a steady-state flux distribution involving a minimal set of reactions. This concept is similar to that of a *minimal T-invariant* in Petri net theory [21,22] and has also been used for analyzing signaling and regulatory networks [23]. It has been shown that elementary modes span the steady-state flux cone. In other words, each steady-state flux distribution can be expressed as a non-negative linear combination of elementary modes. For a more detailed explanation of the similarities and differences between the three inner descriptions, we refer to [24,25,11,12,13]. Tab. 1 summarizes the characteristics of the different inner descriptions.

Elementary modes are defined in the original n-dimensional flux space. In contrast, to define extreme pathways (resp. extremal currents), the dimension of the flux space is increased by p (resp. q), the number of internal reversible reactions (resp. the number of all reversible reactions). In the following, we formally characterize the relationships between the three inner descriptions.

Let Rev_{int} be the set of reversible internal reactions. Suppose $Rev_{int} = \{j_1, \ldots, j_p\}$ and $Rev = \{j_1, \ldots, j_q\}$. Let $\pi : C \to \mathbb{R}^{n+p}$ (resp. $\phi : C \to \mathbb{R}^{n+q}$) be the function that maps each flux vector $v \in C$ to $v' = \pi(v)$ (resp. $v' = \phi(v)$) such that $v'_j = v_j$ for all $j \in \{1, \ldots, n\} \setminus Rev_{int}$ (resp. $j \in \{1, \ldots, n\} \setminus Rev$), and for each $k \in \{1, \ldots, p\}$ (resp. $k \in \{1, \ldots, q\}$)

$$v'_{j_k} = v_{j_k} \text{ and } v'_{n+k} = 0 \quad \text{if } v_{j_k} \geq 0,$$
$$v'_{j_k} = 0 \quad \text{and } v'_{n+k} = -v_{j_k} \text{ if } v_{j_k} < 0.$$

The function π (resp. ϕ) formally defines the reconfiguration of a flux vector $v \in C$ by splitting each free variable v_{j_k} with $j_k \in Rev_{int}$ (resp. $j_k \in Rev$) into two non-negative variables v'_{j_k} and v'_{n+k} with $v_{j_k} = v'_{j_k} - v'_{n+k}$. This operation is similar to standard form transformation in linear programming. To define the reverse operation, let $\pi^r : \mathbb{R}^{n+p} \to C$ (resp. $\phi^r : \mathbb{R}^{n+q} \to C$) be the function that maps each vector $v' \in \mathbb{R}^{n+p}$ (resp. $v' \in \mathbb{R}^{n+q}$) to $v = \pi^r(v')$ (resp. $v = \phi^r(v')$) such that $v_j = v'_j$ for all $j \in \{1, \ldots, n\} \setminus Rev_{int}$ (resp. $j \in \{1, \ldots, n\} \setminus Rev$) and $v_{j_k} = v'_{j_k} - v'_{n+k}$ for all $k \in \{1, \ldots, p\}$ (resp. $k \in \{1, \ldots, q\}$).

Finally, let $\Pi \subseteq \mathbb{R}^{n+p}$ and $\Phi \subseteq \mathbb{R}^{n+q}$ be the sets of 2-cycles corresponding to the split reversible reactions, i.e.,

$$\Pi = \{x \in \mathbb{R}^{n+p} \mid x_j = 0 \text{ for all } j \in \{1, \ldots, n+p\} \setminus \{j_k, n+k\}$$
$$\text{and } x_{j_k} = x_{n+k} = 1, \text{ for some } j_k \in Rev_{int}\},$$

$$\Phi = \{x \in \mathbb{R}^{n+q} \mid x_j = 0 \text{ for all } j \in \{1, \ldots, n+q\} \setminus \{j_k, n+k\}$$
$$\text{and } x_{j_k} = x_{n+k} = 1, \text{ for some } j_k \in Rev\}.$$

The following proposition reformulates the relationship between extreme pathways and elementary modes given in [24]. Except the 2-cycles corresponding to the split reactions, each extreme pathway completely defines a unique elementary mode.

Proposition 1 ([24]). *If $x \notin \Pi$ is an extreme pathway, then there exists a unique elementary mode $e \in C$ such that $x = \pi(e)$ and $e = \pi^r(x)$.*

According to the proposition above, the set of extreme pathways corresponds to a subset of elementary modes. Next, we restate the equivalence between elementary modes and extremal currents given in [26].

Proposition 2 ([26]). *Let $e \in C$ be a steady-state flux distribution. The following are equivalent:*

- *e is an elementary mode.*
- *There exists a unique extremal current $x \notin \Phi$ such that $x = \phi(e)$ and $e = \phi^r(x)$.*

It follows that up to the 2-cycles corresponding to the split reactions, extremal currents and elementary modes are equivalent. Accordingly, an algorithm for computing extremal currents could also be used to calculate elementary modes and vice versa.

Thus all three approaches are concerned with describing a pointed reconfigured flux cone C' by means of its extreme rays. There may exist many generating vectors of the reconfigured flux cone C' lying in the interior of the original flux cone C. This observation is important because the number of these generators may be very large, making a complete analysis of the whole metabolic network impossible and limiting the practical applicability of these methods.

5 Outer Description of the Flux Cone

In [10,27,28], the authors proposed an outer description of the flux cone, based on sets of non-negativity constraints. This approach defines a *metabolic behavior* as a set of *irreversible* reactions $D \subseteq Irr, D \neq \emptyset$, such that there exists a flux distribution $v \in C$ with $D = \{i \in Irr \mid v_i \neq 0\}$. A metabolic behavior D is *minimal (MMB)*, if there is no metabolic behavior $D' \subsetneq D$ strictly contained in D.

The set of flux distributions involving only reversible reactions defines the *reversible metabolic space (RMS)*,

$$RMS = \text{lin.space}(C), \tag{9}$$

which corresponds to the lineality space of the flux cone C.

The minimal metabolic behaviors (MMBs) are closely related to the *minimal proper faces* of the flux cone C, i.e., the faces of dimension $t + 1$, where $t = \dim(RMS)$. According to [10], each minimal proper face is described by its *characteristic set* $D = \{j \in Irr \mid v_j > 0, \text{ for some } v \in G\}$. Indeed, G is given by

$$G = \{v \in C \mid v_i = 0, \text{ for all } i \in Irr \setminus D\}. \tag{10}$$

The next theorem shows that each MMB completely defines a corresponding minimal proper face of the flux cone C and vice versa.

Theorem 1 ([10]). *Let $D \subseteq Irr$ be a set of irreversible reactions. Then, the following are equivalent:*

- *D is a minimal metabolic behavior.*
- *D is the characteristic set of a minimal proper face of the flux cone.*

By the theorem above, the MMBs are in a 1-1 correspondence with the minimal proper faces of the flux cone. Accordingly, the set of MMBs is minimal in the sense that no strict subset of MMBs could completely describe the flux cone. We conclude that there are two minimality properties that hold for minimal metabolic behaviors: the minimality of each MMB and the minimality of the set of MMBs. Furthermore, for each MMB D, there exists at least one EM e involving exactly the irreversible reactions from D, i.e., $D = \{i \in Irr \mid e_i \neq 0\}$.

If $G^1, \ldots, G^s$ are the minimal proper faces of the flux cone C, the corresponding MMBs $D^1, \ldots, D^s$ together with the RMS completely describe C, see [10] for additional details. Note that finding such a minimal and unique outer description is different from eliminating the stoichiometric and thermodynamic constraints that are redundant in the flux cone definition.

Based on the concepts of MMBs and the RMS, a refined classification of reactions has been proposed [10]. A reversible reaction $j \in Rev$ is called *pseudo-irreversible* if $v_j = 0$, for all $v \in RMS$. A reversible reaction that is not pseudo-irreversible is called *fully reversible*. In the following, $Prev_0$ and $Frev$ denote the sets of pseudo-irreversible and fully reversible reactions, respectively.

Inside each minimal proper face, the (pseudo-) irreversible reactions take a unique direction. More precisely, we have the following properties.

Theorem 2 ([10]). *Let G be a minimal proper face of the flux cone C and let $j \in \{1, \ldots, n\}$ be a reaction.*

- *If $j \in Irr$ is irreversible, then $v_j > 0$, for all $v \in G \setminus \text{lin.space}(C)$, or $v_j = 0$, for all $v \in G$. Furthermore, $v_j = 0$, for all $v \in \text{lin.space}(C)$.*

- If $j \in Prev_0$ is pseudo-irreversible, then the flux v_j through j has a unique sign in $G \setminus \text{lin.space}(C)$, i.e., either $v_j > 0$, for all $v \in G \setminus \text{lin.space}(C)$, or $v_j = 0$, for all $v \in G \setminus \text{lin.space}(C)$, or $v_j < 0$, for all $v \in G \setminus \text{lin.space}(C)$. For all $v \in \text{lin.space}(C)$, we have again $v_j = 0$.
- If $j \in Frev$ is fully reversible, there exists $v \in \text{lin.space}(C)$ such that $v_j \neq 0$. We can then find pathways $v^+, v^-, v^0 \in G \setminus \text{lin.space}(C)$ with $v_j^+ > 0$, $v_j^- < 0$ and $v_j^0 = 0$.

6   Outer Description of the Reconfigured Flux Cone

Now we analyze the impact of reconfiguring the network. The effects include an increase in the size of the outer description of the reconfigured cone and changes in the reversibility type of reactions. Here, we define the *size* of an outer description of a flux cone as the sum of the number of its minimal proper faces and the dimension of its lineality space.

Let SR $\subseteq$ *Rev* be the set of split reactions. The network reconfiguration can be seen as an iterative procedure that consists of |SR| iterations, each splitting some reversible reaction. As will be shown, each iteration increases the description of the flux cone depending on the reversibility type of the split reaction. The increase is significant when the split reaction is pseudo-irreversible. Note that there are at most t iterations where the split reaction can be fully reversible, with $t = \dim(\text{lin.space}(C))$.

In this section, we consider the case of splitting one reaction, which is denoted by j. The reconfigured flux cone C', which contains all possible steady-state flux distributions in the reconfigured network, is given by

$$C' = \{(v, w) \in \mathbb{R}^{n+1} \mid Sv = w \cdot S_{*j}, \ v_i \geq 0, \ \text{for } i \in Irr, \ v_j \geq 0, \ w \geq 0\}. \quad (11)$$

According to equation (11), splitting reaction j increases the number of variables and constraints by 1 and 2, respectively. Indeed, the reconfigured network contains one more reaction denoted by $n + 1$. The set of irreversible reactions in the reconfigured network is $Irr' = Irr \cup \{j, n + 1\}$. Accordingly, the lineality space of the reconfigured flux cone C' is given by

$$\text{lin.space}(C') = \{(v, 0) \in \mathbb{R}^{n+1} \mid Sv = 0, \ v_i = 0, \ \text{for all } i \in Irr, \ v_j = 0\},$$

or equivalently,

$$\text{lin.space}(C') = \{(v, 0) \in \mathbb{R}^{n+1} \mid v \in \text{lin.space}(C \cap \{v \in \mathbb{R}^n \mid v_j = 0\})\}. \quad (12)$$

Since $\text{lin.space}(C \cap \{v \in \mathbb{R}^n \mid v_j = 0\}) = \text{lin.space}(C) \cap \{v \in \mathbb{R}^n \mid v_j = 0\}$, it follows from equation (12) that

$$\dim(\text{lin.space}(C')) = \dim(\text{lin.space}(C) \cap \{v \in \mathbb{R}^n \mid v_j = 0\}).$$

Lemma 3. *If* $j \in Prev_0$ *is pseudo-irreversible, then* $\dim(\text{lin.space}(C')) = \dim(\text{lin.space}(C))$. *If* $j \in Frev$ *is fully reversible, then* $\dim(\text{lin.space}(C')) = \dim(\text{lin.space}(C)) - 1$.

Proof. Suppose $j \in Prev_0$ is pseudo-irreversible. Then, $b_j = 0$ for each vector $b \in$ lin.space(C). Hence, lin.space$(C) \subseteq \{v \in \mathbb{R}^n \mid v_j = 0\}$ and so dim(lin.space(C')) $=$ dim(lin.space(C)). Now suppose $j \in Frev$ is fully reversible. Then there exists $b \in$ lin.space(C) such that $b_j \neq 0$ and so lin.space$(C) \not\subseteq \{v \in \mathbb{R}^n \mid v_j = 0\}$. Therefore, dim(lin.space(C')) $=$ dim(lin.space(C)) $- 1$. $\qquad\square$

In the following, we will characterize the minimal proper faces of the reconfigured flux cone C'. We first consider the case of a minimal proper face G' with $v'_j = v'_{n+1} = 0$ for all $v' \in G'$.

Lemma 4. *Let $G' \subseteq C'$ such that $v'_j = v'_{n+1} = 0$ for all $v' \in G'$. Then the following are equivalent:*

- *G' is a minimal proper face of C'.*
- *There exists a minimal proper face G of $C \cap \{v \in \mathbb{R}^n \mid v_j = 0\}$ such that $G' = \{(v, 0) \in \mathbb{R}^{n+1} \mid v \in G\}$.*

If this is the case, G and G' have the same characteristic set.

Proof. "$\Rightarrow$": Suppose G' is a minimal proper face of C' and D' is its characteristic set. Since $v'_j = v'_{n+1} = 0$ for all $v' \in G'$, we get $D' \subseteq Irr$. Let $G = \{v \in C \mid v_j = 0, v_i = 0,$ for all $i \in Irr \setminus D'\}$. We have $G' = \{(v, 0) \in \mathbb{R}^{n+1} \mid v \in G\}$ and so dim$(G) =$ dim(G'). Since G' is a minimal proper face of C' and dim(lin.space(C')) $=$ dim(lin.space$(C \cap \{v \in \mathbb{R}^n \mid v_j = 0\})$), we get dim$(G) =$ dim(lin.space$(C \cap \{v \in \mathbb{R}^n \mid v_j = 0\})$) $+ 1$ and the claim follows.
"$\Leftarrow$": Immediate. $\qquad\square$

We now will study the minimal proper faces of the reconfigured flux cone C', depending on the reversibility type of the split reaction.

6.1   Splitting a Fully Reversible Reaction

If $j \in Frev$ is fully reversible, there exists a flux distribution in the reconfigured network that involves either reaction j or $n+1$ and no other irreversible reactions. Accordingly, as will be stated in the following proposition, reactions j and $n+1$ define two trivial minimal proper faces of C' given by

$$G^j = \{(v, 0) \in \mathbb{R}^{n+1} \mid Sv = 0, v_i = 0, \text{ for all } i \in Irr, v_j \geq 0\},$$
$$G^{n+1} = \{(v, w) \in \mathbb{R}^{n+1} \mid Sv = w \cdot S_{*j}, v_i = 0, \text{ for all } i \in Irr, v_j = 0, w \geq 0\}.$$

Proposition 5. *If $j \in Frev$ is fully reversible, then G^j and G^{n+1} are two minimal proper faces of C' whose characteristic sets are $D^j = \{j\}$ and $D^{n+1} = \{n+1\}$, respectively.*

Proof. Suppose $j \in Frev$ is fully reversible. Then there exists $b \in$ lin.space(C) such that $b_j > 0$. Let $I_j = Irr \cup \{n+1\}$. We have $G^j = \{v' \in C' \mid v'_j \geq 0, v'_i = 0,$ for all $i \in I_j\}$ and lin.space$(C') = \{v' \in C' \mid v'_j = 0, v'_i = 0,$ for all $i \in I_j\}$. In addition, we have $(b, 0) \in G^j \setminus$ lin.space(C') and so $\dim(G^j) = \dim(\text{lin.space}(C')) + 1$. Therefore, G^j is a minimal proper face characterized by reaction j. Similarly, let $I_{n+1} = Irr \cup \{j\}$. We have $G^{n+1} = \{v' \in C' \mid v'_{n+1} \geq 0, v'_i = 0, \text{ for all } i \in I_{n+1}\}$ and $\text{lin.space}(C') = \{v' \in C' \mid v'_{n+1} = 0, v'_i = 0, \text{ for all } i \in I_{n+1}\}$. Define $u \in \mathbb{R}^{n+1}$ by $u_j = 0$, $u_{n+1} = b_j$ and $u_i = -b_i$ for all $i \in \{1, \ldots, n\} \setminus \{j\}$. We have $u \in G^{n+1} \setminus \text{lin.space}(C')$ and so $\dim(G^{n+1}) = \dim(\text{lin.space}(C')) + 1$. Accordingly, G^{n+1} is a minimal proper face characterized by reaction $n + 1$. Since for each $v' \in G^j$ (resp. $v' \in G^{n+1}$), $v'_i = 0$ for all $i \in Irr' \setminus \{j\}$ (resp. $i \in Irr' \setminus \{n+1\}$), the claim follows.   □

Next we are interested in non-trivial minimal proper faces of C'. Here, we get the following result.

Proposition 6. Let $G' \subseteq C'$ such that $G' \neq G^j$ and $G' \neq G^{n+1}$. If $j \in Frev$ is fully reversible, then the following are equivalent:

- G' is a minimal proper face of C'.
- There exists a minimal proper face G of C such that $G' = \{(v, 0) \in \mathbb{R}^{n+1} \mid v \in G \cap \{v \in \mathbb{R}^n \mid v_j = 0\}\}$.

Proof. "$\Rightarrow$": According to Lemma 4, there exists a minimal proper face G'' of $C \cap \{v \in \mathbb{R}^n \mid v_j = 0\}$ such that $G' = \{(v, 0) \in \mathbb{R}^{n+1} \mid v \in G''\}$. Let D be the characteristic set of G'' and let $G = \{v \in C \mid v_i = 0, \text{ for all } i \in Irr \setminus D\}$. We have $G'' = G \cap \{v \in \mathbb{R}^n \mid v_j = 0\}$. Let $G^0 \subseteq G$ be a minimal proper face of C and $D^0 \subseteq D$ its characteristic set. Since $j \in Frev$, there exists $g \in G^0 \setminus \text{lin.space}(C)$ such that $g_j = 0$. Therefore, $g \in G^0 \cap \{v \in \mathbb{R}^n \mid v_j = 0\} \setminus \text{lin.space}(C)$. Suppose there exists $k \in D \setminus D^0$. Then $v_k = 0$ for all $v \in G^0$ and $G^0 \subseteq G \cap \{v \in \mathbb{R}^n \mid v_k = 0\}$. Since G'' is a minimal proper face of $C \cap \{v \in \mathbb{R}^n \mid v_j = 0\}$, we have $G'' \cap \{v \in \mathbb{R}^n \mid v_k = 0\} = \text{lin.space}(C \cap \{v \in \mathbb{R}^n \mid v_j = 0\})$. It follows that $G^0 \cap \{v \in \mathbb{R}^n \mid v_j = 0\} \subseteq \text{lin.space}(C \cap \{v \in \mathbb{R}^n \mid v_j = 0\})$, in contradiction to $g \in G^0 \cap \{v \in \mathbb{R}^n \mid v_j = 0\} \setminus \text{lin.space}(C)$. We conclude that $D^0 = D$ and so G is a minimal proper face of C.
    "$\Leftarrow$": Immediate.   □

In summary, if reaction $j \in Frev$ is fully reversible, the minimal proper faces of C' are G^j, G^{n+1} and those which are in a 1-1 correspondence with the minimal proper faces of C. The dimension of the lineality space of C' decreases by one. Accordingly, the size of the flux cone description increases by one after splitting a fully reversible reaction.

6.2   Splitting a Pseudo-irreversible Reaction

If $j \in Prev_0$ is pseudo-irreversible, there is no flux distribution in the reconfigured network that involves reaction j (resp. $n+1$) and no other irreversible reactions. The following proposition shows that both (and only) reactions j and $n + 1$ characterize a trivial minimal proper face of C' given by

$$G^c = \{(v, w) \in \mathbb{R}^{n+1} \mid Sv = w \cdot S_{*j}, v_i = 0, \text{ for all } i \in Irr, v_j \geq 0, w \geq 0\}.$$

The minimal proper face G^c contains all the (2-cycle) flux distributions in the reconfigured network that involve only the forward and backward reactions j and $n + 1$.

Proposition 7. *If $j \in Prev_0$ is pseudo-irreversible, then G^c is a minimal proper face of C' whose characteristic set is $D^c = \{j, n + 1\}$.*

Proof. We have $G^c = \{v' \in C' \mid v'_j \geq 0, v'_i = 0, \text{ for all } i \in Irr\}$ and $\text{lin.space}(C') = \{v' \in C' \mid v'_j = 0, v'_i = 0, \text{ for all } i \in Irr\}$. Let $u \in \mathbb{R}^{n+1}$ with $u_j = u_{n+1} = 1$ and $u_i = 0$ for all $i \in \{1, \ldots, n\} \setminus \{j\}$. We have $u \in G^c \setminus \text{lin.space}(C')$ and so $\dim(G^c) = \dim(\text{lin.space}(C')) + 1$. Therefore, G^c is a minimal proper face characterized by reaction j. Since $u_{n+1} \neq 0$ and $u_i = 0$ for all $i \in Irr' \setminus \{j, n + 1\}$, $D^c = \{j, n + 1\}$ is the characteristic set of G^c. $\qquad\square$

Let $G^1, \ldots, G^s$ be the minimal proper faces of C and $D^1, \ldots, D^s$ their characteristic sets, respectively. Starting from [29,10] and using that reaction j is pseudo-irreversible, we partition the set $J = \{G^1, \ldots, G^s\}$ of minimal proper faces of C into three parts:

$$J^0 = \{G \in J \mid v_j = 0 \text{ for all } v \in G\},$$
$$J^+ = \{G \in J \mid v_j > 0 \text{ for all } v \in G \setminus \text{lin.space}(C)\},$$
$$J^- = \{G \in J \mid v_j < 0 \text{ for all } v \in G \setminus \text{lin.space}(C)\}.$$

From each of the sets J^0, J^+, J^- we will obtain different minimal proper faces of C'. We start by characterizing minimal proper faces G' with $v'_j = v'_{n+1} = 0$ for all $v' \in G'$. As will be stated in the next proposition, in addition to minimal proper faces $G \in J^0$, some minimal proper faces of C' are obtained by combining pairs $(G^k, G^l) \in J^+ \times J^-$ of adjacent minimal proper faces in C. The set Φ of these pairs is given by

$$\Phi = \{(G^k, G^l) \in J^+ \times J^- \mid D^i \not\subseteq D^k \cup D^l \text{ for all } i \in \{1, \ldots, s\} \setminus \{k, l\}\}.$$

Each pair $(G^k, G^l) \in \Phi$ defines a minimal proper face of C'

$$\zeta(G^k, G^l) = \{v \in C \mid v_j = 0, v_i = 0, \text{ for all } i \in Irr \setminus D^k \cup D^l\}.$$

Finally, let Adj be the set given by

$$Adj = \{\zeta(G^k, G^l) \mid (G^k, G^l) \in \Phi\}. \tag{13}$$

Proposition 8. *Let $G' \subseteq C'$ such that $v'_j = v'_{n+1} = 0$ for all $v' \in G'$. If $j \in Prev_0$ is pseudo-irreversible, then the following are equivalent:*

- *G' is a minimal proper face of C'.*
- *There exists $G \in J^0 \cup Adj$ such that $G' = \{(v, 0) \in \mathbb{R}^{n+1} \mid v \in G\}$.*

Proof. According to Lemma 4, G' is a minimal proper face of C' if and only if there is a minimal proper face G of $C \cap \{v \in \mathbb{R}^n \mid v_j = 0\}$ such that $G' = \{(v, 0) \in \mathbb{R}^{n+1} \mid v \in G\}$. We show that G is a minimal proper face of $C \cap \{v \in \mathbb{R}^n \mid v_j = 0\}$

if and only if $G \in J^0 \cup Adj$. Since $j \in Prev_0$, we have lin.space$(C \cap \{v \in \mathbb{R}^n \mid v_j = 0\}) = $ lin.space(C).

"$\Rightarrow$": Let $G = \{v \in C \cap \{v \in \mathbb{R}^n \mid v_j = 0\} \mid v_i = 0$, for all $i \in Irr \setminus D\}$. There exists $g \in G \setminus $ lin.space(C) such that $D = \{i \in Irr \mid g_i \neq 0\}$. Let $g^i \in G^i \setminus $ lin.space(C) for $i = 1, \ldots, s$. Since $g \in C$, g can be written in the form $g = \sum_{i=1}^s \alpha_i g^i + b$, for some $\alpha_i \geq 0$ and $b \in $ lin.space(C). Since $g \notin $ lin.space(C), there exists $k \in \{1, \ldots, s\}$ such that $\alpha_k \neq 0$. Accordingly, $D^k \subseteq D$. We have the following cases:

1. $G^k \in J^0$: Since $D^k \subseteq D$ and $G^k \subseteq C \cap \{v \in \mathbb{R}^n \mid v_j = 0\}$, we get $G^k \subseteq G$. Since G is a minimal proper face of $C \cap \{v \in \mathbb{R}^n \mid v_j = 0\}$ and lin.space$(C \cap \{v \in \mathbb{R}^n \mid v_j = 0\}) \subsetneq G^k$, we get $G^k = G$ and $G \in J^0$.

2. $G^k \in J^+$: Suppose $\alpha_i \neq 0$ implies $G^i \in J^+$ for all $i = 1, \ldots, s$. We get $g_j = \alpha_k g_j^k + \sum_{i \neq k} \alpha_i g_j^i > 0$, contradicting $g_j = 0$. Then there exists $l \in \{1, \ldots, s\}$ such that $\alpha_l \neq 0$ and $G^l \in J^-$. It follows that $D^l \subseteq D$ and $D^k \cup D^l \subseteq D$. Let $g' = g^l - (g_j^l/g_j^k) \cdot g^k$ and $G' = \{v \in C \mid v_j = 0, \ v_i = 0$, for all $i \in Irr \setminus (D^k \cup D^l)\}$. We have $g' \in G' \setminus $ lin.space$(C \cap \{v \in \mathbb{R}^n \mid v_j = 0\})$ and $G' \subseteq G$. Since G is a minimal proper face of $C \cap \{v \in \mathbb{R}^n \mid v_j = 0\}$, we get $G' = G$ and $D = D^k \cup D^l$. Suppose there exists $i \in \{1, \ldots, s\}$ such that $D^i \subseteq D^k \cup D^l$. If $G^i \in J^+$ (resp. $G^i \in J^-$), we prove in a similar way that $D^i \cup D^l = D^k \cup D^l$ (resp. $D^k \cup D^i = D^k \cup D^l$) and so $D^i = D^k$ (resp. $D^i = D^l$). It follows that $(G^k, G^l) \in \Phi$, $G = \zeta(G^k, G^l)$ and $G \in Adj$.

3. $G^k \in J^-$: The proof is similar to that of the case above.

"$\Leftarrow$": We can easily see that if $G \in J^0$, then G is a minimal proper face of $C \cap \{v \in \mathbb{R}^n \mid v_j = 0\}$. Suppose $G = \zeta(G^k, G^l)$ for some $(G^k, G^l) \in \Phi$. Let $G' \subseteq G$ be a minimal proper face of $C \cap \{v \in \mathbb{R}^n \mid v_j = 0\}$ and $D' \subseteq D^k \cup D^l$ its characteristic set. Accordingly, $G' \in J^0 \cup Adj$. Suppose $G' \in J^0$. It follows from $D' \subseteq D^k \cup D^l$ and $(G^k, G^l) \in \Phi$ that $D' = D^k$ or $D' = D^l$, contradicting $v_j = 0$ for all $v \in G'$. We conclude that $G' = \zeta(G^{k'}, G^{l'})$ for some $(G^{k'}, G^{l'}) \in \Phi$ and $D' = D^{k'} \cup D^{l'}$. Since $D' \subseteq D^k \cup D^l$, we get $D^{k'} \subseteq D^k \cup D^l$ and $D^{l'} \subseteq D^k \cup D^l$. Therefore, $D^{k'} = D^k$ and $D^{l'} = D^l$. We get $G' = G$ and so G is a minimal proper face of $C \cap \{v \in \mathbb{R}^n \mid v_j = 0\}$. $\square$

Next, we characterize non-trivial minimal proper faces G' of C' with $v_j' > 0$ for all $v' \in G' \setminus $ lin.space(C').

Proposition 9. Let $G' \subseteq C'$ such that $G' \neq G^c$. If $j \in Prev_0$ is pseudo-irreversible, then the following are equivalent:

- G' is a minimal proper face of C' such that $v_j' > 0$ for all $v' \in G' \setminus $ lin.space(C').
- There exists $G \in J^+$ such that $G' = \{(v, 0) \in \mathbb{R}^{n+1} \mid v \in G\}$.

Proof. Suppose $j \in Prev_0$. Then, dim(lin.space(C')) = dim(lin.space(C)).

"$\Rightarrow$": Suppose G' is a minimal proper face of C' such that $v_j' > 0$ for all $v' \in G' \setminus $ lin.space(C') and let D' be its characteristic set. Since $G' \neq G^c$ and

$j \in D'$, we have $n + 1 \notin D'$ and $D' \setminus \{j\} \subseteq Irr$. Let $(g, 0) \in G' \setminus \text{lin.space}(C')$, $D = D' \setminus \{j\}$ and $G = \{v \in C \mid v_i = 0, \text{ for all } i \in Irr \setminus D\}$. We have $g \in G \setminus \text{lin.space}(C)$ and $g_j > 0$. Suppose there exists $v \in G \setminus \text{lin.space}(C)$ such that $v_j \leq 0$ and let $w = v - (v_j/g_j) \cdot g$. We have $(w, 0) \in G' \setminus \text{lin.space}(C')$ and $w_j = 0$, in contradiction to $v'_j > 0$ for all $v' \in G' \setminus \text{lin.space}(C')$. We conclude that $v_j > 0$ for all $v \in G \setminus \text{lin.space}(C)$ and $G' = \{(v, 0) \in \mathbb{R}^{n+1} \mid v \in G\}$. Accordingly, $\dim(G) = \dim(G')$. Since G' is a minimal proper face of C', we have $\dim(G') = \dim(\text{lin.space}(C \cap \{v \in \mathbb{R}^n \mid v_j = 0\})) + 1 = \dim(\text{lin.space}(C)) + 1$ and so $G \in J^+$.

"$\Leftarrow$": Let $G \in J^+$ such that $G' = \{(v, 0) \in \mathbb{R}^{n+1} \mid v \in G\}$. Since $\dim(G') = \dim(G)$ and $\dim(G) = \dim(\text{lin.space}(C)) + 1 = \dim(\text{lin.space}(C \cap \{v \in \mathbb{R}^n \mid v_j = 0\})) + 1$, we conclude that G' is a minimal proper face of C'. Since $v_j > 0$ for all $v \in G \setminus \text{lin.space}(C)$, it follows that $v'_j > 0$ for all $v' \in G' \setminus \text{lin.space}(C')$. $\square$

Finally, we characterize non-trivial minimal proper faces $G' \neq G^c$ of C' with $v'_{n+1} > 0$ for all $v' \in G' \setminus \text{lin.space}(C')$. In such a case, $v'_j = 0$ for all $v' \in G'$ and the characteristic set of G' is $D \cup \{n + 1\}$ for some $D \subseteq Irr$.

Proposition 10. *Let $D \subseteq Irr$ be a set of irreversible reactions. If $j \in Prev_0$ is pseudo-irreversible, then the following are equivalent:*

- *There exists a minimal proper face G' of C' whose characteristic set is $D \cup \{n + 1\}$.*
- *There exists $G \in J^-$ whose characteristic set is D.*

Proof. Suppose $j \in Prev_0$. Then, $\dim(\text{lin.space}(C')) = \dim(\text{lin.space}(C))$.

"$\Rightarrow$": Suppose G' is a minimal proper face of C' whose characteristic set is $D \cup \{n + 1\}$. Let $g' \in G' \setminus \text{lin.space}(C')$ and $g \in \mathbb{R}^n$ such that $g_i = g'_i$ for all $i \in \{1, \ldots, n\} \setminus \{j\}$ and $g_j = -g'_{n+1}$. Let $G = \{v \in C \mid v_i = 0, \text{ for all } i \in Irr \setminus D\}$. We have $g \in G \setminus \text{lin.space}(C)$ and $g_j < 0$. Suppose there exists $v \in G \setminus \text{lin.space}(C)$ such that $v_j \geq 0$ and let $w = v - (v_j/g_j) \cdot g$. We have $(w, 0) \in G' \setminus \text{lin.space}(C')$, in contradiction to $v'_{n+1} > 0$ for all $v' \in G' \setminus \text{lin.space}(C')$. We conclude that $v_j < 0$ for all $v \in G \setminus \text{lin.space}(C)$. To show $G \in J^-$, let $F \subseteq G$ be a minimal proper face of C and $D' \subseteq D$ its characteristic set. Let $f \in F \setminus \text{lin.space}(C)$ and $f' \in \mathbb{R}^{n+1}$ with $f'_i = f_i$ for all $i \in \{1, \ldots, n\} \setminus \{j\}$, $f'_j = 0$ and $f'_{n+1} = -f_j > 0$. Since $f' \in C'$ and $\{i \in Irr' \mid f'_i > 0\} = D \cup \{n + 1\}$, we have $f' \in G' \setminus \text{lin.space}(C')$. Suppose there exists $k \in D \setminus D'$. Then $v_k = 0$ for all $v \in F$ and $F \subseteq G \cap \{v \in \mathbb{R}^n \mid v_k = 0\}$. Accordingly, $f \in G \cap \{v \in \mathbb{R}^n \mid v_k = 0\} \setminus \text{lin.space}(C)$ and $f' \in G' \cap \{v \in \mathbb{R}^n \mid v_k = 0\} \setminus \text{lin.space}(C')$. Since G' is a minimal proper face of C' and $k \in D$, $G' \cap \{v \in \mathbb{R}^n \mid v_k = 0\} = \text{lin.space}(C')$, contradicting $f' \in G' \cap \{v \in \mathbb{R}^n \mid v_k = 0\} \setminus \text{lin.space}(C')$. We conclude that $D' = D$, $F = G$ and so the claim follows.

"$\Leftarrow$": Let $G \in J^-$ such that D is its characteristic set. Let $G' = \{(v, w) \in \mathbb{R}^{n+1} \mid Sv = w \cdot S_{*j}, v_i = 0, \text{ for all } i \in Irr \setminus D, v_i \geq 0, \text{ for all } i \in D, v_j = 0, w \geq 0\}$. Let $F' \subseteq G'$ be a minimal proper face of C' and $D' \subseteq D \cup \{n + 1\}$ its characteristic set. Suppose $n + 1 \notin D'$. Since $j \notin D'$, by Proposition 8, there exists $F \in J^0 \cup Adj$ such that $F' = \{(v, 0) \in \mathbb{R}^{n+1} \mid v \in F\}$. The characteristic set of F is D'. Then either $D' = D^i$ with $G^i \in J^0$ or $D' = D^k \cup D^l$ with $(G^l, G^k) \in \Phi$. Since $D' \subseteq D$, both cases are contradicting $G \in J^-$. We conclude that $n + 1 \in D'$. Since $j \notin D'$, $F' \neq G^c$ and its characteristic set is $(D' \setminus \{n + 1\}) \cup \{n + 1\}$. There exists then $K \in J^-$ whose characteristic set is $D' \setminus \{n + 1\}$. Since $D' \setminus \{n + 1\} \subseteq D$ and both G and K are minimal proper faces of C, it follows that $K = G$, $D' = D \cup \{n + 1\}$ and $F' = G'$. We conclude that G' is a minimal proper face of C' whose characteristic set is $D \cup \{n + 1\}$. $\square$

To summarize, a non-trivial minimal proper face G' of C' is given either by

$$G' = \{(v, 0) \in \mathbb{R}^{n+1} \mid v \in G\}, \text{ for some } G \in J^0 \cup J^+ \cup Adj,$$

or by

$$G' = \{v' \in C' \mid v_i' = 0 \text{ for all } i \in (Irr \cup \{j\}) \setminus D^k\}, \text{ for some } G^k \in J^-.$$

Since $\dim(\text{lin.space}(C')) = \dim(\text{lin.space}(C))$, it follows that the size of the flux cone description increases by $|Adj| + 1$ after splitting a pseudo-irreversible reaction. Note that the set Adj can be quite large (cf. Sect. 7).

6.3   Changes in the Reversibility Type of Reactions

Another consequence of the network reconfiguration is the change in the reversibility type of reactions. Indeed, possibly many fully reversible reactions in the original network may become pseudo-irreversible in the reconfigured network. Let $Frev'$ and $Prev_0'$ be the sets of fully and pseudo-irreversible reversible reactions in the reconfigured network, respectively, i.e.,

$$Frev' = \{i \in Rev \setminus \{j\} \mid b_i' \neq 0, \text{ for some } b' \in \text{lin.space}(C')\},$$
$$Prev_0' = Rev \setminus (Frev_0' \cup \{j\}).$$

Since $\text{lin.space}(C') = \{(v, 0) \in \mathbb{R}^{n+1} \mid v \in \text{lin.space}(C), v_j = 0\}$, we have $Frev' \subseteq Frev \setminus \{j\}$ and $Prev_0 \setminus \{j\} \subseteq Prev_0'$. Let Δ be the set of fully reversible reactions of the original network which become pseudo-irreversible in the reconfigured network, i.e.,

$$\Delta = Frev \setminus (Frev' \cup \{j\}).$$

We can easily see that $\Delta = \{i \in Frev \setminus \{j\} \mid b_i = 0 \text{ for each } b \in \text{lin.space}(C) \cap \{v \in \mathbb{R}^n \mid v_j = 0\}\}$. The following proposition further characterizes the set Δ using a basis of the lineality space of C.

Proposition 11. *Let* $B = (b^1, \ldots, b^t)$ *be a basis of* $\text{lin.space}(C)$. *Then*

$$\Delta = \{i \in Frev \setminus \{j\} \mid \text{there exists } \lambda \neq 0 \text{ with } b_i^k = \lambda b_j^k \text{ for all } k = 1, \ldots, t\}.$$

Proof. Let $\Omega = \{i \in Frev \setminus \{j\} \mid \text{there exists } \lambda \neq 0 \text{ such that } b_i^k = \lambda b_j^k \text{ for all } k = 1, \ldots, t\}$. Then $\Omega \subseteq \Delta$. To show the reverse inclusion, suppose $i \in \Delta$. Since $i \in Frev$, there exists $b \in B$ such that $b_i \neq 0$. Since $i \in \Delta$, we have $b_j \neq 0$. Let $b' \in B$ and let $w = b' - (b_j'/b_j) \cdot b$. We have $w \in \text{lin.space}(C) \cap \{v \in \mathbb{R}^n \mid v_j = 0\}$ and $w_i = b_i' - (b_i/b_j)b_j'$. Since $i \in \Delta$, we get $w_i = 0$ and so $b_i'/b_j' = b_i/b_j \overset{\text{def}}{=} \lambda \neq 0$, independently from b'. $\square$

Corollary 12. *If* $j \in Prev_0$ *is pseudo-irreversible, then* $Frev' = Frev$ *and* $Prev'_0 = Prev_0 \setminus \{j\}$.

Proof. Suppose $j \in Prev_0$. Then, $b_j^k = 0$ for all $k = 1, \ldots, t$. Consider $i \in Frev \setminus \{j\}$. There exists $b \in B$ such that $b_i \neq 0$. Since $b_j = 0$, it follows that $i \notin \Delta$. Therefore, $\Delta = \emptyset$ and the claim follows.   □

7 From Outer to Inner Descriptions

The results in Sect. 6 allow for obtaining an outer description of the reconfigured flux cone after splitting one reversible reaction. Now we are seeking for an inner description of the reconfigured flux cone after splitting a set SR $= \{j_1, \ldots, j_p\}$ of reversible reactions. We propose an iterative procedure that splits, in each iteration k, a reversible reaction, and obtains a minimal generating set of the reconfigured flux cone using the following scheme. Let (R^0, B^0) be a minimal generating set of the original flux cone. The set (R^0, B^0) can be computed using an existing software for polyhedral computations such as cdd [29]. For $1 \leq k \leq p$, let j_k be the reversible reaction to be split in iteration k and let Irr^{k-1}, $Prev_0^{k-1}$ and $Frev^{k-1}$ be the set of irreversible, pseudo-irreversible and fully reversible reactions after splitting reactions $j_1, \ldots, j_{k-1}$, respectively. Set $Irr^0 := \{1, \ldots, n\} \setminus Rev$, $Prev_0^0 := \{i \in Rev \mid b_i = 0 \text{ for all } b \in B^0\}$ and $Frev^0 := Rev \setminus Prev_0^0$. Iteration k comprises two basic steps. First, we deduce a minimal generating set (B^k, R^k) from (B^{k-1}, R^{k-1}) based on the results given in Sect. 6. This step is straightforward if $j_k \in Frev^{k-1}$. In such a case, the inner description of the reconfigured flux cone increases by one. However, if $j_k \in Prev_0^{k-1}$, in addition to the generators we can directly deduce from (B^{k-1}, R^{k-1}), the inner description of the reconfigured flux cone includes a subset $\Psi \subseteq R^{k-1} \times R^{k-1}$ that contains possibly many generators. In this case, the increase in the inner description is equal to $|\Psi| + 1$. In the second step, we update the reversibility type of reactions using Proposition 11. The deduction procedure terminates in iteration p and an inner description of the reconfigured flux cone is (B^p, R^p). For a more detailed description, see Algorithm 1.

Note that, for the reconfigured flux cone to be pointed, we must have $p \geq t$, where t is the dimension of the lineality space of the original flux cone, i.e., $t = \dim(\text{lin.space}(C))$. This is typically the case for the extreme pathway and extremal current approaches. In such a case, we have $\eta := |\{k \in \{1, \ldots, p\} \mid j_k \in Prev_0^{k-1}\}| \geq p - t$. Accordingly, the above procedure contains at least η iterations where the increase in the inner description of the flux cone is significant. This explains why, for large-scale metabolic networks, the size of the inner descriptions may be several orders of magnitude larger than that of the outer description.

Tab. 2 shows the sizes of the inner and outer descriptions of the flux cone of some typical metabolic networks. The computation of the extreme pathways, extremal currents, the minimal metabolic behaviors and the reversible metabolic space was done using the software cdd [29]. For computing the elementary flux modes, we used METATOOL [34]. We can see that the size of the outer description, given as the sum of the number of MMBs and dim(RMS), is typically much

Input   : Set of reversible reactions to be split $SR = \{j_1, \ldots, j_p\}$;
              Set J of minimal proper faces and lineality space L of the flux cone.
Output   : Minimal generating set GenSet of the reconfigured cone.
Initialization: $R^0 :=$ set of generators $g \in G \setminus L$, one for each $G \in J$;
              $B^0 := (b^1, \ldots, b^t)$ vector basis of L;
              $Prev_0^0 := \{i \in Rev \mid b_i = 0 \text{ for all } b \in B^0\}$;
              $Frev^0 := Rev \setminus Prev_0^0$, $Irr^0 := \{1, \ldots, n\} \setminus Rev$.
foreach $k \in \{1, \ldots, p\}$ **do**
  if $j_k \in Frev^{k-1}$ **then**
    Choose $u \in B^{k-1}$ such that $u_{j_k} > 0$, add$((u, 0), R^k)$,
    Let $w \in \mathbb{R}^{n+k}$ such that $w_i' := -u_i$ for all $i \in \{1, \ldots, n+k-1\} \setminus \{j_k\}$,
    $w_{j_k} := 0$ and $w_{n+k} := u_{j_k}$, add(w, R^k),
    foreach $g \in R^{k-1}$ **do**
      | add$((g - (g_{j_k}/u_{j_k}) \cdot u, 0), R^k)$,
    end
    foreach $b \in B^{k-1} \setminus \{u\}$ **do**
      | add$((b - (b_{j_k}/u_{j_k}) \cdot u, 0), B^k)$,
    end
    $\Delta := \{i \in Frev^{k-1} \setminus \{j_k\} \mid$ there exists $\lambda \neq 0$ with $b_i = \lambda b_{j_k}$ for all $b \in B^{k-1}\}$,
    $Frev^k := Frev^{k-1} \setminus (\Delta \cup \{j_k\})$, $Prev_0^k := Prev_0^{k-1} \cup \Delta$.
  end
  else
    Let $w \in \mathbb{R}^{n+k}$ such that $w_i' := 0$ for all $i \in \{1, \ldots, n+k-1\} \setminus \{j_k\}$,
    $w_{j_k} := w_{n+k} := 1$, add(w, R^k),
    $P := \{g \in R^{k-1} \mid g_{j_k} > 0\}$, $N := \{g \in R^{k-1} \mid g_{j_k} < 0\}$,
    $Z := \{g \in R^{k-1} \mid g_{j_k} = 0\}$.
    foreach $g \in P \cup Z$ **do**
      | add$((g, 0), R^k)$,
    end
    foreach $g \in N$ **do**
      Let $g' \in \mathbb{R}^{n+k}$ such that $g_i' := g_i$ for all $i \in \{1, \ldots, n+k-1\} \setminus \{j_k\}$,
      $g_{j_k}' := 0$ and $g_{n+k}' := -g_{j_k}$, add(g', R^k).
    end
    $\Psi := \{(g^1, g^2) \in P \times N \mid \{i \in Irr^{k-1} \mid g_i > 0\} \nsubseteq \{i \in Irr^{k-1} \mid g_i^1 + g_i^2 > 0\}$,
    for all $g \in R^{k-1} \setminus \{g^1, g^2\}\}$,
    foreach $(g^1, g^2) \in \Psi$ **do**
      | add$((g^2 - (g_{j_k}^2/g_{j_k}^1) \cdot g^1, 0), R^k)$,
    end
    foreach $b \in B^{k-1}$ **do**
      | add$((b, 0), B^k)$,
    end
    $Frev^k := Frev^{k-1}$, $Prev_0^k := Prev_0^{k-1} \setminus \{j_k\}$.
  end
  $Irr^k := Irr^{k-1} \cup \{j_k, n+k\}$.
end
GenSet $:= (B^p, R^p)$.

Algorithm 1. Deducing an inner description from an outer description

Table 2. Metabolic networks, with the number of internal metabolites (Met), the number of irreversible (Irr) reactions, the number of reversible internal (Rev-Int) and external (Rev-Ext) reactions, the number of minimal metabolic behaviors (MMB), the dimension of the reversible metabolic space (RMS), the number of elementary modes (EM), extreme pathways (EP), and extremal currents (EC). "?" indicates that the existing implementation of cdd has failed in the computation of the inner description. This is not the case for the computation of the outer description, illustrating that the network reconfiguration renders more complex the constraint system that defines the reconfigured flux cone. Except the 2-cycles corresponding to the split reactions, the set of EPs corresponds to a subset of the set of EMs, which is equivalent to the set of ECs.

Metabolic network	Network size				Outer description size		Inner description size		
	Met	Irr	Rev-Int	Rev-Ext	RMS	MMB	EM	EP	EC
Chloroplast stroma [30]	19	9	12	3	0	11	15	27	30
Human red blood cell [31]	38	18	17	15	1	48	3557	127	3590
S. cerevesiae [32]	48	30	17	0	0	657	8726	8743	8743
Escherichia coli [33]	90	83	27	1	0	3560	507632	?	?
Purple bacteria [33]	77	61	24	3	2	12	393524	?	?

smaller than the number of elementary flux modes, extreme pathways and extremal currents. This observation holds even if the flux cone is pointed. In such a case, the MMBs correspond to the set of extreme rays of the flux cone. The extreme pathways and extremal currents are extreme for the only reason that the split reversible reactions have been decomposed into forward and backward reactions. In the initial cone, these extreme rays are conically dependent and their numbers are much larger than the number of MMBs.

8 Conclusion

In this paper, we have studied the relationship between inner and outer descriptions of the steady-state flux cone. By distinguishing two types of reversible reactions (pseudo-irreversible, fully reversible), we have analyzed the impact of reconfiguring the metabolic network in terms of the size of the description of the reconfigured flux cone as well as the reversibility type of reactions. This leads to a generic procedure for computing inner descriptions from the outer one. This procedure makes clear why the size of the inner descriptions may be several orders of magnitude larger than that of the outer description.

References

1. Heinrich, R., Schuster, S.: The Regulation of Cellular Systems. Chapman and Hall, New York (1996)
2. Covert, M., Famili, I., Palsson, B.: Identifying constraints that govern cell behavior: a key to converting conceptual to computational models in biology? Biotechnol. Bioeng. 84(7), 763–772 (2003)

3. Palsson, B.: The challenges of in silico biology. Nat. Biotechnol. 18(11), 1147–1150 (2000)

4. Price, N., Reed, J., Palsson, B.: Genome-scale models of microbial cells: evaluating the consequences of constraints. Nat. Rev. Microbiol. 2(11), 886–897 (2004)

5. Bonarius, H., Schmid, G., Tramper, J.: Flux analysis of underdetermined metabolic networks: the quest for the missing constraints. Trends Biotechnol. 15(8), 308–314 (1997)

6. Kauffman, K., Prakash, P., Edwards, J.: Advances in flux balance analysis. Curr. Opin. Biotechnol. 14(5), 491–496 (2004)

7. Lee, J., Gianchandani, E., Papin, J.: Flux balance analysis in the era of metabolomics. Brief. Bioinformatics 7(2), 140–150 (2006)

8. Schuetz, R., Kuepfer, L., Sauer, U.: Systematic evaluation of objective functions for predicting intracellular fluxes in Escherichia coli. Mol. Syst. Biol. 3, 119 (2007)

9. Lee, S., Phalakornkule, C., Grossmann, I., Domach, M.: Recursive MILP model for finding all the alternate optima in LP models for metabolic networks. Comput. Chem. Eng. 24, 711–716 (2000)

10. Larhlimi, A., Bockmayr, A.: A new constraint-based description of the steady-state flux cone of metabolic networks. Discrete Applied Mathematics (to appear, 2008)

11. Papin, J., Price, N., Wiback, S., Fell, D., Palsson, B.: Metabolic pathways in the post-genome era. Trends Biochem. Sci. 28(5), 250–258 (2003)

12. Papin, J., Stelling, J., Price, N., Klamt, S., Schuster, S., Palsson, B.: Comparison of network-based pathway analysis methods. Trends Biotechnol. 22(8), 400–405 (2004)

13. Wagner, C., Urbanczik, R.: The geometry of the flux cone of a metabolic network. Biophys. J. 89(6), 3837–3845 (2005)

14. Schrijver, A.: Theory of Linear and Integer Programming. Wiley, Chichester (1986)

15. Clarke, B.: Stability of complex reaction networks. In: Prigogine, I., Rice, S. (eds.) Advances in Chemical Physics, vol. 43, pp. 1–216. John Wiley & Sons, Chichester (1980)

16. Schilling, C., Letscher, D., Palsson, B.: Theory for the systemic definition of metabolic pathways and their use in interpreting metabolic function from a pathway-oriented perspective. J. Theor. Biol. 203(3), 229–248 (2000)

17. Dandekar, T., Moldenhauer, F., Bulik, S., Bertram, H., Schuster, S.: A method for classifying metabolites in topological pathway analyses based on minimization of pathway number. BioSystems 70(3), 255–270 (2003)

18. Schuster, S., Hilgetag, C.: On elementary flux modes in biochemical reaction systems at steady state. J. Biol. Syst. 2(2), 165–182 (1994)

19. Clarke, B.: Complete set of steady states for the general stoichiometric dynamical system. J. Chem. Phys. 75(10), 4970–4979 (1981)

20. Schuster, S., Hilgetag, C., Woods, J., Fell, D.: Reaction routes in biochemical reaction systems: algebraic properties, validated calculation procedure and example from nucleotide metabolism. J. Math. Biol. 45(2), 153–181 (2002)

21. Heiner, M., Koch, I., Voss, K.: Analysis and simulation of steady states in metabolic pathways with Petri nets. In: Workshop and Tutorial on Practical Use of Coloured Petri Nets and the CPN Tools, CPN 2001, Aarhus University, Denmark, pp. 15–34 (2001)

22. Schuster, S., Pfeiffer, T., Moldenhauer, F., Koch, I., Dandekar, T.: Structural analysis of metabolic networks: elementary flux modes, analogy to Petri nets, and application to Mycoplasma pneumoniae. In: German Conference on Bioinformatics, GCB 2000, Heidelberg, Germany, pp. 115–120. Logos Verlag (2000)

23. Klamt, S., Saez-Rodriguez, J., Lindquist, J., Simeoni, L., Gilles, E.: A methodology for the structural and functional analysis of signaling and regulatory networks. BMC Bioinformatics 7, 56 (2006)
24. Klamt, S., Stelling, J.: Two approaches for metabolic pathway analysis? Trends Biotechnol. 21, 64–69 (2003)
25. Palsson, B., Price, N., Papin, J.: Development of network-based pathway definitions: the need to analyze real metabolic networks. Trends Biotechnol. 21(5), 195–198 (2003)
26. Gagneur, J., Klamt, S.: Computation of elementary modes: a unifying framework and the new binary approach. BMC Bioinformatics 5, 175 (2004)
27. Larhlimi, A., Bockmayr, A.: A new approach to flux coupling analysis of metabolic networks. In: Berthold, M.R., Glen, R.C., Fischer, I. (eds.) CompLife 2006. LNCS (LNBI), vol. 4216, pp. 205–215. Springer, Heidelberg (2006)
28. Larhlimi, A., Bockmayr, A.: Minimal direction cuts in metabolic networks. In: Computational Life Sciences III, CompLife 2007, Utrecht, The Netherlands. American Institute of Physics Conference Series, vol. 940, pp. 73–86 (2007)
29. Fukuda, K., Prodon, A.: Double description method revisited. In: Deza, M., Manoussakis, I., Euler, R. (eds.) CCS 1995. LNCS, vol. 1120, pp. 91–111. Springer, Heidelberg (1996)
30. Poolman, M., Fell, D., Raines, C.: Elementary modes analysis of photosynthate metabolism in the chloroplast stroma. Eur. J. Biochem. 270(3), 430–439 (2003)
31. Wiback, S., Palsson, B.: Extreme pathway analysis of human red blood cell metabolism. Biophys. J. 83(2), 808–818 (2002)
32. Cakir, T., Tekir, D., Önsan, Z., Kutlu, U., Nielsen, J.: Effect of carbon source perturbations on transcriptional regulation of metabolic fluxes in Saccharomyces cerevisiae. BMC Syst. Biol. 18(1) (2007)
33. Klamt, S., Saez-Rodriguez, J., Gilles, E.: Structural and functional analysis of cellular networks with cellnetanalyzer. BMC Syst. Biol. 1, 2 (2007)
34. von Kamp, A., Schuster, S.: Metatool 5. 0: fast and flexible elementary modes analysis. Bioinformatics 22(15), 1930–1931 (2006)

A Combinatorial Approach to Reconstruct Petri Nets from Experimental Data

Markus Durzinsky, Annegret Wagler, and Robert Weismantel

Magdeburg Center of Systems Biology (MaCS),
Otto-von-Guericke Universität Magdeburg,
Universitätsplatz 2,
39106 Magdeburg, Germany

Abstract. For many aspects of health and disease, it is important to understand different phenomena in biology and medicine. To gain the required insight, experimental data are provided and need to be interpreted, thus the challenging task is to generate all models that explain the observed phenomena. In systems biology the framework of Petri nets is often used to describe models for the regulatory mechanisms of biological systems. The aim of this paper is to present an exact combinatorial approach for the reconstruction of such models from experimental data.

1 Introduction

Models of biological systems and phenomena are of high scientific interest and practical relevance, but not always easy to obtain, as the underlying biological systems are typically complex and, thus, their structure and function in most cases non-obvious. One fundamental question in this context is to detect the local mechanisms of interaction starting from the experimentally observed global behavior of a biological system.

Structure and function of the studied system can be probed by stimulating one or several of its elements and by measuring the values of a set of elements as a function of time to see how this stimulation propagates through the system. The task is to reconstruct all possible models from such experimental data, i.e., to determine the network topology (telling in which way the measured elements interact with each other) and to describe the dynamic behavior of the system (by presenting rules how the states of the system change).

In Section 2, we briefly outline an approach proposed in [7] that, starting with the given experimental data, finally yields all solutions that account for the time-dependent mass or signal flux in the system. The practical impact of this approach was shown in [2], where the approach was applied to reconstruct models from experimental data taken from [4,5,6] describing the light controlled commitment of sporulation of physarum polycephalum plasmodia [4,6] and the photocycles in halobacterial sensory rhodopsins [5].

One main step of this approach is to determine the topology of the network in terms of a Petri net. Thereby, a central issue is to represent the vectors

M. Heiner and A.M. Uhrmacher (Eds.): CMSB 2008, LNBI 5307, pp. 328–346, 2008.

representing the changes of the values of the observed elements as conic integer combinations of yet unknown vectors which have to respect certain biological constraints. The main contribution of this paper is to present an efficient way for this step using experimental time series data of different quality as input data of the problem (Section 3).

2   An Approach to the Network Reconstruction Problem

Regulatory mechanisms of biological systems are often modeled as certain bipartite graphs related to Petri nets, see [3,8,9]. In this paper we use the Petri net terminology to present our reconstruction approach.

Let $G = (N \cup T, A, w)$ be a weighted directed bipartite graph with two kinds of nodes, so-called places and transitions. The places represent the set $N = \{1, \ldots, n\}$ of studied elements (as proteins or their conformational states, enzymes ect.), the transitions t in $T = \{1, \ldots, \tau\}$ reactions (as chemical reactions, activations, causal dependencies ect.). The arcs a in A link a place with a transition (or vice versa) and have some integral weights w_a to reflect the stoichiometric coefficients of the reactions. To keep things simple we assume without loss of generality that there is no pair of place and transition linked in both directions.

Each place $i \in N$ can be marked with an integral number of tokens and each labeling of the places with tokens defines a state $x \in \mathbb{Z}_+^n$ of the system. In biological systems, all elements can be considered to be bounded, as the value x_i of any state refers to the measured concentration of the studied element, which can only increase up to a certain maximum u_i. Indeed, many of the elements are binary ($u_i = 1$) to distinguish between the presence ($x_i = 1$) and absence ($x_i = 0$) of the element only. Thereby the set of all potential states of the system can be restricted to

$$\mathcal{X} = \{x \in \mathbb{Z}^n : 0 \le x_i \le u_i \, \forall i \in N\}.$$

This leads to a bounded *Petri net* $P = (G, u)$.

The system can change its state by switching a transition. We associate with each transition $t \in T$ a reaction vector r^t with

$$r_i^t = \begin{cases} -w_{it} & \text{if } it \in A \\ w_{ti} & \text{if } ti \in A \\ 0 & \text{else} \end{cases}$$

and say that t is applicable to a state $x \in \mathcal{X}$ if $x + r^t \in \mathcal{X}$ holds [1]. The dynamic of a network can be described in terms of reachability (by constructing all possible switching sequences starting in an initial state) or by means of stronger activation rules (for deterministic systems where each state has a unique successor state), see [7,10] for more details.

---

[1] In the usual context of Petri nets, a switching event is called firing and, if the number of tokens in a place i is not bounded by a capacity u_i, a transition is enabled and can fire if there are enough tokens available in its pre-places.

Reconstructing a network $G = (N \cup T, A, w)$ from experimental data means the following. One chooses a set N of elements which are expected to be crucial for the studied phenomenon. Then one triggers the system (by external stimuli like the change of nutrient concentrations or the exposition to some pathogens), thereby generating an initial state x^0 of the system. Then one observes how the system reacts by changing its states according to the stimulus, thereby measuring the values of the elements in N as a function of time. This generates experimental time series data $\mathcal{X}' = (x^0, x^1, \ldots, x^m)$ with $x^j \in \mathcal{X}$ for $0 \leq j \leq m$. Eventually, this procedure can be repeated for different initial states x^0, see [7,10] for more details.

In the best case, two consecutively measured states $x^j, x^{j+1} \in \mathcal{X}'$ are also consecutive system states, i.e., x^{j+1} can be obtained from x^j by switching a *single* transition. This is, however, in general not the case (and depends on the chosen time points to measure the states in $\mathcal{X}'$). Indeed, x^{j+1} is typically obtained from x^j by a switching sequence

$$x^j = y^1 \xrightarrow{t_1} y^2 \xrightarrow{t_2} \ldots \xrightarrow{t_k} y^{k+1} = x^{j+1}$$

of some length k in a yet unknown network, where the intermediate states $y^2, \ldots, y^k$ are not reported in $\mathcal{X}'$, see Figure 1.

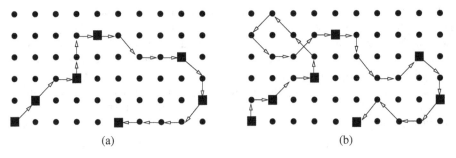

(a)   (b)

Fig. 1. Sequences of states passed through by an experiment (depicted as directed paths in the set of states) where only a subset of the passed states is measured (drawn as black squares)

We say that a network $G = (N \cup T, A, w)$ is *conformal* with the given data $\mathcal{X}'$ if for any two consecutively measured states x^j, x^{j+1} in $\mathcal{X}'$ we have that

- for $d = x^{j+1} - x^j$, the linear equation system $d = R\lambda$ has a solution $\lambda \in \mathbb{Z}_+^{|T|}$ where R is the incidence matrix of G (with columns r^t for all $t \in T$),

- λ is the Parikh vector of a sequence $\sigma_{d,\lambda} = (t_1, \ldots, t_k)$ of switches, i.e., there are states $x^j = y^1, y^2, \ldots, y^{k+1} = x^{j+1} \in \mathcal{X}$ with $y^{l+1} = y^l + r^{t_l}$ for $1 \leq l \leq k$ and λ encodes how often each reaction occurs in the sequence.

To solve this problem of reconstructing a conformal network $G = (N \cup T, A, w)$ for a given set N of elements, a capacity vector $u \in \mathbb{Z}_+^n$, and experimental data $\mathcal{X}'$, the following approach was proposed in [7].

As initial step, extract the observed changes of states from the experimental data. For that, define the set

$$\mathcal{D} := \left\{ d^j = x^{j+1} - x^j \ : \ x^j, x^{j+1} \in \mathcal{X}' \right\}.$$

The goal is to find a matrix R s.t. each $d \in \mathcal{D}$ has an integral conic representation of the form $d = R\lambda$. Every column r^t of R belongs to the set

$$\mathcal{R} = \{ r \in \mathbb{Z}^n : -u_i \leq r_i \leq u_i \ \forall i \in N \} \setminus \mathbb{Z}_+^n$$

as each r^t has to respect the capacity bounds u, be a lattice point (due to the integrality of the states), and must not be positive (as no internal reaction can produce substances without consuming anything).

Remark 1. Note that some elements in N represent external stimuli to the system, say all $i \in N_{\text{ext}} \subseteq N$. We encode such stimuli within the initial state(s) x^0 of the experimental time serie(s), instead of using reactions $r \in \mathbb{Z}_+^n$ with $r_i > 0$ for $i \in N_{\text{ext}}$. Therefore, we can exclude all vectors in $\mathbb{Z}_+^n$ from the set $\mathcal{R}$ of all potential reactions in order to consider for the reconstruction process only vectors corresponding to internal reactions of the system. In addition, for every $i \in N_{\text{ext}}$, the value x_i^0 in an initial state x^0 provides a natural upper bound u_i for the number of tokens in place i for any system state (or the maximum of x_i^0 taken over all initial states x^0, if several time series data are considered simultaneously).

Thus, the set $\mathcal{R}$ contains the vectors corresponding to all potential reactions of the studied network (resp. transitions of the Petri net). We are interested in this (inclusion-wise) minimal subset $\mathcal{B}$ of $\mathcal{R}$ which is necessary to provide for each $d \in \mathcal{D}$ *all* representations in the form

$$d = \sum_{r^t \in \mathcal{R}} \lambda_t r^t, \ \lambda_t \in \mathbb{Z}_+. \tag{1}$$

Indeed, there are typically several solutions λ for this equation system as different sequences can link two consecutively measured states in $\mathcal{X}'$, see Figure 1 again. Note that in both pictures (a) and (b), the sets of measured states (drawn as black squares) are the same, while the sequences of intermediate states passed through by the experiment differ. We have to take into account all possible solutions and denote by $\Lambda(d)$ the set of all integral solutions of the linear equation system (1). For each vector $\lambda \in \Lambda(d)$, let

$$\mathcal{R}_{d,\lambda} = \{ r^t \in \mathcal{R} : \lambda_t \neq 0 \}$$

be the (multi-)set of reactions used for this representation λ, and let

$$\mathcal{R}(d) = \bigcup_{\lambda \in \Lambda(d)} \mathcal{R}_{d,\lambda}$$

be the set of all reactions used for these representations. By construction,

$$\mathcal{B} = \bigcup_{d \in \mathcal{D}} \mathcal{R}(d)$$

holds.

We discuss in Section 3 how the two sets $\Lambda(d)$ and $\mathcal{R}(d)$ can be determined. It turns out that in general (without any restriction on the intermediate states between two consecutively measured states in $\mathcal{X}'$), the set $\Lambda(d)$ is infinite and the set $\mathcal{R}(d)$ equals $\mathcal{R}$ for any $n \geq 3$. Thus, the reconstruction problem is in general a trivial task and the network generated from $\mathcal{B} = \mathcal{R}$ is certainly conformal with $\mathcal{X}'$, but also with any other experimental data within $\mathcal{X}$ (Section 3.1).

The situation changes if the provided experimental data are of a certain quality, called *locally bounded* and *monotone data*, which reflect more or all changes of states that are crucial to describe the studied biological phenomenon. We show that in these cases both sets $\mathcal{R}(d)$ and $\Lambda(d)$ are substantially smaller than in the general case and that the problem can be efficiently solved (Section 3.2 and Section 3.3).

However, it can happen that there exists *no* network being conformal with locally bounded or monotone data. We provide a *feasibility test* to detect this situation directly by inspecting the set $\mathcal{D}$. If our analysis shows that there is no network conformal with the yet considered data, then additional non-observed elements are required to provide us with a meaningful model (which leads to a mixed case as there is no restriction for the values of the artificial elements). We again determine $\mathcal{R}(d)$ and $\Lambda(d)$ and show that there exists always a conformal network using two artificial elements.

In summary, we present for different types of experimental data a complete description of the sets $\mathcal{R}(d)$ and $\Lambda(d)$ and an efficient way to determine the studied conformal network.

3   The Integer Decomposition of the Difference Vectors

The aim of this section is to determine, for each $d \in \mathcal{D}$, the sets $\Lambda(d)$ of integral solutions λ of system (1) and $\mathcal{R}(d) \subseteq \mathcal{R}$ of reactions used for these representations for different types of experimental data.

3.1   The General Case

In order to determine $\Lambda(d)$ in the general case, i.e., without any restriction on the intermediate states between two consecutively measured states in $\mathcal{X}'$, we have also to consider the integral solutions η of the homogeneous system

$$\sum_{r^t \in \mathcal{R}} \eta_t r^t = \mathbf{0}, \ \eta_t \in \mathbb{Z}_+,$$

and denote the set of all homogeneous solutions $\eta \neq \mathbf{0}$ by $\Lambda(\mathbf{0})$. The reason is that every solution $\lambda \in \Lambda(d)$ can be extended by a homogeneous solution $\eta \in \Lambda(\mathbf{0})$

to a new solution $\lambda + \eta \in \Lambda(d)$. More precisely, $\Lambda(d)$ can be represented by two finite and uniquely determined minimal sets $\Lambda_{\text{inhom}}(d)$ and $H \subset \Lambda(\mathbf{0})$ such that

$$\Lambda(d) = \Lambda_{\text{inhom}}(d) + \text{cone}_{\mathbb{Z}}(H),$$

where $\text{cone}_{\mathbb{Z}}(H)$ is the set of all integral conic combinations of vectors in H (the so-called Hilbert basis of $\Lambda(\mathbf{0})$, see [1,10] for more details).

Theorem 1. *The set $\Lambda(d)$ is*

- *empty if and only if $n = 1$, $d > \mathbf{0}$ or $n = 2$, $d > \mathbf{0}$, and $u = 1$,*
- *finite if and only if $n = 1$, $d \leq \mathbf{0}$, and*
- *infinite otherwise.*

Proof. We always have a representation for $d \not\geq \mathbf{0}$ (as $d \in \mathcal{R}$ allows the solution $\lambda = e^t$ with $r^t = d$) and for $d = \mathbf{0}$ (as the empty sequence $\lambda = \mathbf{0}$ is also a solution). Consider now $d > \mathbf{0}$. For $n = 1$, there is clearly no representation as $\mathcal{R} = \{(-1), (-2), \ldots, (-u_1)\}$. For $n = 2$ and $u = 1$ we obtain

$$\mathcal{R} = \left\{ \begin{pmatrix} -1 \\ -1 \end{pmatrix}, \begin{pmatrix} -1 \\ 0 \end{pmatrix}, \begin{pmatrix} -1 \\ 1 \end{pmatrix}, \begin{pmatrix} 0 \\ -1 \end{pmatrix}, \begin{pmatrix} 1 \\ -1 \end{pmatrix} \right\}$$

and it is straightforward to see that (1) has no solution $\lambda \in \mathbb{Z}_+^{|\mathcal{R}|}$. For $n = 2$ and $u \neq 1$, say $u_1 \geq 2$, there is always a representation as any d can be decomposed as

$$d = \begin{pmatrix} d_1 \\ -1 \end{pmatrix} + \begin{pmatrix} -1 \\ d_2 \end{pmatrix} + 2 \cdot \begin{pmatrix} 2 \\ -1 \end{pmatrix} + 3 \cdot \begin{pmatrix} -1 \\ 1 \end{pmatrix}$$

For $n \geq 3$, we can always represent $d = (d_1, d_2, d_3, d')^T$ for any u by

$$d = \begin{pmatrix} d_1 \\ 1 \\ -1 \\ d' \end{pmatrix} + \begin{pmatrix} -1 \\ d_2 \\ 1 \\ 0 \end{pmatrix} + \begin{pmatrix} 1 \\ -1 \\ d_3 \\ 0 \end{pmatrix}.$$

Hence, we have $\Lambda_{\text{inhom}}(d) = \emptyset$ and thus $\Lambda(d) = \emptyset$ if and only if $d > \mathbf{0}$ and $n = 1$ or $n = 2$ and $u = 1$. There always exist homogeneous solutions, except for $n = 1$. Combining the sets $\Lambda_{\text{inhom}}(d)$ and H yields finally the assertion of the theorem. □

Thus, except for some pathological cases, $\Lambda(d)$ is infinite; whether a solution $\lambda \in \Lambda(d)$ has to be taken into account depends, however, on its realization as a sequence $\sigma_{d,\lambda}$ of switches. The inhomogeneous solutions $\lambda \in \Lambda_{\text{inhom}}(d)$ are minimal in the sense that, for any $\eta \in H$, the sets of involved reactions satisfy $\mathcal{R}_{d,\lambda} \subset \mathcal{R}_{d,\lambda+\eta}$. We can discard those non-minimal solutions $\lambda + \eta$ where in all possible sequences $\sigma_{d,\lambda+\eta}$ the reactions corresponding to η are consecutive and, thus, form a cycle (i.e., there is a sequence $\sigma_{d,\lambda}$ realizing the inhomogeneous part λ of the solution, and on some intermediate state y^l, the cycle of reactions corresponding to the homogeneous part η is attached, returning back to the state

y^l). However, a solution $\lambda + \eta$ is feasible if there exists a sequence $\sigma_{d,\lambda+\eta}$ such that the reactions corresponding to η are not consecutive and, thus, the sequence is cycle-free. Indeed, for the sake of completeness we have to consider the subset $\Lambda_\sigma(d)$ of $\Lambda(d)$ consisting of all inhomogeneous solutions $\lambda \in \Lambda_{\text{inhom}}(d)$ and of all feasible combinations $\lambda + \eta_1 + \ldots + \eta_l$ with $\lambda \in \Lambda_{\text{inhom}}(d), \eta_i \in H$. Thus, $\Lambda_\sigma(d)$ is certainly finite (since there are only finitely many cycle-free paths between the finitely many states in $\mathcal{X}$), but still far too large in general and might contain solutions without a meaningful biological interpretation.

$\mathcal{R}(0)$ is the set of reactions to represent the zero vector, so these are the reactions involved in at least one minimal homogeneous solution.

Lemma 1. *We have that $\mathcal{R}(0)$*

- *is empty if and only if $n = 1$,*
- *equals $\left\{ \binom{-1}{1}, \binom{1}{-1} \right\}$ if and only if $n = 2$, $u = 1$, and*
- *equals $\mathcal{R}$ otherwise.*

Proof. If $n = 1$, there is no non-trivial solution of equation (1) using negative reactions only, and $\mathcal{R}(0) = \emptyset$.

If $n = 2$ and $u = 1$, then $\binom{-1}{1} + \binom{1}{-1}$ is the unique minimal cycle, thus $\mathcal{R}(0)$ contains these two reactions only.

If $n = 2$ and $u \neq 1$, say $u_2 \geq 2$, then any $r = \binom{r_1}{r_2} \in \mathcal{R}$ is part of a representation, namely

$$\binom{0}{0} = \binom{r_1}{r_2} + \binom{-r_1}{-1} + \binom{-1}{-r_2} + 3 \cdot \binom{1}{-1} + 2 \cdot \binom{-1}{2}.$$

If $n \geq 3$, there is a homogeneous solution using any $r \in \mathcal{R}$ as well. Let $r = (r_1, r_2, r')$ with $r' \in \mathbb{Z}^{n-2}$, then r is used in the representation

$$\begin{pmatrix} 0 \\ 0 \\ 0 \end{pmatrix} = \begin{pmatrix} r_1 \\ r_2 \\ r' \end{pmatrix} + \begin{pmatrix} -1 \\ -r_2 \\ -r' \end{pmatrix} + \begin{pmatrix} -r_1 \\ -1 \\ 1 \end{pmatrix} + \begin{pmatrix} 1 \\ 1 \\ -1 \end{pmatrix}.$$

$\square$

As clearly $\mathcal{R}(0) \subseteq \mathcal{R}(d)$ holds for any $d \in \mathcal{D}$ with $\Lambda(d) \neq \emptyset$, we obtain:

Theorem 2. *The set $\mathcal{R}(d)$ is*

- *empty if and only if $n = 1$, $d \geq 0$ or $n = 2$, $d > 0$, and $u = 1$,*
- *equals $\{d, d+1, \ldots, -2, -1\}$ if and only if $n = 1$, $d < 0$*
- *equals $\{r \in \mathcal{R} : r_1 + r_2 \geq d_1 + d_2\}$ if and only if $n = 2$, $d \not> 0$, $u = 1$, and*
- *equals $\mathcal{R}$ otherwise.*

Proof. If $n = 1$, then any vector $d < 0$ can be represented with any $r \in \{d, d+1, \ldots, -2, -1\}$ by

$$d = \begin{cases} r & \text{if } r = d \\ r + r' & \text{if } r \neq d \end{cases}$$

where $r' = d - r < 0$. Otherwise no reaction $r < 0$ can be used to represent a vector $d \geq 0$.

If $n = 2$, $u = 1$, we have $\Lambda(d) = \emptyset$ for any $d > \mathbf{0}$ (by Theorem 1), hence $\mathcal{R}(d)$ is empty. In the case of $d \not> \mathbf{0}$, the vector d is either zero or has a negative component. In the former case, we obtain from the previous lemma

$$\mathcal{R}(\mathbf{0}) = \left\{ \begin{pmatrix} -1 \\ 1 \end{pmatrix}, \begin{pmatrix} 1 \\ -1 \end{pmatrix} \right\} = \{ r \in \mathcal{R} : r_1 + r_2 \geq 0 \}.$$

In the latter case, say $d_1 = -1$ and $d_2 \in \{-1, 0, 1\}$, we examine the possible sums $d_1 + d_2$ and $r_1 + r_2$ for any reaction $r \in \mathcal{R}$.

$$d_1 + d_2 \in \{-2, -1, 0\}$$
$$r_1 + r_2 = \begin{cases} -2 & \text{if } r = \begin{pmatrix} -1 \\ -1 \end{pmatrix} \\ -1 & \text{if } r \in \left\{ \begin{pmatrix} -1 \\ 0 \end{pmatrix}, \begin{pmatrix} 0 \\ -1 \end{pmatrix} \right\} \\ 0 & \text{if } r \in \left\{ \begin{pmatrix} -1 \\ 1 \end{pmatrix}, \begin{pmatrix} 1 \\ -1 \end{pmatrix} \right\} \end{cases}$$

For any representation $d = r^1 + r^2 + \cdots + r^k$, also

$$d_1 + d_2 = (r_1^1 + r_2^1) + \cdots + (r_1^k + r_2^k)$$

holds. Hence, $r = \begin{pmatrix} -1 \\ -1 \end{pmatrix}$ can only be used if $d_1 + d_2 = -2$ and the reactions $\begin{pmatrix} -1 \\ 0 \end{pmatrix}$, $\begin{pmatrix} 0 \\ -1 \end{pmatrix}$ can only be used if $d_1 + d_2 \leq -1$. All other combinations are indeed possible.

$$d = \begin{pmatrix} -1 \\ -1 \end{pmatrix} = \begin{pmatrix} -1 \\ 0 \end{pmatrix} + \begin{pmatrix} 0 \\ -1 \end{pmatrix} \Rightarrow \mathcal{R}(d) = \mathcal{R}$$

$$d = \begin{pmatrix} -1 \\ 0 \end{pmatrix} = \begin{pmatrix} 0 \\ -1 \end{pmatrix} + \begin{pmatrix} -1 \\ 1 \end{pmatrix} \Rightarrow \mathcal{R}(d) = \left\{ \begin{pmatrix} -1 \\ 0 \end{pmatrix}, \begin{pmatrix} 0 \\ -1 \end{pmatrix}, \begin{pmatrix} -1 \\ 1 \end{pmatrix}, \begin{pmatrix} 1 \\ -1 \end{pmatrix} \right\}$$

$$d = \begin{pmatrix} -1 \\ 1 \end{pmatrix} = \begin{pmatrix} -1 \\ 1 \end{pmatrix} \qquad \Rightarrow \mathcal{R}(d) = \mathcal{R}(\mathbf{0})$$

These sets contain exactly those reaction vectors with $r_1 + r_2 \geq d_1 + d_2$.

In all other cases, $n = 2$ and $u > 1$ or $n \geq 3$, Theorem 1 shows $\Lambda(d) \neq \emptyset$ and the previous lemma implies $\mathcal{R}(\mathbf{0}) = \mathcal{R}$, hence the final assertion $\mathcal{R}(d) = \mathcal{R}$ follows. $\qquad \square$

Thus, except for some pathological cases, already $\mathcal{R}(d) = \mathcal{R}$ holds and building $\mathcal{B} = \bigcup_{d \in \mathcal{D}} \mathcal{R}(d)$ cannot exclude any vector from $\mathcal{R}$.

Corollary 1. *For general experimental data $\mathcal{X}'$ and $n \geq 3$, the incidence matrix of the conformal network contains all potential reaction vectors from $\mathcal{R}$.*

Hence, the reconstruction problem is a trivial task in the general case as the conformal network can be generated without solving the linear equation system (1) and the generated network is conformal with *any* experimental data on N.

3.2 The Case of Locally Bounded Experimental Data

From our analysis of the general case, the question emerges whether a nontrivial reconstruction process for a given series of experimental data is possible if the data meet a certain quality which restricts the intermediate sequences linking two consecutively measured states.

From a biological point of view it is interesting to consider the case when the network elements have been measured so accurately that an oscillation of their values between two measured states can be *restricted* a priori. This is the setting where we can assume that the chosen level of resolution in time to produce the time series data $\mathcal{X}'$ suffices to guarantee the following property:

Definition 1. *We say that the experimental data $\mathcal{X}'$ are* locally bounded *if for any two consecutively measured states $x^j, x^{j+1} \in \mathcal{X}'$ and $d = x^{j+1} - x^j$, the intermediate states y^l of the sequence*

$$x^j = y^1, y^2, \ldots, y^k, y^{k+1} = x^{j+1}$$

belong to the box

$$\mathcal{X}(d) = \{y \in \mathcal{X} : y_i \in [x_i^j, x_i^{j+1}] \text{ for all } i \in N\}.$$

Figure 2 shows again the two sequences from Figure 1 where for any two consecutively measured states (indicated by black squares) the state boxes are included (drawn as grey-shaded regions). While the sequence (a) is locally bounded, the sequence (b) is not since states outside the boxes are visited.

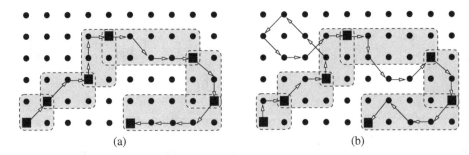

(a)   (b)

Fig. 2. Sequences of states passed through by an experiment (depicted as directed paths in the set of states) where the subset of measured states (drawn as black squares) are (a) and are not (b) locally bounded

Remark 2. Note that the experimental sequence from Figure 1(b) can be turned into locally bounded data by measuring additional states (in Figure 3(b) depicted as white squares) as then the state boxes (again drawn as grey-shaded regions) indeed cover the whole experimental sequence.

As in the case of locally bounded data, in all intermediate states the values of the elements cannot arbitrarily oscillate, we have to choose the columns of the studied matrix R from the set

$$Box(d) = \left\{ r \in \mathbb{Z}^n : \begin{array}{l} -|d_i| \leq r_i \leq |d_i| \text{ if } d_i \neq 0 \\ r_i = 0 \quad \text{if } d_i = 0 \end{array} \right\} \setminus \mathbb{Z}_+^n$$

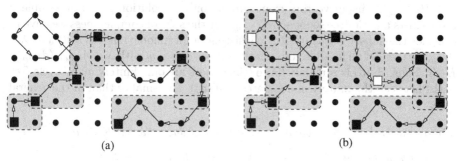

(a)   (b)

Fig. 3. Experimental sequence (depicted as directed path) where the subset of originally measured states (drawn as black squares) in (a) is extended by additional states (depicted as white squares) in order to obtain locally bounded data (b)

in order to represent a vector $d \in \mathcal{D}$ which reduces the problem to find all integral solutions λ of the system

$$\sum_{r^t \in Box(d)} \lambda_t r^t = d, \ \lambda_t \in \mathbb{Z}_+ \tag{2}$$

and we only have to consider homogeneous solutions η of the system

$$\sum_{r^t \in Box(d)} \eta_t r^t = \mathbf{0}, \ \eta_t \in \mathbb{Z}_+$$

as well.

That is, for the representation of any vector $d \in \mathcal{D}$, we are locally in the general setting. More precisely, we restrict the set of considered places to those $i \in \mathrm{supp}(d)$ (i.e., to all $i \in N$ with $d_i \neq 0$), and use $|d_i|$ as capacity bound for any $i \in \mathrm{supp}(d)$ instead of u_i. This implies that we obtain $\mathcal{X}(d)$ instead of $\mathcal{X}$ as state space and $Box(d)$ instead of $\mathcal{R}$ as reaction space.

Hence, we deduce as immediate consequence of Theorem 1:

Theorem 3. *For locally bounded experimental data, the set $\Lambda(d)$ is*

- *empty if and only if $|\mathrm{supp}(d)| = 1$, $d > \mathbf{0}$ or $|\mathrm{supp}(d)| = 2$, $d > \mathbf{0}$, and $d_i = 1$ for all $i \in \mathrm{supp}(d)$,*
- *finite if and only if $|\mathrm{supp}(d)| = 1$, $d \leq \mathbf{0}$, and*
- *infinite otherwise.*

Proof. The set $Box(d)$ is the projection of $\mathcal{R}$ to $\mathrm{supp}(d)$ and with bounds $u_i = |d_i|$ for all $i \in \mathrm{supp}(d)$. Hence, this statement follows directly from Theorem 1 by adjusting the dimension n to $|\mathrm{supp}(d)|$ and the bounds u according to d. $\square$

Thus, except for some cases, $\Lambda(d)$ is infinite; whether a solution $\lambda \in \Lambda(d)$ has to be taken into account depends again on its realization as a sequence $\sigma_{d,\lambda}$ of switches. The inhomogeneous solutions $\lambda \in \Lambda_{\mathrm{inhom}}(d)$ are minimal and for any

$\eta \in H(d) \subseteq H$ using vectors from $Box(d)$ only, a solution $\lambda + \eta$ is feasible if there exists a cycle-free sequence $\sigma_{d,\lambda+\eta}$. For the sake of completeness we have to consider the subset $\Lambda_\sigma(d)$ of $\Lambda(d)$ consisting of all inhomogeneous solutions $\lambda \in \Lambda_{\text{inhom}}(d)$ and of all feasible combinations $\lambda+\eta_1+\ldots+\eta_l$ with $\lambda \in \Lambda_{\text{inhom}}(d), \eta_i \in H(d)$. Thus, $\Lambda_\sigma(d)$ is certainly finite (since there are only finitely many cycle-free paths between the finitely many states in $\mathcal{X}(d)$), and smaller than in the general case.

Theorem 4. *For locally bounded experimental data, the set $\mathcal{R}(d)$ is*

- *empty if and only if $|\text{supp}(d)| = 1$, $d > \mathbf{0}$ or $|\text{supp}(d)| = 2$, $d > \mathbf{0}$, and $d_i = 1$ for all $i \in \text{supp}(d)$,*
- *equals $\{e^i - e^j, e^j - e^i\}$ if and only if $|\text{supp}(d)| = 2$, $d_i = 1$, $d_j = -1$, and*
- *equals $Box(d)$ otherwise.*

Proof. This Theorem can also be derived directly from Theorem 2 by adjusting n and u as follows.

If $|\text{supp}(d)| = 1$, $d > \mathbf{0}$ or $\text{supp}(d) = 2$, $d > \mathbf{0}$, and $d_i = 1$ for all $i \in \text{supp}(d)$, then $\mathcal{R}(d)$ is empty.

If $|\text{supp}(d)| = 1$, $d \leq \mathbf{0}$, then d has exactly one negative entry $d_i < 0$ and $\mathcal{R}(d) = \{d, d + e^i, \ldots, -2e^i, -e^i\}$ which actually equals $Box(d)$.

If $|\text{supp}(d)| = 2$, $d \not> \mathbf{0}$, $|d_i| = 1$ for all $i \in \text{supp}(d)$, say $d_i, d_j \in \{-1, 1\}$, then $\mathcal{R}(d) = \{r \in Box(d) : r_i + r_j \geq d_i + d_j\}$. In the case of $d_i = -1$ and $d_j = -1$, the property $r_i + r_j \geq d_i + d_j = -2$ is always true, such that $\mathcal{R}(d) = Box(d)$ holds. In the other cases, where d_i and d_j have different signs, the property $r_i + r_j \geq d_i + d_j = 0$ is true only for $r \in \{e^i - e^j, e^j - e^i\}$.

In all remaining cases the sets $\mathcal{R}(d)$ and $Box(d)$ are always equal. □

Thus, whenever $d \not> \mathbf{0}$ or $|\text{supp}(d)| \geq 3$ holds, $\mathcal{R}(d)$ is explicitly known and building $\mathcal{B} = \bigcup_{d \in \mathcal{D}} \mathcal{R}(d)$ can be easily done without any computation. However, in two cases with $d > \mathbf{0}$ and $|\text{supp}(d)| \leq 2$, we obtain $\mathcal{R}(d) = \emptyset$ and, thus, d cannot be represented. This implies:

Corollary 2. *For locally bounded experimental data*

- *there is no conformal network if and only if there exists a vector $d \in \mathcal{D}$ with $|\text{supp}(d)| = 1$, $d > \mathbf{0}$ or $|\text{supp}(d)| = 2$, $d > \mathbf{0}$, and $d_i = 1$ for all $i \in \text{supp}(d)$;*

- *if a conformal network exists, then $\mathcal{B} = \bigcup_{d \in \mathcal{D}} \mathcal{R}(d)$ holds where $\mathcal{R}(d) = \{e^i - e^j, e^j - e^i\}$ for any d with $|\text{supp}(d)| = 2$, $d_i = 1$, $d_j = -1$, and $\mathcal{R}(d) = Box(d)$ otherwise.*

Hence, if a conformal network exists, then the reconstruction problem is an easy task in the case of locally bounded experimental data, as the conformal network can be generated even without explicitly solving the linear equation system (2). Moreover, we can deduce an efficient *feasibility test* for the existence of a conformal network by inspecting the set $\mathcal{D}$ only which implies that the problem is not solvable with the considered set N of elements in the two cases with $d > \mathbf{0}$ and $|\mathrm{supp}(d)| \leq 2$. In this situation, we have to extend the set of elements by some *additional elements* and partition the index set accordingly into $N = N_o \cup N_a$ where N_o contains the indices $1, \ldots, n$ of all original elements and N_a the indices $n + 1, \ldots, n + a$ of all additional elements. We choose an upper bound $u_i = 1$ for the capacity of each $i \in N_a$ (as we can only deal with the availability of the additional elements). The n-dimensional vectors $x^j \in \mathcal{X}'$ and $d \in \mathcal{D}$ have to be extended to vectors $\bar{x}^j$ and $\bar{d}$ of dimension $n + a$, starting with unknown values for the additional elements (as those elements were not subject to experimental observation). We call this situation the *mixed locally bounded case* as the values of the additional elements clearly cannot be restricted, but the values of the original elements only. The columns of the matrix R have to be chosen from the set

$$Box(\bar{d}) = \left\{ r \in \mathbb{Z}^{n+a} : \begin{array}{l} -|d_i| \leq r_i \leq |d_i| \text{ if } i \in N_o,\ d_i \neq 0 \\ r_i = \ 0 \ \ \text{ if } i \in N_o,\ d_i = 0 \\ -1 \leq r_i \leq \ 1 \ \ \text{ if } i \in N_a \end{array} \right\} \setminus \mathbb{Z}_+^{n+a}$$

and the system

$$\sum_{r^t \in Box(\bar{d})} \lambda_t r^t = \bar{d},\ \lambda_t \in \mathbb{Z}_+ \tag{3}$$

has to be solved for all possible start values $\bar{d}_i \in \{-1, 0, 1\}$ for all $i \in N_a$. In addition, we have to consider homogeneous solutions η of the system

$$\sum_{r^t \in Box(\bar{d})} \eta_t r^t = \bar{\mathbf{0}},\ \eta_t \in \mathbb{Z}_+.$$

In the mixed locally bounded case with $a = 1$, we obtain:

Theorem 5. *For locally bounded experimental data on N_o and $a = 1$, $\mathcal{R}(\bar{d})$ is*

- *empty if and only if $d = \mathbf{0}$, $\bar{d}_{n+1} \geq 0$ or $|\mathrm{supp}(d)| = 1$, $\bar{d} > \mathbf{0}$ and $d_i = 1$ with $i \in \mathrm{supp}(d)$,*
- *equals $\{\bar{d}\}$ if and only if $d = \mathbf{0}$ and $\bar{d}_{n+1} = -1$,*
- *equals $\{r \in Box(\bar{d}) : r_i + r_{n+1} \geq d_i + \bar{d}_{n+1}\}$ if and only if $|\mathrm{supp}(d)| = 1$, $\bar{d} \not> \mathbf{0}$, $|d_i| = 1$ with $i \in \mathrm{supp}(d)$, and*
- *equals $Box(\bar{d})$ otherwise.*

Proof. The projection of $Box(\bar{d})$ to $\mathrm{supp}(d) \cup N_a$ results in the set of all reactions with an upper bound $u_i = |d_i|$ for the original elements and $u_i = 1$ for the additional elements. This allows us to apply Theorem 2 by adjusting the dimension to $|\mathrm{supp}(d)| + a$ and the bounds as specified.

If $|\mathrm{supp}(d)| = 0$ and $\bar{d}_{n+1} \geq 0$, we have $\mathcal{R}(\bar{d}) = \emptyset$. If $|\mathrm{supp}(d)| = 0$ and $\bar{d}_{n+1} = -1$, the set $\mathcal{R}(\bar{d})$ reduces to the single vector $\bar{d} = \binom{0}{-1}$.

If $|\mathrm{supp}(d)| = 1$, let d_i be the single non-zero entry and $|d_i| = 1$ (implying and upper bound $u = \mathbf{1}$ for Theorem 2). In the case of $\bar{d} \geq \mathbf{0}$, the set $\mathcal{R}(\bar{d})$ is again empty. Otherwise $\bar{d}_i$ has a negative entry and we derive $\mathcal{R}(\bar{d}) = \{r \in Box(\bar{d}) : r_i + r_{n+1} \geq d_i + \bar{d}_{n+1}\}$.

In all other than the above cases, the set $\mathcal{R}(\bar{d})$ equals $Box(\bar{d})$. $\qquad\square$

Thus, for every single vector $d \in \mathcal{D}$ there is an extension $\overline{d}$ which has a representation. The existence of a conformal network depends, therefore, on the sequence of vectors in $\mathcal{D}$. The previous theorem implies that there is no conformal network using one additional element if and only if $\mathcal{X}'$ gives rise to two consecutive positive vectors. In this situation, two additional elements are required:

Theorem 6. *For locally bounded experimental data on N_o and $a = 2$, $\mathcal{R}(\overline{d})$ is*

- *empty if and only if $d = 0$, $\left(\begin{smallmatrix} \overline{d}_{n+1} \\ \overline{d}_{n+2} \end{smallmatrix}\right) > 0$,*
- *equals $\{r \in Box(\overline{d}) : r_{n+1} + r_{n+2} \geq \overline{d}_{n+1} + \overline{d}_{n+2}\}$ if and only if $d = 0$, $\left(\begin{smallmatrix} \overline{d}_{n+1} \\ \overline{d}_{n+2} \end{smallmatrix}\right) \not> 0$, and*
- *equals $Box(\overline{d})$ otherwise (i.e., if $|supp(d)| \geq 1$).*

Proof. This theorem can be derived similarly to Theorem 5 with the only difference that $a = 2$ rises the dimension once more.

If $|supp(d)| = 0$, we use an upper bound $u = 1$ for each reaction. Applying Theorem 2 results in $\mathcal{R}(\overline{d}) = \emptyset$ if $\overline{d} > 0$ and $\mathcal{R}(\overline{d}) = \{r \in Box(\overline{d}) : r_{n+1} + r_{n+2} \geq \overline{d}_{n+1} + \overline{d}_{n+2}\}$ otherwise.

Any $\overline{d}$ with $|supp(d)| \geq 1$ immediately results in $\mathcal{R}(\overline{d}) = Box(\overline{d})$.   □

Thus, for every single vector $d \in \mathcal{D}$ there is an extension $\overline{d}$ which has a representation. In addition, we can find suitable extensions for two consecutive positive vectors $d^j, d^{j+1} \in \mathcal{D}$, for instance

$$\overline{d}^j = \begin{pmatrix} d^j \\ -1 \\ 0 \end{pmatrix} \text{ and } \overline{d}^{j+1} = \begin{pmatrix} d^{j+1} \\ 0 \\ -1 \end{pmatrix}$$

which implies the existence of a network being conformal with any locally bounded experimental data $\mathcal{X}'$ using up to two artificial elements.

3.3   The Case of Monotone Experimental Data

An even better situation occurs if the network elements can be measured so accurately that an oscillation of their values inbetween two measured states can be *excluded* a priori. This is the setting where we can assume that the chosen level of resolution in time to produce the time series data $\mathcal{X}'$ suffices to guarantee the following property:

Definition 2. *We say that the experimental data $\mathcal{X}'$ are* monotone *if for any two consecutively measured states $x^j, x^{j+1} \in \mathcal{X}'$ and $d = x^{j+1} - x^j$, the intermediate states y^l of the sequence*

$$x^j = y^1, y^2, \ldots, y^k, y^{k+1} = x^{j+1}$$

satisfy

- $y_i^1 \leq y_i^2 \leq \ldots \leq y_i^k \leq y_i^{k+1}$ for all $i \in N$ with $x_i^j \leq x_i^{j+1}$ and
- $y_i^1 \geq y_i^2 \geq \ldots \geq y_i^k \geq y_i^{k+1}$ for all $i \in N$ with $x_i^j \geq x_i^{j+1}$.

Monotone data are in particular locally bounded, hence all intermediate states y^l belong to the box $\mathcal{X}(d)$ again.

Recall that Figure 3 shows the sequence from Figure 1(a) and the refined sequence from Figure 2(b) where for any two consecutively measured states (indicated by black and white squares) the state boxes are included (drawn as grey-shaded regions). While the sequence (a) is monotone, the sequence (b) is not since in the last state box of the sequence, the values of one element oscillate (w.r.t. the vertical axis).

Remark 3. Note that the refined experimental sequence from Figure 3(b) can be turned into monotone data by measuring two further states (in Figure 4(b) the two last white squares in the sequence) as then no oscillation of the values within the state boxes (again drawn as grey-shaded regions) occurs anymore.

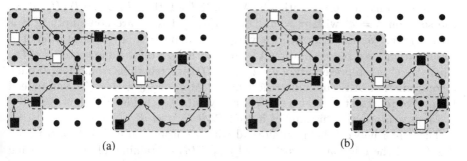

(a)   (b)

Fig. 4. Experimental sequence (depicted as directed path) where the subset of originally measured states (drawn as black squares) in (a) is extended by additional states (the last two white squares in the sequence) in order to obtain monotone data (b).

As in the case of monotone data, in all intermediate states the values of the elements cannot oscillate at all, we have to choose the columns of the studied matrix R from the set

$$Mon(d) = \left\{ r \in \mathbb{Z}^n : \begin{array}{l} 0 \leq r_i \leq d_i \ \text{if} \ d_i > 0 \\ d_i \leq r_i \leq 0 \ \text{if} \ d_i < 0 \\ r_i = 0 \ \text{if} \ d_i = 0 \end{array} \right\} \setminus \mathbb{Z}^n_+$$

in order to represent a vector $d \in \mathcal{D}$ which reduces the problem to find all integral solutions λ of the system

$$\sum_{r^t \in Mon(d)} \lambda_t r^t = d, \ \lambda_t \in \mathbb{Z}_+. \tag{4}$$

In fact, no homogeneous solutions have to be considered in the monotone case and we can characterize $\mathcal{R}(d)$ explicitly:

Theorem 7. *Let $\mathcal{X}'$ satisfy the monotonicity property.*

- *For each $d \in \mathcal{D}$, we have $\Lambda(d) = \Lambda_{\text{inhom}}(d)$ and $\mathcal{R}(d) = Mon(d) \setminus U_0$ with*

$$U_0 = \left\{ r \in \mathbb{Z}^n : \begin{array}{l} r_i = d_i \text{ for all } i \text{ with } d_i \leq 0 \\ r_i \neq d_i \text{ for at least one } i \text{ with } d_i > 0 \end{array} \right\}.$$

  For $d > \mathbf{0}$ it follows that $Mon(d) = \emptyset$ and $\Lambda(d) = \emptyset$.
- *There is a network conformal with $\mathcal{X}'$ if and only if there is no $d \in \mathcal{D}$ with $d > \mathbf{0}$.*

Proof. Due to the monotonicity property, $\mathcal{R}(d) \subseteq \text{Mon}(d)$ clearly holds. As for any $i \in N$ and all $r \in \text{Mon}(d)$, the entries r_i are either all negative, all positive, or all equal to zero, there are clearly no homogeneous solutions, which implies $H = \emptyset$ and $\Lambda(d) = \Lambda_{\text{inhom}}(d)$.

From $d > \mathbf{0}$ it follows $r \geq \mathbf{0}$ for any $r \in \text{Mon}(d)$, thus $\text{Mon}(d)$ is empty. The vector $d = \mathbf{0}$ has a representation by $\lambda = \mathbf{0}$. If $d \not\geq \mathbf{0}$ then $d \in \text{Mon}(d) \setminus U_0$ allows the solution $\lambda = e^t$ with $r^t = d$. Thus, $\Lambda(d)$ and $\mathcal{R}(d)$ are non-empty iff d is not positive. We next characterize which vectors occur in a representation of d. Assume there is a representation $\lambda \in \Lambda(d)$ with $\lambda_t > 0$ for some vector $r^t \in U_0$. Then $\lambda - e^t$ is a representation for $d - r^t > \mathbf{0}$ using reactions from $\text{Mon}(d)$ only, a contradiction. All remaining reactions $r \in \text{Mon}(d) \setminus U_0$ with $r \neq d$ can indeed be used to represent d, as also $r' = d - r \in \text{Mon}(d) \setminus U_0$ holds (and thus $d = r + r'$ is a representation). This implies indeed $\mathcal{R}(d) = Mon(d) \setminus U_0$.

There is a network being conformal with $\mathcal{X}'$ if and only if $\Lambda(d) \neq \emptyset$ for all $d \in \mathcal{D}$ and the minimal network $\mathcal{B} = \bigcup_{d \in \mathcal{D}} \mathcal{R}(d)$ containing all networks being conformal with $\mathcal{X}'$ is explicitly known (without any computation). $\square$

In particular, $\Lambda(d)$ contains minimal inhomogeneous solutions only; it is not necessary to solve equation system (4) in order to determine $\mathcal{B}$ as all sets $\mathcal{R}(d)$ can be listed explicitly.

Moreover, we can deduce an efficient *feasibility test* for both the existence of a representation for $d \in \mathcal{D}$ and a conformal network: whenever a positive vector d occurs in $\mathcal{D}$, the problem is not solvable with the considered set N of elements.

In this situation, we again extend the set of elements by additional elements in $N_a = \{n + 1, \ldots, n + a\}$ with $u_i = 1$ for each of them and extend the n-dimensional vectors in $\mathcal{X}'$ and $\mathcal{D}$ accordingly. We call this situation the *mixed monotone case* as the additional elements clearly do not satisfy the monotonicity property, but the original elements only. The columns of the matrix R have to be chosen from the set

$$Mon(\overline{d}) = \left\{ r \in \mathbb{Z}^{n+a} : \begin{array}{l} 0 \leq r_i \leq d_i \text{ if } i \in N_o, \ d_i > 0 \\ d_i \leq r_i \leq 0 \text{ if } i \in N_o, \ d_i < 0 \\ r_i = 0 \text{ if } i \in N_o, \ d_i = 0 \\ -1 \leq r_i \leq 1 \text{ if } i \in N_a \end{array} \right\} \setminus \mathbb{Z}_+^{n+a}$$

and the system

$$\sum_{r^t \in Mon(\overline{d})} \lambda_t r^t = \overline{d}, \ \lambda_t \in \mathbb{Z}_+ \tag{5}$$

has to be solved for all possible start values $\overline{d}_i \in \{-1, 0, 1\}$ for all $i \in N_a$. In the mixed monotone case, homogeneous solutions can involve additional elements only; thus all of them are integral solutions η of the system

$$\sum_{r^t \in \mathcal{R}(0)} \eta_t r^t = 0, \ \eta_t \in \mathbb{Z}_+$$

using vectors from the following set

$$\mathcal{R}(0) = \left\{ r \in \mathbb{Z}^{n+a} : \begin{array}{l} r_i = 0 \quad \text{if } i \in N_o \\ -1 \leq r_i \leq 1 \text{ if } i \in N_a \end{array} \right\} \setminus \mathbb{Z}_+^{n+a}.$$

In the mixed monotone case with $a = 1$, for every single vector $d \in \mathcal{D}$ there is an extension $\overline{d}$ which has a representation. Thus, the existence of a conformal network again depends on the sequence of vectors in $\mathcal{D}$.

Theorem 8. *Let $\mathcal{X}'$ satisfy the monotonicity property on N_o and let $a = 1$.*

- *For each $d \in \mathcal{D}$, we have $\Lambda(\overline{d}) = \Lambda_{\text{inhom}}(\overline{d})$ and $\mathcal{R}(\overline{d}) \subseteq Mon(\overline{d}) \setminus U_1$ with*

$$U_1 = \left\{ r \in \mathbb{Z}^{n+1} : \begin{array}{l} r_i = d_i \quad \text{for all } i \in N_o \text{ with } d_i \leq 0 \\ r_i \neq d_i \quad \text{for at least one } i \in N_o \text{ with } d_i > 0 \\ r_{n+1} = -1 \end{array} \right\}.$$

Moreover, if $d > 0$, then there is a unique representation

$$\mathcal{R}(\overline{d}) = \left\{ \begin{pmatrix} d \\ -1 \end{pmatrix} \right\} \ \text{and} \ \Lambda(\overline{d}) = \left\{ e^t : r^t = \begin{pmatrix} d \\ -1 \end{pmatrix} \right\}.$$

- *There is a network conformal with $\mathcal{X}'$ if and only if $\mathcal{X}'$ does not contain three consecutive states $x^j, x^{j+1}, x^{j+2} \in \mathcal{X}'$ with $d^j, d^{j+1} > 0$.*

Proof. Due to the monotonicity property on N_o and $u_{n+1} = 1$, $\mathcal{R}(\overline{d}) \subseteq \text{Mon}(\overline{d})$ follows. As $\mathcal{R}(0)$ consists of the vector $(0, \ldots, 0, -1)^T$ only, there are clearly no homogeneous solutions, which implies $H = \emptyset$ and $\Lambda(d) = \Lambda_{\text{inhom}}(d)$ again. For any $d \in \mathcal{D}$, we have $r^t = (d, -1)^T \in \text{Mon}(\overline{d}) \setminus U_1$ and thus $\lambda = e^t$ is always a solution. Thus, for any $d \in \mathcal{D}$ there is an extension $\overline{d}$ such that both sets $\mathcal{R}(\overline{d})$ and $\Lambda(\overline{d})$ are non-empty.

If $d > 0$, then $\text{Mon}(\overline{d})$ only contains vectors r with $r_{n+1} = -1$. This yields $\overline{d}_{n+1} = -1$, and thus a single reaction has to create all changes on N_o at once, which makes the above described representation the unique one in $\Lambda(\overline{d})$. Any representation $\lambda \in \Lambda(\overline{d})$ with $\lambda_t > 0$ for some $r^t \in U_1$ would result in a representation $\lambda - e^t$ of the vector $\overline{d} - r^t > 0$ using only reactions from $\text{Mon}(\overline{d})$, but $\overline{d}_{n+1} - r^t_{n+1} \geq 0$ contradicts the unique representation. Hence, only reactions $r \notin U_1$ can be used.

If there are three consecutive states $x^j, x^{j+1}, x^{j+2} \in \mathcal{X}'$ with $d^j, d^{j+1} > 0$, the unique representations imply $\overline{d}^j_{n+1} = -1$ (and thus $\overline{x}^j_{n+1} = 1$ and $\overline{x}^{j+1}_{n+1} = 0$) and $\overline{d}^{j+1}_{n+1} = -1$ (and thus $\overline{x}^{j+1}_{n+1} = 1$ and $\overline{x}^{j+2}_{n+1} = 0$), a contradiction. Hence, there is no conformal network in this situation, but there is always one otherwise (as we can find appropriate values for the additional element). $\square$

Thus, if $\mathcal{X}'$ gives rise to two consecutive positive vectors in $\mathcal{D}$, at least two additional elements are required. We next obtain that two such elements always suffice to obtain a conformal network. Since in this situation the set $\Lambda(\bar{d})$ is infinite for each $d \in \mathcal{D}$, we consider again the realizations of the solutions in $\Lambda(\bar{d})$ as sequences. For $a = 2$, there is exactly one minimal homogeneous solution $\eta \in H$ using the two vectors $r^t = (0, \ldots, 0, 1, -1)^T$ and $r^{t'} = (0, \ldots, 0, -1, 1)^T$. Thus, any solution in $\Lambda_\sigma(\bar{d})$ is of the form $\lambda + k\,\eta$ with $\lambda \in \Lambda_{\text{inhom}}(\bar{d})$ and the vectors corresponding to λ alternate in the worst case with r^t and $r^{t'}$ in any sequence $\sigma_{\bar{d}, \lambda + k\,\eta}$. This enables us to derive an upper bound on the number of solutions in $\Lambda_\sigma(\bar{d})$.

Theorem 9. *Let $\mathcal{X}'$ satisfy the monotonicity property on N_o and $a = 2$.*
- *For each $d \in \mathcal{D}$, the set $\Lambda(\bar{d})$ is empty iff $d \geq \mathbf{0}$, $\bar{d}_{n+1} + \bar{d}_{n+2} \geq 1$ and infinite otherwise. We have $|\Lambda_\sigma(\bar{d})| \leq \left(1 + |\mathcal{R}(\bar{d})|\right)^l$ with $l \leq 1 + 2\sum_{i \leq n} |d_i|$ and $\mathcal{R}(\bar{d}) \subseteq Mon(\bar{d})$.*
- *There is always a network being conformal with $\mathcal{X}'$.*

Proof. Due to the monotonicity property on N_o and $u_{n+1} = u_{n+2} = 1$, $\mathcal{R}(\bar{d}) \subseteq Mon(\bar{d})$ follows. If $d \geq \mathbf{0}$, then similar arguments as in the general case show that there is no representation if $(\bar{d}_{n+1}, \bar{d}_{n+2})^T > \mathbf{0}$; otherwise we have a valid decomposition $\bar{d} = (d, +1, -1)^T + (\mathbf{0}, -1, +1)^T$ whenever $\bar{d}_{n+1} = \bar{d}_{n+2} = 0$ holds. If $\bar{d} \not\geq \mathbf{0}$, then $\bar{d} = r^t \in Mon(\bar{d})$ follows and $\bar{d}$ can be represented by $\lambda = e^t$. Thus, there is for any $d \in \mathcal{D}$ an extension $\bar{d}$ such that $\Lambda_{\text{inhom}}(\bar{d})$ is non-empty, and the existence of a homogeneous solution $\eta \in H$ implies that $\Lambda(\bar{d})$ is infinite.

The number of representations in $\Lambda_\sigma(\bar{d})$ is at most the number of sequences $\sigma_{d,\lambda}$, which is clearly bounded by $\sum_{i=0}^{l} |\mathcal{R}(\bar{d})|^i \leq (1 + |\mathcal{R}(\bar{d})|)^l$ with l being the maximal length of such a sequence. To also derive a bound on l, consider a subsequence σ' of $\sigma_{d,\lambda}$ with $d' = r^1 + r^2 + \cdots + r^k$ consisting of reactions from

$$\mathcal{R}(\mathbf{0}) = \left\{ \begin{pmatrix} 0 \\ -1 \\ -1 \end{pmatrix}, \begin{pmatrix} 0 \\ -1 \\ 0 \end{pmatrix}, \begin{pmatrix} 0 \\ -1 \\ 1 \end{pmatrix}, \begin{pmatrix} 0 \\ 0 \\ -1 \end{pmatrix}, \begin{pmatrix} 0 \\ 1 \\ -1 \end{pmatrix} \right\}$$

only. For any $r^i \in \mathcal{R}(\mathbf{0})$, $r^i_{n+1} + r^i_{n+2} \leq 0$ holds, thus $d'_{n+1} + d'_{n+2} \leq 0$ follows, which implies either $d' = \mathbf{0}$ (the subsequence is a cycle) or d' has a negative component (thus, $d' \in \mathcal{R}(\mathbf{0})$). Hence, in any representation of $\bar{d}$, each subsequence only modifying the elements N_a can be replaced by a single reaction in $\mathcal{R}(\mathbf{0})$, without changing the behavior of the network. We only have to consider sequences, where each reaction in $\mathcal{R}(\mathbf{0})$ is followed by a reaction $r \in \mathcal{R}$ which modifies an original element $i \in N_o$. Due to the monotonicity property, such changes on N_o can occur at most $\sum_{i \leq n} |d_i|$ times, which leads to the specified bound.

One conformal network can be obtained by alternatively extending the difference vectors

$$\bar{d}^1 = \begin{pmatrix} d^1 \\ -1 \\ +1 \end{pmatrix}, \bar{d}^2 = \begin{pmatrix} d^2 \\ +1 \\ -1 \end{pmatrix}, \ldots, \bar{d}^m = \begin{pmatrix} d^1 \\ \mp 1 \\ \pm 1 \end{pmatrix},$$

and using these vectors as reactions and representations. Thus, for $a = 2$ there is always a conformal network.   □

Hence, our results show that we can efficiently solve the problem to reconstruct a conformal network using up to two additional elements, if the experimental data are monotone.

4 Concluding Remarks

To summarize, we provide in this paper different results for the reconstruction of a conformal network, depending on the quality of the provided experimental time series data $\mathcal{X}'$:

- In the *general case* where no information is known on the intermediate states between two consecutively measured states in $\mathcal{X}'$, we have to consider homogeneous solutions in order to represent a difference vector $d \in \mathcal{D}$. As soon as at least 3 elements are considered in N, all potential reactions from $\mathcal{R}$ occur in one minimal homogeneous solution (Lemma 1), which implies $\mathcal{R}(d) = \mathcal{R}$ for all $d \in \mathcal{D}$ (Theorem 2) and, thus, also $\mathcal{B} = \mathcal{R}$ (Corollary 1). Hence, none of the potential reactions from $\mathcal{R}$ can be excluded as transition of the studied conformal network. As consequence, the network has a huge number of transitions (and is also conformal with any other experimental data using the elements in N).
- In the case of *locally bounded data*, an oscillation of the values of the intermediate states between two consecutively measured states in $\mathcal{X}'$ can be restricted as all such states belong to the box $\mathcal{X}(d)$. Here, the representations of d and the involved homogeneous solutions use exclusively reactions from the restricted set $Box(d)$. This setting corresponds locally to the general case (using d instead of u as capacity bound, $\mathcal{X}(d)$ instead of $\mathcal{X}$ as state space, and $Box(d)$ instead of $\mathcal{R}$ as reaction space) which is reflected by Theorem 3 and Theorem 4. In contrary to the general case, the representability of a vector $d \in \mathcal{D}$ does not depend on the number of elements in N, but in its support. This implies a *feasibility test* for the existence of a conformal network: the problem is not solvable with the considered set N of elements in two cases with $d > 0$ and $|\mathrm{supp}(d)| \leq 2$ (Corollary 2). If, however, a conformal network exists as $d \not> 0$ or $|\mathrm{supp}(d)| \geq 3$ holds for all $d \in \mathcal{D}$, each set $\mathcal{R}(d)$ is explicitly known and $\mathcal{B}$ can be easily constructed without any computation (Corollary 2).
- In the case of *monotone data*, an oscillation of the values of the intermediate states between two consecutively measured states x^j, x^{j+1} in $\mathcal{X}'$ can be excluded. Thus, no homogeneous solutions occur and the representations of d use exclusively reactions from the restricted set $Mon(d)$. We can characterize the set $\mathcal{R}(d)$ for all $d \in \mathcal{D}$ (Theorem 7) and, thus, again explicitly construct the set $\mathcal{B}$ without any computation, provided a conformal network exists. We again have a *feasibility test* for the existence of a conformal network: the problem is not solvable with the considered set N of elements if and only if $\mathcal{D}$ contains a positive vector.

To conclude, we could show for all three cases in which situations a conformal network exists. In particular, it turned out that it is not necessary to determine the set $\Lambda(d)$, as we could characterize all sets $\mathcal{R}(d)$ combinatorially and, thus, the set $\mathcal{B}$ can be explicitly constructed without any computation. While infeasibility occurs in the general setting only in some pathological cases (with $n \leq 2$), we have infeasibility in the locally bounded (and, thus, in the monotone) case if one vector $d \in \mathcal{D}$ is positive. If this happens, we introduce additional elements which enables us to generate conformal networks as there exists a network being conformal with such data with

- $a = 1$ if and only if $\mathcal{D}$ contains no two consecutive vectors $d^j, d^{j+1} > \mathbf{0}$;
- $a = 2$ always.

Thus, we conclude that the proposed approach provides a powerful tool for efficiently reconstructing Petri net models of biological systems.

Acknowledgment. The authors thank Monica Heiner for helpful discussion which helped to improve the presentation of the results.

References

1. Bertsimas, D., Weismantel, R.: Optimization over Integers. Dynamic Ideas, Belmont, MA (2005)
2. Durzinsky, M., Marwan, W., Wagler, A., Weismantel, R.: Automatic reconstruction of molecular and genetic networks from experimental time series data. BioSystems 93, 181–190 (2008)
3. Heiner, M., Koch, I.: Petri net based model validation in systems biology. In: Proceedings of 25th International Conference on Application and Theory of Petri Nets, Bologna. Springer, Berlin (2004)
4. Marwan, W.: Detecting functional interactions in a gene and signalling network by time-resolved somatic complementation analysis. BioEssays 25, 950–960 (2003)
5. Hoff, W.D., Jung, K.-H., Spudich, J.L.: Molecular mechanism of photosignaling by archaeal sensory rhodopsins. Annu. Rev. Biophys. Biomol. Struct. 26, 223–258 (1997)
6. Marwan, W., Sujatha, A., Starostzik, C.: Reconstructing the regulatory network controlling commitment and sporulation in physarum polycephalum based on hierarchical petri net modeling and simulation. J. Theor. Biol. 236, 349–365 (2005)
7. Marwan, W., Wagler, A., Weismantel, R.: A mathematical approach to solve the network reconstruction problem. Math. Methods of Operations Research 67, 117–132 (2008)
8. Matsuno, H., Tanaka, Y., Aoshima, H., Doi, A., Matsui, M., Miyano, S.: Biopathways representation and simulation on hybrid functional petri net. Silico Biol. 3, 389–404 (2003)
9. Pinney, J.W., Westhead, R.D., McConkey, G.A.: Petri net representations in systems biology. Biochem. Soc. Trans. 31, 1513–1515 (2003)
10. Wagler, A., Weismantel, R.: The combinatorics of modeling and analysing biological systems. Natural Computing (submitted)

Analyzing a Discrete Model of *Aplysia* Central Pattern Generator*

Ashish Tiwari and Carolyn Talcott

SRI International, Menlo Park, CA 94025
{tiwari,clt}@csl.sri.com

Abstract. We present a discrete formal model of the central pattern generator (CPG) located in the buccal ganglia of *Aplysia* that is responsible for mediating the rhythmic movements of its foregut during feeding. Our starting point is the continuous dynamical model for pattern generation underlying fictive feeding in *Aplysia* proposed by Baxter et. al. [1]. The discrete model is obtained as a composition of discrete models of ten individual neurons in the CPG. The individual neurons are interconnected through excitatory and inhibitory synaptic connections and electric connections. We used Symbolic Analysis Laboratory (SAL) to formally build the model and analyzed it using the SAL model checkers. Using abstract discrete models of the individual neurons helps in understanding the buccal motor programs generated by the network in terms of the network connection topology. It also eliminates the need for detailed knowledge of the unknown parameters in the continuous model of Baxter et. al. [1].

1 Introduction

Background The last several years have witnessed rapid growth in the amount of detailed high quality experimental data on neural processes underlying behavior. Concurrently, computational neuroscience has also experienced a surge of activity in the formulation of models of increasing complexity. These twin developments present opportunities as well as challenges to neuroinformatics. There are exciting opportunities to describe and simulate neural processes in hitherto unprecedented detail. The challenges are to manage vast amounts of intricate data consisting of experimental and model-derived results, and also to construct tractable models at levels of abstraction that provide useful insights for guiding research. Many current models of neural processes utilize a framework of differential equations such as the Hodgkin-Huxley (H-H) equations [2]. Other modeling techniques include integrate-and-fire methods and artificial neural networks [3, 4]. Detailed models, such as the H-H models, tend to exhibit high sensitivity to system parameters and consequently require very accurate measurements of these parameters. However, such data are frequently

---

* Research supported in part by the National Science Foundation under grants IIS-0513857 and CNS-0720721.

M. Heiner and A.M. Uhrmacher (Eds.): CMSB 2008, LNBI 5307, pp. 347–366, 2008.

unavailable. Furthermore, these models do not scale easily because they rapidly become intractable as the number of cells incorporated increases. The situation is analogous to that in Systems Biology. A complementary approach based on using logical computing formalisms and symbolic techniques [5, 6, 7], called Symbolic Systems Biology, is being successfully pursued. By representing knowledge at an abstract, symbolic level this approach has enabled development of models capable of performing sophisticated queries about large models of biological processes, while not being overly constrained by lack of low-level details.

Neurons and Neural Circuits. Neurons are highly specialized eukaryotic cells capable of communicating with each other by means of electrical and chemical signaling (see, for example, [8]). While there are several kinds of neurons, all possess a cell body, called the soma, from which emerge several tree like structures called dendrites as well as a single long tube called the axon, which ends in several branches, the synaptic terminals. The synaptic terminals of the transmitting (presynaptic) neuron communicate to the dendrites of nearby receiving (postsynaptic) neurons by releasing specialized molecules, the neurotransmitters, such as glutamate, serotonin, and dopamine. A neurotransmitter is released by the presynaptic neuron once a sufficient transmembrane voltage depolarization has reached the synaptic terminal. The buildup of membrane depolarization at the synapse occurs after a nerve impulse, or action potential, travels from the cell body along the axon in a series of depolarizations and repolarizations caused by transfer of sodium and potassium ions between the cytoplasm of the neuron and the extracellular space. The ions are transferred by electrochemical gradients and they cross the membrane through voltage-dependent channels, which are membrane pores that become permeable to ions due to conformational changes induced by transmembrane voltage depolarizations. Chemical synaptic connections between neurons can be of two types, inhibitory, and excitatory. Neurons can also signal each other via electrical synapses (also called gap junctions) in which the membranes of the two neurons contact each other via transmembrane pores through which electrical signals can spread bidirectionally. All of these, and other factors, endow neural signaling with a rich collection of signaling modalities and enormous complexity in communications and signaling behavior.

Feeding Behavior and its Neural Circuit. A major goal of neuroscience is to understand neural processes that underlie the generation of behavior and the modification of behavior induced by learning. A basic tenet of neuroscience is that the ability of the nervous system to generate behaviors arises from the organization of neurons into networks or circuits and that the functional capabilities of these circuits emerge from interactions among: i) the intrinsic biophysical properties of the individual neurons, ii) the pattern of the synaptic connections among these neurons, and iii) the physiological properties of these synaptic connections. To investigate how neural circuits function, a diverse collection of animal model systems has been developed. By virtue of their relatively simple nervous systems, often with large identifiable neurons that are amenable to detailed study, invertebrates are frequent candidates for cellular analyses of neural circuits and their relationship to behavior. A key advantage of invertebrate model systems is that the neural circuits controlling specific behaviors often contain relatively few neurons, which allows the circuit to be described in detail on a cell-by-cell and synapse-by-synapse basis. Another advantage is that many of these neural circuits produce fictive motor patterns when isolated in vitro, which facilitates investigating how behaviorally relevant neural activity is generated and regulated.

One useful animal model system is the marine mollusk Aplysia. Feeding behavior in Aplysia has been an important focus of study and research. The neural circuitry that mediates feeding behavior is located primarily in the cerebral and buccal ganglia. The structure of this circuit is relatively well understood [9]. A major component is the central pattern generator (CPG) in the buccal ganglia (Figure 1) that generates the motor activity for controlling the rhythmic movements of the odontophore and radula. Although a great deal is known about the individual components of the feeding circuitry, how this collection of cells and synapses functions as a circuit is not well understood. The nonlinear, diverse and dynamic nature of neural circuits provide formidable challenges to systematically analyzing and understanding how circuits operate and adapt. Symbolic system models can help in this endeavor. We believe that properties of a neuron network can be understood as properties of the *connections* between the neurons in the network. Such abstract models can also be used to test the plausibility of hypotheses and to manipulate components of the system in a manner that may not be feasible in the real nervous system.

2 Biology

We describe the biology of feeding in *Aplysia* and its neural control. The material here is taken mainly from Baxter et. al. [10, 1].

The feeding cycle in *Aplysia* consists of ingestion, which brings food into the buccal cavity, and egestion, which ejects unwanted material from the buccal cavity. These two stages of feeding involve rhythmic movements of structures in the foregut that can be classified into two phases: a protraction phase, where the jaws open and the odontophore rotate forward and a retraction phase, the odontophore retracts and the jaws close. During ingestion, the two halves of the radula (grasping surface of the odontophore) are separated and open during protraction, and closed during retraction. This causes the food to enter the buccal cavity. On the other hand, during egestion, the radula is closed during protraction and open during retraction. This causes ejection of unwanted material.

The buccal ganglia of *Aplysia* contain a central pattern generator (CPG) that controls the rhythmic movements of the foregut during feeding. The CPG generates two corresponding buccal motor programs (BMP), one for ingestion and another for egestion. Based on extensive intracellular recordings of action potentials, Baxter et.al. [10, 1] have proposed a speculative type of model of the CPG. This model contains continuous dynamical models of ten neuron cells: B4, B8, B31, B34, B35, B51, B52, B63, B64, and a hypothetical neuron Z, and their connections.

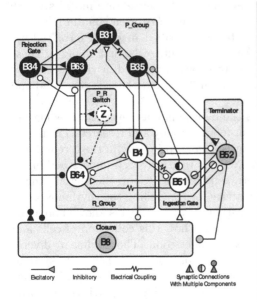

This diagram shows the model of the central pattern generator. The circles represent neurons and the edges represent the interconnections. Subsets of neurons are collected into groups to indicate their function/role. The P_Group neurons are active during protraction and the R_Group neurons during retraction. The rejection gate neuron participates in egestion, whereas the ingestion gate neuron plays a role in the ingestion BMP. The circle-terminated edge represents inhibitory synaptic connection, the triangle-terminated edge denotes an excitatory synaptic connection, and the wiggly edge denotes an electrical coupling. An edge terminating on a group indicates that it effects all members of the group.

Fig. 1. Central Pattern Generator: Neurons and their interconnections that are responsible for generating the BMPs associated with ingestion and egestion [1]

We give a brief description of the role of these neurons. B8 is a radula-closer motor neuron and the timing of its activity can distinguish ingestion from egestion. Neurons B31, B35, and B63 are active during the radula protraction phase whereas B4 and B64 are active during the radula retraction phase. The neuron B52 terminates the retraction phase. The hypothetical neuron Z mediates transition from the protraction to the retraction phase. The neurons B34 and B51 control B8 and are part of the system that switches mode from egestion to ingestion and vice-versa.

The ten identified neurons are connected to each other forming a complex network. Each interconnection is either an excitatory synaptic connection, an inhibitory synaptic connection, or an electric coupling. The connections between the ten neurons used in the model of Baxter et.al. [10, 1] are reproduced in Figure 1.

3   Related Work

Baxter et.al. [1] presented a continuous dynamical system model of the ten neuron interconnected network. They used continuous differential equations, in the form of Hodgkin-Huxley-type models, for each of the ten cells and their electrical and synaptic connections. Unfortunately, such a model requires inferring, either experimentally or computationally, several parameters that describe the details of membrane currents. Specifically, there are about 18 parameters for each ion channel (Na, K) of each neuron. In addition, there are three parameters for each synaptic connection, two for synaptic plasticity, and two for every electric coupling. Estimating these parameters is a challenge and after the parameters have been discovered, the result is an immensely complicated model that can only be analyzed by simulation.

In [11] a simple rewriting logic model of a two neuron subsystem (B31,B63) of the Aplysia bucchal ganglia was presented. This work demonstrated that essential features of the two neuron system could be modeled by appropriate choice of a small number of parameters. This preliminary success lead us to look for a more principled way of determining the parameters that control a neuron's behavior. After studying the various proposed neuron models [4, 12], including the "simple model" of [4], we built a highly abstract qualitative model of a single neuron and used it as a starting point.

Hybrid systems have been used as a modeling language for System Biology [13] in general and for modeling single neurons [14] in particular. Due to their high expressiveness, hybrid models can more closely approximate HH models, but hybrid analysis does not scale to studying a large collection of cells and is restricted to studying a single cell [14]. Our interest was in analyzing a large neuron network and this motivated looking at discrete models.

4   Discrete Formal Model

In this section we present a simple discrete model of the central pattern generator derived from the continuous model of Baxter et.al. [1]. The model of a single neuron is a simple generic qualitative "integrate and fire" model. We use the *same* model for each of the ten neurons, but specialize the generic model for some of the "special" neurons in the network. The electrical and synaptic interconnections are also modeled at a highly abstract qualitative level. There are no parameters that need instantiation in our model. The properties exhibited by the model are solely a consequence of the abstract model of neuron behavior and the positive and negative *interconnections*.

We will complement our informal presentation of the discrete model with a formal description in the Symbolic Analysis Laboratory (SAL) language [15]. SAL is a formal language for describing discrete state transition systems. We do not present a detailed introduction to the syntax and semantics of SAL here. Since SAL syntax uses inspiration from standard imperative and functional languages, readers unfamiliar with SAL can still possibly understand and appreciate the formal description. The full SAL model is available online [16].

4.1   Discrete Model of a Single Neuron

A generic neuron is a simple input/output automaton. It receives an input `i`, changes its internal state `level` in response to it, and optionally produces an output `o`. In our model, the input `i` can be one of three qualitative values: `pos`, `neg`, and `zero`, collectively referred to as `SIGS`. The value of `i` indicates whether the

```
N: NATURAL = 4;
LEVELS: TYPE = [0 .. N];
SIGS: TYPE = { pos, neg, zero };
NEURONS: TYPE = { B31, B35, B63, B34, B64, B4, B51, B52, B8, Z };
PHASES: TYPE = { protraction, retraction, termination };
GNEURONS(n: NEURONS): BOOLEAN = ( n /= B31 AND n /= B64 );
```

Fig. 2. SAL Global Declarations. N is a constant. LEVELS, SIGS, NEURONS, PHASES are types denoting finite sets. GNEURONS is a function that returns true when its argument is a generic neuron, and false when it is a specific neuron (B31, B64).

neuron receives a positive (pos), negative (neg), or no (zero) impulse (from its neighbors). The internal state of the neuron stores a value in the range $[0, \ldots, N]$ in a variable called level[1]. A value of 0 indicates that the neuron is at its resting potential, and N indicates that the neuron at its highest membrane potential. The values in between represent abstractions of the concrete intermediate membrane potentials. The output o is a Boolean-valued variable. A value of TRUE indicates that the neuron emits an impulse, whereas FALSE indicates that it does not do so.

Figure 2 shows the declarations for the parameter N[1] and types used to describe the complete discrete CPG model. The type NEURONS denotes the set of all ten neurons, while the function GNEURONS identifies the eight that are generic. The type PHASES denotes the set of three phases.

The dynamics of single generic neurons are given by the following intuitive rules. The rules abstractly capture the "integrate and fire" behavior of neurons.

Integrate Positive Input Impulse (IPII): If the input impulse i is positive (pos) and the level level is less-than N, then the level level is incremented. The amount of increment is given by a parameter sens. Again, the properties of the model are robust to changes in the value of sens. We use sens $= 2$ in our experiments.

Integrate Negative Input Impulse (INII): If the input impulse i is negative (neg) and the level level is less-than N, then the level level is decremented. The amount of decrement is fixed to 2 units. The increments and decrements are always saturated so as to force the value of level to remain in the range $[0, \ldots, N]$.

Fire: If the level level is equal to N, then it is reset to $N - 2 * \text{sens}$ (which is 0 for the choices made above). Additionally, the output o is set to True whenever level $= N$ to indicate that the neuron fired.

If none of the conditions are applicable, then the state of the neuron remains unchanged. In particular, this means that we do not model decay of the membrane potential (level) with time. This is because in the time intervals of interest, decay does not play an important role in determining the overall behavior

---

[1] We could use any positive value for N. We used the value 4 in our model, but the choice is really arbitrary and we could have used a different value, say 5 or 3: the properties of the final CPG model do *not* dependent on the exact value used.

```
generic[n: NEURONS]: MODULE = BEGIN
  INPUT i: SIGS
  OUTPUT o: BOOLEAN, level: LEVELS
  LOCAL sens: [1 .. 2], pir: BOOLEAN
  INITIALIZATION
     pir = FALSE; level = 0
  DEFINITION
     sens = IF (n = Z) THEN 1 ELSE 2 ENDIF;
     o = (level = N)
  TRANSITION
     [
     FIRE: level = N AND GNEURONS(n) AND NOT(pir) -->
              level' = N - 2*sens
     []
     IPII: level < N AND i' = pos -->
              level' = IF (level > N - sens) THEN N ELSE level + sens ENDIF
     []
     INII: (NOT(GNEURONS(n)) OR level < N) AND i' = neg -->
              level' = IF (level > 1) THEN level - 2 ELSE 0 ENDIF
     []
     SPIR: n = B52 AND level = 0 AND i = neg AND i' = neg -->
              pir' = TRUE; level' = N
     []
     ELSE -->
     ]
END;
```

Fig. 3. SAL Model of a Generic Neuron. The model is parameterized by the name n of the neuron. The model is specialized for (a) neurons that exhibit a plateau-like potential (identified by NOT(GNEURONS(n)) in the code above), (b) the neuron B52 that exhibits postinhibitory rebound (PIR). The ELSE clause says that when all the above conditions fail, do not change the state of the system. The neuron model responds to the value of i in the "next" step (i') because the module computing i (aplysia_wiring described later) already introduces a one-step delay.

of the system. Also note that the model of a single neuron is *deterministic*. The conditions for the three cases above are mutually exclusive and given an input i and level level, the next state of the neuron is deterministically specified.

The formal description of the above neuron model is given in SAL in Figure 3. Since some of the neurons exhibit behavior that can not be entirely captured by an "integrate and fire" model, the generic model has a couple of specializations. The first one is for neurons that exhibit a plateau-like potential. In the model of the CPG, neurons B31 and B64 fall in this category [17, 18]. They are modeled to behave just like the other neurons, except that they do not fire; that is, the membrane potential (level) is not reset (to 0) when it reaches its highest value N). As we shall later discuss, this distinction is important for the CPG to exhibit the observed behavior.

```
aplysia_neurons: MODULE =
  (WITH INPUT ins: ARRAY NEURONS OF SIGS
   WITH OUTPUT outs: ARRAY NEURONS OF BOOLEAN
   WITH OUTPUT levels: ARRAY NEURONS OF LEVELS
    (|| (n: NEURONS): (RENAME o to outs[n], i to ins[n],
                              level to levels[n] IN generic[n]))));
```

Fig. 4. SAL model of the collection of the ten neurons in the CPG. For example, ins[B63], outs[B63], and levels[B63] will now denote the input signal for B63, the output generated by B63 and the internal level of B63.

The second specialization is for neuron B52 that exhibits postinhibitory rebound (PIR). PIR is defined as membrane depolarization (activation) occurring at the offset of a hyperpolarizing stimulus. The mechanism for PIR is not well understood. In our abstract model, we modeled it in the form of a special transition (labeled SPIR in Figure 3) that depolarizes B52 when it is at its resting potential and it receives an inhibitory pulse. Again, this special behavior of B52 is important for ensuring termination of the ingestion and egestion neural programs. Neither of these special behaviors can be exhibited by a simple "integrate and fire" model [4].

4.2 Modeling a Collection of Neurons

We get a model of each of the ten neurons in the CPG network (enumerated in Figure 2) by instantiating the model of a generic neuron described above ten times. Thus, we get a collection of ten identical neurons. The few subtle distinctions between these ten neurons have already been captured in the description of the generic neuron described above.

Figure 4 describes the SAL code for generating models of the ten neurons. The ten instantiations of the generic neuron are synchronously composed (|| is the synchronous composition operator). The input i, output o, and the level level variables are renamed to avoid a naming conflict.

The result is a collection of ten neurons, but there is no interconnection between them yet. In the next section, we will model the interconnections.

4.3 Modeling the Interconnects

As mentioned before, there are three types of connections between the neurons in our model: excitatory synaptic connection, inhibitory synaptic connection, and electrical connection.

Excitatory Synaptic Connection. This is a *directed* connection between a source and a target neuron. In this connection, the source neuron produces an excitatory postsynaptic potential (EPSP) in the target neuron. This connection is modeled as follows: whenever the source neuron fires (that is, generates a TRUE signal on its output port o), the target neuron receives a pos input on its input port; otherwise, it receives only a zero input.

```
epsp(x: BOOLEAN): SIGS = IF x THEN pos ELSE zero ENDIF;
```

Inhibitory Synaptic Connection. This is a *directed* connection between a source and a target neuron. In this connection, the source neuron produces an inhibitory postsynaptic potential (IPSP) in the target neuron. This connection is modeled as follows: whenever the source neuron fires (that is, generates a TRUE signal on its output port o), the target neuron receives a neg input on its input port; otherwise, it receives only a zero input.

```
ipsp(x: BOOLEAN): SIGS = IF x THEN neg ELSE zero ENDIF;
```

Electrical Coupling. Electrical connections are undirected, that is, they effect both neurons that are coupled by an electrical connection. Electrical coupling indicates that there can be flow of current between the two coupled neurons that is proportional to the *difference* in the membrane potentials of the two cells. Since the membrane potentials are abstracted to qualitative levels (level), we model electrical coupling using the difference in the levels of the two neurons. Specifically, if the levels of electrically-coupled neurons, say A and B, are $level_A$ and $level_B$ respectively, then

(a) neuron A receives a pos input and neuron B receives a neg input if $level_B > level_A$;

(b) neuron A receives a neg input and neuron B receives a pos input if $level_B < level_A$; and

(c) both neurons receive a zero input if $level_B = level_A$.

```
diff(x1: LEVELS, x2: LEVELS): SIGS =
IF (x1>x2) THEN pos ELSIF (x1<x2) THEN neg ELSE zero ENDIF;
```

Accumulating Effects from Multiple Connections. The above three cases describe the input each neuron receives from each of its neighboring neurons. Each neuron now has to accumulate all its input signals and map the result to one value: pos, neg, or zero, that it will use as its actual input signal.

The accumulation process is modeled using the following simple rules: If pp denotes the total number of pos inputs and nn denotes the total number of neg inputs received by the neuron, then

1) if pp > nn, then the result is a pos signal.

2) if pp < nn, then the result is a neg signal.

3) Otherwise, the result is a zero signal.

```
integrate(x1: SIGS, x2: SIGS, ..., x7: SIGS): SIGS =
    LET pp:[0..7] = count(pos, x1,x2,x3,x4,x5,x6,x7),
        nn:[0..7] = count(neg, x1,x2,x3,x4,x5,x6,x7) IN
    IF (pp > nn) THEN pos
    ELSIF (nn > pp) THEN neg
    ELSE zero ENDIF;
```

```
aplysia_wiring: MODULE =
BEGIN
  INPUT b63i: SIGS, outs: ARRAY NEURONS OF BOOLEAN
  INPUT levels: ARRAY NEURONS OF LEVELS
  OUTPUT ins: ARRAY NEURONS OF SIGS
  INITIALIZATION ins = [ [i: NEURONS] zero ]
  TRANSITION
  [ TRUE -->
  ins' = [ [n:NEURONS] LET
    ec:[[NEURONS,NEURONS]->SIGS]=LAMBDA(a,b:NEURONS):diff(levels[a],levels[b]),
    ep:[NEURONS->SIGS] = LAMBDA(x:NEURONS): epsp(outs[x]),
    ip:[NEURONS->SIGS] = LAMBDA(x:NEURONS): ipsp(outs[x]) IN
    IF (n=B31) THEN integrate31(ip(B64),ep(B34),ep(B63),ep(B35),
                                ep(B4),ec(B63,B31),ec(B35,B31))
    ELSIF (n=B34) THEN integrate(ep(B63),ip(B64),zero,...,zero)
    ELSIF (n=B63) THEN integrate(ip(B64),ec(B31,B63),ep(B34),b63i,...)
    ELSIF (n=B35) THEN integrate(ip(B64),ec(B31,B35),ip(B52),...)
    ELSIF (n=B64) THEN integrate64(ep(Z),ip(B52),ip(B34),ip(B63),
                                ip(B4),ep(B51),ec(B51,B64))
    ELSIF (n=B4) THEN integrate(ip(B52),ep(B35),ec(B51,B4),ep(B64),...)
    ELSIF (n=B52) THEN integrate(ep(B35),ip(B64),ip(B51),ip(B4),...)
    ELSIF (n=B8) THEN integrate8(ep(B51),ip(B52),ip(B63),ip(B4),ep(B34))
    ELSIF (n=Z) THEN integrate(ep(B63),...) ]
    ELSE integrate51(ip(B35),ip(B52),ip(B4),ec(B64,B51),ec(B4,B51))
  ]
END;
```

Fig. 5. SAL Model of the Interconnections of the ten neurons in the CPG. Neurons B31, B64, B8 and B51 have their own special integrate functions. Missing arguments, indicated by ..., are all zero.

The Wiring Diagram. Figure 5 contains the final wiring diagram for the ten neurons. For each neuron, the `integrate` function described above is used to collect all inputs for that neuron. Depending on the type of connection, each input is obtained using either `epsp`, `ipsp` or the `diff` function.

Some of the neurons, namely B31, B64, B51 and B8, use a specialization of the `integrate` function described above. (Hence the names of the function used in Figure 5 for these neurons are different.) The specialization captures *preference* of some neurons to some excitatory or inhibitory inputs; that is, strength of some connections is stronger than others. The `integrate` function treats all connections equally. If we use the `integrate` function to accumulate inputs *for all* the neurons, then the resulting model's behavior does not match the observed behavior. We therefore specialized `integrate` for neurons B31, B64, B8 and B51 as follows: B31 requires inhibition from B64 to see a negative input, B64 gives high priority to signals from Z and B52, B8 gives lower priority to the inhibitory signal from B63, and B51 gives higher priority to its synaptic connections and lower to its electrical coupling with B4. These priorities can be

```
observer: MODULE =
BEGIN
  OUTPUT b63i: SIGS, phase: PHASES
  INPUT levels: ARRAY NEURONS OF LEVELS
INITIALIZATION b63i = pos
DEFINITION
  phase = IF (b63i=pos OR levels[B31]>=N-1) THEN protraction
          ELSIF (levels[B64]=N OR levels[B4]=N) THEN retraction
          ELSE termination ENDIF
TRANSITION
  [ levels[B35]=N --> b63i' = zero
  [] ELSE --> ]
END;
```

Fig. 6. SAL Model of the Observer: It generates the input trigger and observes phase changes

captured by weighting the inputs and doing a weighted sum in the `integrate` function. These specializations are discussed further in Section 6.

4.4   Exciting the System and Observing the Phases

We wish to study the behavior of the system elicited by a brief depolarization of B63. We add a separate component to our model to inject an abstract depolarizing current pulse to B63. Figure 6 contains the SAL description of this additional "observer" module. The module begins by sending a `pos` input to B63, but as soon as B35 reaches its firing threshold (value N), the external input to B63 is reset to zero. This simulates the effect of giving B63 a brief depolarizing current pulse.

The protraction and retraction phases are characterized by activity in, respectively, the P_Group and the R_Group neurons (Figure 1). The "observer" module observes these phase changes: depending on the levels of B31 (a P_Group neuron) and B64 (a R_Group neuron), it decides whether the phase is protraction, retraction, or termination.

4.5   The Complete Model: Putting It All Together

We have completely described the models of all components of the model, namely the ten neurons, the connections between the neurons, and the observer. The final complete model is simply a synchronous composition of the three components, which is described in SAL as:

```
aplysia: MODULE = aplysia_wiring || aplysia_neurons || observer;
```

Synchronous composition means that at each abstract time step, each of the components (the ten neurons, the wiring, and the observer) simultaneously take a transition.

We analyzed the above discrete abstract model using model checking tools (described in detail in Section 5) and found that the CPG generates an egestion-like behavior. As suggested in the literature [1, 19], external modulatory influences, mostly through the neurotransmitter dopamine, can alter the CPG and cause it to exhibit an ingestion-like behavior. Following the approach of [1], we modeled the effect of dopamine by changing the strength of some of the interconnections. Specifically, we decreased the excitability of B34 and B4. We achieved this by specializing the `integrate` functions for B34 and B4. This resulted in a different model that is revealed to have ingestion-like behavior by model checking.

5   Analysis

The discrete model described above is appealing for its simplicity of description and the lack of any requirement for hundreds of parameters. Moreover, we can use model checking to systematically explore the system for desired behaviors.

As a basic sanity check, we can verify that if there is no input trigger to B63, then all neurons in the system indeed remain in their resting potentials. Excitation of B63 by an external pulse initiates either the ingestion, or the egestion buccal motor program. Figure 7 shows plots generated by the continuous model of [1]. This bursting pattern is typical of the egestion BMP. We can capture the salient features of the patterns in Figure 7 in the form of Linear Temporal Logic (LTL) formulas. The LTL properties can then be model checked against the developed abstract qualitative model of Section 4. This gives us a way to validate our model against (experimental) observations.

We classify the properties into different groups depending on their pertinent phase.

5.1   The Protraction Phase

The protraction phase is characterized by activity in B63, B31, and B35. We first make sure that the input pulse to B63 is modeled correctly by checking that it is initially present (`pos`), and then terminated (`zero`) before the system reaches the retraction phase. Property p0 in Figure 8 formally states this fact in LTL.

An important feature in Figure 7 is that B31 (eventually) reaches a plateau and stays there all through the protraction phase and until the start of retraction. This is captured in Figure 8 Property p1.

The neurons B63 and B35 show periodic firings in the protraction phase. Property p2 partly encodes this fact as follows: at all points until two steps before retraction starts, it holds that B35 eventually fires. The same can be stated and verified for B63. Note that non-plateau neurons, such as B35 and B63, reset upon firing, and hence Property p2 says that B35 *repeatedly* (and not necessarily periodically) fires. In the egestion scenario, the same is also true for B34 and Property p3 captures this. Also during egestion, B8 fires during the protraction phase (Property p4).

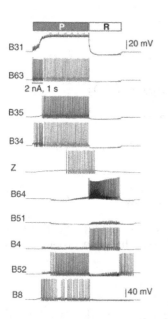

The plots show the activity in each of the ten neurons of the modeled CPG when B63 is excited by a short pulse. The excitation causes activity in the neurons in the protraction group (P-group) neurons (B31, B35, B63) . In the egestion mode, B34 and B8 are active in the protraction phase. Towards the end of the protraction phase, Z starts firing that in turn causes activity in B64. This immediately causes the P-group neurons to switch "off", while the retraction group (R-group) neurons (B4, B64) remain active. In this retraction phase, B8 is inactive, which indicates that this pattern corresponds to egestion. Activity in B52 signals an end of the retraction phase. B51 shows little activity throughout the two phases. In the ingestion mode (not shown here), the plots of B34, B8, B4, and B51 are different. Essentially, during ingestion, B34 remains mostly inactive, B8 shows no activity during protraction, but is active during retraction, B4 shows reduced activity, and B51 shows higher activity.

Fig. 7. [Figure 5 from [1]] Simulating the continuous model of the CPG [1] generates patterns characteristic of egestion. These plots are formalized as LTL properties in Figures 8, 9,10, and 11.

```
p0: THEOREM
    aplysia ⊢ U(b63i = pos, U(b63i = zero, phase = retraction));
p1: THEOREM
    aplysia ⊢ F(levels[B31]=N AND U(levels[B31] >= N-1, phase=retraction));
p2: THEOREM
    aplysia ⊢ U( F(levels[B35] = N), X(X(phase = retraction)) );
p3: THEOREM
    aplysia ⊢ U( F(levels[B34] = N), X(X(levels[B64] = N)) );
p4: THEOREM
    aplysia ⊢ F(phase = protraction AND levels[B8] = N);
```

Fig. 8. Temporal Properties for the Protraction Phase in SAL. $F(A)$ means that A holds eventually, $U(A, B)$ means that eventually B holds and until then A holds, and $X(A)$ means that A holds in the next time step.

We successfully verified all the above properties on our discrete model. Property **p3** and Property **p4** are verified for the egestion scenario, but, as expected, they fail when the model is modified for ingestion. Instead, for the ingestion model, the following property is verified, which states that B8 remains inactive all through the protraction phase.

```
p5: THEOREM aplysia ⊢ G(phase=protraction => levels[B8] < N);
```

```
t1: THEOREM
    aplysia ⊢ F( levels[Z] = N );
t2: THEOREM
    aplysia ⊢ F( b63i = zero AND levels[B64] = N );
t3: THEOREM
    aplysia ⊢ F( levels[B64] = N AND G( levels[B63] < N ) );
t4: THEOREM
    aplysia ⊢ F( phase = retraction AND levels[B64] = N );
```

Fig. 9. Temporal Properties for the Transition Phase in SAL. The notation $G(A)$ means that A holds always from that instance onwards.

5.2   Transitioning from Protraction to Retraction

Figure 9 contains LTL properties pertaining to the switching off of the protraction phase and transitioning into the retraction phase. The main events in the transition from protraction to retraction are:

(a) the hypothetical neuron Z becomes active (Property **t1**),
(b) this causes the depolarization (activation) of B64, (Property **t2**),
(c) this, in turn, simultaneously causes the hyperpolarization (deactivation) of the protraction group neurons, such as B63, (Property **t3**), and
(d) eventually the retraction phase neurons, such as B64, are active (Property **t4**).
Property **t3** also states that B63 remains deactivated ever after. The same can also be said for the other protraction group neurons.

The properties above verify that the system eventually transitions from the protraction to the retraction phase. These properties hold true for the egestion, as well as the ingestion, model.

5.3   The Retraction Phase

Figure 10 contains LTL properties pertaining to the retraction phase. During retraction, B64 remains active (Property **r0**). The protraction phase neurons remain inactive at all instances during the retraction phase. Property **r1** states this for neuron B35, but the property can be stated and verified for the other P-group neurons as well. The same is also true for the neuron B34 (Property **r2**). Furthermore, the neuron B4 repeatedly fires during retraction. Again, this fact is partly encoded in LTL as follows: (Property **r3**) at all instances until two steps before the termination phase starts, it is true that the neuron B4 eventually fires.

The next two properties are specific to egestion. The neuron B51 is not part of the egestion behavior, but participates only during ingestion. Property **r4** states that B51 always remains inactive. Finally, Property **r5** states that the radula-closer motor neuron, B8, is inactive during the retraction phase (in fact, at any non-protraction state).

All the properties described above, except **r4**, are verified to be valid for egestion. Properties **r0**, **r1** and **r2** remain valid even when the model is specialized

```
r0: THEOREM
    aplysia ⊢ G( phase = retraction => levels[B64] = N );
r1: THEOREM
    aplysia ⊢ G( phase = retraction => levels[B35] < N );
r2: THEOREM
    aplysia ⊢ G( phase = retraction => levels[B34] < N );
r3: THEOREM
    aplysia ⊢ W( F(levels[B4] = N), X(X(phase = termination)) );
r4: THEOREM
    aplysia ⊢ G( levels[B51] < N );
r5: THEOREM
    aplysia ⊢ G( phase /= protraction => levels[B8] < N );
```

Fig. 10. Temporal Properties for the Retraction Phase in SAL. The LTL operator W is the "weak until" operator. $W(A, B)$ says that A continues to hold until B becomes true. Unlike $U(A, B)$, here B may never become true.

```
e1: THEOREM
    aplysia ⊢ G(phase=termination => F(G(levels[B64]=0 AND levels[B4]=0)));
e2: THEOREM
    aplysia ⊢ G(phase=termination => G(levels[B31]<N AND levels[B63]<N));
e3: THEOREM
    aplysia ⊢ U(phase=protraction, W(phase=retraction, phase=termination));
```

Fig. 11. Temporal Properties for the Termination Phase in SAL

to the ingestion case. However, the remaining properties, Properties r3, r4 and r5 are, as expected, invalid for ingestion. While the plots in Figure 7 suggest that Property r4 should be valid for egestion, it is not so for our model. We discuss this further in Section 6.

5.4 Termination

Figure 11 contains LTL properties pertaining to the termination of the ingestion/egestion Buccal Motor Program. As is evident from Figure 7, the termination phase is characterized by the hyperpolarization (deactivation) of all the protraction group neurons and the retraction group neurons. Property e1 says that when the system enters the termination phase, eventually B64 and B4 return to their resting levels. Note that B64 and B4 may not be at their resting potential when the termination phase begins, and hence the eventuality operator (F) is important. The same fact can be stated for the P-group neurons, B31, B35, and B63. Property e2 states that B31 and B63 neurons always remain inactive during the termination phase. Since we are stating that the neurons are inactive (levels $< N$), and not asking for levels to be 0, we do not need the eventuality operator (F) here.

Finally, we state one of the most important properties of the CPG network: the protraction phase is followed by the retraction phase, which in turn is (optionally) followed by the termination phase – exactly in that order. This is stated in Property e3, which basically says that the three phases occur sequentially.

The three termination phase properties are model checked and verified to hold for both the egestion and the ingestion model.

6   Results and Discussion

The main observations from the study of the discrete model are described below. Most of the observations made here are similar to those obtained by working with the continuous model based on using Hodgkin-Huxley-type models for the neurons and several hundreds of parameters. Thus, by just looking at the interconnections at a fairly abstract level, it is still possible to build and analyze useful and interesting models that help in refining our understanding of the way a neuron network works.

Specialized neurons. B31 and B64 are different from the other neurons because they exhibit a plateau-like potential [17, 18]. This difference is important to sustain the activity in the protraction (B31) and retraction (B64) phases. If B31 and B64 are modeled in the same way as the other neurons, then the modified model does not exhibit the sustained activity of B31, B35, and B63 during protraction and that of B64 and B4 during retraction.

Electrical couplings. The effect of electrical coupling needs to be necessarily asymmetric to enable the model to have the desired behavior. In a symmetric scenario, if two neurons A and B are electrically coupled and if level$(A) >$ level(B), then level(A) decrements as level(B) increments. In an asymmetric scenario, we are allowed to have, say, level(B) increase while level(A) remains unchanged. In the model, using a symmetric effect on the B31 - B35 coupling and the B31 - B63 coupling would cause the protraction phase to prematurely end. Similarly, B64 - B51 coupling can cause the retraction phase to terminate earlier if the effect is symmetric. We note that electrical coupling is asymmetric in the continuous model [1].

Robust protraction phase. The positive feedback loops between the protraction-phase neurons (B31, B35, B63, B34) lead to a very robust protraction phase, that is, once the system settles into the protraction phase (characterized by periodic firing of these neurons), it is not "easy" to drive the system out of that phase. In fact, we had to make the inhibition of the protraction-phase neurons by B64 a very "strong" signal to really cause the protraction phase to terminate. The transition from protraction to the retraction phase is not very well understood and had led to the hypothetical neuron Z in the model [1]. Our observation here suggests that apart from the unknown component Z, there is another important aspect related to the strengths of the synaptic connections between the P-group and the R-group neurons that is required to ensure transition into retraction.

Retraction phase. Unlike the protraction-phase neurons that are connected in a strong positive feedback, the neurons active in the retraction phase do not form any positive feedback cycle. As a result, the retraction phase is not robust and it can be "easily" terminated. Again, its continued sustainability depends crucially on the intrinsic ability of B64 to maintain a plateau-like potential, *despite the negative (inhibitory) feedback from its inter-connections.* Note that in the speculative model used here (proposed in [1]), B64 gets only inhibitory inputs in the retraction phase of egestion.

Ingestion. The default model exhibits *egestion* behavior, that is, the radula motor neuron (B8) is active (radula is closed) during protraction and inactive (radula is open) during retraction. As suggested in the literature [1, 19], if the excitability of cells B34 and B4 is reduced (characteristic of dopaminergic modulation), then the situation changes and B8 is inactive (radula is open) during protraction and active (radula is closed) during retraction. This corresponds to the ingestion BMP. The explanation for this change is as follows: reduced excitability of B34 causes it to remain inactive during the protraction phase, and hence B8, which depended on B34 for excitatory pulse, remains inactive too. On the other hand, during retraction, the low excitability of B4 causes it to remain relatively inactive, which allows B51 to activate and cause B8 to activate as well. Note that the mode change is solely explained by the network and the connections and does not depend on the detailed physical modeling of the single neurons or their synaptic connections.

B51. In contrast to the prediction made by the continuous model of [1], B51 is activated during egestion in our model. However, it does not affect the activation pattern of B8. It is possible to reduce the responsiveness of B51 to positive signals and have it remain inactive during egestion. However, in that case, it also tends to remain inactive during ingestion: the reduced inhibition by B4 (during ingestion) is not enough to overcome the reduced responsive of B51 to positive signals. These results indicate that the unmodeled components – the sensory input neurons and the motor neurons – may have a significant effect on firing of B51.

Specialization versus Robustness. One important feature of biological networks is that they are robust to minor changes. The system continues to have the same behavior (equivalently, satisfy the same set of properties) even when certain changes are made to it. The positive feedback in the P-group neurons that sustains the protraction phase is such an example. Most properties that describe the ingestion and egestion behaviors are, in fact, robust to minor changes in the discrete model proposed in this paper. This robustness is the reason why abstract discrete models are useful. However, some of the properties of our model are quite sensitive to the strengths of certain synaptic connections. For examples, the interconnections of B64, and to some extent those of B51, B8, and B31 (note that we had to use specialized accumulator functions for these neurons, see Figure 5), appear to influence the overall behavior. The specializations indicate the sensitive parts of the model. This suggests that we may have to refine our current understanding of the CPG.

Weighted Integrate Function. Biological data shows that certain connections between neurons are stronger than others. Our definition of the weighted integrate functions is chosen to reflect this fact. The exact integrate function is, however, not important and other qualitatively similar functions produce the same behavior.

Model Checking Effort. Though we have used our in-house model checking environment SAL for experiments, we could have used any other temporal logic model checker. We used the symbolic model checker, sal-smc, which uses Boolean Decision Diagrams (BDDs) to represent states. Each model checking run (one for each property) takes about 10 seconds. The total state space of the system is of the order of 4^{10}; however, the set of reachable states must be significantly less. Adding more non-determinism to the model increases the model checking effort.

We also note here that, for any reasonable choice for the parameters, N, increment, and decrement, the model satisfies the same set of LTL properties. Our specific choice, such as N=4, was a reasonable compromise between being small and giving enough qualitative states (N=0,1,2,3) to model details (of other kinds of neurons we foresee adding to the network.)

7   Conclusions

We presented an abstract discrete model of the central pattern generator responsible for generating the egestion and ingestion buccal motor programs during the feeding cycle in *Aplysia*. We formalized the neural activation patterns, observed during egestion and ingestion BMPs, using LTL formulas. We verified that our model satisfies the LTL formulas using a symbolic model checker. While many properties are a direct consequence of the excitatory, inhibitory, and electrical connections between the various neurons, some of the properties – especially those related to transitioning from protraction phase to the retraction phase and finally to the termination phase – depend on the relative strengths of the various synaptic and electrical connections.

We plan to expand the model to include other missing elements from the feeding neural circuit, such as the sensory neurons in the cerebral ganglion and the cerebral-buccal interneurons. Simultaneously, we will need to expand the collection of LTL properties to capture more functions and behaviors.

A second direction for future work is generalizing the present model and making it nondeterministic. The present model is deterministic, except for the initiation of the termination phase. We can drop assumptions and make our model more nondeterministic, while still preserving its properties. For example, when a neuron receives both positive and negative inputs, it presently behaves as if it received no input – implicitly assuming that all signals are of "equal strength".

Instead, we can drop this assumption and let the neuron nondeterministically behave as if it received a positive, negative, or zero input.

References

[1] Cataldo, E., Byrne, J.H., Baxter, D.A.: Computational model of a central pattern generator. In: Priami, C. (ed.) CMSB 2006. LNCS (LNBI), vol. 4210, pp. 242–256. Springer, Heidelberg (2006)

[2] Baxter, D., Canavier, C., Byrne, J.: Dynamical properties of excitable membranes. In: [8], pp. 161–196

[3] Abbott, L.: Single neuron dynamics: an introduction. In: Ventriglia, F. (ed.) Neural Modeling and Neural Networks, pp. 57–78. Pergamon Press, Oxford (1994)

[4] Izhikevich, E.M.: Which model to use for cortical spiking neurons? IEEE Trans. on Neural Networks 15(5) (2004)

[5] Eker, S., Knapp, M., Laderoute, K., Lincoln, P., Meseguer, J., Sonmez, K.: Pathway logic: Symbolic analysis of biological signaling. In: Proc. Pacific Symposium on Biocomputing, pp. 400–412 (2002)

[6] Lincoln, P., Tiwari, A.: Symbolic systems biology: Hybrid modeling and analysis of biological networks. In: Alur, R., Pappas, G.J. (eds.) HSCC 2004. LNCS, vol. 2993, pp. 660–672. Springer, Heidelberg (2004)

[7] Tiwari, A., Talcott, C., Knapp, M., Lincoln, P., Laderoute, K.: Analyzing pathways using sat-based approaches. In: Anai, H., Horimoto, K., Kutsia, T. (eds.) Ab 2007. LNCS, vol. 4545, pp. 155–169. Springer, Heidelberg (2007)

[8] Byrne, J.H., Roberts, J.L. (eds.): From Molecules to Networks: An Introduction to Cellular and Molecular Neuroscience. Elsevier, Amsterdam (2004)

[9] Elliott, C., Susswein, A.: Comparative neuroethology of feeding control in molluscs. J. Exp. Biol. 205, 877–896 (2002)

[10] Susswein, A.J., Hurwitz, I., Thorne, R., Byrne, J.H., Baxter, D.A.: Mechanisms underlying fictive feeding in aplysia: Coupling between a large neuron with plateau potentials activity and a spiking neuron. J. Neurophysiology 87(5), 2307–2323 (2002)

[11] Iyengar, S.M., Talcott, C., Mozzachiodi, R., Cataldo, E., Baxter, D.A.: Executable symbolic models of neural processes. In: Network Tools and Applications in Biology NETTAB 2007 (2007)

[12] Herz, A.V.M., Gollisch, T., Manchens, C.K., Jaeger, D.: Modeling single-neuron dynamics and computations: A balance of detail and abstraction. Science 314 (October 2006)

[13] Ghosh, R., Tiwari, A., Tomlin, C.: Automated symbolic reachability analysis with application to delta-notch signaling automata. In: Maler, O., Pnueli, A. (eds.) HSCC 2003. LNCS, vol. 2623, pp. 233–248. Springer, Heidelberg (2003)

[14] Grosu, R., Mitra, S., Ye, P., Entcheva, E., Ramakrishnan, I.V., Smolka, S.A.: Learning cycle-linear hybrid automata for excitable cells. In: Bemporad, A., Bicchi, A., Buttazzo, G. (eds.) HSCC 2007. LNCS, vol. 4416, pp. 245–258. Springer, Heidelberg (2007)

[15] de Moura, L., Owre, S., Shankar, N.: The SAL intermediate language, Computer Science Laboratory, SRI International, Menlo Park, CA (2003), http://sal.csl.sri.com/

[16] Tiwari, A.: SAL model of aplysia central pattern generator (2008), http://www.csl.sri.com/~tiwari/html/cmsb08.html

[17] Hurwitz, I., Goldstein, R., Susswein, A.: Compartmentalization of pattern-initiation and motor function in the B31 and B32 neurons of the buccal ganglia of aplysia californica. J. Neurophysiol. 71, 1514–1527 (1994)

[18] Susswein, A.J., Byrne, J.H.: Identification and characterization of neurons initiating patterned neural activity in the buccal ganglia of aplysia. J. Neurosci. 8, 2049–2061 (1988)

[19] Kabotyanski, E.A., Baxter, D.A., Cushman, S.J., Byrne, J.H.: Modulation of fictive feeding by Dopamine and Serotonin in aplysia. J. Neurophysiol. 83, 378–392 (2000)

Stochastic Analysis of Amino Acid Substitution in Protein Synthesis[*]

D. Bošnački[1], H.M.M. ten Eikelder[1], M.N. Steijaert[1], and E.P. de Vink[2]

[1] Dept. of Biomedical Engineering, Eindhoven University of Technology
[2] Dept. of Mathematics and Computer Science, Eindhoven University of Technology

Abstract. We present a formal analysis of amino acid replacement during mRNA translation. Building on an abstract stochastic model of arrival of tRNAs and their processing at the ribosome, we compute probabilities of the insertion of amino acids into the nascent polypeptide chain. To this end, we integrate the probabilistic model checker Prism in the Matlab environment. We construct the substitution matrix containing the probabilities of an amino acid replacing another. The resulting matrix depends on various parameters, including availability and concentration of tRNA species, as well as their assignment to individual codons. We draw a parallel with the standard mutation matrices like Dayhoff and PET91, and analyze the mutual replacement of biologically similar amino acids.

1 Introduction

The transfer of genetic information from DNA to mRNA to protein happens with very high precision. Errors can have dramatic consequences for the organism as a whole. In this paper we analyze the second stage of this information pathway—the translation from mRNA to protein, i.e., the protein biosynthesis, in the light of translation errors and the factors of potential influence.

An mRNA molecule can be considered as a string of codons, each of which codes for a specific amino acid. The codons of an mRNA molecule are sequentially read by a ribosome, where each codon is translated using an amino acid specific transfer-RNA (aa-tRNA). This way, one-by-one, a chain of amino acids, i.e. a protein is built. In this setting, aa-tRNA can be seen as molecules containing a so-called anticodon, and carrying a specific amino acid. Arriving by Brownian motion, an aa-tRNA, docks into the ribosome and may succeed in adding its amino acid to the chain under construction, or alternatively dissociates in an early or later stage of the translation. This depends on the pairing of the codon under translation with the anticodon of the aa-tRNA, as well as on the stochastic influences such as the changes in the conformation of the ribosome.

Thanks to the vast amount of research during the last thirty years, the overall process of translation is reasonably well understood from a qualitative perspective. The process can be divided into around twenty small steps/reactions, a

[*] Partially supported by EU FP6-project ESIGNET.

number of them being reversible. Relatively little is known exactly about the kinetics of the translation. Over the past several years, Rodnina and collaborators have measured kinetic rates for various steps in the translation process for a small number of specific codons and anticodons [16,18,21,9]. Using various advanced techniques, they were able to show that in several of those steps the rates strongly depend on the degree of matching between the codon and the anticodon. Additionally, in [6] the average concentrations (amounts) of aa-tRNAs per cell have been collected for the model organism *Escherichia coli*. Based on these results, Viljoen and co-workers started from the assumption that the rates found by Rodnina et al. can be used in general, for all codon-anticodon pairs as estimates for the reaction dynamics. In [7], a complete detailed model is presented for all 64 codons and all 48 aa-tRNA classes for *E. coli*, on which extensive Monte Carlo experiments are conducted. In particular, using the model, codon insertion times and frequencies of erroneous elongations are established. A strong correlation of the translation error and the ratio of the concentrations of the so-called near-cognate and cognate aa-tRNA species was observed. Consequently, one can argue that the competition of aa-tRNAs, rather than their availability, decides both speed and fidelity of codon translation.

In the present paper, we model the translation kinetics via the modelchecking of continuous-time Markov chains (CTMCs) using the tool Prism [15,10]. The tool provides built-in performance analysis algorithms and a formalism (Computational Stochastic Logic, CSL) to reason about various properties of the CTMCs, removing the burden of extensive mathematical calculations from the user. Additionally, in our case, the Prism tool provides much shorter response times compared to Gillespie simulation.

We present an improvement of the stochastic model from [1], integrated in a Matlab environment. We use our model in the context of an original case study. To this end, we define the notion of a *translation substitution matrix*. The columns and rows of this matrix are labeled with amino acids. The element of the matrix indexed by amino acids a_1 and a_2 is the probability that a_1 is substituted by a_2 in the polypeptide chain. The translation substitution matrix can be used to check the error resistance capabilities of the translation process and the genetic code in general. It can also be used as an alternative similarity measure between amino acids from a point of view of translation.

The flexibility of our integrated Matlab-Prism model makes it possible to relatively easily investigate possible factors that can influence the probabilities in the substitution matrix. Here, we consider two of them:

- *Concentrations of the tRNAs.* To check this, we modify our model by assuming artificial conditions where all tRNAs have concentrations that deviate from the realistic *E. coli* model.
- *An alternative set of tRNAs.* Instead of the standard tRNAs that are confirmed by the experiments in *E. coli* we use 'synthetic' tRNAs assuming different number of tRNA species and their distributions over codons.

The obtained results indicate that the overall translation error is dependent on the tRNA set.

We also use the matrix to check the hypothesis that similar amino acids substitute for one another with higher probability than dissimilar ones. As a measure of similarity we use mutation data matrices, like Dayhoff [5] and PET91 [12] which are used for a similar purpose in sequence alignment tools, like BLAST. Our results confirm this hypothesis by showing that to a great extent the similarity patterns implied by the mutation data matrices emerge also in the translation substitution matrix.

Related work. We did not find other work on translation substitution matrices and investigating the hypothesis that similar amino acids tend to substitute each other. The model that is used in this paper builds upon [1], which was inspired by the simulation experiments of mRNA translation reported in [7]. There, only insertion errors per codon are considered, rather than amino acid-amino acid substitution probabilities.[1] A similar model, based on ordinary differential equations, was developed in [11]. Although probabilistic, it is used to compute insertion times, but no translation errors. The model of mRNA translation in [8] assumes insertion rates that are directly proportional to the mRNA concentrations, but assigns the same probability of translation error to all codons.

There exist numerous applications of formal methods to biological systems. A selection of recent papers from modelchecking and process algebra includes [17,3,4]. More specifically pertaining to the current paper, [2] applies the Prism modelchecker to analyze stochastic models of signaling pathways. Their methodology is presented as a more efficient alternative to ordinary differential equations models, including properties that are not of probabilistic nature. Also, [10] employs Prism on various types of biological pathways, showing how the advanced features of the tool can be exploited to tackle large models.

2 Biological Background

Proteins are essential building blocks for the production and regulation in cellular life. In fact, proteins take care of the major part of the functioning of the cell. Proteins are produced in two stages from the genetic information carried by the DNA: From a gene at the DNA, several copies of mRNA can be generated involving RNA-polymerases. Subsequently, from each mRNA, many identical proteins can be produced with the help of a ribosome. In the present case study, we will focus on the latter aspect of expression, the generation of proteins from mRNA, the process generally referred to as translation.

In effect, proteins are long, typically folded, strains composed from amino acids. Grossly, there are only twenty different types of amino acid present in living material. On the other hand, a string of mRNA can be interpreted as a sequence of nucleotides, molecules out of four types only, A, C, G and U, short for adenine, cytosine, guanine and uracil. Each three consecutive nucleotides form a codon. So, codons are essentially triplets of nucleotides. One type of codon

---

[1] The implementation of [7] was not available to us, impeding a direct comparison with our implementation. However, the overall insertion error per codon is comparable.

prescribes exactly one type of amino acid, but not vice versa. Generally, in an organism, several codons may code for the same amino acid. The correspondence of codons and amino acids is the same for all organisms but for a few exceptions, and is called the genetic code. So, an mRNA, as sequence of codons, specifies precisely a protein, as sequence of amino acids.

Basically, the translation of an mRNA into a protein takes effect as follows: A ribosome attaches to the mRNA. Next, the codons of the mRNA are processed one-by-one, stepwise building up a chain of amino acids, the protein in nascent. The amino acids used are brought to the mRNA-ribosome complex by aa-tRNA, tRNA charged with an amino acid. Characteristic for an aa-tRNA is a specific triplet of nucleotides, called the anticodon. It turns out that the anticodon decides which amino acid can be charged at the tRNA. As for codons, there is exactly one amino acid that corresponds to an anticodon. Now, an aa-tRNA arriving at the mRNA-ribosome complex, docks into the A-site of the ribosome. The aa-tRNA may either (i) immediately dissociate during the initial binding or codon recognition phase, (ii) be rejected after reconfiguration of the elongation factor Tu (EF-Tu) or (iii) successfully finish the second translocation phase and have its amino acid added to the protein under construction. The particular codon of the mRNA is then considered to be processed; the mRNA-ribosome complex shifts one position with a new codon for translation, if available.

The binding of a codon and an anticodon differs from pair to pair. Given a codon, we distinguish between cognate, near-cognate and non-cognate aa-tRNA, dependent on the match of the codon and the anticodon of the particular aa-tRNA. For a cognate aa-tRNA the binding of the codon and anticodon is strong, for near-cognate the binding is less strong, for a non-cognate the binding is weak. In fact, the codon-anticodon binding influences the success (going through none, one or two phases) and the actual speed of a translation attempt.

Next, we give a brief overview of the individual steps of the translation mechanism. Our explanation is based on [18,20,13].

An aa-tRNA arrives at the ribosome-mRNA complex in a ternary complexation with EF-Tu and GTP with a rate determined by the interaction of EF-Tu and the ribosome [20]. The initial binding is relatively weak. Codon recognition comprises (i) the establishing of contact between the anticodon of the aa-tRNA and the current codon in the ribosome-mRNA complex, and (ii) subsequent conformational changes of the ribosome, that are different for cognate and near-cognates. The overall rates are similar for cognates and near-cognates. Note that, non-cognates are not selected during codon recognition. GTPase-activation of the elongation factor is largely favored by the conformational changes in the ribosome induced by a cognate aa-tRNA, while for near-cognate aa-tRNA GTPase-activation is lessened [19,9]. During GTP-hydrolysis that takes place next, inorganic phosphate P_i and GDP are produced. It is assumed that P_i is released, EF-Tu reconfigures and that aa-tRNA dissociates from the complex with EF-Tu and GDP, see [14]. The accommodation step that follows happens rapidly for cognate aa-tRNA, whereas for near-cognate aa-tRNA this proceeds slower and the aa-tRNA is likely to be rejected, also because of the instability of the binding to the ribosome-mRNA complex.

The translocation phase that follows, is unidirectional except for its reversible first step involving the elongation factor EF-G. In short, during the translocation phase, another GTP-hydrolysis catalyzed by EF-G, produces GDP and P_i and results in unlocking and movement of the aa-tRNA to the P-site of the ribosome. The latter step is preceded or followed by P_i-release. Translocation of the ribosome, with dramatic shifts in the positioning of ribosomal subunits, and release of EF-G moves the tRNA, that has transferred its amino acid to the polypeptide chain, into the E-site of the ribosome. Further rotation eventually leads to dissociation of the used tRNA.

3   Abstract Model

In this section, we present an abstract model of the translation mechanism sketched in the previous section. Our aim is to capture various combined steps by a probabilistic automaton. The grouping of multiple steps results in a limited number of states and, subsequently, in a smoother analysis and quicker response times for the Prism experiments. Figure 1 depicts the probabilistic automaton obtained.

Given a particular codon under translation at the ribosome, we distinguish the following states:

- State 1, initial binding. An aa-tRNA binds, in an arrival process, at the mRNA-ribosome complex.
- State 2, recognition. The weak binding of aa-tRNA at the complex either stabilizes (transition to state 3), or the binding breaks and the tRNA dissociates (transition to state 0).
- State 3, conformation. Again, one out of two options may happen. A number of steps related to the processing of GTP may take place (modeled by the transition to state 4). Alternatively, the binding of the aa-tRNA at the mRNA-ribosome complex may lose strength (modeled by the transition back from state 3 to state 2).
- State 4, proofreading. The aa-tRNA can either be rejected, resulting in a dissociation from the mRNA-ribosome complex (transition to state 5), or the aa-tRNA shifts to P-site of the ribosome.
- State 6, accommodation. A reversible reaction involving the elongation factor EF-G, prepares for translocation (transition to state 7).
- State 7, translocation. Translocation may take place, as well as a number of other unidirectional steps in the translation process, and the amino acid is successfully added to the peptidyl chain (transition to state 8), or the binding of EF-G does not lead to reconformation of the ribosome (transition back form state 7 to state 6).

Further, we have a number of auxiliary states, that do not have a concrete, biological counterpart. However, the states are introduced for modelchecking purposes, to discriminate between the various 'exit modes' of an aa-tRNA.

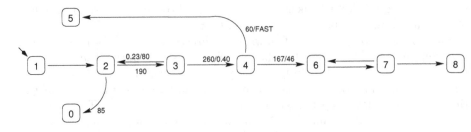

Fig. 1. Abstract automaton representing translation

- State 0, dissociation. The initial binding of the aa-tRNA and the mRNA-ribosome complex does not stabilize and the aa-tRNA floats away.
- State 5, rejection. The aa-tRNA dissociates from the mRNA-ribosome complex while being transferred from the A-site to the P-site of the ribosome.
- State 8, elongation. The amino acid carried by the aa-tRNA is added to the peptide chain. The uncharged tRNA flows back into the cytosol. (aa-tRNA synthesis, the recharging of a tRNA with an amino acid is not modeled here.)

The choices in the abstract automaton of Figure 1 can all be considered to be probabilistic. Of relevance are those in states 2, 3 and 4. In states 1 and 6 the choice is degenerated, in auxiliary states 0, 5 and 8 it does not occur. The probabilistic choice in state 7 does not influence the eventual exit state. In the past decade, Rodnina and co-workers have collected various kinetic parameters of a number of steps in the peptidyl transfer phase of the translation mechanism [16,18,21,9]. Additionally, using advanced fluorescent and crystallographic techniques, they showed that the binding of the codon under translation, on the one hand, and the anticodon of the aa-tRNA that has attached to the mRNA-ribosome complex, on the other hand, decisively influences the success or failure of some of the steps. Their rates are incorporated in the automaton, that can be interpreted as a continuous-time Markov chain.[2]

In state 2, the transition to state 3 has rate 190 whereas the transition to state 0 is taken with rate 85. These two rates are equal for cognate and near-cognate aa-tRNA alike. The transition back from state 3 to state 2 has rate 0.23 for cognate aa-tRNA against 80 for near-cognate aa-tRNA. Conversely, the transition forward from state 3 to state 4 is of rate 260 for cognate aa-tRNA and of rate 0.40 for near-cognate aa-tRNA. Note the difference between the two classes of aa-tRNA. A similar phenomenon can be observed for state 4. A transition from state 4 to state 6 of rate 167 and to state 5 of rate 60 for cognate aa-tRNA, compared to a rate 46 from state 4 to state 6 for near-cognate aa-tRNA and a fast rate (chosen to be 1000 in our experiments) for the alternative transition to state 5.

Clearly, based on the rates provided, the probabilities of ending up in state 8 differ significantly for cognate aa-tRNA compared to near-cognate aa-tRNA.

---

[2] All rates in this paper are of dimension s^{-1}.

Non-cognate aa-tRNA probabilistic choices do not apply, as stable association to the mRNA-ribosome complex is considered to be negligible.

The substitution error for a codon is the probability that another amino acid is added to the nascent protein than is coded for by the codon under translation. The average insertion time is the expected time it takes for any amino acid to be added for the particular codon. The model given above has been used previously in [1] to analyze both the substitution error for a codon and its average insertion time. To this end, the automaton of Figure 1 is combined with an arrival process of cognate, near-cognates and non-cognate aa-tRNAs, the separation in classes of aa-tRNA depending on the active codon. In the current paper, we refine the analysis of the substitution error, by considering the actual amino acid that is inserted.

We define $P(\text{aa}|\text{cd})$ as the probability for an elongation of the peptidyl chain with an amino acid aa, given a codon cd. Let cd code for amino acid bb. Then, we distinguish five categories of aa-tRNA: xx-cognate, yy-cognate, xx-near-cognate, yy-near-cognate and non-cognate aa-tRNA. A cognate aa-tRNA is classified as an xx-cognate if it carries the amino acid aa of interest (i.e. if aa equals bb); the cognate aa-tRNA is considered an yy-cognate if it carries an amino acid different from the amino acid aa (i.e. if we are interested in an amino acid aa that is different from the amino acid bb coded for by cd). Similarly, a near-cognate aa-tRNA is referred to as an xx-near-cognate if its amino acid is aa, and is called an yy-near-cognate if its amino acid is different from aa. All aa-tRNA that do not belong to any of the other classes are considered non-cognates. Thus, the selected amino acid determines the xx-type or yy-type for a cognate or near-cognate aa-tRNA; the active codon decides whether an aa-tRNA is cognate, near-cognate or non-cognate. In our model, non-cognates never lead to insertion of an amino acid. The binding of a non-cognate anticodon and the codon under translation is considered too weak for the tRNA to move to state 3. Non-cognate aa-tRNA will always exit at state 0.

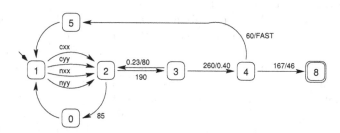

Fig. 2. Adapted automaton

Given the above definitions of aa-tRNA, we end up with the automaton as given in Figure 2. The rates cxx, cyy, nxx and nyy depend on the amount of the aa-tRNAs that are classified as xx-cognate, yy-cognate, xx-near-cognate and yy-near-cognate, respectively, for the considered codon and amino acid. The transitions from state 0 and state 5 to state 1 are added, since the process

Table 1. Molecular reactions underlying the adapted model

$$C + R1 \xrightleftharpoons[85]{cxx/cyy} CR2 \xrightleftharpoons[190]{0.23} CR3 \xrightarrow{260} CR4$$

$$CR4 \xrightarrow{167} CR8 \qquad CR4 \xrightarrow{60} C + R1$$

$$N + R1 \xrightleftharpoons[85]{nxx/nyy} NR2 \xrightleftharpoons[190]{80} NR3 \xrightarrow{0.40} NR4$$

$$NR4 \xrightarrow{46} NR8 \qquad NR4 \xrightarrow{FAST} NN + R1$$

continues as long as state 8 has not been reached, i.e. until an amino acid has been transferred successfully. States 6 to 8 have been identified, as from state 6 the final state 8 will always be reached eventually. The rates have been kept deliberately, although they could have been replaced by probabilities. In the set-up of our experiments, it comes in handy to deal with CTMCs to which we can feed numbers of aa-tRNAs, avoiding to calculate their relative fractions. An overview of the reactions involved are collected in Table 1.

4   Amino Acid Substitution

In this section, we discuss how the amino acid-codon insertion probability matrix AC and the amino acid-amino acid translation substitution matrix TS are computed for the *Escherichia coli* bacterium. The abstract model of the previous section is, per codon, supplemented with relevant concentrations of the various types of aa-tRNAs, to establish the probability for an amino acid to be inserted with the particular codon being active. When this information is available for all codons, by proper grouping, the probability for an amino acid to replace the amino acid that is actually coded for, can be obtained.

To calculate $P(\text{aa}|\text{cd})$, the probability of elongation of the growing protein chain with the amino acid **aa**, when codon **cd** is under translation, we proceed as follows: We provide the Prism modelchecker with four input parameters, viz. **xx_cogn**, **yy_cogn**, **xx_near** and **yy_near**, each representing the amount of available cognate and near cognate tRNAs, carrying either the amino acid of interest (indicated by **xx**) or an other amino acid (indicated by **yy**). This instantiates the arrival process of an aa-tRNA at state 1 of the abstract automaton. The probabilities for success, i.e. reaching state 8, and failure, dissociation via state 0 or rejection via state 5 followed by re-activation of the arrival process at state 1, are not independent from the arrival process itself. Therefore, we have to consider the CTMC simultaneous representing the arrival process and translation mechanism. With this done in Prism, we only need to establish the CSL formula

$$\text{P=?} \ [\ (\text{s!=0 \& s!=5}) \ \text{U} \ (\text{s=8}) \ ] .$$

See Appendix B for the complete Prism code.

In order to facilitate the construction of the insertion probability matrix, we have combined the Prism modelchecker with Matlab. We make use of Matlab's extern call mechanism and Prism's command line option for parameter instantiation. In two nested loops, the outer loop iterating over codons, the inner loop iterating over amino acids, we call from within Matlab the Prism tool, for example by

```
prism ourmodel.sm ourformula.csl -const xx_cogn=1037, ...
```

communicating the values of parameter xx_cogn and others to the modelchecker. The negligible value $1.0e-6$ is used instead of 0, as only positive values are allowed as rates of an exponential distribution. Apart from directory management, the interaction of the two tools quite conveniently suited our purposes.

For the usual *E. coli* aa-tRNA set with standard concentrations, referred to as *real*, we illustrate how the parameter values are obtained for the codon UUU and amino acids Phe, Leu, and Ile.

Table 2. Some input values for the realistic model

codon	amino acid	xx_cogn	yy_cogn	xx_near	yy_near
UUU	Phe	1037	0	0	2944
UUU	Leu	0	1037	2944	0
UUU	Ile	0	1037	0	2944

In case of calculating $P(\text{Phe}|\text{UUU})$, where the amino acid under consideration coincides with the amino acid of the codon, we assign to xx_cogn the total amount of cognate tRNAs. From Table 7 in Appendix A, we see that tRNA 28 is the only cognate, so xx_cogn is set to 1037, the number of tRNA 28 in Table 8. In this case, there are no cognates coding for an amino acid other than Phe, hence yy_cogn = 0. Next, we check the near-cognates of UUU. According to Table 7, only tRNAs 22 and 23 act as near-cognates. Both code for Leu, the amino acid leucine. Therefore, we put xx_near = 0, as no near-cognate codes for Phe and yy_near = 1913 + 1031 = 2944 the sum of all counts of near-cognates coding for an amino acid different from Phe. See the first row of Table 2.

To establish $P(\text{Leu}|\text{UUU})$, we put xx_cogn to zero (or rather 10^{-6}), as there are no cognates of the codon UUU carrying Leu, and yy_cogn to 1037, the number of molecules of the non-Leu cognate aa-tRNA 28. Since, the near-cognates aa-tRNAs 22 and 23 both carry Leu, xx_near is set to 2944, the sum of their counts, and yy_near is zero as there are no near-cognate with other amino acids.

To compute $P(\text{Ile}|\text{UUU})$, the substitution probability for the amino acid isoleucine with respect to the codon UUU, we have that xx_cogn and xx_near are both nihil. No cognate or near-cognate aa-tRNAs will insert Ile, hence the substitution probability will be zero. Because of this, no further calculation is required. The Matlab script automatically puts $P(\text{Ile}|\text{UUU}) = 0$.

Having the amino acid-codon insertion matrix $AC = (P(\text{aa}|\text{cd}))_{\text{aa,cd}}$ available, we derive the amino acid-amino acid translation substitution matrix TS.

Each item $TS_{aa,bb}$ of TS denotes the probability that amino acid bb is inserted while the current translation codes for the amino acid aa. To compensate for the differences of occurrence in the *E. coli* genome of the various codons of an amino acid, we balance the sum of amino acid-codon insertion probabilities, by the relative frequencies of the codons:

$$TS_{aa,bb} = \sum_{cd \in codons(aa)} rf(cd, aa) \cdot P(aa|cd)$$

with *codons*(aa) the set of codons coding for the amino acid aa according to the genetic code, and *rf*(cd, aa) the relative frequency of the codon cd with respect to *codons*(aa).

5   Alternative aa-tRNA Sets

We investigate how substitution probabilities are affected by the aa-tRNA population. We consider both the effect of the concentration of aa-tRNA molecules of a specific species, and the influence of the composition of the aa-tRNA set. Our starting point is the *E. coli* model from the previous sections with an aa-tRNA set containing 48 different tRNA species and numbers of molecules from each species as reported in [6]. Alternative models, with an aa-tRNA set different from usual and with concentrations deviating from standard values, are examined as well. The experimental set-up is rather flexible in this respect, only different aa-tRNA populations with other amounts of available molecules need to be supplied. We consider eight different models, listed in Table 3.

Table 3. Alternative models: aa-tRNA sets of 48, 64 or 25 species, aa-tRNA concentrations based on real measurements, flat, or proportional to codon matching

Name	species	aa-tRNA counts
Model *48R*	48	based on Table 8
Model *48F*	48	1000 per aa-tRNA
Model *48C*	48	1000 for per codon recognized
Model *64R*	64	based on Table 8
Model *64FC*	64	1000 per aa-tRNA / codon
Model *25R*	25	based on Table 8
Model *25F*	25	1000 per aa-tRNA
Model *25C*	25	1000 per codon recognized

48 Species aa-tRNA Set

Model *48R*, the so-called realistic model with parameters based on [7,6], has 48 aa-tRNA species and molecule counts based on physical measurements. In order to assay the stability of amino acid substitution, we have run similar experiments with varying parameters. In one model, we use for each tRNA species the same amount of molecules, arbitrarily chosen as 1000. This model with flat aa-tRNA concentrations is referred to as *48F*, with *F* for flat. Another variation is a model in which we assign the equal amounts (again 1000 molecules) to each codon. If the same codon is recognized by multiple aa-tRNAs, each of them is assigned a proportional part. So, the count of 1000 is equally split over the number of cognate tRNA species. If an aa-tRNA is cognate to several codons, it will be allotted accordingly. We refer this model as *48C*, with *C* for codon. Note that, for models *48F* and *48C*, the arbitrarily chosen value of 1000 does not influence the outcome, as the error probabilities are determined by the fractions of cognate and near-cognate species and not by the values themselves.

64 Species aa-tRNA Set

Apart from variations in the concentrations of aa-tRNAs, one can also modify, in silico, the sets of tRNA species. An obvious choice, is the model with the maximal number of 64 aa-tRNA species. In this model, each of aa-tRNA is considered to recognize exactly one codon. Thus, under this assumption, each codon has exactly one cognate tRNA and nine near-cognate aa-tRNA.

For the model *64R*, the count for each aa-tRNA species is equal to the sum of tRNA species in the original model that recognize the corresponding codon. For aa-tRNAs in the original model that recognize more than one codon, the new aa-tRNAs get an equal share of the original amount. Analogously to the models with a 48 species aa-tRNA set, we also define flat and cognate models for the 64 aa-tRNA case, with equal amounts of molecules per tRNA species and per codon. However, for this specific case these two models coincide, each with an amount of 1000 molecules for each of the 64 aa-tRNA species. Consequently, the model is named *64FC*.

25 Species aa-tRNA Set

The other obvious choice of aa-tRNA set, is the opposite case, where the number of species is minimal. However, instead of the theoretical minimum of aa-tRNA species, where the choice of cognates seems arbitrarily, we decide to have exactly one tRNA for each 'genomic box or block', i.e. a group of codons that codes for the same amino acid and only differ in the third nucleotide. This model contains 25 aa-tRNA species, reminiscent to to fine-tuned genetic code in eukaryote mitochondria. As before, *25F* denotes the flat model variant, each tRNA species having 1000 molecules. For *25R*, the amount aa-tRNA is set to the total of the tRNAs in the *48R* model that belong to the block. The one exception being release factor RF1, for which we assign the full amount to the block UAA, UAG although it recognizes both UAA and UAG (which belong to separate blocks). Finally, for the model *64C*, the amount of molecules for each aa-tRNA species is calculated as 1000 times the number of codons in the corresponding block. See Table 4.

For all the above models we have computed the substitution probability matrix as sketched in Section 4. The diagonal of the matrices denotes the probability

Table 4. The 25 species aa-tRNA set

aa-tRNA	recognized codons	aa-tRNA	recognized codons
1 (Phe)	UUU, UUC	14 (Gln)	CAA, CAG
2 (Leu)	UUA, UUG	15 (Asn)	AAU, AAC
3 (Leu)	CUU, CUC, CUA, CUG	16 (Lys)	AAA, AAG
4 (Ile)	AUU, AUC, AUA	17 (Asp)	GAU, GAC
5 (Met)	AUG	18 (Glu)	GAA, GAG
6 (Val)	GUU, GUC, GUA, GUG	19 (Cys)	UGU, UGC
7 (Ser)	UCU, UCC, UCA, UCG	20 (End)	UGA
8 (Pro)	CCU, CCC, CCA, CCG	21 (Trp)	UGG
9 (Thr)	ACU, ACC, ACA, ACG	22 (Arg)	CGU, CGC, CGA, CGG
10 (Ala)	GCU, GCC, GCA, GCG	23 (Ser)	AGU, AGC
11 (Tyr)	UAU, UAC	24 (Arg)	AGA, AGG
12 (End)	UAA, UAG	25 (Gly)	GGU, GGC, GGA, GGG
13 (His)	CAU, CAC		

for peptide elongation with the amino acid that was coded for. The average substitution error for the model, shown in Table 5, is obtained by taking the average over all amino acid for the probability of a substitution by a non-coded amino acid. The table displays in parentheses a weighted average of errors, obtained by scaling the errors for individual amino acids with the relative occurrence of their codons in the *E. coli* genome. Remarkably, the errors for the 48 tRNA models are always smaller than those of the synthesized 25 and 64 tRNA models. The errors for individual amino-acids are shown in Figure 3.

Table 5. Mean substitution error and occurrence-weighted substitution error (within parentheses) for all eight models (model *64RF* twice)

model	'real'	'flat'	'codon'
48 aa-tRNA set	0.48% (0.45%)	0.45% (0.45%)	0.39% (0.36%)
64 aa-tRNA set	0.93% (0.85%)	0.73% (0.69%)	0.73% (0.69%)
25 aa-tRNA set	1.16% (0.83%)	0.66% (0.64%)	0.76% (0.68%)

Notably, the 'real' *48R* model has a striking low probability for errors for stop-codons. None of the other models reaches such low value. Moreover, the probability for other codons to accidentally act as a stop-codon is lowest for *48R* model as well. See the red/grey bars in Figure 4.

6   Groups of Related Amino Acids

In this section, we check the hypothesis that biologically similar amino acids substitute for each other during translation. Its rationale is that such substitutions would lower the chance that the resulting protein is non-functional. Namely, it

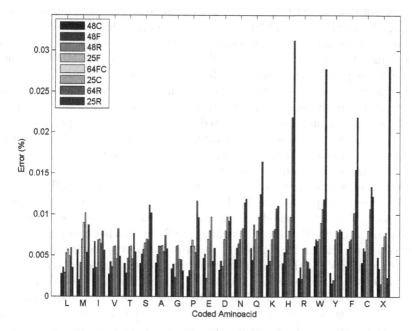

Fig. 3. Substitution errors for individual amino acids and the for stop-codo n

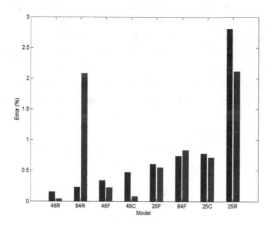

Fig. 4. Percentage of erroneous substitution of a stop-codon for an amino acid (blue/dark), and percentage of erroneous substitution of any other codon by a stop-codon (red/gray)

has been observed that if in a polypeptide chain an amino acid is substituted by another biochemically similar amino acid, then it is likely that the modified protein has biochemical properties similar to the original one. Therefore, it seems plausible to assume that, under evolutionary pressure, an error robustness mechanism has been developed also exploited at translational level.

Our measure of similarity of amino acids is based on so-called mutation data matrices, like Dayhoff [5] and PET91 [12]. Essentially, these matrices give for each pair of amino acids an evolutionary substitution probability. Mutation data matrices and their derivatives are widely used as an amino acid similarity measure, e.g. in sequence alignment tools like BLAST. At first sight it may look strange that we compare mutation errors, which introduce protein changes that are permanent for the organism, with translation errors which only affect one copy of the protein and are not inherited. Thus, it is worth emphasizing that we use the matrices only as a similarity measure 'approved' by the evolution without trying to draw any further analogy between these two phenomena.

Based on the mutation data matrices, we divide amino acids in four groups. There are different groupings of amino acids in the literature usually based on their biochemical properties [23]. In this paper, we use the partitioning proposed by Swanson [22] which is based on mutation data matrices, but also rather well in agreement with the classifications based purely on biochemical properties.

The two main criteria for Swanson's partitioning are the size (small vs. large) of the amino acids and their positioning in the proteins based on their affinity for water (hydrophobic/inner vs. hydrophilic/outer) The four groups are 1) the *small* group: Pro, Gly, Ala, Ser, Thr, 2) the *outer* group: Glu, Asp, Asn, Gln, Lys, His, 3) the *large* group: Arg, Trp, Tyr, Phe, and 4) the *inner* group: Cys, Val, Ile, Met, Leu. A natural cyclic arrangement of those groups can be made such that neighboring groups are small-outer, outer-large, large-inner and inner-small. Thus, the groups small and large are considered as opposite groups and the same holds for the groups inner and outer. See [22].

During translation three types of errors can be made. The least serious misreading results in adding to the protein an amino acid of the same group as the intended amino acid. A greater error would be a substitution with an amino acid of a neighboring group. Finally, potentially greatest consequences have substitutions with an amino acid of the opposite group. A completely different error is the misinterpretation of a codon as the stop codon, which means that the forming of the protein is terminated.

The probability GSP_{G_1, G_2} that an amino acid in group G_1 will be substituted with an amino acid from group G_2 is given by

$$GSP_{G_1, G_2} = \sum_{\text{aa} \in G_1, \text{bb} \in G_2} rf(\text{aa}, G_1) \cdot TS_{\text{aa, bb}}$$

where aa and bb are amino acids in groups G_1 and G_2, respectively, $rf(\text{aa}, G_1)$ is the relative occurrence frequency of amino acids aa within G_1, and $TS_{\text{aa,bb}}$ is the probability that bb substitutes aa, i.e., the corresponding element of the substitution matrix TS as defined in Section 4. The relative frequency of amino acid aa within its group G_1 is obtained as a ratio between the sum of the occurrence frequencies of all codons of aa and the sum of the occurrence frequencies of all codons of the amino acids in G_1. (Again we deal with the so-called codon bias, i.e. within the genome of an organism not all codons and amino acids are used with equal frequency.)

Table 6. Realistic model, percentage of erroneous amino acids, stop codon and the correct amino acid for amino acids from the four groups

	to group 1	to group 2	to group 3	to group 4	stop-codon	correct
group 1	**0.22 %**	0.077 %	0.047 %	0.11 %	0.0006 %	99.5 %
group 2	0.076 %	**0.25 %**	0.081 %	0.10 %	0 %	99.7 %
group 3	0.065 %	0.027 %	**0.043 %**	0.21 %	0.0019 %	99.7 %
group 4	0.069 %	0.029 %	0.055 %	**0.17 %**	0.0015 %	99.7 %

In Table 6 we give the above defined *GSP* probabilities for the realistic model. Each row gives the error probabilities for amino acids of the corresponding group. For all groups, the probability of generating the correct amino acid is more than 0.995. The probability of generating a wrong amino acid from the *correct* group is indicated in bold. Note that for groups 1, 2 and 4 this probability is larger than the probability of generating a wrong amino acid of another group. Somewhat surprisingly, this does not hold for group 3. Furthermore, the probability of an unexpected termination (stop codon) is extremely small.

A graphical representation of the error probabilities is given in Figure 5. In this figure only the error probabilities, except the very small probability of an erroneous stop codon, are shown. The probabilities of generating an erroneous amino acid of the original group are indicated in white. The probabilities of indicating an amino acid of the opposite group are shown in black. Clearly, the probability of generating erroneous amino acids in the correct (white) group is higher than the probability of generating amino acids in the opposite (black) group, except for group 3.

We performed analogous analyses of the other models and in all cases the outcome was similar. This indicates that the relatively small probability of out-of-group amino acid substitutions is due to the distribution of the amino acids over codons, i.e., the genetic code, rather than to the tRNA species and their concentrations.

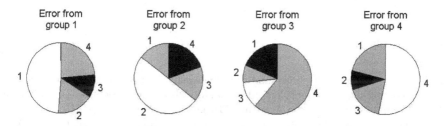

Fig. 5. Realistic model, probabilities of erroneous amino acid from another group (own group: white, opposite group: black)

7   Conclusions and Future Work

We described a formal analysis of codon misreading errors during translation of mRNA to protein, caused by a mismatch between codons and tRNAs. To this end, we presented a method based on probabilistic modelchecking, in particular the modelchecker Prism in combination with Matlab.

Inspired by mutation data matrices, we introduced the notion of a translation substitution matrix. Using our model, we computed the elements of this matrix which are the probabilities that amino acids replace each other in the protein as a result of codon misreading. Further, we investigated the influence of some parameters, like tRNA concentrations and different tRNA species, to the misreading probabilities. It turned out that the translation mechanism is quite robust. The mean substitution error for the realistic model is in line with experimental findings, cf. [7]. Remarkably, for the realistic model it is smaller than for our synthesized models. We also showed that biologically similar amino acid replace each other with higher probabilities than dissimilar ones. Experiments as described in Section 5 can easily be done in silico, but will require substantial effort, if not impossible on such rigorous scale, in a wetlab. Additionally, our case studies confirm that probabilistic modelchecking has advantage over simulation regarding reliability and running times. Preliminary experiments indicate that our modelchecking approach is about 10 times faster than our Gillespie simulations.

In the future, we plan to apply our translation model to further investigate the robustness of the translation mechanism and the genetic code. The translation substitution belongs to a class of case studies for which the essential properties are of a probabilistic nature. It would be interesting to employ the methodology of this paper to similar problems, like the precision of DNA repair and antibody recognition.

Acknowledgements. We are indebted to Timo Breit, Christiaan Henkel, Erik Luit, Jasen Markovski, Tessa Pronk and Hendrik Viljoen for fruitful discussions and constructive feedback. We gratefully acknowledge the contribution of the students of the 8P135 Bioinformatics project at TU/e.

References

1. Bosnacki, D., et al.: In Silico modelling and analysis of ribosome kinetics and aatrna competition. In: Proc. Computational Models for Cell Processes. Turku Centre for Computer Science, Åbo Academia, Turku, 16 p. (2008)
2. Calder, M., et al.: Analysis of signalling pathways using continuous time Markov chains. In: Priami, C., Plotkin, G. (eds.) Transactions on Computational Systems Biology VI. LNCS (LNBI), vol. 4220, pp. 44–67. Springer, Heidelberg (2006)
3. Chabrier, N., Fages, F.: Symbolic model checking of biochemical networks. In: Priami, C. (ed.) CMSB 2003. LNCS, vol. 2602, pp. 149–162. Springer, Heidelberg (2003)

4. Danos, V., et al.: Rule-based modelling of cellular signalling. In: Caires, L., Vasconcelos, V.T. (eds.) CONCUR 2007. LNCS, vol. 4703, pp. 17–41. Springer, Heidelberg (2007)

5. Dayhoff, M.O.: Suplements 1, 2 and 3. Atlas of Protein Sequence and Structure 5 (1978)

6. Dong, H., et al.: Co-variation of tRNA abundance and codon usage in Escherichia coli at different growth rates. Journal of Molecular Biology 260, 649–663 (1996)

7. Fluitt, A., et al.: Ribosome kinetics and aa-tRNA competition determine rate and fidelity of peptide synthesis. Computational Biology and Chemistry 31, 335–346 (2007)

8. Gilchrist, M.A., Wagner, A.: A model of protein translation including codon bias, nonsense errors, and ribosome recycling. Journal of Theoretical Biology 239, 417–434 (2006)

9. Gromadski, K.B., Rodnina, M.V.: Kinetic determinants of high-fidelity tRNA discrimination on the ribosome. Molecular Cell 13(2), 191–200 (2004)

10. Heath, J., et al.: Probabilistic model checking of complex biological pathways. In: Priami, C. (ed.) CMSB 2006. LNCS (LNBI), vol. 4210, pp. 32–47. Springer, Heidelberg (2006)

11. Heyd, A.W., Drew, D.A.: A mathematical model for elongation of a peptide chain. Bulletin of Mathematical Biology 65, 1095–1109 (2003)

12. Jones, D.T., et al.: The rapid generation of mutation data matrices from protein sequences. CABIOS 3, 275–282 (1992)

13. Karp, G.: Cell and Molecular Biology, 5th edn. Wiley, Chichester (2008)

14. Knudsen, C., et al.: The importance of structural transitions of the switch II region for the functions of elongation factor Tu on the ribosome. Journal of Biological Chemistry 276, 22183–22190 (2001)

15. Kwiatkowska, M., et al.: Probabilistic symbolic model cheking with Prism: a hybrid approach. Journal on Software Tools for Technology Transfer 6, 128–142 (2004), http://www.prismmodelchecker.org/

16. Pape, T., et al.: Complete kinetic mechanism of elongation factor Tu-dependent binding of aa-tRNA to the A-site of E. coli. EMBO Journal 17, 7490–7497 (1998)

17. Priami, C., et al.: Application of a stochastic name-passing calculus to represent action and simulation of molecular processes. Information Processing Letters 80, 25–31 (2001)

18. Rodnina, M.V., Wintermeyer, W.: Ribosome fidelity: tRNA discrimination, proofreading and induced fit. TRENDS in Biochemical Sciences 26(2), 124–130 (2001)

19. Rodnina, M.V., et al.: Codon-dependent conformational change of elongation factor Tu preceding GTP hydrolysis on the ribosome. EMBO Journal 14, 2613–2619 (1995)

20. Rodnina, M.V., et al.: Recognition and selection of tRNA in translation. FEBS Letters 579, 938–942 (2005)

21. Savelsbergh, A., et al.: An elongation factor G-induced ribosome rearrangement precedes tRNA–mRNA translocation. Molecular Cell 11, 1517–1523 (2003)

22. Swanson, R.: A unifying concept for the amino acid code. Bulletin of Mathematical Biology 46(2), 187–203 (1984)

23. Taylor, W.R.: The classification of amino acid conservation. Journal of Theoretical Biology 119, 205–218 (1986)

A   Additional Tables

Table 7. Codons and their cognate and near-conate tRNAs. Derived form Table 7

Codon	Cognates	Near-Cognates	Codon	Cognates	Near-Cognates
UUU	28	22,23	GUU	44,45,46	
UUC	28	9,17,20,22,23,36,42,43,45,46	GUC	45,46	2,8,15,17,20,28,44
UUG	22,23	18,19,25,26,27,28,34,41	GUG	44	13,18,19,22,25,26,27,45,46
UUA	23	21,22,28,32,33,44	GUA	44	1,12,14,21,23,45,46
UCU	33,36	34	GCU	1	2
UCC	36	2,9,28,30,33,34,37,39,42,43	GCC	2	1,8,15,30,36,37,39,45,46
UCG	33,34	22,29,36,38,41	GCG	1	2,13,29,34,38
UCA	33	1,23,31,32,34,36,40	GCA	1	2,12,14,31,33,40,44
UGU	9	3,32,41	GGU	15	3,13,14
UGC	9	15,28,32,35,36,41,42,43	GGC	15	2,8,9,13,14,35,45,46
UGG	41	4,6,9,13,22,32,34	GGG	13,14	4,6,15,41
UGA	32,48	5,9,14,23,33,41	GGA	14	1,5,12,13,15,32,44
UAU	42,43		GAU	8	12
UAC	42,43	7,8,9,16,28,36	GAC	8	2,7,12,15,16,42,43,45,46
UAG	47	11,22,34,41,42,43	GAG	12	8,11,13
UAA	47,48	10,12,23,24,32,33,42,43	GAA	12	1,8,10,14,24,44
CUU	20	3,19,21	AUU	17	18,25,26,27
CUC	20	16,17,19,21,28,30,45,46	AUC	17	7,18,20,25,26,27,28,35,37,39,45,46
CUG	19,21	4,11,18,20,22,25,26,27,29	AUG	25,26,27	6,17,18,19,22,38
CUA	21	10,19,20,23,31,44	AUA	18	5,17,21,23,24,25,26,27,40,44
CCU	30,31	3,29	ACU	37,39,40	38
CCC	30	2,16,20,29,31,36,37,39	ACC	37,39	2,7,17,30,35,36,38,40
CCG	29,31	4,11,19,30,34,38	ACG	38,40	6,18,25,26,27,29,34,37,39
CCA	31	1,10,21,29,30,33,40	ACA	40	1,5,24,31,33,37,38,39
CGU	3	4	AGU	35	3,5,6
CGC	3	4,9,15,16,20,30,35	AGC	35	5,6,7,9,15,17,37,39
CGG	4	3,6,11,13,19,29,41	AGG	6	4,5,13,18,25,26,27,35,38,41
CGA	3	4,5,10,14,21,31,32	AGA	5	6,14,24,32,35,40
CAU	16	3,10,11	AAU	7	24
CAC	16	7,8,10,11,20,30,42,43	AAC	7	8,16,17,24,35,37,39,42,43
CAG	11	4,10,16,19,29	AAG	24	6,7,11,18,25,26,27,38
CAA	10	11,12,16,21,24,31	AAA	24	5,7,10,12,40

Table 8. tRNA species in *E. coli*, data from [6] and [7]

	tRNA	Amino acid	Anticodon	Recognized Codons	Molecules/cell
1	Ala1	A	UGC	GCU, GCA, GCG	3250
2	Ala2	A	GGC	GCC	617
3	Arg2	R	ACG	CGU, CGC, CGA	4752
4	Arg3	R	CCG	CGG	639
5	Arg4	R	UCU	AGA	867
6	Arg5	R	CCU	AGG	420
7	Asn	N	GUU	AAC, AAU	1193
8	Asp1	D	GUC	GAC, GAU	2396
9	Cys	C	GCA	UGC, UGU	1587
10	Gln1	Q	UUG	CAA	764
11	Gln2	Q	CUG	CAG	881
12	Glu2	E	UUC	GAA, GAG	4717
13	Gly1	G	CCC	GGG	1068.5
14	Gly2	G	UCC	GGA, GGG	1068.5
15	Gly3	G	GCC	GGC, GGU	4359
16	His	H	GUG	CAC, CAU	639
17	Ile1	I	GAU	AUC, AUU	1737
18	Ile2	I	CAU	AUA	1737
19	Leu1	L	CAG	CUG	4470
20	Leu2	L	GAG	CUC, CUU	943
21	Leu3	L	UAG	CUA, CUG	666
22	Leu4	L	CAA	UUG	1913
23	Leu5	L	UAA	UUA, UUG	1031
24	Lys	K	UUU	AAA, AAG	1924
25	Met f1	M	CAU	AUG	1211
26	Met f2	M	CAU	AUG	715
27	Met m	M	CAU	AUG	706
28	Phe	F	GAA	UUC, UUU	1037
29	Pro1	P	CGG	CCG	900
30	Pro2	P	GGG	CCC, CCU	720
31	Pro3	P	UGG	CCA, CCU, CCG	581
32	Sec	X	UCA	UGA	219
33	Ser1	S	UGA	UCA, UCU, UCG	1296
34	Ser2	S	CGA	UCG	344
35	Ser3	S	GCU	AGC, AGU	1408
36	Ser5	S	GGA	UCC, UCU	764
37	Thr1	T	GGU	ACC, ACU	104
38	Thr2	T	CGU	ACG	541
39	Thr3	T	GGU	ACC, ACU	1095
40	Thr4	T	UGU	ACA, ACU, ACG	916
41	Trp	W	CCA	UGG	943
42	Tyr1	Y	GUA	UAC, UAU	769
43	Tyr2	Y	GUA	UAC, UAU	1261
44	Val1	V	UAC	GUA, GUG, GUU	3840
45	Val2A	V	GAC	GUC, GUU	630
46	Val2B	V	GAC	GUC, GUU	635
47	RF1	X		UAA, UAG	1200
48	RF2	X		UAA, UGA	6000

B   Prism Code

```
stochastic

// constants
const double ONE=1;
const double FAST=1000;

// tRNA rates, precalculated
const double c_xx_cogn ;
const double c_yy_cogn ;
const double c_xx_near ;
const double c_yy_near ;
const double c_nonc ;

const double k1f =  140;
const double k2b  =   85;
const double k2bx=2000;
const double k2f =  190;
const double k3bc=   0.23;
const double k3bn=   80;
const double k3fc=  260;
const double k3fn=    0.40;
const double k4rc=   60;
const double k4rn=FAST;
const double k4fc= 166.7;
const double k4fn=  46.1;
const double k6f =  150;
const double k7b =  140;
const double k7f = 145.8;
```

```
module ribosome

s_rib : [0..8] init 1 ;
cogn : bool init false ;
near : bool init false ;
nonc : bool init false ;
xx : bool init false ;

// initial binding
[ ] (s_rib=1) -> k1f * c_xx_cogn : (s_rib'=2) & (xx'=true) & (cogn'=true) ;
[ ] (s_rib=1) -> k1f * c_yy_cogn : (s_rib'=2) & (cogn'=true) ;
[ ] (s_rib=1) -> k1f * c_xx_near : (s_rib'=2) & (xx'=true) & (near'=true) ;
[ ] (s_rib=1) -> k1f * c_yy_near : (s_rib'=2) & (near'=true) ;
[ ] (s_rib=1) -> k1f * c_nonc : (s_rib'=2) & (nonc'=true) ;
[ ] (s_rib=2) & ( cogn | near ) -> k2b :
        (s_rib'=0) & (cogn'=false) & (near'=false) & (xx'=false) ;
[ ] (s_rib=2) &  nonc -> k2bx : (s_rib'=0) & (nonc'=false) ;

// codon recognition
[ ] (s_rib=2) & ( cogn | near ) -> k2f : (s_rib'=3) ;
[ ] (s_rib=3) & cogn -> k3bc : (s_rib'=2) ;
[ ] (s_rib=3) & near -> k3bn : (s_rib'=2) ;

// GTPase activation, GTP hydrolysis, EF-Tu conformation change
[ ] (s_rib=3) & cogn -> k3fc : (s_rib'=4) ;
[ ] (s_rib=3) & near -> k3fn : (s_rib'=4) ;

// rejection
[ ] (s_rib=4) & cogn -> k4rc : (s_rib'=5) & (cogn'=false) & (xx'=false);
[ ] (s_rib=4) & near -> k4rn : (s_rib'=5) & (near'=false) & (xx'=false);

// accommodation, peptidyl transfer
[ ] (s_rib=4) & cogn -> k4fc : (s_rib'=6) ;
[ ] (s_rib=4) & near -> k4fn : (s_rib'=6) ;

// EF-G binding
[ ] (s_rib=6) -> k6f : (s_rib'=7) ;
[ ] (s_rib=7) -> k7b : (s_rib'=6) ;

// GTP hydrolysis, unlocking, tRNA movement and Pi release,
// rearrangements of ribosome and EF-G, dissociation of GDP
[ ] (s_rib=7) -> k7f : (s_rib'=8) ;

// no entrance, re-entrance at state 1
[ ] (s_rib=0) -> FAST*FAST : (s_rib'=1) ;
// rejection, re-entrance at state 1
[ ] (s_rib=5) -> FAST*FAST : (s_rib'=1) ;
// elongation
[ ] (s_rib=8) -> FAST*FAST : (s_rib'=8) ;


endmodule
```

A Stochastic Single Cell Based Model of BrdU Measured Hematopoietic Stem Cell Kinetics

Richard C. van der Wath and Pietro Lio'

Computer Laboratory, University of Cambridge, William Gates Building,
15 JJ Thomson Avenue, Cambridge CB3 0FD, UK

Abstract. The therapeutic potential of stem cells due to their ability to build and maintain tissues and organs is widely recognised. Much can be learned by studying stem cell turnover dynamics and Bromodeoxyuridine (BrdU) is often used for this purpose. Good computational models are however needed for a full understanding of BrdU data and in this paper we present such a model. Our approach is to model single cells as well as their chromosomes as agents which make probabilistic decisions over fixed intervals of time. We demonstrate the power of our model by comparing its performance to a deterministic BrdU model used in a recently published study on asymmetric chromosome segregation in Hematopoietic stem cells.

1 Introduction

Stem cells can be defined as cells that have the ability to self-renew (produce copies of themselves) or differentiate into mature specialised progeny (cells that the body needs to grow and maintain itself) [1]. Much can be learned by studying the turnover kinetics of stem cells and there are several experimental techniques available to do this. One of the most widely used approaches is 5-bromo-2-deoxyuridine (BrdU) assays. BrdU is a synthetic nucleotide and an analogue of thymine and can thus be incorporated in synthesised DNA by substituting for thymine (Fig. 1). Fluorescently marked antibodies that attaches to BrdU are used to detect DNA strands that are BrdU$^+$. BrdU are usually applied by adding it to the drinking water of animals and/or by injection. Tracking the proportion of BrdU$^+$ cells during both the uptake and loss (chase) period provides a mechanism to measure the turnover kinetics of a given population of cells. BrdU data can be misleading however when interpreted directly. Much more certainty about turnover rates can be gained by using computational and mathematical models to simulate BrdU dynamics [2,3]. Note that BrdU uptake and loss are monotonic systems: increasing during the administration phase and decreasing during the chase phase.

A recent study by Kiel et al. [4] uses an ODE-based model to fit BrdU data on murine Hematopoietic stem cells (HSCs). The authors use the result of their model to refute the asymmetrical segregation of chromosomes hypothesis in favour of the random segregation hypothesis. The asymmetrical segregation or immortal strand hypothesis was first proposed in 1975 [5] and suggests that adult

M. Heiner and A.M. Uhrmacher (Eds.): CMSB 2008, LNBI 5307, pp. 387–401, 2008.

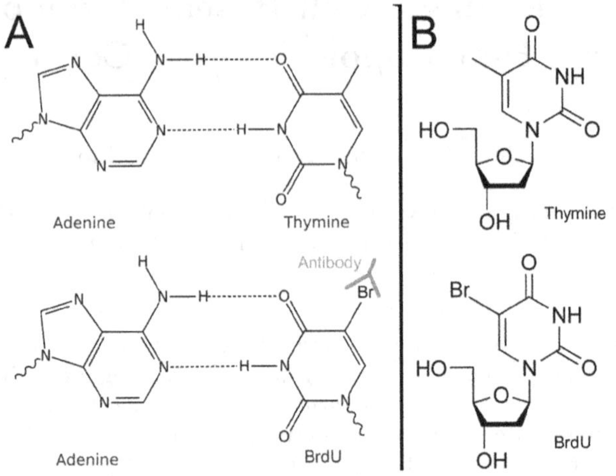

Fig. 1. A: Normal hydrogen bond between Adenine and Thymine (top) and BrdU substituting for Thymine (bottom). **B:** Comparing the chemical structure of Thymine and BrdU.

stem cells retain older DNA strands during mitotic cell division and newly synthesised DNA strands is asymmetrically segregated to differentiating daughter cells. The older DNA strands act as templates for all divisions thereby providing a mechanism to limit the accumulation of DNA mutations. We will describe the equations of the Kiel et al. model and give their solutions in Sect. 2.2, after which we show how the system of uptake equations for random segregation can be improved. In Sect. 2.4 we present a description of a stochastic single cell based BrdU model. We use this model to repeat the study of Kiel et al. in Sect. 3 as a case study to compare the results of the two approaches.

2   Methods

2.1   Model Evaluation

Two key criteria when evaluating a model's suitability is complexity and how well the model describes empirical data (goodness-of-fit). The quest is thus to find the simplest model that fits experimental data at a satisfactory level, known as the *Occam's razor* principle. Often visual inspection of the plot comparing model prediction with empirical data can be a quick and accurate goodness-of-fit estimate, especially in the monotonic system we are considering. We will use a single statistic, the Residual Sum of Squares (RSS), to quantify the visual goodness-of-fit measure. The RSS statistic we use are calculated as follows

$$RSS = \sum_{i=1}^{n} (e(t_i) - m(t_i))^2 \ , \tag{1}$$

where $e(t_i)$ is the experimental value observed at time t_i and $m(t_i)$ is the predicted value at time t_i. Smaller values for RSS indicate a better fit. The RSS statistic is suitable in this study since it is very difficult if not impossible to define more complex model evaluation statistics (likelihood based criterions for example) for the single cell model. In addition, the RSS statistic enables us to compare two very different modelling approaches.

2.2   The Deterministic Model of Kiel et al.

The model of Kiel et al. consists of sets of coupled ordinary differential equations (ODEs) that describe the rate of change of the fraction of BrdU labelled cells over time. Different sets of equations describe the dynamics during BrdU uptake and loss for the random and asymmetric segregation case separately. For both random and asymmetric segregation, the rate of change of the proportion of BrdU$^+$ cells at time t during BrdU application, $y_1(t)$, is simply equivalent to the proportion of BrdU$^-$ cells that has divided at time t, $y_0(t)$:

$$\frac{dy_1}{dt} = \alpha y_0 \ , \tag{2}$$

where α is the proliferation rate of the cells. Because cells leave the y_0 population when they divide, the equation for y_0 in turn is

$$\frac{dy_0}{dt} = -\alpha y_0 \ . \tag{3}$$

The solution of $y_1(t)$ under the initial condition of $y_0(0) = 1$ turns out to be the Cumulative Distribution Function (CDF) of an exponential distribution with parameter α:

$$y_1(t) = 1 - e^{-\alpha t} = 1 - y_0(t) \ . \tag{4}$$

For modelling loss of labelling when BrdU is removed at day T, asymmetric segregation is the simplest case since only one DNA strand can take up and consequently lose BrdU. All BrdU$^+$ cells will thus lose their labelling after a single division, so the rate equation is:

$$\frac{dy_{10}}{dt} = -\alpha y_{10} \ , \tag{5}$$

where y_{10} represent the fraction of cells with one labelled strand that hasn't divided yet after day T. The initial condition is $y_{10}(0) = y_1(T)$.

In the case of random segregation however, BrdU dilution is a more complex process since the model needs to take both DNA strands and BrdU detection sensitivity into account. Kiel et al. model BrdU detection sensitivity by defining separate coupled equations for each (decreasing) level of BrdU labelling. With each division it is assumed that labelling is halved and cells move from level k to level $k + 1$. A cell with 2 labelled strands would need one extra division to be on the same level as a cell with only 1 labelled strand. The equations to model this process are:

$$\frac{dy_{10}}{dt} = -\alpha y_{10}$$

$$\frac{dy_{11}}{dt} = -\alpha y_{11} + \alpha y_{10}$$

$$\vdots$$

$$\frac{dy_{1N}}{dt} = \alpha y_{1(N-1)}$$

$$\frac{dy_{20}}{dt} = -\alpha y_{20}$$ (6)

$$\frac{dy_{21}}{dt} = -\alpha y_{21} + \alpha y_{20}$$

$$\vdots$$

$$\frac{dy_{2N}}{dt} = \alpha y_{2(N-1)} \ ,$$

where $y_{ij} : i \in \{1, 2\}, j \in \{0, 1, .., N\}$ is the fraction of cells with i labelled DNA strand(s) that has divided j times after day T.

The equations in 6 are easiest solved if they are written in matrix form as:

$$y_1' = K \cdot y_1, \text{ and } \quad y_2' = K \cdot y_2 \ ,$$

$$\text{with} \quad y_1 = \begin{bmatrix} y_{10} \\ y_{11} \\ \vdots \\ y_{1(N-1)} \end{bmatrix}, \quad y_2 = \begin{bmatrix} y_{20} \\ y_{21} \\ \vdots \\ y_{2(N-1)} \end{bmatrix}, \text{ and } \quad K = \begin{bmatrix} -\alpha & 0 & 0 & \cdots & 0 \\ \alpha & -\alpha & 0 & \cdots & 0 \\ 0 & \alpha & -\alpha & \ddots & 0 \\ \vdots & \ddots & \ddots & \ddots & 0 \\ 0 & \cdots & 0 & \alpha & -\alpha \end{bmatrix}.$$

Here y_1' and y_2' are the vector derivatives of y_1 and y_2 respectively. The solution can now be written in analogy to that of first order differential equations:

$$y_1(t) = exp(K \cdot t) \cdot y_1(0), \text{ and}$$
$$y_2(t) = exp(K \cdot t) \cdot y_2(0) \ ,$$

$$\text{with} \quad y_1(0) = \begin{bmatrix} y_{10}(0) \\ 0 \\ \vdots \\ 0 \end{bmatrix}, \text{ and } \quad y_2(0) = \begin{bmatrix} y_{20}(0) \\ 0 \\ \vdots \\ 0 \end{bmatrix}.$$

These initial conditions imply that the uptake equation (2) for the random segregation case needs to be extended to predict both the proportion of cells with

1 labelled strand and 2 labelled strands during BrdU uptake. For this purpose Kiel et al. introduced y_1 to represent the fraction of cells with one DNA strand BrdU$^+$ after only one division, and y_2 to represent the fraction of cells with both DNA strands BrdU$^+$ after two or more divisions. The updated uptake equations now define a two step process:

$$\frac{dy_1}{dt} = -\alpha y_1 + \alpha y_0 \tag{7}$$

$$\frac{dy_2}{dt} = \alpha y_1 \ . \tag{8}$$

The initial conditions for (6) then becomes $y_{10}(0) = y_1(T)$ and $y_{20}(0) = y_2(T)$. Note that the total uptake dynamics $\frac{dy_1}{dt} + \frac{dy_2}{dt} = \alpha y_0$ remains unchanged.

2.3 Improved Uptake Equations

Careful study of (8) reveals that it is based on an assumption that a cell has 100% of its DNA labelled after only 2 divisions in the presence of BrdU. A more accurate continuous approximation would be that *daughter.labelling* = 0.5 + (0.5 × *parent.labelling*), with *labelling* = 1.0 indicating all DNA (i.e. both strands) are BrdU$^+$. Equation (8) thus over-predicts $y_{20}(0)$, the proportion of cells that has both strands labelled when BrdU is removed (see Fig. 3 in Sect. 3). This can be improved upon by rather letting y_1 represent the fraction of cells that has divided once or twice, and then letting y_2 represent the fraction of cells that has divided three or more times, making provision for an intermediate level of labelling in between y_1 and y_2. Instead of modelling the number of labelled strands explicitly, we rather define equations for two classes of labelling, allowing cells to stay in the first class for one round of divisions:

$$\frac{dy_{1_1}}{dt} = -\alpha y_{1_1} + \alpha y_0 \tag{9}$$

$$\frac{dy_{1_2}}{dt} = -\alpha y_{1_2} + \alpha y_{1_1} \tag{10}$$

$$\frac{dy_2}{dt} = \alpha y_{1_1} \ , \tag{11}$$

where $y_1 = y_{1_1} + y_{1_2}$. There is no need to define a third class in this study, since we will use $T = 10$ and *alpha* = 0.6 (see Sect. 3), and hence we can expect few cells to divide four or more times during BrdU uptake.

2.4 Stochastic Single Cell Model

Since stem cells are rare entities with relatively low numbers compared to other cells, it is quite feasible to simulate large stem cell systems with single cell based methods in a reasonable amount of time. Single cell based models have a one-to-one correspondence between a real cell and a software implementation of a cell (which we will refer to as an agent). The result is that biologists can more easily relate to such models as opposed to more mathematically complex ones. Furthermore, agent-based approaches are lately gaining more widespread use in modelling complex biological systems [6]. A single cell based BrdU model is surprisingly simple since all that it relies on are representation of time, cells, chromosomes and cell division. We will describe our model under each of these components.

Time. We follow a discrete time approach where we evaluate each individual cell at fixed intervals of time. Each instance of time is called a *tick*. A *tick* can be set to represent any level of granularity (seconds, hours, days, etc.) with smaller values producing more exact results but requiring more computational power.

Cells. Cells are implemented as agents. Agents can be of different types to represent distinct cell populations. Each agent is member of a *masterlist* which is used to keep track of the total cell population and its relevant properties, in our case status of BrdU labelling. Each different type of agent has a certain probability of dividing at each *tick*. Concurrency is simulated by traversing the *masterlist* in random order and then performing each agent's division action with its corresponding probability. There is no need for the agents to have spatial properties or to interact and communicate with each other.

Chromosomes. Chromosomes are implemented as collections of boolean pairs. Each unit of the boolean pair represents one of the DNA strands of a cell's chromosome (3-prime or 5-prime strand) and is set to `true` if the strand has taken up BrdU and `false` otherwise. Depending on the organism we are simulating, the number of chromosomes as well as the ploidy of cells can be set. A global boolean variable `brdu_present` indicates whether BrdU is applied or not, so that both BrdU uptake and loss can be simulated. With our agent-based approach we can count the explicit number of strands that are BrdU positive at each time point and this provides a very accurate mechanism to model the sensitivity of BrdU detection.

Algorithm 1. Asymmetric segregation

$\quad$ **for** $i = 1$ to $(num\_chromosomes * ploidy)$ **do**

$\qquad daughter1.chr(i).5prime \Leftarrow mother.chr(i).5prime$

$\qquad daughter1.chr(i).3prime \Leftarrow brdu\_present$

$\qquad daughter2.chr(i).5prime \Leftarrow brdu\_present$

$\qquad daughter2.chr(i).3prime \Leftarrow mother.chr(i).3prime$

$\quad$ **end for**

$\quad masterlist.remove(mother)$

$\quad masterlist.add(daughter1)$

$\quad masterlist.add(daughter2)$

Cell division. When a stem cell agent divides it spawns two new daughter agents and itself is deleted from the *masterlist*. The way in which the hereditary information (i.e. chromosomes) gets transferred to the two daughter cells depends on whether we simulate asymmetric or random chromosome segregation (Algorithm 1 and 2).

Algorithm 2. Random segregation

---

  for $i = 1$ to $(num\_chromosomes * ploidy)$ **do**
    $rnd \Leftarrow Random.uniform(0, 1.0)$
    if $rnd > 0.5$ **then**
      $daughter1.chr(i).5prime \Leftarrow mother.chr(i).5prime$
      $daughter1.chr(i).3prime \Leftarrow brdu\_present$
      $daughter2.chr(i).5prime \Leftarrow brdu\_present$
      $daughter2.chr(i).3prime \Leftarrow mother.chr(i).3prime$
    else
      $daughter1.chr(i).5prime \Leftarrow brdu\_present$
      $daughter1.chr(i).3prime \Leftarrow mother.chr(i).3prime$
      $daughter2.chr(i).5prime \Leftarrow mother.chr(i).5prime$
      $daughter2.chr(i).3prime \Leftarrow brdu\_present$
    end if
  end for
  $masterlist.remove(mother)$
  $masterlist.add(daughter1)$
  $masterlist.add(daughter2)$

We have implemented this model as an Object Oriented application in Java using the Repast Agent Simulation Toolkit [7] and in the following section we evaluate the usefulness of our model as a stochastic BrdU kinetics simulator over the deterministic models described above (based on how well experimental data is explained - see Sect. 2.1).

3 Results

The experimental data we model for BrdU uptake and loss are taken from Figures 2d and 3c (HSCs) respectively in Kiel et al. [4]. Their ODE model prediction is presented in Figure 3a of [4] where they explicitly indicate the prediction values for day 70 only. Kiel et al. conclude that their data are most consistent with random chromosome segregation and that BrdU detection is lost after approximately three divisions. It seems that this conclusion is based solely on the prediction of day 70. To evaluate the model based on all the experimental data we implemented the equations of Kiel et al. as given in Sect. 2.2 and plotted two versions of the predictions (together with the observed values and their standard deviation bars). Figure 2A shows the prediction using the original random segregation uptake equations (7) and (8), and corresponds to Figure 3a of [4]. Figure 2B shows the model prediction when the improved random segregation uptake equations (9), (10) and (11) are used instead. We also show the predicted distribution of y_1 and y_2 during BrdU uptake for the two versions in Fig. 3. From these figures the following observations can be made:

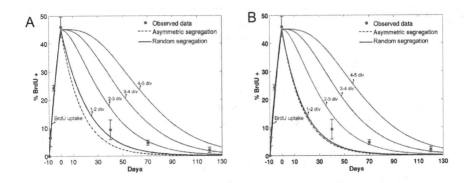

Fig. 2. The deterministic model by Kiel et al., showing the model prediction for: BrdU uptake, asymmetric segregation BrdU chase, and random segregation BrdU chase on different detection thresholds. **A:** The prediction when using the original random segregation uptake equations (7 & 8). **B:** The prediction when using the improved random segregation uptake equations (9 - 10).

- The predictions for BrdU uptake are very accurate. The BrdU administration phase is a very robust system, independent of chromosome segregation mechanism and BrdU detection threshold. The only real free parameter is thus the proliferation rate, and it seems the estimate of 6% per day is very good. In contrast, both models performs poorly in the chase period (days 0-130).
- For the model in Fig. 2A, if all experimental data are taken into account, the 1-2 division threshold has the best fit (see Table 1) which is different to what Kiel et al. concluded. This model thus predicts that BrdU will not be detected unless each and every chromosome of a cell has at least one BrdU$^+$ strand.
- For the model in Fig. 2B, the 2-3 division threshold has the best fit (Table 1) which seems more realistic than what is the case for Fig. 2A. For this model the 1-2 division threshold prediction is very similar to the asymmetric segregation curve. This is due to the smaller predicted proportion of y_2 as can be seen in Fig. 3B, which we think is a more realistic prediction than Fig. 3A for cells with a daily turnover of 6%.
- From this perspective the model in Fig. 2B can thus be regarded as an improved version, but nevertheless, prediction in general remain poor. At any threshold, either the early (day 40), middle (day 70) or late stage (day 120) chase data can be fitted, but the models fails to satisfactorily describe all stages simultaneously. One possible cause is that the poor fit is an artifact of a continuous deterministic model describing a discrete stochastic process. It can also imply the existence of a small more quiescent cell population that retains BrdU for a longer period. Using our single cell model, we can take both of these two possibilities into account.

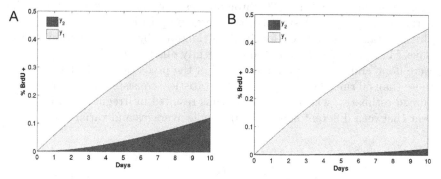

Fig. 3. The proportion of y_2 and y_1 during BrdU uptake. **A:** The prediction of the original random segregation uptake equations (7 & 8). **B:** The prediction of the improved random segregation uptake equations (9 - 10).

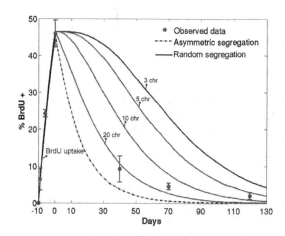

Fig. 4. Single cell model simulation results of 50 runs, showing the average predicted percentage of 400 cells that are BrdU positive after an uptake period of 10 days. A threshold of *20 chr* means at least 20 chromosomes needs to be BrdU labelled before the agent is regarded as BrdU positive.

3.1 Homogenous Population

The prediction of our stochastic single cell model (assuming one homogeneous cell population) is shown in Fig. 4 and Fig. 5. To improve comparison, parameter values were chosen to match the values reported in Kiel et al.:

- Every *tick* represents one day. Division probability per *tick* is 0.06. We used a homogenous population of 400 agents each with 40 chromosomes.

Mus musculus (mice) have 20 chromosomes, but since they are diploid the actual number of DNA strands that can take up BrdU is 80.

– BrdU uptake is simulated for 10 days, followed by a chase period of 130 days. Since each simulation produces slightly different results, 50 repetitions were done and the location and spread of the predictions calculated. Using more than 50 runs did not improve the statistical consistency of location and spread estimates, whilst using fewer runs resulted in irregular and inconsistent (between different simulations) sample averages and variances.

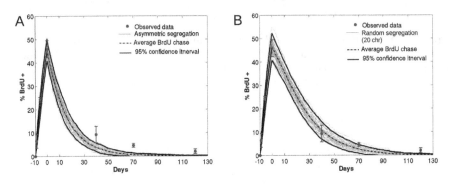

Fig. 5. The 50 individual runs of the single cell model (homogeneous population) with their average and 95% confidence interval. **A:** Asymmetric segregation model predictions. **B:** Random segregation model predictions on a 20 chromosomes detection threshold.

Like the model of Kiel et al., this version of our model also supports the random segregation hypothesis (Comparing Fig. 5A to Fig. 5B). The best fit suggests BrdU is detectable at a 20 chromosomes threshold (Table 1). But from Fig. 5B we see that even with our stochastic simulator, the $2.0 \pm 1.0\%$ observed rate at day 120 falls outside the 95% confidence interval of prediction. There is thus strong reason to believe that the data was generated by a heterogeneous cell population.

Note that our model is much more precise in terms of simulating the BrdU detection threshold, since the model of Kiel et al. simulates the number of divisions N before BrdU becomes undetectable rather than the labelled chromosomes itself. However, N depends on whether one or both DNA strands are labelled, causing the 'overlap' effect between different detection thresholds as apparent from the legend in Figure 3a of [4]. In addition, as pointed out in Sect. 2.2, their model assumes that both DNA strands are BrdU$^+$ after just two divisions during BrdU application under random segregation (equation 3 in [4]). In reality, at the single cell level, this is only one possible outcome in the $2^{\#chromosomes}$ ways in which the chromosomes can segregate during the second division. Our discrete model successfully captures this process and hence have a much more intuitive and accurate way of simulating the BrdU detection sensitivity.

Table 1. RSS values for the two versions of the Kiel et al. model and our stochastic single cell model (homogeneous population). The best RSS for each model are indicated in bold.

Kiel et al. model			Single cell model	
	original	improved		
immortal	55.9	55.9	immortal	59.6
1-2 div	**29.7**	49.7	20 chr	**15.5**
2-3 div	73.7	**40.8**	10 chr	194.8
3-4 div	421.4	324.9	5 chr	721.6
4-5 div	1095	883.7	3 chr	1263.8

3.2   Heterogeneous Populations

For this version of our model we assumed the existence of a smaller quiescent population of cells which will retain BrdU for longer periods of time. We will refer to the other group of cells as the active population. Quiescence and smaller population size is associated with stronger stemness properties [8], so we further assume that the quiescent cells are precursors of the active cells. Differentiated daughter cells of the quiescent population are thus accrued in the active population. The complexity of our model is increased in that the size and division rate of the quiescent population are now free parameters. Fortunately they are constrained since the quiescent population size must be less than 50% of the total population size, and the combined daily turnover rate must still be 6% to accurately simulate BrdU uptake. We found that a 30% (120 out of 400) quiescent population with a 0.0156 probability of dividing per day (0.079 for the active population) results in better RSS values than the homogeneous population model (Table 1 and 2). The predictions based on random segregation are shown in Fig. 6A. The most likely threshold is again 20 chromosomes and the average with 95% confidence interval of this threshold is shown in Fig. 7A. The assumption of heterogeneity has a dramatic effect on the predictions under the immortal strand hypothesis, with two possible assumptions:

Immortal-immortal segregation. The first option is to assume that both populations segregate their chromosomes asymmetrically. In this case a scenario will be created where a small subset of cells (differentiated progeny of the quiescent population during BrdU application) in the active population will have immortal BrdU$^+$ DNA strands. This model is thus extremely efficient in describing long term BrdU retaining data, with a lower RSS than the random segregation model (Fig. 8A and Table 2).

Immortal-random segregation. The second option is for the quiescent population to have asymmetrical segregation, and the active population to have random segregation. This might be a biological more acceptable assumption but the model is not as effective as the previous option in describing the observed data (Fig. 9A). There are nevertheless a significant reduction in RSS compared to the immortal strand option of the homogeneous model (Table 1 and 2).

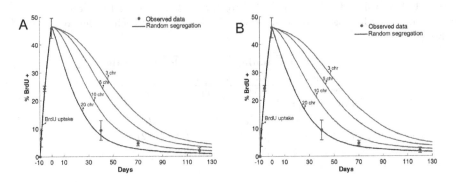

Fig. 6. Prediction of the heterogeneous population model assuming random chromosome segregation and: single-phase turnover (**A**); or bi-phase turnover (**B**)

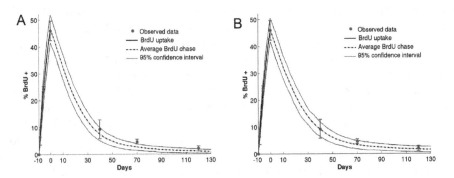

Fig. 7. Average and 95% confidence interval of 50 simulations (heterogeneous population) assuming 20 chromosomes detection threshold random segregation and: single-phase turnover (**A**); or bi-phase turnover (**B**)

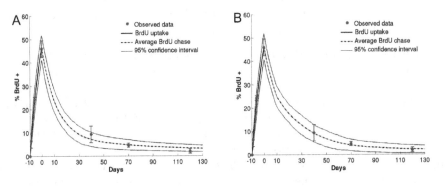

Fig. 8. Average and 95% confidence interval of 50 simulations (heterogeneous population) assuming immortal-immortal chromosome segregation and: single-phase turnover (**A**); or bi-phase turnover (**B**)

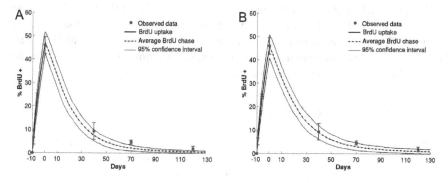

Fig. 9. Average and 95% confidence interval of 50 simulations (heterogeneous population) assuming a 20 chromosomes detection threshold on immortal-random chromosome segregation and: single-phase turnover (**A**); or bi-phase turnover (**B**)

Table 2. RSS values for the stochastic single cell model (heterogeneous population). Comparing single phase vs bi-phase version. The best RSS values for each model are indicated in bold.

	1-phase	2-phase
immortal-immortal	**7.5**	5.9
immortal-random	13.7	5.7
20 chr	9.3	**4.3**
10 chr	47.3	87.9
5 chr	295.1	401.2
3 chr	598	746.9

Toxicity of BrdU. BrdU has been reported to be toxic for some cell types [9], inducing an injury signal that causes the cells to increase their proliferation rate in response. The proliferation rate return to its steady state (healthy) rate soon after BrdU is removed. To take this possibility into account we simulated a bi-phase heterogeneous population whose division probability during BrdU uptake and the first 10 days of chase are higher than the division probability during chase days 10 - 130. Parameters that produced good results are: (in the order of population fraction, first phase division probability, second phase division probability) for the random and immortal-random segregation case: quiescent population (30%,0.0156,0.0096), active population(70%,0.079,0.072); for the immortal-immortal segregation case: quiescent population (10%,0.0156,0.008), active population(90%,0.0649,0.04).

The results are shown in Fig. 6B and Fig. 7B for the random segregation case, and in Fig. 8B and Fig. 9B for the asymmetric segregation case with the corresponding RSS values in Table 2. All three cases (random segregation with 20 chromosomes detection threshold, immortal-random segregation with 20 chromosomes threshold, and immortal-immortal segregation) provides nearperfect goodness of fit statistics. Although at 4.3 the random segregation model has the lowest RSS, no real preference can be given to any of the three hypothesis based only on the RSS. For instance, the immortal-immortal model seems to be the most effective in capturing the large variance observed at day 40 (Fig. 8B). Both versions of the heterogeneous population model thus have equal support for both random and asymmetric segregation.

4   Discussion and Conclusion

The model of Kiel et al. and that of other well known BrdU models in literature [10,11,12] are all ODE-based and thus deterministic and continuous in nature. There are several disadvantages in modelling biological systems as continuous deterministic processes, the most obvious being the fact that a deterministic model needs to assume complete knowledge of the biological system under consideration [13]. This is not possible for most biological systems that researchers are interested in (due to the mere complexity of the spatial position, size, velocity, etc. of billions of molecules). Hence deterministic models invariably have to adopt a higher level view, representing actual biomolecular reactions as some form of aggregate. For systems where all interacting components are present in high numbers individual fluctuations get subsumed in the population average and deterministic models are very effective. However, if some entities are present only in low numbers, the system dynamics behaves in a stochastic manner and needs to be modelled as such. A second advantage of stochastic models over deterministic models is that it provides for much more detailed statistical analysis as we have shown in this study. The disadvantage is that stochastic models are usually much more computationally intensive, and parameter estimates and inference are much harder for these models and not as well established as is the case for deterministic models [13].

We have demonstrated how a single cell based approach, with probabilistic decision making, resulted in a stochastic simulator of BrdU kinetics which can generate Monte Carlo samples from the underlying biological process. Analysis of these samples in turn provides statistical information on the process, whilst being able to take account of each individual cell and its chromosomes provides considerable advantages over conventional approaches, as we have shown in this study.

We evaluated three versions of our model, systematically increasing its complexity and also model prediction accuracy. Taking the results of all three versions of our model together, we conclude that

1. The data of Kiel et al. are unsuitable for making conclusions about the immortal strand hypothesis. More experimental data for the chase period 0-40 days and post 120 days might rectify this situation.
2. It is very likely that the data were generated by a heterogenous population, supporting the possibility of a smaller quiescent group of cells that are the precursors of the larger active cell population.
3. There is further support for the possibility that application of BrdU induces an injury signal, increasing turnover rates.
4. The method used by Kiel et al. to detect BrdU$^+$ cells (immunofluorescence microscopy) has a maximum sensitivity threshold of 20 chromosomes.

Acknowledgements

Part of this work was supported by the Bradlow Foundation Scholarship. We are grateful to Andreas Trumpp, Anne Wilson and Elisa Laurenti for helpful discussions about BrdU and the biology of HSCs. We thank the reviewers for their comments on improving the manuscript.

References

1. Potten, C.S., Loeffler, M.: Stem cells: attributes, cycles, spirals, pitfalls and uncertainties. Lessons for and from the crypt. Development 110(4), 1001–1020 (1990)
2. Cheshier, S.H., Morrison, S.J., Liao, X., Weissman, I.L.: In vivo proliferation and cell cycle kinetics of long-term self-renewing hematopoietic stem cells. Proc. Natl. Acad. Sci. USA 96(6), 3120–3125 (1999)
3. MacKey, M.C.: Cell kinetic status of haematopoietic stem cells. Cell Prolif. 34(2), 71–83 (2001)
4. Kiel, M.J., He, S., Ashkenazi, R., Gentry, S.N., Teta, M., Kushner, J.A., Jackson, T.L., Morrison, S.J.: Haematopoietic stem cells do not asymmetrically segregate chromosomes or retain brdu. Nature 449(7159), 238–242 (2007)
5. Cairns, J.: Mutation selection and the natural history of cancer. Nature 255(5505), 197–200 (1975)
6. Merelli, E., Armano, G., Cannata, N., Corradini, F., d'Inverno, M., Doms, A., Lord, P., Martin, A., Milanesi, L., Müller, S., Schroeder, M., Luck, M.: Agents in bioinformatics, computational and systems biology. Brief Bioinform. 8(1), 45–59 (2007)
7. North, M.J., Collier, N.T., Vos, J.R.: Experiences creating three implementations of the repast agent modeling toolkit. ACM Trans. Model. Comput. Simul. 16(1), 1–25 (2006)
8. Wilson, A., Oser, G.M., Jaworski, M., Blanco-Bose, W.E., Laurenti, E., Adolphe, C., Essers, M.A., Macdonald, H.R., Trumpp, A.: Dormant and self-renewing hematopoietic stem cells and their niches. Ann. N. Y. Acad. Sci. 1106, 64–75 (2007)
9. Caldwell, M.A., He, X., Svendsen, C.N.: 5-bromo-2'-deoxyuridine is selectively toxic to neuronal precursors in vitro. Eur. J. Neurosci. 22(11), 2965–2970 (2005)
10. Grossman, Z., Herberman, R.B., Dimitrov, D.S., Rouzine, I.M., Coffin, J.M., Perelson, A.S., Bonhoeffer, S., Mohri, H., Ho, D.D.: T Cell Turnover in SIV Infection. Science 284(5414), 555a (1999)
11. Bonhoeffer, S., Mohri, H., Ho, D., Perelson, A.S.: Quantification of cell turnover kinetics using 5-bromo-2'-deoxyuridine. J. Immunol. 164(10), 5049–5054 (2000)
12. De Boer, R., Mohri, H., Ho, D.D., Perelson, A.S.: Estimating average cellular turnover from 5-bromo-2'-deoxyuridine (brdu) measurements. Proc. Biol. Sci. 270(1517), 849–858 (2003)
13. Wilkinson, D.J.: Stochastic Modelling for Systems Biology. Chapman & Hall/CRC, Boca Raton (2006)

Analyzing a Discrete Model of *Aplysia* Central Pattern Generator

Ashish Tiwari and Carolyn Talcott

SRI International, Menlo Park, CA 94025
{tiwari,clt}@csl.sri.com

M. Heiner and A.M. Uhrmacher (Eds.): CMSB 2008, LNBI 5307, pp. 347–366, 2008.
© Springer-Verlag Berlin Heidelberg 2008

DOI 10.1007/978-3-540-88562-7_27

In the original online version, the metadata is incorrect.

The original online version for this chapter can be found at
http://dx.doi.org/10.1007/978-3-540-88562-7_24

Author Index

Lecture Notes in Bioinformatics

Vol. 3909: A. Apostolico, C. Guerra, S. Istrail, P.A. Pevzner, M. Waterman (Eds.), Research in Computational Molecular Biology. XVII, 612 pages. 2006.

Vol. 3886: E.G. Bremer, J. Hakenberg, E.-H.(S.) Han, D. Berrar, W. Dubitzky (Eds.), Knowledge Discovery in Life Science Literature. XIV, 147 pages. 2006.

Vol. 3745: J.L. Oliveira, V. Maojo, F. Martín-Sánchez, A.S. Pereira (Eds.), Biological and Medical Data Analysis. XII, 422 pages. 2005.

Vol. 3737: C. Priami, E. Merelli, P. Gonzalez, A. Omicini (Eds.), Transactions on Computational Systems Biology III. VII, 169 pages. 2005.

Vol. 3695: M. R. Berthold, R.C. Glen, K. Diederichs, O. Kohlbacher, I. Fischer (Eds.), Computational Life Sciences. XI, 277 pages. 2005.

Vol. 3692: R. Casadio, G. Myers (Eds.), Algorithms in Bioinformatics. X, 436 pages. 2005.

Vol. 3680: C. Priami, A. Zelikovsky (Eds.), Transactions on Computational Systems Biology II. IX, 153 pages. 2005.

Vol. 3678: A. McLysaght, D.H. Huson (Eds.), Comparative Genomics. VIII, 167 pages. 2005.

Vol. 3615: B. Ludäscher, L. Raschid (Eds.), Data Integration in the Life Sciences. XII, 344 pages. 2005.

Vol. 3594: J.C. Setubal, S. Verjovski-Almeida (Eds.), Advances in Bioinformatics and Computational Biology. XIV, 258 pages. 2005.

Vol. 3500: S. Miyano, J. Mesirov, S. Kasif, S. Istrail, P.A. Pevzner, M. Waterman (Eds.), Research in Computational Molecular Biology. XVII, 632 pages. 2005.

Vol. 3388: J. Lagergren (Ed.), Comparative Genomics. VII, 133 pages. 2005.

Vol. 3380: C. Priami (Ed.), Transactions on Computational Systems Biology I. IX, 111 pages. 2005.

Vol. 3370: A. Konagaya, K. Satou (Eds.), Grid Computing in Life Science. X, 188 pages. 2005.

Vol. 3318: E. Eskin, C. Workman (Eds.), Regulatory Genomics. VII, 115 pages. 2005.

Vol. 3240: I. Jonassen, J. Kim (Eds.), Algorithms in Bioinformatics. IX, 476 pages. 2004.

Vol. 3082: V. Danos, V. Schachter (Eds.), Computational Methods in Systems Biology. IX, 280 pages. 2005.

Vol. 2994: E. Rahm (Ed.), Data Integration in the Life Sciences. X, 221 pages. 2004.

Vol. 2983: S. Istrail, M.S. Waterman, A. Clark (Eds.), Computational Methods for SNPs and Haplotype Inference. IX, 153 pages. 2004.

Vol. 2812: G. Benson, R.D.M. Page (Eds.), Algorithms in Bioinformatics. X, 528 pages. 2003.

Vol. 2666: C. Guerra, S. Istrail (Eds.), Mathematical Methods for Protein Structure Analysis and Design. XI, 157 pages. 2003.